Computational Aspects of Polynomial Identities

Volume 1
Kemer's Theorems
2nd Edition

MONOGRAPHS AND RESEARCH NOTES IN MATHEMATICS

Series Editors

John A. Burns
Thomas J. Tucker
Miklos Bona
Michael Ruzhansky

Published Titles

Application of Fuzzy Logic to Social Choice Theory, John N. Mordeson, Davender S. Malik and Terry D. Clark

Blow-up Patterns for Higher-Order: Nonlinear Parabolic, Hyperbolic Dispersion and Schrödinger Equations, Victor A. Galaktionov, Enzo L. Mitidieri, and Stanislav Pohozaev

Computational Aspects of Polynomial Identities: Volume l, Kemer's Theorems, 2nd Edition Alexei Kanel-Belov, Yakov Karasik, and Louis Halle Rowen

Cremona Groups and Icosahedron, Ivan Cheltsov and Constantin Shramov

Difference Equations: Theory, Applications and Advanced Topics, Third Edition, Ronald E. Mickens

Dictionary of Inequalities, Second Edition, Peter Bullen

Iterative Optimization in Inverse Problems, Charles L. Byrne

Line Integral Methods for Conservative Problems, Luigi Brugnano and Felice Iaverno

Lineability: The Search for Linearity in Mathematics, Richard M. Aron, Luis Bernal González, Daniel M. Pellegrino, and Juan B. Seoane Sepúlveda

Modeling and Inverse Problems in the Presence of Uncertainty, H. T. Banks, Shuhua Hu, and W. Clayton Thompson

Monomial Algebras, Second Edition, Rafael H. Villarreal

Nonlinear Functional Analysis in Banach Spaces and Banach Algebras: Fixed Point Theory Under Weak Topology for Nonlinear Operators and Block Operator Matrices with Applications, Aref Jeribi and Bilel Krichen

Partial Differential Equations with Variable Exponents: Variational Methods and Qualitative Analysis, Vicenţiu D. Rădulescu and Dušan D. Repovš

A Practical Guide to Geometric Regulation for Distributed Parameter Systems, Eugenio Aulisa and David Gilliam

Signal Processing: A Mathematical Approach, Second Edition, Charles L. Byrne

Sinusoids: Theory and Technological Applications, Prem K. Kythe

Special Integrals of Gradshteyn and Ryzhik: the Proofs – Volume l, Victor H. Moll

Special Integrals of Gradshteyn and Ryzhik: the Proofs – Volume II, Victor H. Moll

Forthcoming Titles

Actions and Invariants of Algebraic Groups, Second Edition, Walter Ferrer Santos and Alvaro Rittatore

Analytical Methods for Kolmogorov Equations, Second Edition, Luca Lorenzi

Complex Analysis: Conformal Inequalities and the Bierbach Conjecture, Prem K. Kythe

Geometric Modeling and Mesh Generation from Scanned Images, Yongjie Zhang

Groups, Designs, and Linear Algebra, Donald L. Kreher

Handbook of the Tutte Polynomial, Joanna Anthony Ellis-Monaghan and Iain Moffat

Line Integral Methods and Their Applications, Luigi Brugnano and Felice Iaverno

Microlocal Analysis on R^n and on NonCompact Manifolds, Sandro Coriasco

Practical Guide to Geometric Regulation for Distributed Parameter Systems, Eugenio Aulisa and David S. Gilliam

Reconstructions from the Data of Integrals, Victor Palamodov

Stochastic Cauchy Problems in Infinite Dimensions: Generalized and Regularized Solutions, Irina V. Melnikova and Alexei Filinkov

Symmetry and Quantum Mechanics, Scott Corry

MONOGRAPHS AND RESEARCH NOTES IN MATHEMATICS

Computational Aspects of Polynomial Identities

Volume 1
Kemer's Theorems
2nd Edition

Alexei Kanel-Belov
Bar-Ilan University, Israel

Yakov Karasik
Technion, Israel

Louis Halle Rowen
Bar-Ilan University, Israel

CRC Press
Taylor & Francis Group
Boca Raton London New York

CRC Press is an imprint of the
Taylor & Francis Group, an **informa** business

A CHAPMAN & HALL BOOK

CRC Press
Taylor & Francis Group
6000 Broken Sound Parkway NW, Suite 300
Boca Raton, FL 33487-2742

© 2016 by Taylor & Francis Group, LLC
CRC Press is an imprint of Taylor & Francis Group, an Informa business

First issued in paperback 2019

No claim to original U.S. Government works

ISBN 13: 978-0-367-44580-5 (pbk)
ISBN 13: 978-1-4987-2008-3 (hbk)

Version Date: 20150910

**Visit the Taylor & Francis Web site at
http://www.taylorandfrancis.com**

**and the CRC Press Web site at
http://www.crcpress.com**

Contents

Foreword

The motivation of this second edition is quite simple: As proofs of PI-theorems have become more technical and esoteric, several researchers have become dubious of the theory, impinging on its value in mathematics. This is unfortunate, since a closer investigation of the proofs attests to their wealth of ideas and vitality. So our main goal is to enable the community of researchers and students to absorb the underlying ideas in recent PI-theory and confirm their veracity.

Our main purpose in writing the first edition was to make accessible the intricacies involved in Kemer's proof of Specht's conjecture for affine PI-algebras in characteristic 0. The proof being sketchy in places in the original edition, we have undertaken to fill in all the details in the first volume of this revised edition.

In the first edition we expressed our gratitude to Amitai Regev, one of the founders of the combinatoric PI-theory. In this revision, again we would like to thank Regev, for discussions resulting in a tighter formulation of Zubrilin's theory. Earlier, we thanked Leonid Bokut for suggesting this project, and Klaus Peters for his friendly persistence in encouraging us to complete the manuscript, and Charlotte Henderson at AK Peters for her patient help at the editorial stage.

Now we would also like to Rob Stern and Sarfraz Khan of Taylor and Francis for their support in the continuation of this project. Mathematically, we are grateful to Lance Small for the more direct proof (and attribution) of the Wehrfritz–Beidar theorem and other suggestions, and also for his encouragement for us to proceed with this revision. Eli Aljadeff provided much help concerning locating and filling gaps in the proof of Kemer's difficult PI-representability theorem, including supplying an early version of his write-up with Belov and Karasik. Uzi Vishne went over the entire draft and provided many improvements. Finally, thanks again to Miriam Beller for her invaluable assistance in technical assistance for this revised edition.

This research of the authors was supported by the Israel Science Foundation (Grant no. 1207/12).

Preface

An **identity** of an associative algebra A is a noncommuting polynomial that vanishes identically on all substitutions in A. For example, A is commutative iff $ab - ba = 0$, $\forall a, b \in A$, iff $xy - yx$ is an identity of A. An identity is called a **polynomial identity** (PI) if at least one of its coefficients is ± 1. Thus in some sense PIs generalize commutativity.

Historically, PI-theory arose first in a paper of Dehn [De22], whose goal was to translate intersection theorems for a Desarguian plane to polynomial conditions on its underlying division algebra D, and thereby classify geometries that lie between the Desarguian and Pappian axioms (the latter of which requires D to be commutative). Although Dehn's project was only concluded much later by Amitsur [Am66], who modified Dehn's original idea, the idea of PIs had been planted.

Wagner [Wag37] showed that any matrix algebra over a field satisfies a PI. Since PIs pass to subalgebras, this showed that every algebra with a faithful representation into matrices is a PI-algebra, and opened the door to representation theory via PIs. In particular, one of our main objects of study are **representable** algebras, i.e., algebras that can be embedded into an algebra of matrices over a suitable field.

But whereas a homomorphic image of a representable algebra need not be representable, PIs do pass to homomorphic images. In fact, PIs also can be viewed as the atomic universal elementary sentences satisfied by algebras. Consider the class of all algebras satisfying a given set of identities. This class is closed under taking subalgebras, homomorphic images, and direct products; any such class of algebras is called a **variety** of algebras. Varieties of algebras were studied in the 1930s by Birkhoff [Bir35] and Mal'tsev [Mal36], thereby linking PI-theory to logic, especially through the use of constructions such as ultraproducts.

In this spirit, one can study an algebra through the set of all its identities, which turns out to be an ideal of the free algebra, called a T-**ideal**. Specht [Sp50] conjectured that any such T-ideal is a consequence of a finite number of identities. Specht's conjecture turned out to be very difficult, and became the hallmark problem in the theory. Kemer's positive solution [Kem87] (in characteristic 0) is a tour de force that involved most of the theorems then known in PI-theory, in conjunction with several new techniques such as the use of superidentities. But various basic questions remain, such as finding an explicit set of generators for the T-ideal of 3×3 matrices!

Another very important tool, discovered by Regev, is a way of describing identities of a given degree n in terms of the group algebra of the symmetric group S_n. This led to the asymptotic theory of codimensions, one of the most active areas of research today in PI-theory.

Motivated by an observation of Wagner [Wag37] and M. Hall [Ha43] that the polynomial $(xy - yx)^2$ evaluated on 2×2 matrices takes on only scalar values, Kaplansky asked whether arbitrary matrix algebras have such "central" polynomials; in 1972, Formanek [For72] and Razmyslov [Raz72] discovered such polynomials on arbitrary $n \times n$ matrices. This led to the introduction of techniques from commutative algebra to PI-theory, culminating in a beautiful structure theory with applications to central simple algebras, and (more generally) Azumaya algebras.

While the interplay with the commutative structure theory was one of the main focuses of interest in the West, the Russian school was developing quite differently, in a formal combinatorial direction, often using the polynomial identity as a tool in word reduction. The Iron Curtain and language barrier impeded communication in the formative years of the subject, as illustrated most effectively in the parallel histories of Kurosh's problem, whether or not finitely generated (i.e., affine) algebraic algebras need be finite dimensional. This problem was of great interest in the 1940's to the pioneers of the structure theory of associative rings — Jacobson, Kaplansky, and Levitzki — who saw it as a challenge to find a suitable class of algebras which would be amenable to their techniques. Levitzki proved the result for algebraic algebras of bounded index, Jacobson observed that these are examples of PI-algebras, and Kaplansky completed the circle of ideas by solving Kurosh's problem for PI-algebras. Meanwhile Shirshov, in Russia, saw Kurosh's problem from a completely different combinatorial perspective, and his solution was so independent of the associative structure theory that it also applied to alternative and Jordan algebras. (This is evidenced by the title of his article, "On some nonassociative nil-rings and algebraic algebras," which remained unread in the West for years.)

A similar instance is the question of the nilpotence of the Jacobson radical J of an affine PI-algebra A, demonstrated in Chapter 2. Amitsur had proved the local nilpotence of J, and had shown that J is nilpotent in some cases. There is an easy argument to show that J is nilpotent when A is representable, but the general case is much harder to resolve. By a brilliant but rather condensed analysis of the properties of the Capelli polynomial, Razmyslov proved that J is nilpotent whenever A satisfies a Capelli identity, and Kemer [Kem80] verified that any affine algebra in characteristic 0 indeed satisfies a Capelli identity. Soon thereafter, Braun found a characteristic-free proof that was mostly structure theoretical, employing a series of reductions to Azumaya algebras, for which the assertion is obvious.

There is an analog in algebraic geometry. Whereas affine varieties are the subsets of a given space that are solutions of a system of algebraic equations, i.e., the zeroes of a given ideal of the algebra $F[\lambda_1, \dots, \lambda_n]$ of commutative polynomials, PI-algebras yield 0 when substituted into a given T-ideal of non-commutative polynomials. Thus, the role of radical ideals of $F[\lambda_1, \ldots, \lambda_n]$ in commutative algebraic geometry is analogous to the role of T-ideals of the free algebra, and the coordinate algebra of algebraic geometry is analogous to the relatively free PI-algebra. Hilbert's Basis theorem says that every ideal of the polynomial algebra $F[\lambda_1, \ldots, \lambda_n]$ is finitely generated as an ideal, so Specht's conjecture is the PI-analog viewed in this light.

The introduction of noncommutative polynomials vanishing on A intrinsically involves a sort of noncommutative algebraic geometry, which has been studied from several vantage points, most notably the coordinate algebra, which is an affine PI-algebra. This approach is described in the seminal paper of Artin and Schelter [ArSc81].

Starting with Herstein [Her68] and [Her71], many expositions already have been published about PI-theory, including a book [Ro80] and a chapter in [Ro88b, Chapter 6] by one of the coauthors (relying heavily on the structure theory), as well as books and monographs by leading researchers, including Procesi [Pro73], Jacobson [Jac75], Kemer [Kem91], Razmyslov [Raz89], Formanek [For91], Bakhturin [Ba91], Belov, Borisenko, and Latyshev [BelBL97], Drensky [Dr00], Drensky and Formanek [DrFor04], and Giambruno and Zaicev [GiZa05].

Our motivation in writing the first edition was that some of the important advances in the end of the 20th century, largely combinatoric, still remained accessible only to experts (at best), and this limited the exposure of the more advanced aspects of PI-theory to the general mathematical community. Our primary goal in the original edition was to present a full proof of Kemer's solution to Specht's conjecture (in characteristic 0) as quickly and comprehensibly as we could.

Our objective in this revision is to provide further details for these breakthroughs. The motivating result is Kemer's solution of Specht's conjecture in characteristic 0; the first seven chapters of this book are devoted to the theory needed for its proof, including the featured role of the Grassmann algebra and the translation to superalgebras (which also has considerable impact on the structure theory of PI-algebras). From this point of view, the reader will find some overlap with [Kem91]. Although the framework of the proof is the same as for Kemer's proof, based on what we call the **Kemer index** of a PI-algebra, there are significant divergences; in the proof given here, we also stay more within the PI context. This approach enables us to develop Kemer polynomials for arbitrary varieties, as a tool for proving diverse theorems in later chapters, and also lays the groundwork for analogous theorems that have been proved recently for Lie algebras and alternative algebras, to be handled in Volume II. ([Ilt03] treats the Lie case.) In this revised edition, we add more explanation and detail, especially concerning Zubrilin's theory in Chapter 2 and Kemer's PI-representability theorem in Chapter 6. In Chapter 9, we present counterexamples to Specht's conjecture in characteristic p, as well as their underlying theory.

More recently, positive answers to Specht's conjecture along the lines of Kemer's theory have been found for graded algebras (Aljadeff-Belov [AB10]), algebras with involution, graded algebras with involution, and, more generally, algebras with a Hopf action, which we include in Volume II.

Other topics are delayed until after Chapter 9. These topics include Noetherian PI-algebras, Poincaré–Hilbert series, Gelfand-Kirillov dimension, the combinatoric theory of affine PI-algebras, and description of homogeneous identities in terms of the representation theory of the general linear group GL. In the process, we also develop some newer techniques, such as the "pumping procedure." Asymptotic results are considered more briefly, since the reader should be able to find them in the book of Giambruno and Zaicev [GiZa05].

Since most of the combinatorics needed in these proofs do not require structure theory, there is no need for us to develop many of the famous results of a structural nature. But we felt these should be included somewhere in order to provide balance, so we have listed them in Section 1.6, without proof, and with a different indexing scheme (Theorem A, Theorem B, and so forth). The proofs are to be found in most standard expositions of PI-theory.

Although we aim mostly for direct proofs, we also introduce technical machinery to pave the way for further advances. One general word of caution is that the combinatoric PI-theory often follows a certain Heisenberg principle — complexity of the proof times the manageability of the quantity computed is bounded below by a constant. One can prove rather quickly that affine PI-algebras have finite Shirshov height and satisfy a Capelli identity (thereby leading to the nilpotence of the radical), but the bounds are so high as to make them impractical for making computations. On the other hand, more reasonable bounds now available are for these quantities, but the proofs become highly technical.

Our treatment largely follows the development of PI-theory via the following chain of generalizations:

1. Commutative algebra (taken as given)

2. Matrix algebras (references quoted)

3. Prime PI-algebras (references usually quoted)

4. Subrings of finite dimensional algebras

5. Algebras satisfying a Capelli identity

6. Algebras satisfying a sparse system

7. Algebras satisfying R-Z identities

8. PI-algebras in terms of Kemer polynomials (the most general case)

The theory of Kemer polynomials, which is embedded in Kemer's proof of Specht's conjecture, shows that the techniques of finite dimensional algebras are available for all affine PI-algebras, and perhaps the overriding motivation of this revision is to make these techniques more widely accepted.

Another recurring theme is the Grassmann algebra, which appears first in Rosset's proof of the Amitsur-Levitzki theorem, then as the easiest example of a finitely based T-ideal (generated by the single identity $[[x_1, x_2], x_3]$), later in the link between algebras and superalgebras, and finally as a test algebra for counterexamples in characteristic p.

Enumeration of Results

The text is subdivided into chapters, sections, and at times subsections. Thus, Section 9.4 denotes Section 4 of Chapter 9; Section 9.4.1 denotes subsection 1 of Section 9.4. The results are enumerated independently of these subdivisions. Except in Section 1.6, which has its own numbering system, all results are enumerated according to chapter only; for example, Theorem 6.13 is the thirteenth item in Chapter 6, preceded by Definition 6.12. The exercises are listed at the end of each chapter. When referring in the text to an exercise belonging to the same chapter we suppress the chapter number; for example, in Chapter 9, Exercise 9.12 is called "Exercise 12," although in any other chapter it would have the full designation "Exercise 9.12."

Symbol Description

Due to the finiteness of the English and Greek alphabets, some symbols have multiple uses. For example, in Chapters 2 and 11, μ denotes the Shirshov height, whereas in Chapter 6 and 7, μ is used for the number of certain folds in a Kemer polynomial. We have tried our best to recycle symbols only in unambiguous situations. The symbols are listed in order of first occurrence.

Chapter 1

p. 4: $\mathbb{N}$	The natural numbers (including 0)
$\mathbb{Z}/n$	The ring $\mathbb{Z}/n\mathbb{Z}$ of integers modulo n
$\mathrm{Cent}(A)$	The center of an algebra A
$[a,b]$	The ring commutator $ab - ba$
S_n	The symmetric group
$\mathrm{sgn}(\pi)$	The sign of the permutation π
p. 5: $C[\lambda]$	The commutative algebra of polynomials over C
$C[a]$	The C-subalgebra of A generated by a
$M_n(A)$	The algebra of $n \times n$ matrices over A
p. 6: δ_{ij}	The Kronecker delta
tr	The trace
A^{op}	The opposite algebra
p. 7: $\mathrm{Jac}(A)$	The Jacobson radical of A
p. 9: $S^{-1}A$	The localization of A at a central submonoid A
p. 12: $\sqrt{S}$	The radical of a subset S of A
p. 13: $\mathcal{M}\{X\}$	The word monoid on the set of letters X
$f(x_1,\ldots,x_m)$, $f(\vec{x})$	The polynomial f in indeterminates $x_1,\ldots,x_m$
p. 14: $f(A)$	The set of evaluations of a polynomial f in an algebra A
$\mathrm{id}(A)$	The set of identities of A
p. 15: $\deg f$	The degree of a polynomial f
$\mathrm{UT}(n)$	The set of upper triangular $n \times n$ matrices
p. 16: $\Delta_i f$	The multilinearization step of f in x_i
p. 18: $\tilde{s}_n$	The symmetric polynomial in n letters
$A_1 \sim_{\mathrm{PI}} A_2$	A_1 and A_2 satisfy the same identities
p. 20: s_t	The standard polynomial (on t letters)
c_t	The Capelli polynomial (on t letters)
p. 22: πf	The left action of a permutation π on a polynomial f
p. 23: $f_{A(i_1,\ldots,i_t;X)}$	The alternator of f with respect to the indeterminates $x_{i_1},\ldots,x_{i_t}$
$\tilde{f}$	The symmetrizer of a multilinear polynomial f
p. 24: A_g	The g-component of the graded algebra A

p. 25: $F[\Lambda], F[\lambda_1, \ldots, \lambda_n]$	The commutative polynomial algebra in several indeterminates
$T(V)$	The tensor algebra of a vector space V
$T^n(V)$	The n-homogeneous component of $T(V)$
p. 29 G	The Grassmann algebra, usually in an infinite set of letters
$e_1, e_2, \ldots$	The standard base of the Grassmann algebra G
G_0	The odd elements of G
G_1	The even elements of G
p. 38: $\mathrm{Nil}(A)$	The sum of the nil left ideals of A
p. 39: $\mathcal{M}_{n,F}$	The identities of $M_n(F)$
$\mathcal{M}_n$	The identities of $M_n(\mathbb{Q})$
p. 45: $\mathrm{UT}(n_1, \ldots, n_q)$	The $(n_1, \ldots, n_q)$-block upper triangular matrices
p. 54: $\mathrm{id}(\mathcal{S})$	The identities common to a class $\mathcal{S}$ of algebras
p. 56: $U_{\mathcal{I}}$	The relatively free algebra of a T-ideal $\mathcal{I}$
p. 57: U_A	The relatively free algebra of an algebra A
p. 59: $F\{Y\}_n$	The algebra of generic $n \times n$ matrices
$F(\Lambda)$	The field of fractions of $F[\Lambda]$
$\mathrm{UD}(n, F)$	The generic division algebra of degree n
$A *_C B$	The free product of A and B over C
$A\langle X \rangle$	The free product $A *_C C\{X\}$
$A\langle X \rangle_{\mathcal{I}}$	The relatively free product modulo a T-ideal

Chapter 2

p. 78: $|w|$	The length of a word w
p. 79: $\widehat{W}_\mu$	The Shirshov words of height $\leq \mu$ over W
$\succ$	The lexicographic order on words
$\bar{w}$	The image of a word w in $C\{a_1, \ldots, a_\ell\}$, under the canonical specialization $x_i \mapsto a_i$
p. 80: $\mu = \mu(A)$	The Shirshov height of an affine PI-algebra
p. 81: $\beta(\ell, k, d)$	The Shirshov bound for an affine algebra $C\{a_1, \ldots, a_\ell\}$ of PI-degree d
p. 83: u^∞	The infinite periodic hyperword with period u
p. 84: $\beta(\ell, k, d, h)$	The Shirshov bound for a given hyperword h evaluated on the algebra A
p. 88, 92: $\hat{A}$	The trace ring of a representable algebra A
p. 96: $\delta(xv)$	The cyclic shift
p. 99: $\bar{h} = 0$	The image of a hyperword being 0
p. 108–110: $\Omega, B^p(i), L(j), \psi(p)$	Used in the proof of Theorem 2.8.3
p. 111–112: $\Omega', C^q(i), \phi(q)$	Used in the proof of Theorem 2.8.4

Chapter 10

Chapter 11

Chapter 12

Part I

Basic Associative PI-Theory

Chapter 1

Basic Results

In this chapter, we introduce PI-algebras and review some well-known results and techniques, most of which are associated with the structure theory of algebras. In this way, the tenor of this chapter is different from that of the subsequent chapters. The emphasis is on matrix algebras and their subalgebras (called **representable** PI-algebras) .

1.1   Preliminary Definitions

$\mathbb{N}$ denotes the natural numbers (including 0). $\mathbb{Z}/n$ denotes the ring of integers modulo n. Throughout, C denotes a commutative ring (often a field). Finite dimensional algebras over a field are so important that we often use the abbreviation **f.d.** for them. For any algebra A, Cent(A) denotes the center of A. Given elements a, b of an algebra A, we define $[a, b] = ab - ba$. S_n denotes the symmetric group, i.e., the permutations on $\{1, \ldots, n\}$, and we denote typical permutations as σ or π. We write sgn(π) for the sign of a permutation π.

We often quote standard results about commutative algebras from [Ro05]. We also assume that the reader is familiar with prime and semiprime algebras, and prime ideals. Although the first edition dealt mostly with algebras over a field, the same proofs often work for algebras over a commutative ring C, so we have shifted to that generality.

Remark 1.1.1. *There is a standard way of adjoining* 1 *to a C-algebra A without* 1, *by replacing A by the C-module* $A_1 := A \oplus C$, *made into an algebra by defining multiplication as*

$$(a_1, c_1)(a_2, c_2) = (a_1 a_2 + c_1 a_2 + c_2 a_1, c_1 c_2).$$

We can embed A as an ideal of A_1 *via the identification* $a \mapsto (a, 0)$, *and likewise every ideal of A can be viewed as an ideal of* A_1.

This enables us to reduce most of our major questions about associative algebras to algebras with 1. *Occasionally, we will discuss this procedure in more detail, since one could have difficulties with rings without* 1; *clearly, if* $A^2 = 0$ *we do not have* $A_1^2 = 0$.

In this volume, unless otherwise indicated, an algebra A over C is assumed to be associative with a unit element 1. We will be more discriminating in Volume II, which deals with nonassociative algebras such as Lie algebras.

An element $a \in A$ is **algebraic** (over C) if a is a root of some nonzero polynomial $f \in C[\lambda]$; we say that $a \in A$ is **integral** if f can be taken to be monic. In this case $C[a]$ is a finite module over C. The algebra A is **integral** over C if each element of A is integral.

An element $a \in A$ is **nilpotent** if $a^k = 0$ for some $k \in \mathbb{N}$. An ideal $\mathcal{I}$ of A is **nil** if each element is nilpotent; $\mathcal{I}$ is **nilpotent** of **index** k if $\mathcal{I}^k = 0$ with $\mathcal{I}^{k-1} \neq 0$. One of the basic questions addressed in ring theory is which nil ideals are nilpotent.

Definition 1.1.2. *An element* $e \in A$ *is **idempotent** if* $e^2 = e$; *the **trivial idempotents** are* $0, 1$.

Idempotents e_1 *and* e_2 *are **orthogonal** if* $e_1 e_2 = e_2 e_1 = 0$. *An idempotent* $e = e^2$ *is **primitive** if* e *cannot be written* $e = e_1 + e_2$ *for orthogonal idempotents* $e_1, e_2 \neq 0$.

Remark 1.1.3. *Given a nontrivial idempotent* e *of* A, *and letting* $e' = 1 - e$, *we recall the **Peirce decomposition***

$$A = eAe \oplus eAe' \oplus e'Ae \oplus e'Ae'. \tag{1.1}$$

Note that eAe, $e'Ae'$ *are algebras with respective multiplicative units* e, e'. *If* $eAe' = e'Ae = 0$, *then* $A \cong eAe \times e'Ae'$.

The Peirce decomposition can be extended in the natural way, when we write $1 = \sum_{i=1}^t e_i$ as a sum of orthogonal idempotents, usually taken to be primitive. Now $A = \oplus_{i=1}^t e_i A e_j$. The Peirce decomposition is formulated for algebras without 1 in Exercises 1 and 6.8.

1.1.1 Matrices

$M_n(A)$ denotes the algebra of $n \times n$ matrices with entries in A, and e_{ij} denotes the **matrix unit** having 1 in the i, j position and 0 elsewhere. The set of $n \times n$ matrix units $\{e_{ij} : 1 \leq i, j \leq n\}$ satisfy the properties:

$$\sum_{i=1}^{n} e_{ii} = 1,$$

$$e_{ij} e_{k\ell} = \delta_{jk} e_{i\ell},$$

where δ_{jk} denotes the **Kronecker delta** (which is 1 if $j = k$, 0 otherwise). Thus, the e_{ii} are idempotents.

One of our main tools in matrices is the trace function.

Definition 1.1.4. *For any C-algebra A, and fixed n, a **trace function** is a C-linear map* $\mathrm{tr} : A \to \mathrm{Cent}(A)$ *satisfying*

$$\mathrm{tr}(ab) = \mathrm{tr}(ba), \qquad \mathrm{tr}(a\,\mathrm{tr}(b)) = \mathrm{tr}(a)\,\mathrm{tr}(b), \qquad \forall a, b \in A.$$

It follows readily that

$$\mathrm{tr}(a_1 \ldots a_k) = \mathrm{tr}((a_1 \ldots a_{k-1})a_k) = \mathrm{tr}(a_k a_1 \ldots a_{k-1})$$

for any k.

Of course the main example is $\mathrm{tr} : M_n(C) \to C$ given by $\mathrm{tr}((c_{ij})) = \sum c_{ii}$; here $\mathrm{tr}(1) = n$.

Remark 1.1.5. *The trace satisfies the "nondegeneracy" property that if $\mathrm{tr}(ab) = 0$ for all $b \in A$, then $b = 0$.*

Definition 1.1.6. *Over a commutative ring C, the **Vandermonde matrix** of elements $c_1, \ldots, c_n \in C$ is the matrix (c_i^{j-1}).*

Remark 1.1.7. *When $c_1, \ldots, c_n$ are distinct, the Vandermonde matrix is nonsingular, with determinant $\prod_{1 \leq i < k \leq n}(c_k - c_i)$, cf. [Ro05, Example 0.9]. This gives rise to the famous **Vandermonde argument**, which says that if $\sum_{j=0}^{n-1} c_i^j a_j = 0$ for each $1 \leq i \leq n$, then each $a_j = 0$. The Vandermonde argument occurs repeatedly in proofs in PI theory.*

A^{op} denotes the **opposite algebra**, which has the same algebra structure except with the new multiplication $\cdot$ in A reversed, i.e., $a \cdot b = ba$. In particular, $C^{\mathrm{op}} = C$, and $M_n(C) \cong M_n(C)^{\mathrm{op}}$ via the transpose map.

1.1.2   Modules

We assume the basic properties of modules. We often consider the submodule of an A-module M **spanned** or **generated** by a given subset of M. We say that M is **finitely generated**, denoted by **f.g.**, if $M = \sum_{i=1}^{t} A w_i$ for suitable $w_i \in M$, $t \in \mathbb{N}$. In this case, to avoid confusion with other notions of "generated," we usually say that M is **finite over** A. A module is **finitely presented** over A if it has the form M/N, where M and N are both finite over A.

For C-algebras A_1 and A_2, an A_1, A_2 **bimodule** is a (left) A_1-module M which is also a right A_2-module and a module over C, satisfying the extra associativity condition

$$(a_1 y) a_2 = a_1 (y a_2), \quad \forall a_i \in A_i, \, y \in M,$$

as well as the scalar condition

$$cy = (c1)y = y(c1), \quad \forall c \in C, \, y \in M.$$

Thus, the A_1, A_2 bimodules correspond to the $A_1 \otimes_C A_2^{\mathrm{op}}$-modules. In particular, the sub-bimodules of an algebra A are precisely its ideals.

1.1.3 Affine algebras

Our main interest arises in the following important class of algebras:

Definition 1.1.8. *An algebra A is **affine** over the commutative ring C if A is generated as an algebra over C by a finite number of elements $a_1, \ldots, a_\ell$; in this case we write $A = C\{a_1, \ldots, a_\ell\}$. A commutative affine algebra is notated $C[a_1, \ldots, a_\ell]$.*

In most cases, we shall be considering affine algebras over a field F, so unless specified otherwise, "affine" will mean "affine over a field."

Commutative affine algebras are precisely the coordinate algebras of affine algebraic varieties, and thus play a crucial role in classical algebraic geometry. One of the main thrusts of PI-theory is to generalize commutative affine theory to affine PI-algebras.

1.1.4 The Jacobson radical and Jacobson rings

Definition 1.1.9. *The **Jacobson radical** $\mathrm{Jac}(A)$ of an algebra A is the intersection of the "primitive" ideals of A. (These are the maximal ideals when A is commutative; also see Corollary 1.5.1.)*

Remark 1.1.10. $\mathrm{Jac}(A/J) = \mathrm{Jac}(A)/J$, *whenever* $J \subseteq \mathrm{Jac}(A)$, *cf. [Ro08, Exercise 15.28]*.

We quote a celebrated result of Amitsur [Ro05, Theorem 2.5.23]:

Theorem 1.1.11. *If A has no nonzero nil ideals, then $\mathrm{Jac}(A[\lambda]) = 0$.*

Lemma 1.1.12. *If $\mathrm{Jac}(C) = 0$ and A is a commutative integral domain affine and faithful over C, then $\mathrm{Jac}(A) = 0$.*

Proof. Write $A = C[a_1, \ldots, a_\ell]$, and let $C_1 = C[a_\ell]$. It is enough to show that $\mathrm{Jac}(C_1) = 0$, since then we apply induction on ℓ.

So write $a = a_\ell$ and assume that $A = C[a]$. If a is transcendental over C, then the assertion is clear by Theorem 1.1.11 (since $C[a]$ is isomorphic to a polynomial ring); an easy direct argument is given in Exercise 2.

Thus we may assume that a is algebraic over C, so A is algebraic over C, and by [Ro05, Lemma 6.29] it is enough to show that $C \cap \text{Jac}(A) = 0$. Write $\sum_{i=0}^{t} c_i a^i = 0$ for $c_t \neq 0$, and let $S = \{c_t^i : i \in \mathbb{N}\}$. Let $\mathcal{P}$ be the set of maximal ideals of C not containing c_t, and $J = \cap\{P \in \mathcal{P}\}$. Then $c_t J$ is contained in every maximal ideal of C and thus is 0, implying $J = 0$. On the other hand $S^{-1}A$ is integral over $S^{-1}C$. If $P \in \mathcal{P}$, then $S^{-1}P$ is a prime ideal of $S^{-1}C$, which then is contained in a prime ideal $S^{-1}Q$ of $S^{-1}A$, for some prime ideal Q of A containing P (in view of [Ro05, Proposition 8.11]), implying the integral domain A/Q is a finite extension of the field C/P, and is thus a field. Hence Q is a maximal ideal of A whose intersection with C is P, implying that $C \cap \text{Jac}(A) \subseteq J = 0$, as desired. $\qquad \square$

Definition 1.1.13. *An integral domain C is **local** if it has a unique maximal ideal, which thus is $\text{Jac}(C)$.*

An equivalent formulation [Ro05, Corollary 8.20]: If $a + b = 1$, then either a or b is invertible. One key notion in commutative algebra is localization, treated in [Ro05, Chapter 8].

Definition 1.1.14. *A ring is **Jacobson** (called **Hilbert** in [Kap70b]) if the Jacobson radical of every prime homomorphic image is 0.*

In other words, in a Jacobson ring, any prime ideal is the intersection of primitive ideals of A. Obviously any field is Jacobson, since its only prime ideal 0 is maximal.

Lemma 1.1.15. *Suppose a field $K = C[a_1, \ldots, a_t]$ is affine over a commutative Jacobson subring C. Then C also is a field, and $[K : C] < \infty$.*

Proof. C is an integral domain, and thus $\text{Jac}(C) = 0$. The field K is affine over the field of fractions L of C, implying K is algebraic over C, by [Ro05, Theorem 5.11]. Letting c_i be the leading coefficient of the minimal polynomial of a_i over C, and $c = c_1 \cdots c_t$, we see that each a_i is integral over $C[c^{-1}]$, and thus K is integral over $C[c^{-1}]$, implying $C[c^{-1}]$ is a field, by the easy [Ro05, Proposition 5.31]. Hence any nonzero prime ideal of C contains a power of c, and thus c, implying $c \in \text{Jac}(C) = 0$, a contradiction unless C is already a field, i.e., $L = C$ and thus K is finite over C. $\qquad \square$

We also have a result in the opposite direction.

Lemma 1.1.16. *Any commutative affine algebra $A = C[a_1, \ldots, a_t]$ over a commutative Jacobson ring C is Jacobson.*

Proof. For any prime ideal P of A, $\text{Jac}(A/P) = 0$ by Lemma 1.1.12. $\qquad \square$

This often is called the "weak Nullstellensatz."

1.1.5 Central localization

The localization procedure can be generalized directly from the commutative situation to $S^{-1}A$ whenever S is a (multiplicative) submonoid of Cent(A). In particular the ideals of $S^{-1}A$ are precisely those subsets $S^{-1}\mathcal{I}$ where $\mathcal{I} \lhd A$. We say that an element $s \in A$ is **regular** when $sa, as \neq 0$ for all $a \neq 0$ in A. When A is prime, then every submonoid of Cent(A) is regular. Here is an easy but useful result.

Proposition 1.1.17. *Suppose S is a submonoid of* Cent(A) *which is regular in A. Then $S^{-1}A$ is prime iff A is prime.*

Proof. ($\Rightarrow$) If $\mathcal{I}_1, \mathcal{I}_2 \lhd A$ with $\mathcal{I}_1\mathcal{I}_2 = 0$, then $(S^{-1}\mathcal{I}_1)(S^{-1}\mathcal{I}_2) = 0$, implying $S^{-1}\mathcal{I}_1 = 0$ or $S^{-1}\mathcal{I}_2 = 0$, so $\mathcal{I}_1 = 0$ or $\mathcal{I}_2 = 0$.

($\Leftarrow$) If $S^{-1}\mathcal{I}_1, S^{-1}\mathcal{I}_2 \lhd S^{-1}A$ with $S^{-1}\mathcal{I}_1\mathcal{I}_2 = 0$, then $\mathcal{I}_1 = 0$ or $\mathcal{I}_2 = 0$, implying $S^{-1}\mathcal{I}_1 = 0$ or $S^{-1}\mathcal{I}_2 = 0$, so $\mathcal{I}_1 = 0$ or $\mathcal{I}_2 = 0$. $\square$

Corollary 1.1.18. *Suppose A is a prime algebra, and S is a submonoid of* Cent(A), *and $A \subseteq B \subseteq S^{-1}A$. Then B is prime.*

Proof. $S^{-1}A$ is prime, but $S^{-1}A = S^{-1}B$, implying B is prime. $\square$

1.1.6 Chain conditions

A partially ordered set $\mathcal{S}$ is said to satisfy the ACC (**ascending chain condition**) if every infinite ascending chain

$$S_1 \subseteq S_2 \subseteq \dots$$

stabilizes in the sense that there is some k such that $S_i = S_{i+1}$ for all $i \geq k$. In particular, a commutative ring is **Noetherian** if it satisfies the ACC on ideals. The Hilbert Basis Theorem implies that every commutative affine algebra over a Noetherian ring (in particular, over a field) is Noetherian, thereby elevating the Noetherian theory to a central role in algebra and geometry.

Recall three noncommutative generalizations of "Noetherian," in increasing strength:

Definition 1.1.19. *(i) A ring R is **weakly Noetherian** if it satisfies the ACC on two-sided ideals. (Equivalently, R is a Noetherian $R \otimes R^{\mathrm{op}}$-module.)*

*(ii) A ring R is **left Noetherian** if it satisfies the ACC (ascending chain condition) on left ideals.*

*(iii) R is **Noetherian** if it is left and right Noetherian, i.e., satisfies the ACC on left ideals and also satisfies the ACC on right ideals.*

Any finite module over a left Noetherian ring is left Noetherian. Any weakly Noetherian ring obviously has a unique maximal nilpotent ideal, which is the intersection of its prime ideals.

Remark 1.1.20. *We recall the important technique of "Noetherian induction": To prove a theorem about weakly Noetherian rings, we suppose on the contrary that we have a counterexample R, and take an ideal $\mathcal{I}$ maximal with respect to the theorem failing for $R/\mathcal{I}$. Replacing R by $R/\mathcal{I}$, we may assume that R is a counterexample, but $R/\mathcal{J}$ is not a counterexample for every $0 \neq \mathcal{J} \lhd R$.*

Noetherian induction can also be used for proving theorems about Noetherian modules, in an analogous fashion.

We can pass the Noetherian property to the center by means of the following result.

Proposition 1.1.21 (Artin-Tate Lemma). *Suppose that A is an affine C-algebra, finite over its center Z. If C is Noetherian, then Z is affine, and thus is Noetherian.*

Proof. For the reader's convenience, we reproduce the easy proof given in [Ro88b, Proposition 6.2.5]. Namely, write $A = C\{a_1, \ldots, a_t\}$ and $A = \sum_{\ell=1}^{q} Z b_\ell$. Writing $b_i b_j = \sum_{m=1}^{q} z_{ijm} b_m$ for $z_{ijm} \in Z$, and $a_k = \sum_{\ell=1}^{q} z'_{k\ell} b_\ell$, we let

$$Z_1 = C[z_{ijm}, z'_{k\ell} : 1 \leq i, j, \ell, m \leq t, \ 1 \leq k \leq q],$$

which is affine over C, and thus Noetherian. But $\sum_{\ell=1}^{q} Z_1 b_\ell$ is an algebra over Z_1 containing $C\{a_1, \ldots, a_t\} = A$, and thus is a Noetherian Z_1-module, proving that its submodule Z is finite over Z_1, and thus is affine as an algebra. $\square$

A related result due to Eakin-Formanek (Exercise 3) is that if a ring is Noetherian and finite over its center Z, then Z is Noetherian.

Definition 1.1.22. *Suppose that some set S acts on an algebra A from the right. For any subset $T \subset S$ one defines the **left annihilator***

$$\operatorname{Ann} T = \{a \in A : aT = 0\},$$

*a left ideal of A. **ACC(Left annihilators)** denotes the ACC on $\{$left annihilators$\}$. When $\operatorname{Ann} T$ is a 2-sided ideal of A, we call $\operatorname{Ann} T$ an **annihilator ideal**. In this case, $\operatorname{Ann} T$ is the left annihilator of a 2-sided ideal, namely of its right annihilator.*

Lemma 1.1.23 (Fitting-type Lemma). *Given a module M over a commutative ring Z, with $z \in Z$ and $k \in \mathbb{N}$, let $N = \{a \in M : z^k a = 0\}$. If N satisfies the property that $z^{2k} a = 0$ implies $a \in N$, then $z^k M \cap N = 0$.*

Proof. If $z^k a \in N$, then $z^{2k} a = 0$, implying $a \in N$, so $z^k a = 0$. $\square$

1.1.7 Subdirect products and irreducible algebras

Definition 1.1.24. *A is a **subdirect product** of the algebras $\{A_i : i \in I\}$ if there is an injection $\psi : A \to \prod A_i$ for which $\pi_j \psi : A \to A_j$ is onto for each $j \in I$, where π_j denotes the natural projection $\prod A_i \to A_j$.*

In this case, $\cap \ker \pi_j = 0$. Conversely, if $A_i = A/\mathcal{I}_i$ for each $i \in I$ and $\cap_i \mathcal{I}_i = 0$, then A is a subdirect product of the A_i in the obvious way.

The following concept often fits in with Noetherian.

Definition 1.1.25. *An algebra A is **irreducible** if the intersection of two nonzero ideals is always nonzero.*

By induction, the intersection of finitely many nonzero ideals of an irreducible algebra is always nonzero.

Lemma 1.1.26. *Any weakly Noetherian algebra A is a finite subdirect product of irreducible algebras.*

Proof. The usual Noetherian induction argument. Otherwise, take a counterexample A, and take $\mathcal{I} \lhd A$ maximal with respect to $A/\mathcal{I}$ not being a counterexample. Passing to $A/\mathcal{I}$, we may assume that A is a counterexample to the lemma, but $A/\mathcal{I}$ is not a counterexample, for all $0 \neq \mathcal{I} \lhd A$.

In particular, A itself is reducible, so has nonzero ideals $\mathcal{I}_1, \mathcal{I}_2$ such that $\mathcal{I}_1 \cap \mathcal{I}_2 = 0$. But by hypothesis $A/\mathcal{I}_1$ is a finite subdirect product of irreducible algebras $A/\mathcal{I}_{1,1}, \dots, A/\mathcal{I}_{1,t}$ and $A/\mathcal{I}_2$ is a finite subdirect product of irreducible algebras $A/\mathcal{I}_{2,1}, \dots, A/\mathcal{I}_{2,u}$, implying A is a subdirect product of $A/\mathcal{I}_{1,1}, \dots, A/\mathcal{I}_{1,t}, A/\mathcal{I}_{2,1}, \dots, A/\mathcal{I}_{2,u}$. $\square$

1.1.7.1 ACC for classes of ideals

This subsection contains basic material about chain conditions on classes of ideals of a given ring R, with an eye on applications to ideals of noncommutative algebras. The reason we include it is that Kemer's solution of Specht's problem, given in Chapters 6 and 7, has thrust open the door to a new application of this material, and we might as well present it here to have it available for other purposes (such as for the structure of affine PI-algebras). We skip some proofs, when they are formal and in direct analogy to the well-known proofs in commutative algebra. Throughout, we fix a monoid $\mathcal{S}$ of ideals of R, satisfying the following properties:

(i) The intersection of members of $\mathcal{S}$ is in $\mathcal{S}$;

(ii) If $\mathcal{I}, \mathcal{J} \in \mathcal{S}$, then $\mathcal{I} + \mathcal{J} \in \mathcal{S}$.

Definition 1.1.27. *Given $S \subseteq \mathcal{S}$, the member of $\mathcal{S}$ **generated** by S is defined as $\cap \{\mathcal{I} \in \mathcal{S} : S \subseteq \mathcal{I}\}$. $\mathcal{I} \in \mathcal{S}$ is **finitely generated** in $\mathcal{S}$ if $\mathcal{I}$ is generated by some finite set S.*

(This generalizes the notion of a finite module.)

Remark 1.1.28. *The following are equivalent:*

(i) S satisfies the ACC.

(ii) Every member of S is finitely generated in S.

(iii) Every subset of S has a maximal member.

Definition 1.1.29. *A member P of S is **prime** if, for all $\mathcal{I}, \mathcal{J} \in S$ not contained in P, we have $\mathcal{I}\mathcal{J} \not\subseteq P$. For any $S \subseteq A$, a prime P of S containing S is **minimal**prime over S if P does not properly contain a prime of S containing S.*

Lemma 1.1.30. *Every prime of S containing S contains a minimal prime containing S.*

Proof. In view of Zorn's lemma, we need to show that for any chain $\mathcal{P}$ of primes, that $P = \cap\{P \in \mathcal{P}\}$ is also prime. But this is standard: If $\mathcal{I}\mathcal{J} \subseteq P$ with $\mathcal{I} \not\subseteq P$, then $\mathcal{I} \not\subseteq P_{j_0}$ for some P_{j_0} in $\mathcal{P}$, implying $\mathcal{J} \subseteq P_j$ for each $P_j \subset P_{j_0}$ in $\mathcal{P}$, implying $\mathcal{J} \subseteq P$. $\square$

Theorem 1.1.31. *Suppose that S satisfies the ACC. Then for any $\mathcal{I} \in S$, there are only finitely many primes $P_1, \ldots, P_n$ in S minimal over $\mathcal{I}$, and some finite product of the P_i is contained in $\mathcal{I}$.*

Proof. By Noetherian induction. Otherwise, there is $\mathcal{I} \in S$ maximal with respect to being a counterexample. Certainly $\mathcal{I}$ is not itself prime, so take $\mathcal{J}_1, \mathcal{J}_2 \supset I$ in S such that $\mathcal{J}_1\mathcal{J}_2 \subseteq \mathcal{I}$. (We can replace $\mathcal{J}_i$ by $\mathcal{J}_i + \mathcal{I}$ if necessary.) By hypothesis, the conclusion of the theorem holds for $\mathcal{J}_1$ and $\mathcal{J}_2$, i.e., there are primes P_{ik} minimal over $\mathcal{J}_k$ with some finite product contained in $\mathcal{J}_k$. But then the product together is contained in $\mathcal{J}_1\mathcal{J}_2$ and thus, in $\mathcal{I}$. Any prime P containing $\mathcal{J}_1\mathcal{J}_2$ contains some minimal prime over $\mathcal{J}_1\mathcal{J}_2$, which in turn must contain some P_{ik} and thus must equal P_{ik}. $\square$

Definition 1.1.32. *The **radical** $\sqrt{S}$ of $S \subseteq A$ is the intersection of all primes of S containing S.*

The foregoing results did not involve associativity of the multiplication of S, although the subsequent ones do, in order that $P_1 \cdots P_n$ is well-defined. (The subtleties of the nonassociative case are treated in Volume II.)

Corollary 1.1.33. *Suppose that S satisfies the ACC. If $\mathcal{I} \in S$, then $\sqrt{\mathcal{I}}$ is a finite intersection of primes of S, each minimal over $\sqrt{\mathcal{I}}$.*

Corollary 1.1.34. *If S satisfies the ACC, then $\sqrt{\mathcal{I}}^t \subseteq \mathcal{I}$ for some t.*

Proof. Write $\sqrt{\mathcal{I}} = P_1 \cap \cdots \cap P_n$, and then note that some product of t of the P_i are in $\mathcal{I}$, implying

$$(\sqrt{\mathcal{I}})^t \subseteq P_1 \cdots P_t \subseteq \sqrt{\mathcal{I}}.$$

$\square$

Corollary 1.1.35. *Suppose that $\mathcal{S}$ satisfies the ACC, and $0 \in \mathcal{S}$. If $\mathcal{I} \in \mathcal{S}$ is contained in every prime, then $\mathcal{I}$ is nilpotent.*

Proof. $\mathcal{I} \subseteq \sqrt{0}$, so apply the previous corollary. $\square$

Corollary 1.1.36. *Any nil subset N of a commutative (associative) Noetherian ring C is nilpotent.*

Proof. N is contained in every prime ideal P, since C/P is an integral domain.

$\square$

(This fails for noncommutative rings, even for $\{e_{12}, e_{21}\} \subset M_2(F)$.)

1.2   Noncommutative Polynomials and Identities

In order to get to our subject, we need the noncommutative analog of polynomials.

1.2.1   The free associative algebra

Recall that the free (associative) monoid $\mathcal{M}\{X\}$ in $X = \{x_i : i \in I\}$ is the monoid of **words** $\{x_{i_1} x_{i_2} \cdots x_{i_t} : t \in \mathbb{N}\}$ permitting duplication of subscripts, and whose unit element is the blank word $\emptyset$; the monoid operation is given in terms of juxtaposition of words.

$C\{X\}$, often denoted $C\langle X \rangle$ in the literature, denotes the free associative algebra (with 1) in the set $X = \{x_i : i \in I\}$ of noncommuting indeterminates. (Usually $I = \mathbb{N}$, but often I is taken to be finite.) In other words, $C\{X\}$ is the monoid algebra of $\mathcal{M}\{X\}$. The elements of $C\{X\}$ are called **polynomials**. $C\{X\}$ is free as a C-module, with base consisting of $\mathcal{M}\{X\}$, the set of words; thus, any $f \in C\{X\}$ is written uniquely as $\sum c_j h_j$ where $h_j \in \mathcal{M}(X)$. We call these $c_j h_j$ the **monomials** of f.

Given $f \in C\{X\}$ we write $f(x_1, \ldots, x_m)$ to denote that $x_1, \ldots, x_m$ are the only indeterminates occurring in f. Sometimes we write $f(\vec{x})$ for short. Later, when the notation becomes more cumbersome, we shall have occasion to use Y (and at times Z) to denote extra sets of indeterminates that do not enter the computations as actively as the x_i. In this case we write $C\{X, Y\}$ or $C\{X, Y, Z\}$ in place of $C\{X\}$, and we write $f(\vec{x}, \vec{y})$ or $f(\vec{x}, \vec{y}, \vec{z})$ accordingly.

The main feature of $C\{X\}$ is the following.

Remark 1.2.1. *Given a C-algebra A and elements $\{a_i : i \in I\} \subseteq A$, there is a unique algebra homomorphism $\phi : C\{X\} \to A$, called the **substitution homomorphism**, such that $\phi(x_i) = a_i$, $\forall i \in I$. Indeed, one defines*

$$\phi(x_{i_1} \cdots x_{i_m}) = a_{i_1} \cdots a_{i_m}$$

and extends this linearly to all of $C\{X\}$.

The **evaluation** $f(a_1, \ldots, a_m)$ denotes the image of f under the homomorphism of Remark 1.2.1. We also say that f **specializes** to $f(a_1, \ldots, a_m)$, and $a_1, \ldots, a_m$ are **substitutions** in f.

1.2.2   Polynomial identities

We write $f(A)$ for the set of evaluations $\{f(a_1, \ldots, a_m) : a_i \in A\}$.

Definition 1.2.2. *An element $f \in C\{X\}$ is an **identity** of a C-algebra A if $f(A) = 0$, i.e., $f \in \ker \phi$ for every homomorphism $\phi : C\{X\} \to A$.*

Identities pass to related algebras as follows.

Remark 1.2.3. *If f is an identity of an algebra A, then f is an identity of any homomorphic image of A and also of any subalgebra of A. Furthermore if f is an identity of each C-algebra A_i, $i \in I$, then f is an identity of $\prod_{i \in I} A_i$.*

Remark 1.2.3 provides an alternate approach to identities, cf. §1.7 below.

Definition 1.2.4. *For a monomial h we define $\deg_i h$ to be the number of occurrences of x_i in h, and the **degree** $\deg h = \sum_i \deg_i h$; for a polynomial f, we define $\deg f$ to be the maximum degree of the monomials of f. For example $\deg(x_1 x_2 + x_3 x_4) = 2$.*

One needs some way of excluding the identity px_1, which only says that A has characteristic p. Toward this end, we formulate the main definition of this book.

Definition 1.2.5. *An identity f is a **PI** (polynomial identity) for A if at least one of its coefficients is 1. An algebra A is a **PI-algebra** of PI-**degree** d if A satisfies a PI of degree d.*

This definition might seem restrictive, but in fact is enough to encompass the entire PI-theory, cf. [Am71]. Since PI-algebras are the subject of our study, let us address a subtle distinction in terminology. A ring R is a PI-**ring** when it is a PI-algebra for $C = \mathbb{Z}$. Although most of the general structure theory holds for PI-rings in general, our focus in this book is usually on a particular base ring C, sometimes a field which we denote as F rather than C; often we require $\mathrm{char}(F) = 0$, for reasons to be discussed shortly.

Definition 1.2.6. *We write id(A) for the set of identities of A.*

Here is a notion closely related to PI.

Definition 1.2.7 (Central polynomials). *A polynomial $f(x_1,\ldots,x_n)$ is A-central if $0 \neq f(A) \subseteq \mathrm{Cent}(A)$.*

In other words, $f(x_1,\ldots,x_n)$ is A-central iff $[y,f]$ (but not f) is in $\mathrm{id}(A)$.

The most basic examples of PI-algebras are the matrix algebra $M_n(C)$ for arbitrary n, f.d. algebras over a field, and the Grassmann algebra G, cf. Definition 1.3.26. Since these examples require a bit more theory, we first whet the reader's appetite with some easier examples.

Example 1.2.8.

 (i) *The polynomial x is central for any commutative algebra.*

 (ii) *Let $\mathrm{UT}(n)$ denote the algebra of upper triangular matrices over a given base ring C. Any product of n strictly upper triangular $n \times n$ matrices is 0. Since $[a,b]$ is strictly upper triangular, for any upper triangular matrices a,b, we conclude that the algebra $\mathrm{UT}(n)$ satisfies the identity*

$$[x_1,x_2][x_3,x_4]\cdots[x_{2n-1},x_{2n}].$$

 (iii) *(Wagner's identity) If F is a field, then $M_2(F)$ satisfies the identity $[[x,y]^2,z]$ or, equivalently, the central polynomial $[x,y]^2$, cf. Exercise 19.*

 (iv) *Fermat's Little Theorem translates to the fact that any field F of n elements satisfies the identity x^n-x. (See Exercise 27 for a generalization.)*

 (v) *Any Boolean algebra satisfies the identity $x^2 - x$.*

When dealing with arbitrary PIs it is convenient to work with certain kinds of polynomials. We say that a polynomial $f(x_1,\ldots,x_m)$ is **homogeneous** in x_i if x_i has the same degree in each monomial of f. We say that f is **homogeneous** if f is homogeneous in every indeterminate. (Sometimes this is called "completely homogeneous" or "multi-homogeneous" in the literature.) In this case, if x_i has degree d_i in f_i for $1 \leq i \leq m$, we say that f has **multidegree** $(d_1,\ldots,d_m)$, where $\deg f = d_1 + \cdots + d_m$. Here is a very important special case.

Definition 1.2.9. *A monomial h is **linear** in x_i if $\deg_i h = 1$. A polynomial f is **linear** in x_i if each monomial of f is linear in x_i; f is t-**linear** if f is linear in each of $x_1,\ldots,x_t$.*

*A polynomial $f(x_1,\ldots,x_m)$ is **multilinear** if f is m-linear. In other words, each indeterminate of f appears with degree exactly 1 in each monomial of f.*

Thus, $x_1x_2-x_2x_1$ is multilinear. However, $x_1x_2x_3-x_2x_1$ is not multilinear, since x_3 does not appear in the second monomial.

Given a multilinear polynomial $f(x_1, \ldots, x_m)$, we pick any nonzero monomial h, and renaming the indeterminates appropriately, we may assume that $h = cx_1x_2 \ldots x_m$ for some $c \in C$. Thus, the general form for a multilinear polynomial is

$$f(x_1, \ldots, x_m) = c_1 x_1 x_2 \cdots x_m + \sum_{1 \neq \sigma \in S_m} c_\sigma x_{\sigma(1)} x_{\sigma(2)} \cdots x_{\sigma(m)}. \qquad (1.2)$$

Furthermore, if C is a field, then we can divide by c_1 and assume that $c_1 = 1$. The main reason we focus on multilinear identities is because of Proposition 1.2.18 below. However, the linearity property already is quite useful:

Remark 1.2.10. *If f is linear in x_i, then*

$$f\left(a_1, \ldots, \sum_j c_j a_{ij}, \ldots, a_m\right) = \sum_j c_j f(a_1, \ldots, a_{ij}, \ldots, a_m)$$

for all $c_j \in C$, $a_{ij} \in A$.

Lemma 1.2.11. *Suppose A is spanned over C by a set B. Then a multilinear polynomial f is an identity of A iff f vanishes on all substitutions to elements of B; f is A-central iff every substitution of f on B is in $\mathrm{Cent}(A)$ but some substitution on B is nonzero.*

Proof.

$$f\left(\sum_{i_1} c_{i_1} b_{i_1}, \ldots, \sum_{i_m} c_{i_m} b_{i_m}\right) = \sum_{i_1, \ldots, i_m} c_{i_1} \cdots c_{i_m} f(b_{i_1}, \ldots, b_{i_m}),$$

in view of Remark 1.2.10.   □

1.2.3   Multilinearization

These observations raise the question of how to go back and forth from arbitrary identities (or central polynomials) to multilinear ones. The answer is in the process of **multilinearization**, also called **polarization**. This will be tied to group actions in §3.5 (also cf. Exercise 6), but can be described briefly as follows:

Definition 1.2.12 (Multilinearization). *Suppose the polynomial $f(x_1, \ldots, x_m)$ has degree $n_i > 1$ in x_i. We focus on one of the indeterminates, x_i, and define the **partial linearization***

$$\Delta_i f(x_1, \ldots, x_{i-1}, x_i, x_i', x_{i+1}, \ldots, x_m) \qquad (1.3)$$
$$= f(\ldots, x_i + x_i', \ldots) - f(\ldots, x_i, \ldots) - f(\ldots, x_i', \ldots)$$

where x_i' is a new indeterminate. Clearly $\Delta_i f$ remains an identity for A when $f \in \mathrm{id}(A)$, but all monomials of degree n_i in x_i cancel out in $\Delta_i f$. The remaining monomials have x_i' replacing x_i in some (but not all) instances, and thus have degree $< n_i$ in x_i, the maximum degree among them being $n_i - 1$.

Remark 1.2.13. *Since this procedure is so important, let us rename the indeterminates more conveniently, writing x_1 for x_i and y_j for the other indeterminates.*

(i) *Now our polynomial is $f(x_1; \vec{y})$ and our partial linearization may be written as*

$$\Delta_1 f(x_1, x_2; \vec{y}) = \\ f(x_1 + x_2; \vec{y}) - f(x_1; \vec{y}) - f(x_2; \vec{y}), \tag{1.4}$$

where x_2 is the new indeterminate.

(ii) *Before we get started, we must cope with the situation in which x_1 does not appear in each monomial. For example, if we want to multilinearize $f = x_1 y + y$ in x_1, then the only way would be to apply Δ_1, but*

$$\Delta_1 f = (x_1 + x_2)y + y - (x_1 y + y) - (x_2 y + y) = -y,$$

and we have lost x_1 altogether. This glitch could complicate subsequent proofs.

*Fortunately, we can handle this situation by defining $g = f(0; \vec{y})$, the sum of those monomials in which x_1 does not appear. If $f \in \mathrm{id}(A)$, then also $f - g \in \mathrm{id}(A)$, so we can replace f by $f - g$ and thereby assume that any indeterminate appearing in f appears in each monomial of f, as desired. We call such a polynomial **blended**.*

(iii) *Let $n = \deg_1 f$. Iterating the linearization procedure $n - 1$ times (each time introducing a new indeterminate x_i) yields an n-linear polynomial $\bar{f}(x_1, \ldots, x_n; \vec{y})$ which preserves only those monomials h originally of degree n in x_1. For each such monomial h in f we now have $n!$ monomials in $\bar{f}$ (according to the order in which $x_1, \ldots, x_n$ appears), each of which specializes back to h when we substitute x_1 for each x_i. Thus, when f is homogeneous in x_1, we have*

$$\bar{f}(x_1, \ldots, x_1; \vec{y}) = n!f. \tag{1.5}$$

*We call $\bar{f}$ the **linearization** of f in x_1. In characteristic 0 this is about all we need, since $n!$ is invertible and we have recovered f from $\bar{f}$. This often makes the characteristic 0 PI-theory easier than the general theory.*

(iv) *Repeating the linearization process for each indeterminate appearing in f yields a multilinear polynomial, called the **multilinearization**, or **total multilinearization**, of f.*

(v) *There is a refinement of (iii) which may help us in nonzero character-istic. Write*

$$\Delta_1 f = \sum_{j=1}^{n-1} f_j(x_1, x_2; \vec{y})$$

where $\deg_1 f_j = j$ *(and thus,* $\deg_2 f_j = n - j$*). Then, for any* j*, we get* $\bar{f} = \bar{f}_j$*, i.e., we get* $\bar{f}$ *by performing the multilinearization procedure on any* f_j*, first on* x_1 *and then on* x_2*, and (1.5) becomes*

$$\bar{f}(x_1, \ldots, x_1, x_2, \ldots, x_2; \vec{y}) = j!(n - j)! f. \tag{1.6}$$

Example 1.2.14. *The multilinearizations of Example 1.2.8:*
(i), (ii) are already multilinear.
(iii) multilinearizes to the central polynomial

$$[x_1, x_2][x_3, x_4] + [x_3, x_4][x_1, x_2] + [x_1, x_4][x_3, x_2] + [x_3, x_2][x_1, x_4].$$

The multilinearization of x^n *is called the* **symmetric polynomial**

$$\tilde{s}_n = \sum_{\pi \in \mathrm{Sym}_n} x_{\pi(1)} \cdots x_{\pi(n)}. \tag{1.7}$$

$\tilde{s}_n$ *also is the multilinearization of (iv), since the (partial) linearization of* x *in the first step is 0.*

The multilinearization of (v) is $x_1 x_2 + x_2 x_1$*. Specializing back* $x_2 \mapsto x_1$ *yields* $2x_1^2$*, which is weaker than the original identity and only says that* A *has characteristic 2.*

1.2.4   PI-equivalence

Often we shall be interested in extracting information from $\mathrm{id}(A)$. This leads us to

Definition 1.2.15. *Algebras* A_1, A_2 *are called PI-**equivalent** if* $\mathrm{id}(A_1) = \mathrm{id}(A_2)$*; in this case we write* $A_1 \sim_{\mathrm{PI}} A_2$*.*

It is convenient to start with multilinear identities.

Proposition 1.2.16. *If* f *is a multilinear identity of* A *and* H *is a commu-tative* C*-algebra, then* f *is an identity of* $A \otimes_C H$*.*

Proof. $A \otimes 1$ spans $A \otimes H$ over H.   □

Thus, the class of PI-algebras is closed under commutative tensor exten-sions. We progress next to homogeneous identities.

Lemma 1.2.17. *If* F *is an infinite field, then every identity of an algebra* A *is a consequence of homogeneous identities.*

Proof. This is a well-known application of the Vandermonde argument, cf. Remark 1.1.7 and Exercise 8. □

Proposition 1.2.18. *If A is an algebra over a field F of characteristic zero, then* id(A) *is determined by the multilinear identities.*

Proof. We look a bit closer at the linearization procedure Δ_i. We want to recover f from its various multilinearizations. By Lemma 1.2.17 we may assume that f is homogeneous, so we conclude with Remark 1.2.13(iii). □

Thus, PI-equivalence in characteristic 0 is established by checking the multilinear identities. For this reason, some of our main theorems only hold for an algebra over a field of characteristic 0.

The situation for characteristic $p \neq 0$ is more delicate, because an identity need not be a consequence of its multilinearizations; the simplest example is the Boolean identity (see Example 1.2.8(v) and Example 1.2.14). When F is a finite field, the proposition fails miserably, cf. Exercise 27. Nevertheless, with some care, one can treat algebras over arbitrary infinite fields. Let us record the following result for further use.

Proposition 1.2.19. *If C is an infinite integral domain and K is a commutative C-algebra without C-torsion, then $A \otimes_F K \sim_{\mathrm{PI}} A$, for any F-algebra A.*

Proof. We need only consider homogeneous polynomials, say of degree d_i, in x_i.

$$f(a_1 \otimes k_1, \ldots, a_m \otimes k_m) = f(a_1, \ldots, a_m) \otimes k_1^{d_1} \ldots k_m^{d_m}.$$

Thus one side is 0 iff the other side is 0. Hence id($A \otimes K$) ⊆ id(A).

To prove (⊇), we need to evaluate f to be 0 on sums of simple tensors, which is seen as a modification of the linearization argument; see [Ro80, Theorem 2.3.29], for details. □

In particular, the hypothesis of Proposition 1.2.19 holds over any infinite field.

Remark 1.2.20. *If $\mathcal{I}, \mathcal{J} \lhd A$ with $\mathcal{I} \cap \mathcal{J} = 0$, then $A \sim_{\mathrm{PI}} A/\mathcal{I} \times A/\mathcal{J}$. (Indeed, there is a natural injection $A \to A/\mathcal{I} \times A/\mathcal{J}$; but if $f \in$ id($A/\mathcal{I} \times A/\mathcal{J}$), then checking components we see that $f(A) \subseteq \mathcal{I} \cap \mathcal{J} = 0$.)*
Thus, by induction, if A is a subdirect product of $A_1, \ldots, A_m$ then

$$A \sim_{\mathrm{PI}} A_1 \times \cdots \times A_m. \tag{1.8}$$

Remark 1.2.20 leads us to a slightly weaker definition:

Definition 1.2.21. *A f.d. algebra A is **PI-reducible** if there are f.d. algebras A_1 and A_2 with* id(A_j) ⊃ id(A) *and* id(A_1) ∩ id(A_2) = id(A).

In other words, A is PI-equivalent to a finite subdirect product of f.d. algebras, each satisfying "more" identities than A.

1.2.5   Alternating polynomials

Definition 1.2.22. *An n-linear polynomial $f(x_1, \ldots, x_m)$ is n-**alternating** if f becomes 0 when we specialize x_j to x_i for any $1 \leq i < j \leq n$; i.e.,*

$$f(x_1, \ldots, x_i, \ldots, x_i, \ldots, x_m) = 0 \qquad (1.9)$$

Our next result is clear in characteristic $\neq 2$, but in fact holds in general.

Proposition 1.2.23. *A multilinear n-linear polynomial f is n-alternating iff f satisfies the condition*

$$f(x_1, \ldots, x_i, \ldots, x_j, \ldots, x_m) + f(x_1, \ldots, x_j, \ldots, x_i, \ldots, x_m) = 0 \qquad (1.10)$$

for all $1 \leq i < j \leq n$.

Proof. Write $\tau = (i, j)$. Given a polynomial f, we write

$$\bar{f} := f(\ldots, x_i, \ldots, x_j, \ldots) + f(\ldots, x_j, \ldots, x_i, \ldots).$$

If f satisfies (1.9), then we get (1.10) by linearizing. To wit, as in Definition 1.2.12(i), the left side of $\bar{f}$ can be rewritten as

$$f(\ldots, x_i + x_j, \ldots, x_i + x_j, \ldots) - f(\ldots, x_i, \ldots, x_i, \ldots) - f(\ldots, x_j, \ldots, x_j, \ldots),$$

all of whose terms are 0 by definition.

Conversely, suppose f satisfies (1.10), i.e., $\bar{f} = 0$. Let S denote the subset $\{\pi \in S_n : \pi^{-1}i < \pi^{-1}j\}$. Write f_1 to be the sum of those monomials in f in which x_i precedes x_j, and f_2 to be the sum in which x_j precedes x_i. Thus we can write

$$f_1 = \sum_{\pi \in S} \alpha_\pi h_\pi; \qquad f_2 = \sum_{\pi \in S} \beta_\pi h_{\tau\pi},$$

where $h_\pi = x_{\pi(1)} \cdots x_{\pi(n)}$ and τ is the transposition (i, j).

Then $\overline{h_\pi} = \overline{h_{\tau\pi}}$ for each $\pi \in S$, implying

$$0 = \bar{f} = \sum_{\pi \in S} (\alpha_\pi + \beta_\pi) h_\pi,$$

from which we conclude $\alpha_\pi = -\beta_\pi$ for each $\pi \in S$. $\qquad\square$

For example, $[x_1, x_2]$ is 2-alternating. More generally, the **standard polynomial**

$$s_t = \sum_{\pi \in S_t} \operatorname{sgn}(\pi) x_{\pi(1)} \cdots x_{\pi(t)}$$

is t-multilinear and t-alternating, since the terms cancel out when any two x's are the same.

1.2.5.1 Capelli polynomials

Our main example will be the **Capelli polynomial**polynomial!Capelli

$$c_t = \sum_{\pi \in S_t} \operatorname{sgn}(\pi) x_{\pi(1)} x_{t+1} x_{\pi(2)} x_{t+2} \cdots x_{2t-1} x_{\pi(t)} x_{2t},$$

which is $2t$-multilinear and t-alternating for the analogous reason. In order to distinguish the permuted indeterminates from the fixed ones, we also rewrite this as

$$c_t(\vec{x}; \vec{y}) = c_t(x_1, \ldots, x_t, y_1, \ldots, y_t) = \sum_{\pi \in S_t} \operatorname{sgn}(\pi) x_{\pi(1)} y_1 x_{\pi(2)} y_2 \cdots x_{\pi(t)} y_t.$$

For example, for $t = 2$, we have

$$c_4 = x_1 y_1 x_2 y_2 - x_2 y_1 x_1 y_2.$$

Lemma 1.2.11 contains the key for constructing identities for f.d. algebras.

Proposition 1.2.24. *Any t-alternating polynomial f is an identity for every algebra A spanned as C-module by $< t$ elements.*

Proof. It is enough to check that f vanishes on a spanning set B of A. But choosing B to have $< t$ elements, every evaluation of f on B is 0, by definition. $\square$

Corollary 1.2.25. *Any f.d. algebra A is PI, satisfying the Capelli identity c_t for each $t > \dim A$.*

Corollary 1.2.26. *$M_n(C)$ is a PI-algebra, since it is spanned by the n^2 matrix units.*

In particular, s_{n^2+1} and c_{n^2+1} are PIs of $M_n(C)$. The standard polynomial upstaged the Capelli polynomial throughout the early history of PI theory, in view of the celebrated Amitsur-Levitzki Theorem 1.4.11, which shows that s_{2n} is the unique PI of minimal degree of $M_n(C)$. However, as the alternating property of polynomials became more important over time, the Capelli polynomial assumed its proper leading role. Today it is impossible to delve deeply into the PI theory without constant recourse to the Capelli polynomial, and most of this book is devoted to exposing its more subtle properties.

Since these results yield Capelli identities c_n for "large enough" n, it is useful for us to have ways of building c_n from Capelli polynomials of smaller degree.

Lemma 1.2.27.

(i) *Suppose $n = n_1 + n_2 + \cdots + n_t$. If A satisfies the multilinear identity $c_{n_1} \cdots c_{n_t}$ (using distinct indeterminates), then A satisfies the Capelli identity c_n. More precisely, if $c_n(a_1, \ldots, a_n, b_1, \ldots, b_n) \neq 0$, then one can partition $\{1, \ldots, n\}$ into $I_1 \cup \cdots \cup I_t$ with $I_j = \{i_{j,1}, \ldots, i_{j,n_j}\}$ such that*

$$c_{n_j}(a_{i_{j,1}}, \ldots, a_{i_{j,n_j}}, b_{i_{j,1}}, \ldots, b_{i_{j,n_j}}) \neq 0$$

for each $1 \leq j \leq t$.

(ii) *If $A/\mathcal{I}$ satisfies c_m for m odd, with $\mathcal{I} \lhd A$ and $\mathcal{I}^k = 0$, then A satisfies c_{km}.*

(iii) *If $A/\mathcal{I}$ satisfies c_m, with $\mathcal{I} \lhd A$ and $\mathcal{I}^k = 0$, then A satisfies $c_{k(m+1)}$.*

Proof. (i) Viewing the symmetric group $S_{n_1} \times \cdots \times S_{n_t} \hookrightarrow S_n$, we partition S_n into orbits under the subgroup $S_{n_1} \times \cdots \times S_{n_t}$ and match the permutations in c_n.

(ii) This time we note that any interchange of two odd-order sets of letters has negative sign, so we partition S_{km} into k parts each with m letters.

(iii) Any algebra satisfying c_m for m even, also satisfies c_{m+1}, and $m+1$ is odd. □

1.2.6   The action of the symmetric group

Let us see how Capelli polynomials arise so naturally in the theory of alternating polynomials.

Definition 1.2.28. *Anticipating Chapter 3, given a polynomial $f(x_1, \ldots, x_n, \vec{y})$ and $\pi \in S_n$, we write πf for $f(x_{\pi(1)}, \ldots, x_{\pi(n)}, \vec{y})$.*

This defines an action of the symmetric group S_n on the free algebra $C\{X, Y\}$. In this notation,

$$s_n = \sum_{\pi \in S_n} \text{sgn}(\pi) \pi h$$

where $h = x_1 \cdots x_n$;

$$c_n = \sum_{\pi \in S_n} \text{sgn}(\pi) \pi h$$

where $h = x_1 y_1 \cdots x_n y_n$.

Corollary 1.2.29. *The following are equivalent for an n-linear polynomial $f(x_1, \ldots, x_n; \vec{y})$, where $f_{(n)}$ denotes the sum of monomials of f in which $x_1, \ldots, x_n$ appear in ascending order:*

(i) *f is n-alternating.*

(ii) $\pi f = \operatorname{sgn}(\pi)f$ *for every* $\pi \in S_n$.

(iii) $f = \sum_{\pi \in S_n} \operatorname{sgn}(\pi)\, \pi f_{(n)}$.

(iv) $f = \sum_i g_{0,i} c_n(x_1, x_2, \ldots x_n, g_{1,i}, g_{2,i}, \ldots, g_{n,i})$ *for suitable polynomials* $g_{0,i}, g_{1,i}, g_{2,i}, \ldots, g_{n,i}$.

Proof. $(i) \Leftrightarrow (ii)$ Every permutation is a product of transpositions, so we apply Proposition 1.2.23.

$(ii) \Rightarrow (iii)$ by comparing the different monomials of f.

$(iii) \Rightarrow (iv)$ is immediate.

$(iv) \Rightarrow (i)$ is clear, since c_n is n-alternating. $\qquad\square$

Remark 1.2.30. *Thus, by (iv), each n-alternating polynomial can be written in terms of c_n. If $c_n \in \operatorname{id}(A)$, then $\operatorname{id}(A)$ contains every n-alternating polynomial.*

1.2.6.1 Alternators and symmetrizers

Definition 1.2.31. *The* **alternator** *of a multilinear polynomial f in indeterminates* $x_{i_1}, \ldots, x_{i_n}$ *is*

$$f_{\mathcal{A}(i_1,\ldots,i_n;X)} := \sum \operatorname{sgn}(\pi) f,$$

summed over all permutations π of $\{i_1, \ldots, i_n\}$. Up to sign, this is independent of the order of $i_1, \ldots, i_n$. When $i_1 = 1, \ldots, i_n = n$, we call this the **n-alternator**

$$\sum_{\pi \in S_n} \operatorname{sgn}(\pi) \pi f.$$

For example, the n-alternator of $x_1 \cdots x_n$ is s_n, and the n-alternator of $x_1 y_1 \cdots x_n y_n$ is c_n.

Remark 1.2.32. *If f already is alternating, then its n-alternator is $n!f$. On the other hand, if $f = \tau f$ for some odd permutation τ (for example, for a transposition), then its n-alternator is 0. Thus, before applying the alternator, we often ascertain that f lacks such symmetry.*

See Exercise 23 for some examples. There is an analog needed in Chapter 7. We say that a polynomial f is **symmetric** in $x_1, \ldots, x_n$ if $f = \tau f$ for any transposition $\tau = (i_1, i_2)$.

Definition 1.2.33. *Define the* **n-symmetrizer** *$\tilde{f}$ of a n-linear polynomial $f(x_1, \ldots, x_n, \vec{y})$ to be*

$$\sum_{\pi \in S_t} \pi f.$$

For example, the n-symmetrizer of $x_1 \cdots x_n$ is $\tilde{s}_n$ of (1.7). (Note the slight discrepancy of notation; we do not worry about ambiguity since the symmetrizer of s_n is 0.)

Remark 1.2.34.

(i) *If $f(x_1, \ldots, x_{n_1}; \vec{y})$ is the linearization in x_1 of a polynomial $g(x_1; \vec{y})$, each of whose monomials has degree n_1 in x_1, then*

$$f(x_1, \ldots, x_1; \vec{y}) = n_1! g,$$

  cf. Remark 1.2.13.

(ii) *If $f(x_1, \ldots, x_n; \vec{y})$ is n-linear and symmetric in $x_1, \ldots, x_n$, then $n! f$ is the linearization in x_1 of the polynomial $f(x_1, \ldots, x_1; \vec{y})$, cf. Remark 1.2.13(iii).*

(iii) *The symmetrizers of πf and of f are the same, for any $\pi \in S_n$.*

Thus, there is a natural correspondence between symmetric polynomials and linearizations with respect to a single indeterminate.

1.3   Graded Algebras

We turn to another major example.

Definition 1.3.1. *A C-algebra A is $\mathcal{M}$-graded by a monoid $(\mathcal{M}, +)$ if one can write*

$$A = \oplus \{A_g : g \in \mathcal{M}\},$$

*a direct sum of C-modules, such that $A_g A_h \subseteq A_{gh}$ for each $g, h \in \mathcal{M}$. An element belonging to some A_g is called **homogeneous**.*

*The **support** under the grading, denoted $\mathrm{supp}(A)$, is $\{g \in \mathcal{M} : A_g \neq 0\}$.*

Usually the grading monoid $\mathcal{M}$ is understood, so we merely write **graded** instead of $\mathcal{M}$-**graded**.

Given $a \in A$, we can write $a = \sum a_g$ (uniquely) where $a_g \in A_g$, so any element can be written uniquely as a sum of homogeneous elements. In particular, $(a + b)_g = a_g + b_g$ for all $a, b \in A$ and all $g \in \mathcal{M}$.

Write e for the identity element of M. Then A_e is always a subalgebra of A, and each A_g is an A_e-module. Recall that a monoid $\mathcal{M}$ is **cancellative** if $hg = hk$ implies $g = k$. Any group is cancellative. In many of our examples, the grading monoid $\mathcal{M}$ is Abelian, written in additive notation; then $e = 0$. In particular, $\mathbb{N}$ denotes $(N, +)$ and $\mathbb{Z}$ denotes $(Z, +)$.

Remark 1.3.2. *Grades enter the theory in several significant ways:*

(i) *The commutative polynomial algebra $F[\Lambda] = F[\lambda_1, \ldots, \lambda_n]$ can be graded by $\mathbb{N}^{(n)}$, by associating to any monomial f the "multi-degree" $(d_1, \ldots, d_n)$, where d_i is the degree of λ_i in f; then $F[\Lambda]_{(d_1, \ldots, d_n)}$ is spanned by the monomials of multi-degree $(d_1, \ldots, d_n)$. On the other hand, $F[\Lambda]$ is $\mathbb{N}$-graded when we take $F[\Lambda]_d$ to be spanned by all monomials of total degree d. This is a standard tool in the theory of commutative algebra, relating to Hilbert Series which we consider more generally for PI-algebras in Chapter 11.*

(ii) *In analogy to (i), the free algebra $C\{X\}$ is $\mathbb{N}$-graded, seen by taking $C\{X\}_n$ to be the homogeneous polynomials of degree n. This has already been used implicitly (in treating homogeneous polynomials) and enables us to specify many of their important properties.*

(iii) *The free affine associative algebra $A = C\{x_1, \ldots, x_\ell\}$ is $\mathbb{N}^{(\ell)}$-graded, by defining the **multi-degree** of a monomial $f \in A$ as $(d_1, \ldots, d_\ell)$, where $d_i = \deg_i f$.*

(iv) *One can develop a graded structure theory, and thereby obtain grades on other algebras which play a key role in the PI-theory, such as relatively free algebras, cf. §1.8. We present some basic properties below, but only develop the theory in depth in Volume II.*

(v) *One grading monoid of great interest for us here is the group $\mathbb{Z}/2$, in which case we call A a **superalgebra**. This is crucial in the conclusion of the solution of Specht's conjecture in Chapter 7, and also is useful in describing counterexamples in Chapter 9.*

(vi) *The Grassmann algebra turns out to be the most important superalgebra for us, used in Chapter 7 to link affine superalgebras with ungraded algebras.*

We consider these various aspects in turn.

1.3.1   Grading the free algebra

Remark 1.3.3. *If A, B are $\mathbb{N}$-graded algebras over C, then $A \otimes_C B$ can be $\mathbb{N}$-graded via*

$$(A \otimes B)_n = \sum_{j+k=n} A_j \otimes B_k.$$

The following generalization is also needed quite often.

Definition 1.3.4 ([Ro05, Example 18.38]). *Let V be a C-module. Define $T^0(V) = C$ and, for $n \geq 1$,*

$$T^n(V) := V^{\otimes n} = V \otimes \cdots \otimes V \quad (n \text{ times})$$

and

$$T(V) = \bigoplus_{n=0}^{\infty} T^n(V),$$

the **tensor algebra** *of V. We define multiplication*

$$T^m(V) \times T^n(V) \to T^{m+n}(V)$$

by $(a, b) \mapsto a \otimes b$, *which extends naturally to a multiplication on* $T(V)$ *that makes it an* $\mathbb{N}$*-graded algebra.*

Remark 1.3.5. *If V is a free module with base* $\{x_1, \ldots, x_\ell\}$, *then*

$$C\{x_1, \ldots, x_\ell\} \approx T(V),$$

with the submodule of homogeneous polynomials in $x_1, \ldots, x_\ell$ *of total degree n isomorphic to* $T^n(V)$.

1.3.2   Graded modules

In analogy to the ungraded theory, we introduce graded modules in order to study graded algebras.

Definition 1.3.6. *We say that a left A-module M is* $\mathcal{M}$*-**graded** if*

$$M = \bigoplus_{g \in \mathcal{M}} M_g$$

(where each M_g *is a C-module) and* $A_g M_h \subseteq M_{gh}$ *for all* $g, h \in \mathcal{M}$.

A **graded submodule** *is a submodule of M which is graded with respect to the same grading as M.*

Remark 1.3.7. *If* $\varphi : \mathcal{M} \to \mathcal{N}$ *is a homomorphism of monoids, then any* $\mathcal{M}$*-graded module M also is* $\mathcal{N}$*-graded, defining* $M_h = \sum_{\varphi(g)=h} A_g$ *for* $h \in \mathcal{N}$. *(This works also for graded algebras.)*

Remark 1.3.8. *If N is a graded submodule of an* $\mathcal{M}$*-graded module M, then* $M/N \cong \bigoplus_g M_g/N_g$ *is also* $\mathcal{M}$*-graded, i.e., we put* $(a + N)_g = a_g + N$. *This is easily seen to be well-defined.*

Let us generalize this slightly.

Definition 1.3.9. *Let M and N be* $\mathcal{M}$*-graded A*$-$*modules. An A*$-$*module homomorphism* $\varphi : M \to N$ *is* h*-**graded** for a given* $h \in \mathcal{M}$ *if* $\varphi(M_g) \subseteq N_{hg}$ *for every* $g \in \mathcal{M}$.

In particular, $\varphi : M \to N$ *is* e*-**graded** if* $\varphi(M_g) \subseteq N_g$ *for each* g *in* $\mathcal{M}$.

We usually write $\mathcal{M}$ in additive notation, in which case e-graded means 0-graded.

Lemma 1.3.10. *When the grading monoid $\mathcal{M}$ is cancellative, the kernel of any h-graded module homomorphism $\varphi : M \to N$ is a graded submodule of M.*

Proof. Suppose $a = \sum_{g \in G} a_g \in \ker \varphi$ for $a_g \in M_g$. Then

$$\varphi(a) = \sum \varphi(a_g) \in \oplus N_{hg},$$

which is possible only if all $\varphi(a_g) = 0$. $\square$

Lemma 1.3.11 (Generalizing Remark 1.3.8). *If N is a graded submodule of an $\mathcal{M}$-graded module M and $h \in \mathcal{M}$ is invertible, then $M/N \cong \bigoplus_g M_g/N_g$ is also $\mathcal{M}$-graded, where we put $(a + N)_g = a_{h^{-1}g} + N$, and the natural homomorphism $M \to M/N$ is h-graded.*

Proof. One just has to check that we wind up in the correct components when adding and multiplying by elements of A. $\square$

Proposition 1.3.12. *Suppose M is an G-graded module, for G a group. The F-algebra $End_A(M)$ is G-graded, by declaring $End_A(M)_h$ to consist of all h-graded homomorphisms.*

Proof. This indeed defines a G-grading on $End_A(M)$, since the composite of an h-graded homomorphism and an h'-graded homomorphism is an hh'-graded homomorphism. Moreover, given $\varphi \in End_A(M)$ and letting P_g denote the projection onto M_g, we define

$$\varphi_h = \sum_{g \in G} P_{hg} \varphi P_g$$

and have

$$\sum_{h \in G} \varphi_h = \sum_{h,g \in G,} P_{hg} \varphi P_g = \left(\sum_{h \in G} P_h \right) \cdot \varphi \cdot \left(\sum_{g \in G} P_g \right) = \varphi.$$

$\square$

1.3.3 Superalgebras

One grading monoid of great interest for us here is $\mathbb{Z}/2$, in which case we call A a **superalgebra**. Thus, $A = A_0 \oplus A_1$ where $A_0^2, A_1^2 \subseteq A_0$, and $A_0 A_1, A_1 A_0 \subseteq A_1$. The C-submodule A_0 (respectively A_1) is called the **even** (respectively **odd**) part of A.

Remark 1.3.13. *By Remark 1.3.7, any $\mathbb{Z}$-graded or $\mathbb{N}$-graded algebra A becomes a superalgebra where $\sum \{A_m : m \text{ even}\}$ becomes the even part and $\sum \{A_m : m \text{ odd}\}$ becomes the odd part.*

Example 1.3.14. *The free algebra $C\{X\}$ becomes a superalgebra, by taking the even (respectively odd) part to be generated by all homogeneous polynomials of even (respectively odd) degree. This is the grade obtained by applying Remark 1.3.7 to Remark 1.3.2(ii).*

1.3.4   Graded ideals and homomorphic images

Definition 1.3.15. *An ideal $\mathcal{I}$ of A is **graded** if $\mathcal{I}$ is graded as a submodule of A.*

Thus, $\mathcal{I}$ is a graded ideal of A iff $\mathcal{I} = \sum \mathcal{I}_g$ where $\mathcal{I}_g = \mathcal{I} \cap A_g$, i.e., each element of $\mathcal{I}$ is a sum of homogeneous elements of $\mathcal{I}$. It is easy to see that an ideal is graded iff it is generated by homogeneous elements. Remark 1.3.8 also yields:

Remark 1.3.16. *If $\mathcal{I}$ is a graded ideal of A, then $A/\mathcal{I} = \bigoplus_g A_g/\mathcal{I}_g$ is $\mathcal{M}$-graded as an algebra, where we put $(a + \mathcal{I})_g = a_g + \mathcal{I}_g$.*

Thus, graded ideals take on a crucial role in the structure theory, and we would like to relate them to the ideals of A_e.

Obviously, if $\mathcal{I}$ is a graded ideal of A, then $\mathcal{I}_e \lhd A_e$. The other direction is trickier.

Example 1.3.17. *For the superalgebra $A = M_2(F) = A_0 \oplus A_1$ with respect to the checkerboard grading, A_0 is a direct product of two fields, so is not simple.*

We do have:

Lemma 1.3.18. *When A is graded by an Abelian cancellative monoid $\mathcal{M}$, $\mathrm{Cent}(A)$ is also $\mathcal{M}$-graded.*

Proof. If $z = \sum z_g \in \mathrm{Cent}(A)$, then for any homogeneous element $a \in A_h$ we have

$$0 = \left[a, \sum z_g\right] = \sum [a, z_g] \in \bigoplus A_{hg},$$

implying each $[a, z_g] = 0$, and thus $z_g \in Cent(A)$. □

Remark 1.3.19. *If $\mathcal{I} \lhd \mathrm{Cent}(A_e)$, then $A\mathcal{I} = \oplus_g A_g\mathcal{I}$ is a graded ideal of A.*

Here is a way of relating A to A_e.

Lemma 1.3.20. *For any cancellative monoid $\mathcal{M}$ and any elements $g_1, \dots, g_t \in \mathcal{M}$ with $t > |\mathcal{M}|(m + 1)$, there are $i_1 < i_2 < \cdots < i_m$ such that $g_{i_j+1} \cdots g_{i_{j+1}} = e$ for each $1 \le j \le m$.*

Proof. For each $u \le t$ define $h_u = g_1 \cdots g_u$. Then $m + 1$ of these must be the same, i.e., there are $i_1 < i_2 < \cdots < i_m$ such that $h_{i_1} = \cdots = h_{i_{m+1}}$. Writing $h_{i_{j+1}} = h_{i_j}\tilde{h}_{i_j}$ and canceling $h_{i_{j+1}} = h_{i_j}$ yields the desired result $\tilde{h}_{i_j} = e$ for each j. □

Proposition 1.3.21. *If $\mathcal{M}$ is cancellative and $\mathcal{I}$ is a graded ideal of A with $\mathcal{I}_e$ nilpotent, then $\mathcal{I}$ is a nilpotent ideal of A.*

Proof. Suppose $\mathcal{I}_e^m = 0$. We claim that $\mathcal{I}^{|\mathcal{M}|(m+1)} = 0$. It is enough to show that every product of homogeneous elements of $\mathcal{I}$ of length $|\mathcal{M}|(m+1)$ is 0. But by the lemma, each such product contains a product of m elements of $\mathcal{I}_e$, which is 0. □

Corollary 1.3.22. *If $\mathcal{M}$ is cancellative and A is a f.d. graded algebra with no nonzero nilpotent graded ideals, then A_e is semisimple.*

Proof. A_e is f.d. with no nonzero nilpotent ideals, so is semisimple. □

1.3.5   Gradings on matrix algebras

As usual, much of the theory hinges on matrices.

Example 1.3.23.   *(i) We can grade the matrix algebra $A = M_n(C)$ by $\mathbb{Z}$, setting $A_m = \sum\{Ce_{ij} : i - j = m\}$; this gives rise naturally to a $\mathbb{Z}/m$-grading, taking the components modulo m. The most important special case is for $m = 2$, yielding the superalgebra with respect to the **checkerboard grading**. (Here e_{ij} is even iff $i - j$ is even.)*

(ii) Fixing any $k \le n$, $M_n(C)$ becomes a superalgebra where

$$M_n(C)_0 = \left\{\sum \alpha_{ij} e_{ij} : 1 \le i, j \le k \text{ or } k < i, j \le n\right\}; \tag{1.11}$$

$$M_n(C)_1 = \left\{\sum \alpha_{ij} e_{ij} : 1 \le i \le k < j \le n \text{ or } 1 \le j \le k < i \le n\right\}. \tag{1.12}$$

Letting $\ell = n - k$, it is customary to call this superalgebra $M_{k,\ell}(C)$. (This is a generalization of (i), cf. Exercise 15.)

The matrix algebra $M_n(F)$ has several important grades other than those of Example 1.3.23. In particular, the endomorphism algebra of a free graded vector space is graded, by Proposition 1.3.12. These have all been classified by Bakhturin and Zaicev [BaZa02]. We quote their results here, and sketch the proofs in Exercises 11ff. We take $\mathcal{M}$ to be a group G, not necessarily Abelian.

Definition 1.3.24. *Suppose $A = M_n(F)$ is G-graded, i.e., $A = \oplus_{g \in G} R_g$.*

- *The grading is **elementary** if there are $g_1, \ldots, g_n \in G$ such that*

$$e_{ij} \in A_{g_i^{-1} g_j}$$

  for each $1 \le i, j \le n$.

- *The grading is **fine** if $\dim_F A_g \le 1$ for each $g \in G$.*

For example, the grading of Example 1.3.23 is elementary. By definition, $|\operatorname{supp}(M_n(F))| = n^2$ for any fine grading.

Theorem 1.3.25. *Suppose F is algebraically closed, and $A = M_n(F)$ is G-graded. Then $n = n_1 n_2$, where G has a subgroup H of order n_1^2, and elements $\{g_1, \ldots, g_{n_2}\}$, such that there is a graded isomorphism $A \cong A' \otimes A''$, where $A' = M_{n_1}(F)$ has a fine H-grading and $A'' = M_{n_2}(F)$ has an elementary grading corresponding to $\{g_1, \ldots, g_{n_2}\}$ (as in Definition 1.3.24).*

The proof is sketched in Exercise 16.

1.3.6   The Grassmann algebra

Definition 1.3.26. *Let F be a field, and $V = \sum F x_i \subset F\{X\}$. The **Grassmann**, or **exterior**, algebra is $G := F\{X\}/I$ where I is the ideal $\langle v^2 : v \in V \rangle$ of $F\{X\}$. (This is defined more generally in Exercise 17.)*

Since I is generated by even polynomials, we see that $I \cap V = 0$, so we view $V \subseteq G$; from this point of view (which is the one we shall take), G is the algebra generated by V modulo the relations $v^2 = 0$, $\forall v \in V$.

Remark 1.3.27. *Since we shall refer to the Grassmann algebra so often, let us fix some basic notation. We fix a base $e_1, e_2, \ldots$ of V. By definition, for any α_i in F,*

$$0 = \left(\sum \alpha_i e_i\right)^2 = \sum \alpha_i^2 e_i^2 + \sum \alpha_i \alpha_j (e_i e_j + e_j e_i), \qquad (1.13)$$

which clearly is equivalent to the following set of relations:

$$e_i^2 = 0; \qquad e_i e_j = -e_j e_i, \quad \forall i \neq j. \qquad (1.14)$$

Let

$$B = \{1\} \cup \{e_{i_1} \cdots e_{i_m} : i_1 < \cdots < i_m, \quad m = 1, 2, \ldots\}. \qquad (1.15)$$

Clearly, G is spanned by the words $e_{i_1} e_{i_2} \cdots e_{i_m}$. We can use (1.14) to rearrange the indices in ascending order, and any repetition of e_i must be 0; hence G is spanned by elements of B. We claim that B is in fact a base for G. Otherwise given any relation $\sum \alpha_j b_j = 0$ of shortest length (for $b_j \in B$), suppose e_i appears in b_1 but not in b_2. Then $e_i b_1 = 0$, implying the relation $\sum_{j>1} \alpha_j e_i b_j = 0$ is of shorter length, contradiction.

Remark 1.3.28. *Let us now describe the $\mathbb{Z}/2$-grade directly in terms of B. A word $e_{i_1} \cdots e_{i_m} \in B$ is **even** (respectively **odd**) if m is even or odd, respectively. G_0 is spanned by the even words of B, and G_1 is spanned by the odd words of B. Furthermore, since $1 \in G_0$, we have $G_1 = G_0 V$.*

By definition, G_0 is the subalgebra generated by all words $e_{i_1} e_{i_2}$. But for any $v \in V$,

$$e_{i_1} e_{i_2} v = -e_{i_1} v e_{i_2} = v e_{i_1} e_{i_2};$$

thus, $e_{i_1} e_{i_2} \in \text{Cent}(G)$. The same argument also shows that if $a, b \in G$ are homogeneous, then $ab = \pm ba$, the $-$ sign occurring iff both a, b are odd. In particular, $\text{Cent}(G) = G_0$.

G can also be described in the following way: We say two elements a, b of an algebra A **strictly anticommute** if $arb = -bra$ for all $r \in A$. (One also defines "strictly commute" analogously, but we shall not need that notion.) Whereas any even element of G is central, we have

Remark 1.3.29. *(i) Any two odd elements a, b of G strictly anticommute. (Indeed, take any element $r = r_0 + r_1$, the sum of the even and odd parts;*

$$arb = r_0 ab - r_1 ab = -r_0 ba + r_1 ba = -b(r_0 + r_1)a = -bra.)$$

(ii) It follows that $ara = 0$ for any odd a, since writing $a = \sum a_i$ as a sum of odd words a_i, we have

$$ara = \sum_{i,j} a_i r a_j = \sum_i a_i r a_i + \sum_{i<j}(a_i r a_j + a_j r a_i) = 0 + 0 = 0.$$

Remark 1.3.30. *The innocent computation of (i) has far-reaching consequences. Its first application was an observation over half a century ago by P.M. Cohn that*

$$s_m(e_1, \ldots, e_m) = m! e_1 \cdots e_m, \tag{1.16}$$

since

$$e_{\pi(1)} \cdots e_{\pi(m)} = \text{sgn}(\pi) e_1 \cdots e_m \quad in \quad G,$$

and in fact, Remark 1.3.29 analogously implies

$$c_m(e_1, \ldots, e_{2m}) = m! e_1 \cdots e_{2m}. \tag{1.17}$$

Consequently, the Grassmann algebra of an infinite dimensional vector space in characteristic 0 cannot satisfy any Capelli identity!

This observation turns out to have deep significance since, as we shall see, affine PI-algebras do satisfy suitable Capelli identities. It was not until a generation later that Kemer fully utilized the fact that G lies on the opposite side of the coin from matrix algebras. See Exercise 18 for a description of the structure of G. Here is the most important PI of G.

Proposition 1.3.31. *G satisfies the **Grassmann identity** $[[x_1, x_2], x_3]$; equivalently, $[x_1, x_2]$ is G-central.*

Proof. It is enough to check for homogeneous elements a, b that $[a, b]$ is even (and thus, central). If a or b is even, then $[a, b] = 0$, so we may assume that a, b both are odd, in which case $[a, b]$ is even, as desired. $\square$

In fact, all identities of G are consequences of the Grassmann identity, by a fundamental result of Regev, cf. Corollary 6.2.8. The Grassmann algebra has played an increasingly important role in PI-theory, as discussed in Remark 1.4.12 and Chapter 7.

1.4   Identities and Central Polynomials of Matrices

Now we shall prove two of the most basic PI-theorems of matrix algebras: The Amitsur-Levitzki Theorem and the existence of central polynomials. Surprisingly, neither of these theorems is used much in this volume, outside of this chapter. The standard polynomial has been replaced largely by the Capelli polynomial, and the combinatorics of the Capelli polynomial can be used to replace the transition to commutative theory afforded by central polynomials which is emphasized in [Ro80]. Nevertheless, both the results and their proofs are important and instructive, and fully merit inclusion in an in-depth treatment of PIs. Each theorem now has a short proof, which also utilizes an extension of the notion of PI.

The following result enables us to pass to the special case $C = \mathbb{Z}$, when considering identities with integer coefficients.

Remark 1.4.1. *Any multilinear identity of $M_n(\mathbb{Z})$ also is an identity of $M_n(C)$ for any commutative ring C, the reverse direction holding if C has characteristic 0. (Indeed, $M_n(C) \approx M_n(\mathbb{Z}) \otimes_{\mathbb{Z}} C$, so we appeal to Proposition 1.2.16. If $\operatorname{char}(C) = 0$, then $\mathbb{Z} \hookrightarrow C$ so $\operatorname{id}(M_n(C)) \subseteq \operatorname{id}(M_n(\mathbb{Z}))$.)*

This fact actually holds for all identities, cf. Exercise 48. Let us turn to evaluating specific polynomials. Suppose we want to show that a multilinear polynomial $f(x_1, \ldots, x_n)$ is *not* an identity of an algebra A. Writing f as in (1.2), it suffices to find $a_1, \ldots, a_n$ with $ca_1 \cdots a_n \neq 0$ but every other rearrangement $a_{\sigma(1)} \cdots a_{\sigma(n)} = 0$; then the summation in the right side of (1.2) is 0, but the first term is nonzero. (This is accomplished most readily when $a_j a_i = 0$ for all $j > i$.) Let us see how this idea works with some examples.

Remark 1.4.2. *If $f(x_1, \ldots, x_n) \in \operatorname{id}(M_n(C))$ of degree d, then $d \geq 2n$. (Indeed, otherwise, we may assume that f is multilinear; say*

$$f = cx_1 x_2 \ldots x_d + \sum_{\sigma \neq 1} c_\sigma x_{\sigma(1)} x_{\sigma(2)} \ldots x_{\sigma(d)}.$$

Then

$$0 = f(e_{11}, e_{12}, e_{22}, e_{23}, \ldots, e_{k-1,k}, e_{kk}, \ldots) = ce_{1m}$$

where $m = \left[\frac{d}{2}\right] + 1$, a contradiction.)

In particular, the standard polynomial s_{2n-1} is not an identity of $M_n(C)$. For $n = 2$, we already know that s_3 is not a PI, but s_5 is a PI. To check s_4 it is enough to evaluate s_4 on the spanning set of matrix units $e_{11}, e_{12}, e_{21}, e_{22}$, and observe $s_4(e_{11}, e_{12}, e_{21}, e_{22}) = 0$. In general, the famous Amitsur-Levitzki Theorem (to be proved shortly) says s_{2n} is a PI of $M_n(C)$, for any commutative ring C, also cf. Exercise 22.

Remark 1.4.3. *The Capelli polynomial, although more complicated to write than the standard polynomial, is easier to analyze on matrices. It is easy to see that c_{n^2} is not an identity of $M_n(C)$. Indeed, we shall find matrix units $\bar{x}_k$ and $\bar{y}_k$ such that*

$$c_{n^2}(\bar{x}_1, \ldots, \bar{x}_{n^2}, \bar{y}_1, \ldots, \bar{y}_{n^2}) \neq 0.$$

First, we list a set of matrix units in any order, say

$$\bar{x}_1 = e_{11}, \quad \bar{x}_2 = e_{12}, \ldots, \quad \bar{x}_n = e_{1n}, \quad \bar{x}_{n+1} = e_{21}, \ldots, \quad \bar{x}_{n^2} = e_{nn}.$$

Note for any matrix unit a, we have $e_{i_1 j_1} a e_{i_2 j_2} = 0$ unless $a = e_{j_1 i_2}$, in which case $e_{i_1 j_1} a e_{i_2 j_2} = e_{i_1 j_2}$. Thus, there is a unique specialization of the y_k to matrix units $\bar{y}_k$ such that $\bar{x}_1 \bar{y}_1 \bar{x}_2 \bar{y}_2 \ldots \bar{x}_{n^2} \bar{y}_{n^2} = e_{11}$. But the same argument shows that $\bar{y}_{k-1} \bar{x}_{\pi(k)} \bar{y}_k = 0$ unless $\pi(k) = k$; hence

$$\bar{x}_{\pi(1)} \bar{y}_1 \bar{x}_{\pi(2)} \bar{y}_2 \ldots \bar{x}_{\pi(n^2)} \bar{y}_{n^2} = 0$$

for every $\pi \neq (1)$. We conclude that

$$c_{n^2}(\bar{x}_1, \ldots, \bar{x}_{n^2}, \bar{y}_1, \ldots, \bar{y}_{n^2}) = \bar{x}_1 \bar{y}_1 \ldots \bar{x}_{n^2} \bar{y}_{n^2} = e_{11}.$$

Multiplying $\bar{y}_{n^2}$ on the right by e_{1k}, we see that c_{n^2} takes on the value e_{1k}, and hence, by symmetry, any matrix unit is a value of c_{n^2}.

*Let us formalize this idea, for use in Chapter 6: We call a set of matrix units $u_1, \ldots, u_t; v_1, \ldots, v_t$ **compatible with** the monomial $x_1 y_1 x_2 y_2 \ldots x_t y_t$ if*

$$u_1 v_1 u_2 v_2 \cdots u_t v_t = e_{11}$$

but

$$u_{\sigma(1)} v_1 u_{\sigma(2)} v_2 \cdots u_{\sigma(t)} v_t = 0$$

for every permutation $\sigma \neq (1)$. Such a set exists iff $t \leq n^2$; when $t = n^2$ the $u_1, \ldots, u_t$ must be a base, i.e., must run over all the $n \times n$ matrix units. We just saw examples of this phenomenon. Then in this case

$$c_t(u_1, \ldots, u_t; v_1, \ldots, v_t) = u_1 v_1 u_2 v_2 \ldots u_t v_t = e_{11} \neq 0.$$

Example 1.4.4. *A related result: If $c_{n^2+1} \in \mathrm{id}(M_n(H))$ for an algebra H, then H is commutative. Indeed, take any $a, b \in H$ and, as in Remark 1.4.3, we list the set of matrix units*

$$\bar{x}_1 = e_{11}, \quad \bar{x}_2 = e_{12}, \quad \ldots, \quad \bar{x}_{n^2} = e_{nn},$$

and take the unique matrix units $\bar{y}_i$ such that $\bar{x}_1 \bar{y}_1 \bar{x}_2 \bar{y}_2 \ldots \bar{x}_{n^2} \bar{y}_{n^2} = e_{11}$. Note that $\bar{y}_1 = e_{11}$. Then (repeating $\bar{y}_1 = e_{11}$ in the evaluation)

$$0 = e_{11} c_{n^2+1}(a\bar{x}_1, b\bar{x}_1, \bar{x}_2 \ldots, \bar{x}_{n^2}, \bar{y}_1, \bar{y}_1, \bar{y}_2 \ldots, \bar{y}_{n^2})$$
$$= (abe_{11} - bae_{11})\bar{y}_1 \ldots \bar{x}_{n^2} \bar{y}_{n^2}$$
$$= (ab - ba)e_{11},$$

proving $ab = ba$, $\forall a, b \in H$, as desired.

One of the main themes in PI-theory is to utilize the (monic) characteristic polynomial

$$f_a = \det(\lambda I - a)$$

for any $a \in M_n(C)$. The Cayley-Hamilton theorem says $f_a(a) = 0$.

When C is a $\mathbb{Q}$-algebra, the coefficients can be expressed in terms of $\{\mathrm{tr}(a^m) : m \le n\}$, via **Newton's Formulas**, cf. [MacD95, p. 23] or [Jac80, p. 140], which are best formulated in terms of symmetric polynomials in commuting indeterminates $\lambda_1, \ldots, \lambda_n$ (not to be confused with the noncommutative symmetric polynomials defined earlier):

Remark 1.4.5 (Newton's Formulas). *Let*

$$e_m = e_m(\lambda_1, \ldots, \lambda_n) = \sum_{1 \le i_1 < \cdots < i_m \le n} \lambda_{i_1} \cdots \lambda_{i_m}$$

*be the **elementary symmetric polynomials**, and let*

$$p_k = p_k(\lambda_1, \ldots, \lambda_n) = \lambda_1^k + \cdots + \lambda_n^k$$

*be the **power symmetric polynomials** [MacD95]. Then, for $m = 1, 2, \ldots,$*

$$m e_m = \sum_{k=1}^{m} (-1)^{k-1} p_k e_{m-k}.$$

Recursively, Newton's formulas allow us to derive formulas of the form

$$e_m = q_m(p_1, \ldots, p_m),$$

where q is a suitable polynomial with rational coefficients and with constant term 0. (This is where we use characteristic 0; actually, we need only divide by $m!$.)

Thus, $m = 1$ implies $e_1 = p_1$; together with the case $m = 2$, we see that $e_2 = \frac{1}{2}(p_1^2 - p_2)$. Similarly, one deduces that $e_3 = \frac{1}{6}(p_1^3 - 3p_1 p_2 + 2p_3)$, etc.

Remark 1.4.6. *If $p_k = 0$ for $1 \le k \le m$, then $m e_m = 0$, implying (in characteristic 0) that $e_m = 0$.*

Let us translate this fact into matrix theory. Suppose a is an $n \times n$ matrix with characteristic values $\gamma_1, \ldots, \gamma_n$ in a field K. Clearly $\mathrm{tr}(a) = \sum_{i=1}^{n} \gamma_i$; hence

$$p_k(\gamma_1, \ldots, \gamma_n) = \mathrm{tr}(a^k).$$

Furthermore, let $f_a = \det(\lambda I - a)$ be the characteristic polynomial of a and write

$$f_a = \lambda^n + \alpha_1 \lambda^{n-1} + \cdots + \alpha_n.$$

Then $\alpha_m = (-1)^m e_m(\gamma_1, \ldots, \gamma_n)$.

Proposition 1.4.7. *Suppose* $a \in M_n(C)$, *where* C *is a commutative* $\mathbb{Q}$-*algebra, and denote* $t_k = \text{tr}(a^k)$. *Then each coefficient* α_m *can be written as a polynomial in* $t_1, \ldots, t_m$ *with rational coefficients and with constant term 0. In particular, if* $t_k = 0$ *for* $1 \leq k \leq n$, *then* $f_a = \lambda^n$.

Proof. First we assume that C is an integral domain. Write $a = (c_{i,j})$ for $(c_{i,j} \in C)$. Let K be the algebraic closure of the field of fractions of C. Then f_a factors completely over K:

$$f_a = \lambda^n + \alpha_1 \lambda^{n-1} + \cdots + \alpha_n = \prod_{i=1}^{n} (\lambda - \gamma_i).$$

As noted above, $\alpha_m = e_m(\gamma_1, \ldots, \gamma_n)$, so the first assertion follows from Newton's formulas. The assumption for the second assertion is

$$0 = t_k = p_k(\gamma_1, \ldots, \gamma_n)$$

for $1 \leq k \leq n$, so Remark 1.4.6 shows that $\alpha_m = 0$ for $1 \leq m \leq n$, as desired.

To prove the proposition in general, we turn to a trick that comes up rather frequently in PI-theory. Let $\xi_{i,j}$ be commuting indeterminates, and form the $n \times n$ **generic** matrix $\bar{x} = (\xi_{i,j}) \in M_n(D)$ where D is the integral domain $D = F[\xi_{i,j} : 1 \leq i, j \leq n]$. The homomorphism $\varphi : D \to C$, given by $\varphi : \xi_{i,j} \mapsto c_{i,j}$, extends to $\varphi : M_n(D) \to M_n(C)$ with $\varphi : \bar{x} \mapsto a$, and obviously, $\varphi(\text{tr}(\bar{x}^j)) = \text{tr}(a^j)$. Thus, it suffices to prove the proposition for the matrix $\bar{x}$, or, in other words, we may replace C by the integral domain D, for which case the proposition has just been proved. $\qquad\square$

Corollary 1.4.8. *Suppose* C *is a commutative* $\mathbb{Q}$-*algebra. If* $a \in M_n(C)$ *satisfies* $\text{tr}(a^r) = 0$, $r = 1, \ldots, n$, *then* $a^n = 0$.

These considerations lead us to consider identities formal involving traces, thereby giving rise to **trace identities** which will be studied in detail in Chapter 8 and more formally in Volume II; Newton's formulas enable us to translate the Cayley-Hamilton theorem into a trace identity for $M_n(C)$. On the other hand, we can find trace identities which hold for *all* n.

Example 1.4.9. *By definition,* $M_n(C)$ *satisfies the trace identity* $\text{tr}[x_1, x_2]$.

1.4.1 Standard identities on matrices

Generalizing Example 1.4.9, Kostant observed the following fact:

Remark 1.4.10. $\text{tr}\, s_{2k}(x_1, \ldots, x_{2k})$ *is a trace identity of* $M_n(C)$ *for any* k, n. *(Indeed for any matrices* $a_1, \ldots, a_{2k}$ *and any permutation* π, *we see that*

$$\text{tr}(a_{\pi(1)} a_{\pi(2)} \cdots a_{\pi(2k)}) = \text{tr}(a_{\pi(2)} \cdots a_{\pi(2k)} a_{\pi(1)});$$

but the cycle $(\pi(1)\ \pi(2)\ \ldots\ \pi(2k))$ *is odd, so the terms* $a_{\pi(1)} a_{\pi(2)} \cdots a_{\pi(2k)}$ *and* $a_{\pi(2)} \ldots a_{\pi(2k)} a_{\pi(1)}$ *appear in* $s_{2k}(a_1, \ldots, a_{2k})$ *with opposite sign, and their contributions cancel.)*

Combining these basic remarks about trace identities with the Grassmann algebra yields the following basic result:

Theorem 1.4.11 (Amitsur-Levitzki). *s_{2n} is an identity of $M_n(C)$, for every commutative ring C.*

Proof. (Rosset) We need only check $s_{2n} \in \text{id}(M_n(\mathbb{Q}))$, since then clearly $s_{2n} \in \text{id}(M_n(\mathbb{Z}))$, and we would conclude with Remark 1.4.1. Let us work in $M_n(G) \approx M_n(\mathbb{Q}) \otimes_{\mathbb{Q}} G$. Take arbitrary matrices $a_1, \dots, a_n$ in $M_n(\mathbb{Q})$, and consider $a_i \otimes e_i$ in $M_n(G)$. Recalling that $e_{\pi(1)} \dots e_{\pi(2k)} = \text{sgn}(\pi)e_1 \cdots e_{2k}$, we have

$$s_{2k}(a_1, \dots, a_{2k}) \otimes e_1 \cdots e_{2k} = \sum_{\pi \in S_{2k}} (a_{\pi(1)} \otimes e_{\pi(1)}) \dots (a_{\pi(2k)} \otimes e_{\pi(2k)})$$

$$= (a_1 \otimes e_1 + \cdots + a_{2k} \otimes e_{2k})^{2k}.$$

Thus, letting $c = (a_1 \otimes e_1 + \cdots + a_{2n} \otimes e_{2n})^2$, it suffices to show that $c^n = 0$.

Note that $c \in M_n(G_0)$; since G_0 is commutative, using Corollary 1.4.8, it is enough to show that $\text{tr}(c^k) = 0$ for each $k \leq n$. But applying (1.17) in reverse, in conjunction with Remark 1.3.29(ii),

$$\text{tr}((a_1 \otimes e_1 + \cdots + a_{2n} \otimes e_{2n})^{2k})$$

$$= \text{tr}\left(\sum_{1 \leq i_1, \dots, i_{2k} \leq 2n} (a_{i_1} \otimes e_{i_1}) \dots (a_{i_{2k}} \otimes e_{i_{2k}}) \right)$$

$$= \sum_{i_1 < \cdots < i_k} \text{tr}(s_{2k}(a_{i_1}, \cdots, a_{i_{2k}})) \otimes e_{i_1} \cdots e_{i_{2k}} = 0, \qquad (1.18)$$

by Remark 1.4.10.   □

Remark 1.4.12. *This proof anticipates a key idea we shall need in Chapter 7. To check s_{2n} on the algebra $A = M_n(\mathbb{Z})$, we looked at the symmetric polynomial evaluated on the odd part of $A \otimes G$. Procesi [Pro14] has formalized this argument, noting that Rosset's proof of Theorem 1.4.11 boils down to verifying the "superidentity" that any odd matrix (in $M_n(G_1)$) is nilpotent of index $2n$.*

The interplay between $\text{id}(A)$ and certain graded identities of the superalgebra $A \otimes G$ lies at the heart of Kemer's solution of Specht's problem.

1.4.2   Central polynomials for matrices

We turn to central polynomials for matrices. The easiest verification of a central polynomial is obtained by extracting the main idea from Razmyslov's construction, and trivializing the remainder of his proof.

Definition 1.4.13. *A* **k-weak identity** *of $M_n(F)$ is a polynomial $f(x_1, \ldots, x_m)$ that vanishes on all substitutions to matrices $a_1, \ldots, a_m$ for which* $\mathrm{tr}(a_1) = \cdots = \mathrm{tr}(a_k) = 0$.

For example, the Capelli polynomial c_{n^2} is a multilinear n^2-weak identity of $M_n(F)$ (since it is n^2-alternating but the matrices of trace 0 have dimension only $n^2 - 1$), but c_{n^2} is not an identity of $M_n(F)$, by Remark 1.4.3. Here is an example of lower degree found by Halpin, inspired by an example of Amitsur given in Exercise 20.

Lemma 1.4.14. $s_{n-1}([x, y], [x^2, y], \ldots, [x^{n-2}, y], [x^n, y])$ *is a 1-weak identity of $M_n(F)$.*

Proof. Suppose $a \in M_n(F)$ has trace 0. Then the characteristic polynomial of a has coefficient 0 for λ^{n-1}, thereby yielding a linear dependence for $1, a, \ldots, a^{n-2}, a^n$. Commuting with arbitrary b, noting that $[1, b] = 0$, we see that $[a, b], \ldots, [a^{n-2}, b], [a^n, b]$ are linearly dependent, so can be spanned by $n - 2$ elements, implying $s_{n-1}([a, b], [a^2, b], \ldots, [a^{n-2}, b], [a^n, b]) = 0$, i.e., $s_{n-1}([x, y], [x^2, y], \ldots, [x^{n-2}, y], [x^n, y])$ is a 1-weak identity. $\square$

This 1-weak identity has degree $n + \frac{1}{2}(n-2)(n-1) = \frac{1}{2}(n^2 - n + 2)$ in x and $n-1$ in y. Let $m = \frac{1}{2}(n^2 - n + 2)$. Multilinearizing at x_1 yields an m-weak identity $f(x_1, \ldots, x_m, y_1, \ldots, y_{n-1})$. To get a multilinear 1-weak identity we substitute $[x_{i1}, x_{i2}]$ for x_i, $2 \le i \le m$, since these will specialize to matrices of trace 0. This polynomial has total degree

$$m + (m-1) + n - 1 = n^2 - n + 1 + n - 1 = n^2,$$

and is not an identity of $M_n(C)$, cf. Exercise 24.

The easiest way to understand the connection between central polynomials and 1-weak identities is in terms of another generalization of identity.

Definition 1.4.15. *A* **linear generalized identity** *(LGI) for A is an expression $f = \sum_{j=1}^{k} a_j x b_j$ with a_j, b_j in A, for which $\sum a_j r b_j = 0$ for all r in A. Likewise f is* **linear generalized central** *(LGC) if $\sum a_j r b_j \in \mathrm{Cent}(A)$ for all r in A.*

(More generally, a **generalized polynomial** is a noncommutative polynomial with coefficients from A, interspersed throughout the x_i, defined explicitly below in Definition 1.9.3. There is a well-developed theory of generalized polynomial identities, cf. [BeMM96] or [Ro80, Chapter 7]. This is discussed in the context of universal algebras in Volume II. However, for our present purposes, it suffices to consider this very special case.)

Lemma 1.4.16. *Suppose $f = \sum_j a_j x b_j$. Then $\sum b_j x a_j$ is an LGC of $M_n(C)$ iff $f(r) = 0$ for every commutator matrix $r = [v, w]$.*

Proof. In view of Remark 1.1.5, $\sum b_j x a_j$ is an LGC of $M_n(C)$ iff $\mathrm{tr}\left(\left[\sum_j b_j x a_j, u\right] v\right)$ is identically 0, which is

$$\mathrm{tr}\left(\sum_j b_j x a_j uv - \sum_j u b_j x a_j v\right) = \sum_j \mathrm{tr}(x a_j uv b_j) - \sum_j \mathrm{tr}(x a_j vu b_j)$$

$$= \sum_j \mathrm{tr}(x a_j [u,v] b_j) = \mathrm{tr}\left(x \sum_j a_j [u,v] b_j\right), \tag{1.19}$$

iff $\sum_j a_j [u,v] b_j = 0$ for all u, v.   □

We are ready for a major result.

Theorem 1.4.17 (Razmyslov). *There is a 1 : 1 correspondence between the multilinear central polynomials of $M_n(C)$ and the multilinear 1-weak identities that are not identities.*

Proof. Write

$$f(x, x_1, \ldots, x_n) = \sum_{i=1}^{t} f_i(x_1, \ldots, x_n) x g_i(x_1, \ldots, x_n),$$

and let

$$\hat{f}(x, x_1, \ldots, x_n) = \sum_{i=1}^{t} g_i(x_1, \ldots, x_n) x f_i(x_1, \ldots, x_n).$$

Picking matrices $r_1, \ldots, r_n$ arbitrarily, and letting $a_i = f_i(r_1, \ldots, r_n)$ and $b_i = g_i(r_1, \ldots, r_n)$, we see by Lemma 1.4.16 that $f \in \mathrm{id}(M_n(C))$ iff $\hat{f} \in \mathrm{id}(M_n(C))$. Likewise, by Lemma 1.4.16, $\hat{f}$ is central or an identity for $M_n(C)$, iff f is a 1-weak identity. Combining these two observations gives the result.   □

Since we have already constructed 1-weak identities which are not PIs for $M_n(C)$, we have the existence of central polynomials. Surprisingly, the minimal possible degree of a 1-weak identity (and thus, central polynomial) for $M_n(F)$ is not known. Formanek's polynomial [For72] has degree n^2, as does the central polynomial constructed via Theorem 1.4.17 from the Amitsur-Halpin 1-weak identity. However, for $n = 3$, a computer search uncovered a central polynomial of degree 8. The best result known, in [Dr95], is a central polynomial of degree $(n-1)^2 + 4$ for $M_n(\mathbb{Q})$.

1.5   Review of Major Structure Theorems in PI Theory

This section lists those basic theorems which we shall need from structure theory, which can be found in [Ro80] or [Ro88b]. They hold for PI-algebras over an arbitrary commutative ring, and are so basic for further reference in this book that we refer to them with letters rather than numbers, for special distinction.

1.5.1 Classical structure theorems

Theorem A. *(Kaplansky's Theorem) Any primitive algebra A satisfying a PI of degree d has the form $M_t(D)$ where D is a division algebra of dimension m^2 over $\mathrm{Cent}(A)$, with $mt \leq \left[\frac{d}{2}\right]$. Thus, $[A : \mathrm{Cent}(A)] = \left[\frac{d}{2}\right]^2$.*

Corollary 1.5.1. *Every primitive ideal of a PI-algebra A is maximal. In particular, the Jacobson radical $\mathrm{Jac}(A)$ is the intersection of the maximal ideals of A.*

In particular, every prime ideal of a Jacobson PI-algebra is the intersection of maximal ideals.

Theorem B (Amitsur). *Any semiprime PI-algebra A has no nonzero left or right nil ideals.*

In other words, if we define $\mathrm{Nil}(A)$ to be the sum of the nil left ideals of A, the algebra A is semiprime iff $\mathrm{Nil}(A) = 0$.

Corollary 1.5.2. *If A is a semiprime PI-algebra, then $\mathrm{Jac}(A[\lambda]) = 0$.*

Proof. Apply Theorem 1.1.11. □

Theorem C (Rowen). *Every ideal of a semiprime PI-algebra A intersects $\mathrm{Cent}(A)$ nontrivially.*

Corollary D (Posner et al). *If A is prime PI with center C, and $S = C \backslash \{0\}$, then the localization $S^{-1}A$ is simple and f.d. over its center $F = S^{-1}C$, the field of fractions of C.*

Corollary 1.5.3. *If A is prime PI with center C, then, for some n, either C is a finite field with $A \cong M_n(C)$, or C is infinite and $A \otimes_C F \sim_{\mathrm{PI}} M_n(F)$ for any infinite field F containing C.*

Proof. By Corollary D the algebra $\hat{A}$ of central fractions of $A \otimes_C F$ is central simple over the field of fractions F of C. If C is finite, then $F = C$, and $A = M_n(F)$ by a well-known theorem of Wedderburn. Thus, we may assume that C is infinite, in which case $A \subset A \otimes \bar{F}$ where $\bar{F}$ is the algebraic closure of F. $A \otimes \bar{F}$ is simple and f.d. over $\bar{F}$, so is isomorphic to $M_n(\bar{F})$ since $\bar{F}$ is algebraically closed. By Proposition 1.2.19, applied twice,

$$A \sim_{\mathrm{PI}} M_n(\bar{F}) \sim_{\mathrm{PI}} M_n(F).$$

□

This leads us to study algebras that are PI-equivalent to $M_n(F)$. The following definition focuses on the key properties.

Definition 1.5.4. *Write* $\mathcal{M}_{n,F}$ *for* $\mathrm{id}(M_n(F))$, *and* $\mathcal{M}_n$ *for* $\mathcal{M}_{n,\mathbb{Q}} = \mathrm{id}(M_n(\mathbb{Q}))$. *An algebra* A *has PI-class* n *if*

$$\mathcal{M}_n = \mathrm{id}(A) \cap \mathcal{M}_{n-1};$$

in other words, if A *satisfies all identities of* $M_n(\mathbb{Q})$, *but no extra identities of* $M_{n-1}(\mathbb{Q})$.

Thus, $M_n(C)$ has PI-class n, whereas it has PI-degree $2n$ by the Amitsur-Levitzki Theorem. Together with the structure theory, this yields

Corollary E. *Any semiprime PI-algebra has suitable PI-class* n *(and PI-degree* $2n$) *for suitable* n.

Note that a PI-algebra need not have PI-class, cf. Remark 1.3.29(iii).

Explicitly, PIs have the fundamental connection to representation theory. We define an **irreducible representation** of degree n to be an onto homomorphism $\rho_i : A \to M_{n_i}(F)$.

Remark 1.5.5. *Suppose* A *is an affine PI-algebra over an algebraically closed field* F. *By Kaplansky's theorem, each primitive image* $\bar{A}$ *of* A *is simple Artinian, whose center is f.d. over* F *and thus is* F *by Hilbert's Null-stellensatz, implying* $\bar{A}$ *is of the form* $M_{n_i}(F)$. *Thus, we have a homomorphism* $\rho : A \to \prod_{i \in I} M_{n_i}(F)$, *whose kernel is* $\mathrm{Jac}(A)$. *In this way, we see that PI-* $\mathrm{class}(A/\mathrm{Jac}(A))$ *is the maximal degree of the irreducible representations of* A *over* F.

1.5.2   Applications of alternating central polynomials

Remark F. *If* $g(x_1, \ldots, x_d)$ *is multilinear and* $M_n(\mathbb{Q})$-*central, and* A *has PI-class* n, *then*

$$h_n = g(c_{n^2}(x_1, \ldots, x_{n^2}, y_1, \ldots, y_{n^2}), x_{n^2+1}, \ldots, x_{n^2+d-1}) \qquad (1.20)$$

is multilinear, n^2-*alternating, and also* A-*central.*

Indeed, the multilinear and alternating properties are immediate from those of c_{n^2}. Furthermore, $h_n(A) \subseteq g(A)$ and g is either an identity of A or is A-central, so it remains to show that $h_n \notin \mathrm{id}(A)$. Since $c_{n^2} \in \mathcal{M}_{n-1}$, clearly $h_n \in \mathcal{M}_{n-1}$, so, in view of Definition 1.5.4 it suffices to note $h_n \notin \mathcal{M}_n$, which is true by Remark 1.3.29 since c_{n^2} evaluated on $M_n(\mathbb{Q})$ produces all the matrix units.

In order to ease the notation in the future, for $i > n^2$ we rewrite the indeterminate x_i as y_{i-n^2}, in order to differentiate between the alternating *indeterminates (the x_i for $i \leq n^2$) and the other indeterminates. Thus, we notate h_n as*

$$h_n(x_1, \ldots, x_{n^2}, y_1, \ldots, y_m)$$

where $m = n^2 + d - 1$.

Let us reiterate that we have an n^2-alternating central polynomial, despite the fact that c_{n^2+1} is an identity of A since A has PI-class n. Thus, the next remark is relevant, taking $t = n^2$. $\vec{y}$ denotes a collection of indeterminates.

Lemma G. *Suppose $h(x_1, \ldots, x_t, x_{t+1}; \vec{y})$ is any t-alternating polynomial. Then the polynomial*

$$\tilde{h}(x_1, \ldots, x_{t+1}; \vec{y}) = h - \sum_{i=1}^{t} (-1)^i h(x_1, \ldots, x_{i-1}, x_{t+1}, x_{i+1}, \ldots, x_t, x_i; \vec{y})$$

is $(t+1)$-alternating.

Indeed, suppose we set $x_j = x_u$ for some $1 \leq u < j \leq t+1$. Then all terms on the right side are 0 except

$$(-1)^u h(x_1, \ldots, x_{u-1}, x_{t+1}, x_{u+1}, \ldots, x_u, \vec{y})$$

and

$$(-1)^j h(x_1, \ldots, x_{j-1}, x_{t+1}, x_{j+1}, \ldots, x_j; \vec{y})$$
$$= (-1)^j (-1)^{j-i-1} h(x_1, \ldots, x_{u-1}, x_{t+1}, x_{u+1}, \ldots, x_{j-1}, x_u, x_{j+1}, \ldots, x_j; \vec{y})$$
$$= (-1)^{i-1} h(x_1, \ldots, x_{u-1}, x_{t+1}, x_{u+1}, \ldots, x_{j-1}, x_u, x_{j+1}, \ldots, x_j; \vec{y}),$$

which cancel.

In particular, if A satisfies c_{t+1}, then $\tilde{h} \in \mathrm{id}(A)$. Thus, for all $a_1, \ldots, a_{t+1}$, $r_1, \ldots, r_m$ in A, we have

$$(-1)^t h(a_1, \ldots, a_{t+1}; r_1, \ldots, r_m)$$
$$= \sum_{i=1}^{t} (-1)^i h(a_1, \ldots, a_{i-1}, a_{i+1}, \ldots, a_{t+1}, a_i; r_1, \ldots, r_m). \tag{1.21}$$

Thus, we can apply $\tilde{h}$ when A has PI-class n. (We shall need the more general formulation for the Razmyslov-Zubrilin theory of traces in Chapter 2.)

The following result contains the essence of the Artin-Procesi Theorem, as given in [Ro80, Chapter 1]. In fact, A of Theorem H is an Azumaya algebra, thereby making it possible to localize any prime PI-algebra by a single element (namely, any nonzero evaluation of h_n) to obtain an Azumaya algebra. This technique is fundamental in the structure of PI-theory (and for example is used repeatedly in Braun's proof of the Razmyslov-Kemer-Braun theorem).

Theorem H. *Suppose A has PI-class n and $h(x_1,\ldots,x_t,x_{t+1};\vec{y})$ is the t-alternating central polynomial of Remark F. If $h_n(a_1,\ldots,a_{n^2},r_1,\ldots,r_m)=1$, then A is a free $\mathrm{Cent}(A)$-module with base $a_1,\ldots,a_{n^2}$.*

More generally, if $h_n(a_1,\ldots,a_{n^2},r_1,\ldots,r_m)=c$, then Ac is contained in a f.g. free C-submodule of A. (For the proof, take

$$h = h_n(x_1,\ldots,x_t;\vec{y})x_{t+1}$$

and apply (1.21).)

1.5.3 Cayley-Hamilton properties of alternating polynomials

We now turn to one of the most basic tools in the PI-theory, enabling us to encode the Cayley-Hamilton theorem from linear algebra into alternating polynomials, specifically the Capelli polynomial. The underlying philosophy of the subject is contained in the following observation, recorded in [Ro80, Theorem 1.4.12]:

Remark 1.5.6. *Suppose $f(x_1,\ldots,x_t,\vec{y})$ is a t-linear, multilinear, alternating form on t indeterminates on a vector subspace $V = F^{(t)}$ of A, and $T : V \to V$ is a linear transformation of V. Then*

$$f(T(x_1),\ldots,T(x_t),\vec{y}) = \det(T)f(x_1,\ldots,x_t,\vec{y}).$$

We apply this, viewing a t-alternating polynomial (such as the Capelli polynomial) as a multilinear, alternating A-valued form.

This yields:

Theorem I. *Suppose $A \subseteq M_n(K)$, and let V be a t-dimensional K-subspace of $M_n(K)$ with a base $a_1,\ldots,a_t$ of elements of A. Given a K-linear map $T : V \to V$, let*

$$\lambda^t + \sum_{i=1}^{t}(-1)^i\alpha_i\lambda^{t-i}$$

denote the characteristic polynomial of T (as a $t \times t$ matrix over K). Then, for any t-alternating polynomial $f(x_1,\ldots,x_t;y_1,\ldots,y_m)$, and any $0 \le k \le t$, and $\vec{r}=(r_1,\ldots,r_m)$ for $r_i \in A$,

$$\alpha_k f(a_1,\ldots,a_t,r_1,\ldots,r_m) = \sum f(T^{k_1}a_1,\ldots,T^{k_t}a_t,r_1,\ldots,r_m), \quad (1.22)$$

summed over all vectors $(k_1,\ldots,k_t)$ for which each

$$k_i \in \{0,1\}, \quad k_1 + \cdots + k_t = k.$$

In particular,

$$\mathrm{tr}(T)f(a_1,\ldots,a_t,r_1,\ldots,r_m) =$$
$$\sum_{j=1}^{t} f(a_1,\ldots,a_{j-1},Ta_j,a_{j+1},\ldots,a_t,r_1,\ldots,r_m). \quad (1.23)$$

Proof. First we verify (1.22) for $k = t$. The form

$$f(Tx_1, \ldots, Tx_t, \vec{y})$$

is linear and alternating in the $x_1, \ldots, x_t$, so by Remark 1.5.6 is

$$\det(T)f(x_1, \ldots, x_t; \vec{y}),$$

which is the special case of (1.22) for $k = t$. But now replace T by the transformation $\lambda I_t - T$, whose determinant gives the Cayley-Hamilton . Matching coefficients yields (1.22) in general, and in particular, (1.23). □

Let us also record a key special case.

Theorem J. *If A spans $M_n(K)$ over a field K, so that $t = n^2$, and if we let $\mathcal{I}$ denote the additive subgroup of A generated by $h_n(A)$, then not only is $0 \neq \mathcal{I} \triangleleft \mathrm{Cent}(A)$, but $\alpha\mathcal{I} \subseteq \mathcal{I}$ for every characteristic coefficient α of every element of A (viewed as a matrix).*

When interpreting (1.23), we must exercise care even when $t = n$, since we are interpreting T as an $n^2 \times n^2$ matrix and not an $n \times n$ matrix. See Exercises 25 and 26 for a related result.

Theorem J has profound implications in the realization of the "Shirshov program" described in Chapter 2, especially in connection with the "characteristic closure," since Corollary 1.5.3 shows that the hypotheses of Theorem J are satisfied by prime PI-algebras. It also is the key to the Razmyslov-Zubrilin theory of adjoining traces, developed in Chapter 2, and as such is crucial in Kemer's solution of Specht's problem.

1.6   Representable Algebras

The previous two results lead us to the most important class of PI-algebras. In this section, A always denotes an algebra over a commutative ring C, and unless otherwise indicated, $Z = \mathrm{Cent}(A)$.

Definition 1.6.1. *The algebra A is **weakly representable** if A can be embedded into the matrix algebra $M_n(K)$ for some commutative C-algebra K and suitable n. A is **representable** if K can be taken to be a field.*

Sometimes one can obtain a stronger version.

Definition 1.6.2. *A representable algebra A is **strongly representable** if the field K can be taken to be a subalgebra of $S^{-1}Z$ for some multiplicative submonoid S of Z, and the localization $S^{-1}A$ is f.d. over K.*

Digression 1.6.3. *"Weakly representable" can be weakened even further to the condition that*

$$A \hookrightarrow \mathrm{End}_K(M),$$

for a suitable f.g. module M over a commutative ring K, cf. Exercise 33. The ambiguity in terminology has led to confusion; fortunately, by Theorem 1.6.22 given below, all the definitions coincide when A is finite over a central affine algebra. Also, the ambiguity can be resolved in certain important cases by means of Exercise 35.

Representable algebras are more malleable than arbitrary PI-algebras, but provide intuition for the general PI-theory. One basic property of representable algebras is:

Remark 1.6.4. *If A is an algebra over an infinite field and is embedded in $M_n(K)$ for some field K, then KA is a f.d. K-subalgebra of $M_n(K)$ which by Proposition 1.2.19 is PI-equivalent to A. This leads us to:*

Definition 1.6.5. *An algebra R is **PI-representable** if $R \sim_{\mathrm{PI}} A$ for some algebra A which is f.d. over a field.*

PI-representability is a main feature of Kemer's proof of Specht's conjecture, given in Chapters 6 and 7. Having made the preliminary case for representable algebras, let us give some basic examples.

Corollary 1.5.3 shows that prime PI-algebras are representable. On the other hand, by Exercise 4, any representable algebra satisfies ACC(Ann).

According to our definition, the direct product of fields of different characteristics cannot be representable, although it is obviously weakly representable. The story is different for algebras with a common a base field.

Lemma 1.6.6. *If an algebra A over a field F is a finite subdirect product of representable algebras $A_1, \dots, A_m$, then A is representable.*

Proof. Embedding A_i into $M_{n_i}(K_i)$ and letting K be the field compositum of the K_i (over F), we can embed A diagonally into $M_n(K)$, where $n = n_1 + \cdots + n_m$. $\qquad\square$

Remark 1.6.7. *Any semiprime PI-algebra A is weakly representable. (This is seen by embedding A into the direct product of its prime images, each of which can be embedded into matrix algebras $M_{n_i}(K_i)$ for suitable fields K_i and suitable n_i bounded by the PI-class of A, so A is embedded along diagonal blocks into $M_n(\prod K_i)$ where $n = m!$)*

Here is a connection with affine algebras.

Proposition 1.6.8. *If $A = C\{a_1, \dots, a_\ell\}$ is affine over C and weakly representable, then $A \subseteq M_n(H)$ for a suitable commutative affine C-algebra H.*

Proof. For each $1 \leq k \leq \ell$, write each a_k as an $n \times n$ matrix $(a_{ij}^{(k)})$, for $a_{ij}^{(k)} \in K$, and let H be the commutative subalgebra of K generated by these $a_{ij}^{(k)}$. Now the assertion is obvious. $\qquad\square$

Remark 1.6.9. *In Proposition 1.6.8 if C is Noetherian, then H is Noetherian.*

This innocent-looking assertion has some far-reaching consequences, as we soon shall see.

Corollary 1.6.10. *Any semiprime PI-algebra A affine over a field is representable.*

Proof. By Proposition 1.6.8, $A \subseteq M_n(H)$, which is Noetherian. Hence its nilradical N is a finite intersection $P_1 \cap \cdots \cap P_m$ of prime ideals, and $M_n(N) \cap A$ is a nilpotent ideal of A which is thus 0. It follows that A embeds into

$$M_n(H)/M_n(N) \cong M_n(H/N) \subseteqq \prod_{i=1}^{m} M_n(H/P_i),$$

so we conclude with Lemma 1.6.6. $\qquad\square$

1.6.1 Lewin's Theorems

What if our PI-algebra is not semiprime? Lewin [Lew74] proved an interesting theorem, which plays a crucial role in the sequel. For convenience, we take $C = F$, a field.

Definition 1.6.11. *A $(n_1, \ldots, n_q)$-block upper triangular algebra is a subalgebra of $M_n(F)$ in which all nonzero entries are in or above the diagonal $n_k \times n_k$ blocks, for $1 \leq k \leq q$, i.e., blocks of matrices of size $n_i \times n_j$ which occur over the diagonal. $\mathrm{UT}(n_1, \ldots, n_q)$ denotes the $(n_1, \ldots, n_q)$-block upper triangular algebra where $n_1 + \cdots + n_q = n$. Thus $\mathrm{UT}(n) = \mathrm{UT}(1, 1, \ldots, 1)$, taken n times.*

For example, for $q = 2$, $\mathrm{UT}(n_1, n_2)$ looks like $\begin{pmatrix} *_{n_1 \times n_1} & *_{n_1 \times n_2} \\ 0 & *_{n_2 \times n_2} \end{pmatrix}$. This example can be thought of in terms of bimodules over C-algebras A_1 and A_2 where $A_j = M_{n_j}(F)$. Also, it ties in nicely with identities because of our next remark, carried much further in [GiZa03c].

Remark 1.6.12. *$\mathrm{UT}(n_1, \ldots, n_q)$ is a good model for intuition, arising as the critical case in the theory of exponents of Giambruno and Zaicev [GiZa03c], although we shall not use it. Here $s_A = q$.*

Theorem 1.6.13 (Lewin).

(i) *Suppose A is an algebra over a field and $I_1, I_2 \lhd A$. Then there is an $A/I_1, A/I_2$ bimodule M, such that $A/I_1 I_2$ can be embedded into*

$$\begin{pmatrix} A/I_1 & M \\ 0 & A/I_2 \end{pmatrix} \tag{1.24}$$

via $a \mapsto \begin{pmatrix} a + I_1 & \delta(a) \\ 0 & a + I_2 \end{pmatrix}$, where $\delta : A \to M$ is a derivation.

(ii) *If the algebras $F\{X\}/I_i$ are embeddable in $M_{n_i}(C)$, then $F\{X\}/I_1 I_2$ is embeddable in an (n_1, n_2)-block triangular algebra.*

Although we do not prove Lewin's Theorem in this volume, the elegant proof of (i) by Bergman-Dicks [BergmD75] using universal derivations, is given in full in [Ro88b, pp. 136–140]. Lewin's original proof is also given in [GiZa05, pp. 20-23].

Corollary 1.6.14. *If $F\{X\}/I$ is representable, then so is $F\{X\}/I^m$ for any m.*

As we shall see presently, this corollary is false if we try to replace $F\{X\}$ by an arbitrary algebra.

Proposition 1.6.15. *If $W = \begin{pmatrix} W_1 & * \\ 0 & W_2 \end{pmatrix}$, then $\mathrm{id}(W) \supseteq \mathrm{id}(W_1)\,\mathrm{id}(W_2)$.*

Proof. If $f \in \mathrm{id}(W_1)$ then $f(W)$ has the form $\begin{pmatrix} 0 & * \\ 0 & * \end{pmatrix}$; if $g \in \mathrm{id}(W_2)$ then $g(W)$ has the form $\begin{pmatrix} * & * \\ 0 & 0 \end{pmatrix}$. Multiplying these two matrices together gives 0.

$\square$

In fact, Lewin's Theorem gives us equality in its situation, cf. Proposition 1.7.9 and Exercise 46.

1.6.2   Nonrepresentable algebras

Of course, any representable algebra A satisfies all identities $\mathcal{M}_n$, so one might ask whether $\mathcal{M}_n \subseteq \mathrm{id}(A)$ suffices to make A representable, but examples have been found with A nonrepresentable. Other necessary conditions were thrown in (cf. [Ro88b, Remark 6.3.6]), but researchers continued to find counterexamples satisfying these extra conditions, even for the important class of affine algebras.

1.6.2.1 Bergman's example

Bergman [Bergm70, Theorem 3], also cf. [Pas77, pp. 253ff.], presented the following example as a finite ring that is not weakly representable. We need a ring-theoretic preliminary.

Lemma 1.6.16. *Suppose C is a local ring and $ce = 0$ for some idempotent $e \neq 0$ of a ring $R \subseteq M_n(C)$. Then $cR = 0$.*

Proof. We may replace R by $M_n(C)$ without affecting either the hypothesis or conclusion, so we assume that $R = M_n(C)$. Let J be the maximal ideal of C, so C/J is a field K. The image of Re in $W := M_n(K)$ is a projective module over W and thus some power is free, implying some power of Re is free over R (in view of "Nakamaya's Lemma," [Ro88a, Proposition 2.5.24]). Hence Re is faithful, but $c(Re) = Rce = 0$, so $c = 0$. $\square$

Lemma 1.6.17. *Suppose $ce = 0$ for some idempotent e of a C-algebra R. Then for any homomorphism $\varphi : R \to M_n(C')$ where C' is a local C-algebra, writing $\bar{R} = \varphi(R)$ we have $\bar{R}\bar{e}\bar{R} \cap c\bar{R} = 0$.*

Proof. If $\bar{R}\bar{e}\bar{R} \neq 0$, then $c\bar{R} = 0$ by the lemma. $\square$

Example 1.6.18. *Take an arbitrary commutative local ring C with radical $N = Cc$ satisfying $c^2 = 0$ (and thus $N^2 = 0$), and let $M_1 = C/N$ and $M_2 = C$, both viewed as C-modules. There is a natural onto map $f_{12} : M_2 \to M_1$ as well as a monic map $f_{21} : M_1 \to M_2$ given by $[1] \mapsto [c]$.*

Letting $M = M_1 \oplus M_2$, we define $A = \operatorname{End}_C M = \oplus_{i,j=1,2} \operatorname{Hom}(M_i, M_j)$.

For $i = 1, 2$, let e_i denote the projection of M onto M_i, an idempotent of A satisfying $ce_1 = 0$. Then $e_2 = f_{21}e_1 f_{12} \in Ae_1 A$, and thus, by Lemma 1.6.17,

$$cA = ce_2 A = Ae_1 A \cap cA$$

must be in the kernel of any C-algebra homomorphism into matrices over a local ring. Since any commutative ring can be injected into a direct product of local rings ([Ro05, Proposition 8.22]), we conclude that A cannot be representable as a C-algebra.

Bergman takes $C = \mathbb{Z}/p^2\mathbb{Z}$ and $c = p$ for some prime number $p \in \mathbb{N}$. Now A is not representable as a ring.

Elegant as Bergman's example is, if C is an algebra over a field F, R might be representable as an F-algebra, even if not as a C-algebra, so we might ask whether positive results can be had for affine algebras over a field.

Example 1.6.19 (Nonrepresentable affine algebras). *Lewin discovered an elegant cardinality argument to explain why affine algebras over a field usually are not representable. Suppose the base field F has infinite cardinality γ. (For convenience, one could take F countable). Then the following facts hold:*

(i) *Any affine F-algebra has cardinality γ, since its elements are linear combinations of elements of the form $ca_1^{i_1} \cdots a_\ell^{i_\ell}$.*

(ii) *Any affine F-algebra A has at most γ affine subalgebras. Indeed, any such subalgebra is specified by its generators, which we pick a finite number of times from the set A of cardinality γ.*

(iii) *The cardinality of the set of commutative F-affine algebras (up to isomorphism) is at most γ. Indeed any such algebra has the form*

$$F[a_1, \ldots, a_\ell] = F[\lambda_1, \ldots, \lambda_\ell]/\mathcal{I},$$

where $\mathcal{I}$ is an ideal of the commutative polynomial algebra $F[\lambda_1, \ldots, \lambda_\ell]$, for suitable ℓ, and thus, is f.g. as an ideal. For any given ℓ, there are clearly at most γ choices of the finite number of generators, so at most γ possibilities for the algebra; since there are countably many natural numbers ℓ, we get γ choices altogether.

(iv) *The number of representable F-affine algebras (up to isomorphism) is $\le \gamma$. Indeed, by Proposition 1.6.8, any such algebra can be viewed in $M_n(H)$ where H is commutative and F-affine.*

$$M_n(H) \cong M_n(F[\lambda_1, \ldots, \lambda_\ell])/M_n(\mathcal{I}) \cong M_n(F[\lambda_1, \ldots, \lambda_\ell]/\mathcal{I})$$

where $\mathcal{I} \lhd F[\lambda_1, \ldots, \lambda_\ell]$ (since any ideal of $M_n(F[\lambda_1, \ldots, \lambda_\ell])$ has the form $M_n(\mathcal{I})$), so we conclude with (ii) and (iii).

(v) *Any homomorphism between affine algebras is determined by its action on the finite set of generators, so there are at most γ homomorphisms between F-affine algebras.*

Thus, in view of (iv) and (v), in order to prove there are more than γ non-isomorphic, non-representable images of a given affine algebra A, it suffices to show that A has more than γ ideals.

(vi) *The free associative $\mathbb{Q}$-algebra $\mathbb{Q}\{x_1, x_2\}$ is an affine algebra with uncountably many ideals, namely for any subset I of $\mathbb{N}$ we have the ideal $\langle \{x_2 x_1^i x_2 : i \in I\} \rangle$; hence uncountably many of these yield nonrepresentable affine algebras.*

We can improve this observation to get a PI-example.

Lemma 1.6.20. *Suppose $T = F\{a_1, \ldots, a_\ell\}$ is an affine F-algebra and $L = \sum_{j=1}^m T b_j$ for b_i in T. Then*

$$A' = \begin{pmatrix} F + L & T \\ L & T \end{pmatrix}$$

is affine.

Proof.

$$A' = F\{e_{11}, e_{12}, a_i e_{22}, b_j e_{21} : 1 \le i \le \ell,\ 1 \le j \le m\}.$$

For example, $b_j e_{11} = e_{12} b_j e_{21}$.   □

Example 1.6.21. *In Lemma 1.6.20 let $F = \mathbb{Q}$ and $T = \mathbb{Q}[\lambda, \mu]$, for commuting indeterminates λ and μ, and let $L = T\lambda$. In this case T is commutative, so $R \subseteq M_2(T)$ is PI. But for any subset I of $\mathbb{N}$, let*

$$\mathcal{J}_I = L^2 + \sum_{i \in I} L\mu^i \lhd (F + L),$$

and

$$\mathcal{I}_I = \begin{pmatrix} \mathcal{J}_I & \mathcal{J}_I \\ L^2 & L^2 \end{pmatrix} \lhd R.$$

Thus, we see that the affine PI-algebra R has uncountably many ideals $\mathcal{I}_I$. We conclude that uncountably many of the $R/\mathcal{I}_I$ are mutually nonisomorphic.

But the algebraic closure K of $\mathbb{Q}(\lambda, \mu)$ is countable, implying $M_n(K)$ is countable, and thus, has only countably many affine subalgebras (since this involves choosing a finite subset), so therefore uncountably many of the $R/\mathcal{I}_I$ are nonrepresentable affine PI-algebras.

We shall see in Chapter 2 that the Jacobson radical $J = \mathrm{Jac}(A)$ of an affine PI-algebra always is nilpotent. Thus, despite Corollary 1.6.14, when $J^m = 0$, $A = A/J^m$ could be nonrepresentable, although A/J is representable by Corollary 1.6.10.

While this argument does not actually display such an algebra, we later shall see explicit examples in our study of GK-dimension of monomial algebras, cf. Example 11.2.9 and Corollary 11.2.24.

1.6.3   Representability of affine Noetherian PI-algebras

Lewin's argument fails for Noetherian affine algebras, since there are only countably many of them; Anan'in proved that all left Noetherian PI-algebras over a field are representable. In this exposition we prove a somewhat weaker assertion, which suffices for our purposes. Beidar [Bei86], following Wehrfritz [We76], proved that any algebra over a field, finite over a central affine subalgebra, is representable, and this result turns out to be required for our exposition of other important theorems in PI-theory such as Kemer's PI-representability Theorem 6.3.1.

We aim for the following fundamental result.

Theorem 1.6.22 (cf. Wehrfritz [We76] and Beidar [Bei86]). *Suppose C is a commutative Noetherian algebra over a field F. Then any finite algebra over C is representable.*

Remark 1.6.23. *One can reword the hypothesis: If A is left Noetherian and finite faithful over C, then C is Noetherian by the Eakin-Formanek theorem given in Exercise 3, so the conclusion of Theorem 1.6.22 holds.*

We outline the proof, following [RoSm15], since it is comprised of several diverse ideas, and proves a bit more.

- By the Artin-Tate Lemma (Proposition 1.1.21), we can replace C by $Z := \mathrm{Cent}(A)$.

- Using the subdirect decomposition of Lemma 1.1.26, we may assume that A is irreducible.

- Regular elements in Z are regular in A, so we can localize them and assume that every non-nilpotent element of Z is invertible. In particular, the maximal ideal N of Z is nilpotent, and $K = Z/N$ is a field.

- Lifting up the series

$$Z \supset N \supset \cdots \supset N^k = 0,$$

we show that K is isomorphic to a subfield of A, over which A is f.d., clearly implying that A is representable.

We start with the last step, since it is of independent interest, involving a "Hensel Lemma" type argument.

Lemma 1.6.24. *Any ring contains a maximal integral domain.*

Proof. By Zorn's Lemma, since the union of a chain of integral domains is an integral domain. $\square$

Following [Co46, Theorem 9], we say that a commutative local ring Z with maximal ideal J has a **coefficient field** if the field Z/J is isomorphic to a subfield K of Z; i.e., there is a "Mal'cev-Wedderburn" decomposition $Z = K \oplus J$ as K-vector spaces.

Proposition 1.6.25. *Suppose Z is any local commutative algebra over a field, whose (Jacobson) radical J is nilpotent. Then Z contains a coefficient field.*

Proof. Take K to be a maximal integral domain in Z, which by hypothesis is a subfield. We claim that K is the desired coefficient field for Z.

Let $\bar{Z} = Z/J$, a field. Clearly $\bar{Z}$ is algebraic over $\bar{K}$, since any transcendental element would lift to a transcendental element a over K, and then $K[a]$ would be an integral domain, contrary to the maximality of K. It follows that Z is algebraic over K. It remains to show that $\bar{K} = \bar{Z}$.

Otherwise fix an element $\bar{a} \in \bar{Z} \setminus \bar{K}$, of minimal degree, and minimal (irreducible) polynomial $g \in K[\lambda]$. Lift $\bar{a}$ to an element $a \in Z$. Let $f \in K[\lambda]$ be the minimal polynomial of a. If f is irreducible, then $K[a]$ is a field, contrary to the maximality of K, so we assume that f is reducible.

Suppose $J^t = 0$. Clearly $g|f$, and $b := g(a) \in J$ implies $b^t = 0$, implying f divides g^t. Thus there are only finitely many possibilities for f, taken over all representatives a of $\bar{a}$. But we could replace a by $a + \alpha b$ for any $\alpha \in K$, so in particular we may assume that f is the minimal polynomial of $a + \alpha b$ for infinitely many choices of α. Viewing $f(a + \alpha b) = 0$ as a function in α, we conclude by a Vandermonde argument (taking the linear part in α) that

$0 = f'(a + \alpha b)$, contrary to the minimality of f unless the polynomial $f' = 0$. Thus a must be inseparable, and in particular we are done unless K has characteristic $p > 0$.

Take u such that $p^u > t$. Write $g = \sum_{i=0}^{d} \beta_i \lambda^i$ for $\beta_i \in K$, and $\tilde{g} = \sum_{i=0}^{d} \beta_i^{p^u} \lambda^i$. Then

$$g(a)^{p^u} = \tilde{g}(a^{p^u}),$$

so a^{p^u} is a root of $\tilde{g}$. If $\tilde{g}$ is irreducible then we have a contradiction from the previous paragraph unless $a^{p^u} \in K$, but if $\tilde{g}$ is reducible then $\bar{a}^{p^u}$ has lower degree than $\bar{a}$, contrary to assumption. Thus $a^{p^u} \in K$. Taking the smallest u' such that $a^{p^{u'}} \in K$, and replacing a by $a^{p^{u'-1}}$, we may assume that $a \notin K$ but $a^p \in K$. But then a satisfies the polynomial $\lambda^p - \alpha$ where $\alpha = a^p$, which is irreducible, again contrary to the maximality of K. $\qquad\square$

Next, we want to show that regularity in $\mathrm{Cent}(A)$ implies regularity in A. A **principal (left) annihilator** is a left annihilator of an element.

Lemma 1.6.26. *If A is irreducible and satisfies ACC on principal annihilators of the form $\mathrm{Ann}_A z^i$, for $i \in \mathbb{N}$, for fixed $z \in Z := \mathrm{Cent}(A)$ which is not nilpotent, then $\mathrm{Ann}_A z = 0$.*

Proof. Take k such that $\mathrm{Ann}_A z^k = \mathrm{Ann}_A z^{k+1}$. Then Lemma 1.1.23 (taking $M = A$) and the irreducibility of A imply that either $z^k A = 0$ (in which case $z^k = 0$) or $\mathrm{Ann}_A z^k = 0$. $\qquad\square$

For example, when Z is an integral domain, A is torsion-free over Z.

Lemma 1.6.27. *Suppose Z contains a field F, and $z \in Z$ is algebraic over F. Then either z is invertible or nilpotent, or $F[z]$ contains a nontrivial idempotent.*

Proof. This could be seen using the well-known decomposition of a f.d. commutative algebra as a direct product of local algebras, or by the following direct computation: The minimal polynomial of z can be written

$$\sum_{i=t}^{m} \alpha_i \lambda^i := \alpha_t \lambda^t (1 - \lambda f(\lambda)),$$

where $\alpha_t \neq 0$. Thus $z^t = z^{t+1} f(z)$, implying $z^t = z^{t+k} f(z)^k$ for any $k \geq 1$, and in particular $z^t = e z^t$ where $e := (z f(z))^t$ is idempotent, since

$$e^2 = e z^t f(z)^t = z^t f(z)^t = e.$$

If $e = 0$ then $z^t = 0$, and if $e = 1$ then z is invertible. $\qquad\square$

Proposition 1.6.28. *Suppose that A is an irreducible algebra, finite over a central Noetherian F-algebra Z. Let S be the set of regular elements of Z. Then $S^{-1}Z$ is finite over a suitable subfield K of $S^{-1}Z$.*

Proof. By Lemma 1.6.26, all non-nilpotent elements of Z are regular in A, which embeds into $S^{-1}A$, which is where we work. Replacing Z by $S^{-1}Z$, we may assume that each regular element of Z is invertible, and, in view of Lemma 1.6.27, Z is local with nilpotent maximal ideal N.

We conclude with a well-known argument of Levitzki. Take a coefficient subfield K and consider the chain

$$Z \supset N \supset N^2 \supset \cdots \supset N^k = 0.$$

It suffices to prove that each N^i/N^{i+1} is finite over K. But we know that each N^i/N^{i+1} is finite over Z (and thus over $Z/N \cong K$ under the natural action), since Z is Noetherian. □

Proof of Theorem 1.6.22: In view of Lemmas 1.1.26 and 1.6.6, we may assume that A is irreducible. Then every regular element of Z is regular in A. In the notation of Proposition 1.6.28, after localization, Z is finite over K, but we are given A finite over Z, so A is finite over K, implying our original algebra is strongly representable, as desired.

The important related question remains open, generalizing Theorem 1.6.22:

Question 1.6.29. *Is every algebra over a field, that is finite over its center, weakly representable?*

The question also is open whether $\text{End}_C M$ is weakly representable for a module M finite over C, although the answer is positive (Exercise 32) when M is Noetherian.

1.6.4   Nil subalgebras of a representable algebra

As an example of how representable algebras shine light on general PI-theory, we quickly prove certain basic structural results which when proved more generally for affine PI-algebras require much more intricate arguments occupying most of Chapter 4. We start with the classical case, in which A is commutative.

Remark 1.6.30. *Since any commutative affine algebra A over a Noetherian base ring C is Noetherian, the intersection of its prime ideals is nilpotent, by Corollary 1.1.35.*

But for any ideal $\mathcal{I} \lhd A$, the algebra $A/\mathcal{I}$ is also Noetherian, so the intersection of the prime ideals of A containing $\mathcal{I}$ is nilpotent modulo $\mathcal{I}$.

Proposition 1.6.31. *If H is a commutative affine algebra over a Noetherian Jacobson ring (in particular, over a field), then $\text{Jac}(H)$ is nilpotent.*

Proof. H is Jacobson by Lemma 1.1.16, implying $\text{Jac}(H)$ is nilpotent, by Remark 1.6.30. □

To extend Corollary 1.1.36 and Proposition 1.6.31 to noncommutative algebras, we need some other classical results:

Remark 1.6.32.

(i) *[Ro08, Theorem 15.23] (Wedderburn) Any nil subring of an $n \times n$ matrix algebra over a field is nilpotent, of nilpotence index $\leq n$ (in view of [Ro08, Lemma 15.22]).*

(ii) *[Ro08, Theorem 15.18] (Jacobson) The Jacobson radical of an n-dimensional algebra over a field is nilpotent, and thus has nilpotence index $\leq n$, by (i).*

The following result will be useful.

Proposition 1.6.33. *If C is Jacobson, and an PI-algebra A is affine and integral over C, then A also is Jacobson.*

Proof. Passing to prime homomorphic images, we may assume that A is prime, and need to prove that $\mathrm{Jac}(A) = 0$. Note that every nonzero ideal I of A intersects C nontrivially. Indeed, by Theorem C of §1.5, I contains a nonzero element z of $\mathrm{Cent}(A)$, an integral domain, and taking the minimal monic polynomial $f = \lambda^t + \sum_{i=0}^{t-1} c_i \lambda^i$ of z, we see that $c_0 \neq 0$ (since $\mathrm{Cent}(A)$ is an integral domain), and $c_0 \in I \cap C$.

But now we claim that every maximal ideal P of C is contained in a maximal ideal of A. This will prove the assertion, since then

$$C \cap \mathrm{Jac}(A) \subseteq \mathrm{Jac}(C) = 0,$$

and thus $\mathrm{Jac}(A) = 0$. To prove the claim, we take an ideal Q of A maximal with respect to $Q \cap C \subseteq P$. Clearly Q is a prime ideal, so passing to A/Q, we may assume that $Q = 0$. In other words, no nonzero ideal of A intersects C inside P. We want to show in this case that $P = 0$. We conclude by showing that if $0 \neq c \in P$ then $cA \cap C \subseteq P$. Indeed, if $ca \in cA \cap C$ then taking $a^t = \sum_{i=0}^{t-1} c_i a^i$, we have

$$(ca)^t = c \left(c^{t-1} \sum_{i=0}^{t-1} c_i a^i \right) = c \left(c^{t-1} \sum_{i=0}^{t-1} c^{t-1-i} c_i (ca)^i \right) \in cC \subseteq P,$$

implying $ca \in P$. $\square$

Proposition 1.6.34. *Suppose $A = C\{a_1, \ldots, a_\ell\}$ is an affine algebra over a commutative Noetherian ring C, with $A \subseteq M_n(K)$ for a suitable commutative C-algebra K. Then:*

(i) *Any nil subalgebra N of A is nilpotent, of nilpotence index $\leq mn$, where m is given in the proof. When K is reduced, i.e., without nonzero nilpotent elements, then $m = 1$, so $N^n = 0$.*

(ii) If C is also Jacobson, then $\mathrm{Jac}(A)$ is nilpotent.

Proof. By Proposition 1.6.8, $A \subseteq M_n(H)$ for a commutative affine Noetherian algebra H. Since the nilradical of H is nilpotent, we may mod it out, and assume that H is semiprime.

(i) Let $N \subseteq A$ be a nil subalgebra. Now $A \subseteq M_n(H)$, so $N \subseteq M_n(H)$ and is nil. For any prime ideal $P \lhd H$, the homomorphism $H \to H/P$ extends to

$$M_n(H) \to M_n(H/P) \ \ (\cong M_n(H)/M_n(P)).$$

Let $\bar{N}$ be the image of N, so

$$\bar{N} = (N + M_n(P))/M_n(P) \subseteq M_n(H)/M_n(P) \cong M_n(H/P) \subseteq M_n(L),$$

where L is the field of fractions of the domain H/P. By Wedderburn's theorem $\bar{N}^n = 0$ which implies that $N^n \subseteq M_n(P)$.

(ii) By Proposition 1.6.33, H is Jacobson. Thus the intersection of its maximal ideals is 0, and passing to the image of A in $M_n(H/P)$, we may assume that H is a field. But replacing C by its image in H/P, we see that the field H is affine over a Jacobson subring C, so by Proposition 1.1.15, C is a field, and H is finite over C. Thus, A also is finite over C, implying every prime ideal of A is maximal (a standard fact which also follows easily from Corollary 1.5.3), and thus $\mathrm{Jac}(A)$ is nilpotent.

Hence, letting U denote the nilradical of H, we have $N^n \subseteq M_n(U)$. But, in view of Remark 1.6.30, we have $U^m = 0$ for some m. (If K is reduced then $U = 0$, implying $m = 1$.) We conclude that

$$N^{mn} = (N^n)^m \subseteq (M_n(U))^m = M_n(U^m) = 0.$$

$\square$

1.7   Sets of Identities

Until now, we have focused on the properties of a ring satisfying a given PI (whether the PI comes from a central polynomial, or a Capelli identity). At this stage we consider what extra information is available from the set of identities of a given algebra. This turns out to yield a powerful multi-purpose tool, needed in several related contexts. Accordingly, we turn to a general description of identities and their varieties. The collection of identities of an algebra, described as a subset of the free algebra, has a special structure, called a *T*-ideal.

1.7.1 The set of identities of an algebra

Definition 1.7.1. *Fix a commutative ring C. Given a set S of C-algebras, we let $\mathrm{id}(S)$ denote the set of identities holding in each algebra in S i.e., $\cap_{A\in S}\mathrm{id}(A)$.*

The underlying questions of this section are, "Why are we interested in $\mathrm{id}(S)$, and what can we say about it?"

Remark 1.7.2. *If f and g are identities of an algebra A, then so is $f \pm g$, hf, and fh, for all $h \in C\{X\}$, so $\mathrm{id}(A) \lhd C\{X\}$; hence also $\mathrm{id}(S) \lhd C\{X\}$.*

However, there is another crucial property.

Remark 1.7.3. *If $f(x_1,\ldots,x_m) \in \mathrm{id}(A)$ and $h_1,\ldots,h_m \in C\{X\}$, then for any specializations of x_i to $\bar{x}_i$ in A, writing $\overline{h_j}$ for the corresponding value of h_j in A, we clearly have*

$$0 = f(\overline{h_1},\ldots,\overline{h_m}) = \overline{f(h_1,\ldots,h_m)},$$

implying $f(h_1,\ldots,h_m) \in \mathrm{id}(A)$. In other words, an identity remains an identity under any specialization of its indeterminates in $C\{X\}$.

But by Remark 1.2.1, for any algebra R, any set mapping $\psi : X \to R$ extends uniquely to an algebra homomorphism $\psi : C\{X\} \to R$. Hence, taking $R = C\{X\}$ we can restate Remark 1.7.3 more concisely as:

Remark 1.7.4. *If $f \in \mathrm{id}(A)$, then $\psi(f) \in \mathrm{id}(A)$ for all algebra homomorphisms $\psi : C\{X\} \to C\{X\}$.*

1.7.2 *T*-ideals and related notions

In view of Remark 1.7.4 we formulate a basic notion.

Definition 1.7.5. *A T-ideal of an algebra A is an ideal I such that $\psi(I) \subseteq I$ for all homomorphisms $\psi : A \to A$. In this case, we write $I \lhd_T A$.*

Remark 1.7.6. *The T-ideals of $C\{X\}$ are of special interest to us, since $\mathrm{id}(S) \lhd_T C\{X\}$, for any set of algebras S. Furthermore, the homomorphisms $\psi : C\{X\} \to C\{X\}$ are described by the "substitutions" $x_i \mapsto \psi(x_i)$, so in principle the T-ideal generated by a set of polynomials is the ideal generated by all substitutions of these polynomials in $C\{X\}$.*

For example, s_n is in the T-ideal generated by c_n, where we specialize all the $y_i \mapsto 1$. In the same spirit, we have

Remark 1.7.7. *Suppose $\psi : C\{X\} \to A$ is any algebra homomorphism. Then $\mathrm{id}(A) \subseteq \ker\psi$, since*

$$\psi(f(x_1,\ldots,x_m)) = f(\psi(x_1),\ldots,\psi(x_m)) \in f(A) = 0.$$

The most familiar T-ideal is $\mathrm{id}(M_n(F))$, denoted in Definition 1.5.4 as $\mathcal{M}_{n,F}$. It gains in prominence through the following observations.

Remark 1.7.8. $\mathcal{M}_{n,F} \subseteq \mathcal{M}_{m,F}$ *for all* $m \leq n$. *(This is because we can embed* $M_m(F)$ *into* $M_n(F)$ *as an algebra without 1, by adding a fringe of 0's along the last* $n - m$ *rows and columns.)*

T-ideals also connect to Lewin's embeddability theorem 1.6.13.

Proposition 1.7.9. *If* $\mathcal{I}_1, \mathcal{I}_2$ *are* T-*ideals of* $F\{X\}$, *and we put*

$$A = \begin{pmatrix} F\{X\}/\mathcal{I}_1 & M \\ 0 & F\{X\}/\mathcal{I}_2 \end{pmatrix} \tag{1.25}$$

(in the notation of Theorem 1.6.13(ii)), then $\mathrm{id}(A) = \mathcal{I}_1 \mathcal{I}_2$.

Proof. We already showed in Proposition 1.6.15 that $\mathrm{id}(A) \supseteq \mathcal{I}_1 \mathcal{I}_2$. But conversely, Theorem 1.6.13 shows that $F\{X\}/\mathcal{I}_1 \mathcal{I}_2$ is embedded in A, so $\mathrm{id}(A) \subseteq \mathcal{I}_1 \mathcal{I}_2$. $\qquad\square$

One may wonder to what extent the T-ideal $\mathrm{id}(A)$ determines a f.d. algebra A. In general, uniqueness would not hold in view of Proposition 1.7.9 and Exercise 46, but over an algebraically closed field F one has the following easy observation:

Remark 1.7.10. *If* A *is a simple f.d.* F-*algebra, then* $A \cong M_n(F)$ *for some* n, *implying* $\mathrm{id}(A) = \mathrm{id}(M_n(F))$. *But then* A *is determined by* n, *so* $\mathrm{id}(A)$ *determines* A *in this case.*

This observation has led researchers to study nonassociative analogs as well as graded analogs, which we discuss in Volume II.

Being interested in T-ideals, i.e., the set of identities of suitable algebras, we are led at once to PI-equivalence, cf. Definition 1.2.15.

Lemma 1.7.11. *If* R *is a semiprime PI-algebra over an infinite field* F, *then* $R \sim_{\mathrm{PI}} M_n(F)$, *i.e.,* $\mathrm{id}(R) = \mathcal{M}_{n,F}$, *where* n *is the PI-class of* R.

Proof. When R is prime, this is the last sentence of Corollary 1.5.3.

In general, each prime image of R has PI-class $\leq n$, and at least one image has PI-class n, so we conclude by Remark 1.7.8. $\qquad\square$

Thus, the $\mathcal{M}_n$ are the only T-ideals of $\mathbb{Q}\{X\}$ which could be prime as ideals. However, there are other important T-ideals, such as the identities of the Grassmann algebra.

Here are two interesting ways of generalizing T-ideals of the free algebra $F\{X\}$ over a field F.

Definition 1.7.12.

(i) *A* T-**space** *of* $F\{X\}$ *is an* F-*subspace* V *closed under substitutions; i.e., if* $f(x_1, \ldots, x_m) \in V$ *then* $f(h_1, \ldots, h_m) \in V$ *for all* $h_i \in F\{X\}$.

(ii) Suppose A is an F-algebra and V is an F-subspace of A. Then those polynomials vanishing for all substitutions in V are denoted id($A; V$), *and this is also an ideal of* $F\{X\}$, *which Kemer calls an S-**ideal**.*

Note that id($A; V$) was used in proving a basic result, the construction of central polynomials, when we took A to be $M_n(F)$ and V the subspace of matrices of trace 0.

1.7.3 Varieties of algebras

In the other direction, given $\mathcal{I} = $ id($\mathcal{S}$), define the **(algebraic) variety** $\mathcal{V}_\mathcal{I}$ to be the class of algebras for which each $f \in \mathcal{I}$ is an identity. Clearly id($\mathcal{V}_{\text{id}(\mathcal{S})}$) = id($\mathcal{S}$). An abstract characterization of varieties is given in Exercise 43.

Geometrically, any T-ideal $\mathcal{I} \triangleleft_T C\{X\}$ can be considered as the ideal of polynomials of a given (PI) variety ($\mathcal{V}_\mathcal{I}$), in analogy to the ideal of the commutative polynomial ring annihilating a given algebraic variety, so we have the analog with algebraic sets from algebraic geometry, described in the Preface.

1.8 Relatively Free Algebras

It is also enlightening to work backwards.

Definition 1.8.1. *Given a T-ideal $\mathcal{I} \triangleleft_T C\{X\}$, we form the **relatively free** algebra*

$$U_\mathcal{I} = C\{X\}/\mathcal{I}$$

*for $\mathcal{I}$, also called the **universal** PI-algebra for $\mathcal{I}$. We write $\bar{x}_i$ for the image of x_i in $U_\mathcal{I}$. When $\mathcal{I} = $ id(A), we write U_A for $U_\mathcal{I}$.*

The relatively free algebra is the free object of its corresponding variety, and has important structural properties, thereby providing an excellent source of examples, as well as enabling us to unify aspects of the theory.

Lemma 1.8.2. *If* id(A) $\supseteq \mathcal{I}$, *then for any $a_1, a_2, \ldots \in A$, there exists a unique algebra homomorphism $U_\mathcal{I} \to A$ satisfying $\bar{x}_i \to a_i$, $i = 1, 2, \ldots$.*

Proof. We know there is a unique homomorphism $\psi : C\{X\} \to A$ given by $x_i \mapsto a_i$. The kernel contains $\mathcal{I}$, by Remark 1.7.7, so we conclude with Noether's isomorphism theorem. $\qquad\square$

Proposition 1.8.3. id($U_\mathcal{I}$) $= \mathcal{I}$, *for any $\mathcal{I} \triangleleft_T C\{X\}$.*

Proof. $\mathrm{id}(U_{\mathcal{I}}) \subseteq \mathcal{I}$, by Remark 1.7.7. Conversely, for any $f(x_1,\ldots,x_m) \in \mathcal{I}$ and $\overline{h_1},\ldots,\overline{h_m}$ in $U_{\mathcal{I}}$, where $\overline{h_i} = h_i + \mathcal{I}$, we see that

$$f(\overline{h_1},\ldots,\overline{h_m}) = f(h_1,\ldots,h_m) + \mathcal{I} = 0$$

since $\mathcal{I} \lhd_T C\{X\}$, proving $f \in \mathrm{id}(U_{\mathcal{I}})$.   □

Recapitulating, we have started with a set $\mathcal{S}$ of algebras, passed to $\mathcal{I} = \mathrm{id}(\mathcal{S})$, and then to $U_{\mathcal{I}}$, noting that $\mathrm{id}(U_{\mathcal{I}}) = \mathcal{I}$. Thus, we have a 1:1 correspondence between the T-ideals of $F\{X\}$ and relatively free algebras.

Relatively free algebras enable us to encode identities, in the following manner.

Definition 1.8.4. *Elements $y_1, y_2, \cdots \in A$ are* **PI-generic** *for A if they test identities in the following sense:*
$f(x_1,\ldots,x_m) \in \mathrm{id}(A)$ *iff* $f(y_1,\ldots,y_m) = 0$.

Remark 1.8.5. *The elements $\bar{x}_i : i = 1,2,\ldots$ are PI-generic for $U_{\mathcal{I}}$. Indeed $f(\bar{x}_1,\ldots,\bar{x}_m) = 0$ iff $f(x_1,\ldots,x_m) \in \mathcal{I} = \mathrm{id}(U_{\mathcal{I}})$.*

We have the following analog of Lemma 1.8.2.

Lemma 1.8.6. *Suppose an algebra $\hat{A} \in \mathcal{V}_{\mathcal{I}}$ is generated by PI-generic elements $\{r_i : i \in I\}$. Then for any A in $\mathcal{V}_{\mathcal{I}}$ and any $a_i \in A$, there is a (unique) homomorphism $\varphi : \hat{A} \to A$ obtained by sending $r_i \mapsto a_i, \forall i \in I$.*

Proof. Define φ by

$$\varphi(f(r_1,\ldots,r_m)) = f(a_1,\ldots,a_m).$$

This is well-defined since if $f(r_1,\ldots,r_m) = 0$, then $f(x_1,\ldots,x_m) \in \mathrm{id}(A)$, implying $f(a_1,\ldots,a_m) = 0$.   □

The next result is almost trivial but turns out to be surprisingly useful.

Proposition 1.8.7. *Any algebra $\hat{A}$ in $\mathcal{V}_{\mathcal{I}}$ generated by PI-generic elements $\{\hat{r}_i : i \in I\}$ is relatively free for $\mathcal{I}$.*

Proof. By Lemma 1.8.6 we can define a homomorphism $\varphi : \hat{A} \to U_{\mathcal{V}}$ by

$$f(\hat{r}_1,\ldots,\hat{r}_m) \mapsto f(\bar{x}_1,\ldots,\bar{x}_m),$$

and likewise we define a homomorphism $\psi : U_{\mathcal{V}} \to \hat{A}$ by

$$f(\bar{x}_1,\ldots,\bar{x}_m) \mapsto f(\hat{r}_1,\ldots,\hat{r}_m).$$

The composition in either way is the identity, so $\hat{A}$ is isomorphic to $U_{\mathcal{V}}$.   □

Example 1.8.8. *Let U_i be relatively free algebras for $1 \leq i \leq m$, generated by the images of generic matrices which we write as $\bar{x}_{i,1}, \bar{x}_{i,2}, \ldots$ where the $x_{i,j}$ are distinct indeterminates. Let $U = U_1 \times \cdots \times U_m$, and*

$$\hat{x}_j = (\bar{x}_{1,j}, \ldots, x_{m,j}) \in U.$$

Then the $\hat{x}_j$ are PI-generic elements of U, so the subalgebra $\hat{U}$ of U generated by the $\hat{x}_j$ is relatively free. (Indeed, $\mathrm{id}(U) = \cap\, \mathrm{id}(U_j)$. If $f(\hat{x}_1, \ldots \hat{x}_n) = 0$ then specializing all $x_{i,j}$ to 0 except for $i = i_0$ yields $f(\bar{x}_{i_0,1}, \ldots, \bar{x}_{i_0,n}) = 0$, and thus $f \in \mathrm{id}(U_{i_0})$; since this is true for all i_0 we get $f \in \mathrm{id}(U)$.)

We shall have occasion to weaken this property to multilinear polynomials:

Definition 1.8.9. *The elements $y_1, \ldots, y_n$ of an algebra are called ml-generic (ml for "multilinear") for A if they satisfy the following property:*
For any multilinear polynomial $f(x_1, \ldots, x_n)$, $f(x) \in \mathrm{id}(A)$ if and only if $f(y_1, \ldots, y_n) = 0$.

Here is an important example.

Example 1.8.10. *Let A, B be two PI-algebras with corresponding relatively free algebras $U_A = F\{\bar{x}_1, \bar{x}_2, \ldots\}$ and $U_B = F\{\bar{x}_1', \bar{x}_2', \ldots\}$. Then the elements $\bar{x}_1 \otimes \bar{x}_1', \bar{x}_2 \otimes \bar{x}_2', \ldots \in U_A \otimes U_B$ are ml-generic for $A \otimes B$.*
Indeed, assume that $f(\bar{x}_1 \otimes \bar{x}_1', \ldots, \bar{x}_n \otimes \bar{x}_n') = 0$ for f multilinear. We need to show that $f(x) \in \mathrm{id}(A \otimes B)$. It suffices to show that

$$f(a_1 \otimes b_1, \ldots, a_n \otimes b_n) = 0$$

for any $a_1 \ldots, a_n \in A$ and $b_1 \ldots, b_n \in B$. This follows by extending the homomorphisms given by $\bar{x}_i \to a_i$ and $\bar{x}_i' \to b_i$, to the homomorphism given by $\bar{x}_i \otimes \bar{x}_i' \to a_i \otimes b_i$.

Relatively free affine PI-algebras are representable. This is not quite implied by Theorem 1.6.13(ii), but does follow from Kemer's program, to be proved below as Corollary 6.9.7; also see Proposition 7.2.31.

1.8.1 The algebra of generic matrices

Let us collect two very important examples of relatively free algebras, which play a central role both in the general theory and in applications. We start with a famous example. Fixing a field F, we want to describe the relatively free algebra of $\mathcal{M}_{n,F}$. (The case of an arbitrary base ring is treated in Exercise 47.)

Definition 1.8.11. *Take a set $\Lambda = \{\lambda_{ij}^{(k)} : 1 \leq i, j \leq n, k \in \mathbb{N}\}$ of commuting indeterminates over a field F, and define the **algebra of generic $n \times n$ matrices** $F\{\bar{Y}\}_n$ to be the F-subalgebra of $M_n(F[\Lambda])$ generated by the matrices $\bar{y}_k = \left(\lambda_{ij}^{(k)}\right)$ for $k = 1, 2 \ldots$. We call each $\bar{y}_k$ a "generic $n \times n$ matrix."*

Remark 1.8.12. *If $f(\bar{y}_1, \ldots, \bar{y}_m) = 0$, then $f(x_1, \ldots, x_m) \in \mathcal{M}_{n,F}$. (This is seen by specializing the $\lambda_{ij}^{(k)}$ to arbitrary elements of F, thereby specializing the $\bar{y}_k$ to arbitrary matrices in $M_n(F)$.)*

It follows that the generic matrices are PI-generic elements, so by Proposition 1.8.7, $F\{\bar{Y}\}_n$ is relatively free for $\mathcal{M}_{n,F}$.

Let $F(\Lambda)$ denote the field of fractions of the integral domain $F[\Lambda]$.

Proposition 1.8.13. $F\{\bar{Y}\}_n$ *is prime of PI-class n.*

Proof. Using the notation of Definition 1.8.11, let A be the $F(\Lambda)$ subalgebra of $M_n(F(\Lambda))$ spanned by $F\{\bar{Y}\}_n$. Clearly, c_{n^2} is not an identity of $F\{\bar{Y}\}_n$. Hence, $A = M_n(F(\Lambda))$, by Proposition 1.2.24, and thus $F\{\bar{Y}\}_n$ has PI-class n. Furthermore, if $B_1, B_2 \lhd F\{\bar{Y}\}_n$ with $B_1 B_2 = 0$, then $B_1 F(\Lambda) B_2 F(\Lambda) = B_1 B_2 F(\Lambda) = 0$, implying $B_1 = 0$ or $B_2 = 0$ since $M_n(F(\Lambda))$ is simple. $\quad\square$

One can push this result further.

Theorem 1.8.14. *For every field F and every n, $F\{\bar{Y}\}_n$ is a (noncommutative) domain, whose ring of central fractions, denoted $\mathrm{UD}(n, F)$, is a division algebra D of dimension n^2 over its center.*

Proof. In view of Corollary D of §1.5, $\mathrm{UD}(n, F)$ is central simple of degree n, so in view of the Wedderburn-Artin theorem, it suffices to show $\mathrm{UD}(n, F)$ has no nonzero nilpotent elements, or equivalently that $F\{Y\}_n$ has no nonzero nilpotent elements. But in this case, there would be a polynomial $f(x_1, \ldots, x_m)$ not in $\mathcal{M}_{n,F}$, such that $f^t \in \mathcal{M}_{n,F}$ for some $t > 0$. This is false, because there are examples of division algebras D of PI-class n containing any given field (namely, the "generic" symbol algebra). Taking $a_1, \ldots, a_n \in D$ with $a = f(a_1, \ldots, a_n) \neq 0$, we would have $a^t = 0$, contradiction. $\quad\square$

Although we do not pursue this direction here, $\mathrm{UD}(n, F)$ has become a focal example in the theory of division algebras, since it is the test case for many important theorems and conjectures.

1.8.2   Relatively free algebras of f.d. algebras

There is another, less celebrated example, which nevertheless also plays an important role in the PI-theory.

Example 1.8.15. *Suppose A is a f.d. F-algebra with base $b_1, \ldots, b_t$, and let*

$$\Lambda = \{\lambda_{ij} : 1 \le i \le \ell,\ 1 \le j \le t\}$$

be commuting indeterminates over A. Define the **generic element**

$$\bar{y}_i = \sum_{j=1}^{t} b_j \lambda_{ij}, \quad 1 \le i \le \ell,$$

and let $\mathcal{A}$ *be the subalgebra of* $A[\Lambda]$ *generated by these generic elements* $\bar{y}_1, \ldots, \bar{y}_\ell$.

We claim that $\bar{y}_1, \ldots, \bar{y}_\ell$ *are PI-generic, cf. Definition 1.8.4, and there-fore* $\mathcal{A}$ *is isomorphic to the relatively free PI-algebra of A. Indeed, suppose* $f(\bar{y}_1, \ldots, \bar{y}_\ell) = 0$. *Then, given any* $a_1, \ldots, a_\ell$ *in A, write*

$$a_i = \sum_{j=1}^{t} \alpha_{ij} b_j, \quad 1 \le i \le \ell,$$

and define a homomorphism $\psi : A[\lambda] \to A$ *given by* $\lambda_{ij} \mapsto \alpha_{ij}$. *Then* ψ *restricts to a homomorphism* $\hat{\psi} : \hat{A} \to A$ *and*

$$f(a_1, \ldots, a_\ell) = f(\hat{\psi}(\bar{y}_1), \ldots, \hat{\psi}(\bar{y}_\ell)) = \hat{\psi}(f(\bar{y}_1, \ldots, \bar{y}_\ell)) = 0,$$

as desired.

Note that the structure of the relatively free algebra $\mathcal{A}$ can be quite differ-ent from that of the f.d. algebra A; for example, if $A = M_n(F) \times \mathrm{UT}(n-1)$, then its relatively free algebra $\mathcal{A}$ is also the relatively free algebra of $M_n(F)$, and so "lost" the structure of $\mathrm{UT}(n-1)$. Even when $\mathcal{A}$ has nilpotent elements, they might be hard to recover. For example, if A/J has PI-class n, then all values of c_{n^2+1} in $\mathcal{A}$ are nilpotent, but these may be difficult to handle. To get hold of the radical, we offer yet another generic construction in Defini-tion 6.6.24, although it is not quite relatively free.

1.8.3 *T*-ideals of relatively free algebras

Remark 1.8.16. *The T-ideals of an algebra A comprise a sublattice of the lattice of ideals of A; namely, if* $\mathcal{I}, \mathcal{J} \lhd_T A$, *then* $\mathcal{I} + \mathcal{J} \lhd_T A$ *and* $\mathcal{I} \cap \mathcal{J} \lhd_T A$. *(This is immediate since* $\psi(\mathcal{I} + \mathcal{J}) \subseteq \psi(\mathcal{I}) + \psi(\mathcal{J}) \subseteq \mathcal{I} + \mathcal{J}$ *and*

$$\psi(\mathcal{I} \cap \mathcal{J}) \subseteq \psi(\mathcal{I}) \cap \psi(\mathcal{J}) \subseteq \mathcal{I} \cap \mathcal{J},$$

for any homomorphism $\psi : A \to A$).

T-ideals are especially important when viewed in relatively free algebras.

Proposition 1.8.17. *Given a T-ideal* $\mathcal{I}$ *of a relatively free algebra U, there is lattice isomorphism between the T-ideals of U containing* $\mathcal{I}$ *and the T-ideals of* $U/\mathcal{I}$, *given by* $\mathcal{J} \mapsto \mathcal{J}/\mathcal{I}$.

Proof. Suppose $\mathcal{J} \lhd_T U$ contains $\mathcal{I}$. For any homomorphism $\psi : U/\mathcal{I} \to U/\mathcal{I}$ we write $\psi(\bar{x}_i + \mathcal{I}) = y_i + \mathcal{I}$ for suitable $y_i \in U$, By Remark 1.8.6, there is a homomorphism $\hat{\psi} : U \to U$ given by $\hat{\psi}(\bar{x}_i) = y_i$. By definition of T-ideal, $\hat{\psi}(\mathcal{I}) \subseteq \mathcal{I}$ and $\hat{\psi}(\mathcal{J}) \subseteq \mathcal{J}$. Thus, for any $a + \mathcal{I}$ in $\mathcal{J}/\mathcal{I}$,

$$\psi(a + \mathcal{I}) = \hat{\psi}(a) + \mathcal{I} \subseteq \mathcal{J}/\mathcal{I},$$

proving $\psi(\mathcal{J}/\mathcal{I}) \subseteq \mathcal{J}/\mathcal{I}$, thereby implying that $\mathcal{J}/\mathcal{I}$ is a T-ideal.

Conversely, suppose $\mathcal{J}/\mathcal{I} \lhd_T U/\mathcal{I}$. For any $\psi : U \to U$, we have $\psi(\mathcal{I}) \subseteq \mathcal{I}$, so ψ induces a homomorphism of $U/\mathcal{I}$, given by $a + \mathcal{I} \mapsto \psi(a) + \mathcal{I}$. But then, for $a \in \mathcal{J}$, $\psi(a) + \mathcal{I} \subseteq \mathcal{J}/\mathcal{I}$, implying $\psi(a) \in \mathcal{J}$; hence $\mathcal{J}$ is a T-ideal. $\square$

Corollary 1.8.18. *If $\mathcal{J} \lhd_T U$ and U is relatively free, then the algebra $U/\mathcal{J}$ is relatively free.*

Proof. Write $U = C\{X\}/\mathcal{I}$ for $\mathcal{I} \lhd_T C\{X\}$. Then $\mathcal{J} = \mathcal{J}_0/\mathcal{I}$ for some T-ideal $\mathcal{J}_0 \lhd_T C\{X\}$ containing $\mathcal{I}$, and $U/\mathcal{J} = (C\{X\}/\mathcal{I})/(\mathcal{J}_0/\mathcal{I}) \cong C\{X\}/\mathcal{J}_0$ is relatively free. $\square$

1.8.4   Verifying T-ideals in relatively free algebras

A natural question to ask at this stage is, "Which ideals occurring naturally in the structure theory of relatively free algebras are T-ideals?"

Remark 1.8.19. *Suppose U is a relatively free algebra generated by the infinite set $\bar{X} = \{\bar{x}_i : i \in I\}$, and $\mathrm{id}(A) \supseteq \mathrm{id}(U)$.*

(i) *If $r_j = f_j(a_1, \ldots, a_{m_j})$, then there is a homomorphism $\Phi : U \to A$ such that*

$$r_1 = \Phi(f_j(\bar{x}_1, \ldots, \bar{x}_{m_1})), \quad r_2 = \Phi(f_j(\bar{x}_{m_1+1}, \ldots, \bar{x}_{m_1+m_2})), \quad \ldots;$$

the point here is that different $\bar{x}_i$ appear in the different positions.

(ii) *If A is generated by a set of cardinality $\leq |\bar{X}|$, then Φ can be taken to be a surjection. (Indeed, deleting a countable number of letters needed to produce $r_1, r_2, \ldots$, still leaves the same cardinality of letters in $\bar{X}$, enough to produce a surjection on A.)*

(iii) *It follows from (ii) that to check that $\mathcal{I} \lhd U$ is a T-ideal, it is enough to check that $\psi(\mathcal{I}) \subseteq \mathcal{I}$ for all surjections $\psi : U \to U$.*

The point of this is that some of the ideals in structure theory are preserved under surjections but not necessarily under arbitrary homomorphisms, because surjections take ideals to ideals. For example, the Jacobson radical $J = \mathrm{Jac}(U)$ is a quasi-invertible ideal containing all quasi-invertible left ideals. If $\psi : U \to U$ is a surjection, then $\psi(J)$ is an ideal of U, clearly quasi-invertible, and thus contained in J. The same reasoning holds for the nilradical $N = \mathrm{Nil}(U)$: If $\psi : U \to U$ is a surjection, then $\psi(N)$ is a nil ideal of U, implying $\psi(N) \subseteq N$.

Lemma 1.8.20. *If $\mathrm{Jac}(A) = 0$, then $\mathrm{Jac}(U_A) = 0$.*

Proof. Otherwise take $0 \neq f \in \mathrm{Jac}(U_A)$, and elements $a_1, \ldots, a_m$ in A such that $f(a_1, \ldots, a_m) \neq 0$. Using Remark 1.8.19 take a surjection $\psi : U \to A$ such that $\bar{x}_i \mapsto a_i$. Then $0 \neq \psi(\mathrm{Jac}(U_A)) \subseteq \mathrm{Jac}(A) = 0$, contradiction. $\square$

Proposition 1.8.21. *If U is relatively free, then $\mathrm{Jac}(U) = \mathrm{Nil}(U)$ are T-ideals.*

Proof. We just saw that they are T-ideals, so we need to prove equality; we shall do this over an infinite field F. Since $\mathrm{Nil}(U) \subseteq \mathrm{Jac}(U)$, we can pass to $U/\mathrm{Nil}(U)$ and assume that $\mathrm{Nil}(U) = 0$. In particular, U is semiprime and thus has some PI-class n, cf. Corollary E of §1.5. But $\mathrm{Jac}(U[\lambda]) = 0$ by Theorem B of §1.5 and Theorem 1.1.11, and $\mathrm{id}(U[\lambda]) = \mathrm{id}(U)$ by Proposition 1.2.19. Hence, $U = U_{U[\lambda]}$, which has Jacobson radical 0 by Lemma 1.8.20. $\square$

Here is another application of Remark 1.8.19:

Remark 1.8.22. *The relatively free algebra $U = U_A$ is a subdirect product of copies of A. (Indeed, for each surjection $\varphi : U_A \to A$, write A_φ for this copy of A; the homomorphism $U \to \prod A_\varphi$ given by sending $\bar{f}$ to the vector $(\varphi(\bar{f}))$ is an injection, since $\varphi(\bar{f}) = 0$ for all surjections φ iff $f \in \mathrm{id}(A)$, iff $\bar{f} = 0$.*

We shall improve this result in Exercise 7.4.

1.8.5 Relatively free algebras without 1, and their T-ideals

Although we emphasize algebras with 1 in this book, there are important results in the theory for associative algebras without 1, especially concerning nil algebras (which obviously cannot contain 1), such as Theorem 1.8.28 below. So by a slight abuse of terminology, "nil subalgebra" indicates an algebra without 1, whereas all other algebras are with 1. To salve our consciences, let us take a minute to consider the identities of an algebra A without 1.

Remark 1.8.23. *If $f(x_1, \ldots, x_m) \in \mathrm{id}(A)$, then the constant term of f is $f(0, \ldots, 0) = 0$. Thus, writing $C\{X\}^+$ for the polynomials of constant term 0, so that $C\{X\} = C \oplus C\{X\}^+$, we have $\mathrm{id}(A) \subseteq C\{X\}^+$.*

Remark 1.8.24. *Adjoining 1 to any algebra A without 1, via Remark 1.1.1, we could view $\mathrm{id}(A) \lhd C\{X\}$. Furthermore, viewing $\mathrm{id}(A) \lhd C\{X\}^+$, we obtain the decomposition $U_A = C\{X\}/\mathrm{id}(A) = C \oplus (C\{X\}^+/\mathrm{id}(A))$. Then $U_A^+ = C\{X\}^+/\mathrm{id}(A)$ is the **relatively free algebra without 1** of A.*

1.8.6 Consequences of identities

A basic question about T-ideals and identities is, "Does $f \in \mathrm{id}(A)$ imply $g \in \mathrm{id}(A)$?" Alternatively, "When is g in the T-ideal generated by f?" In this case, we say g is a *consequence* of f; we also say f *implies* g.

Example 1.8.25. *The standard polynomial s_n is a consequence of the Capelli polynomial c_n, since we get s_n from c_n by specializing $y_i \mapsto 1$, for $1 < i \le n$. On the other hand, for $m \ge 2$, c_{2m} is not a consequence of s_{2m}, seen by noting that $s_{2m} \in \mathcal{M}_m$ by the Amitsur-Levitzki theorem, whereas $c_{2m} \notin \mathcal{M}_m$ by Remark 1.4.3.*

Also, the multilinearization of any identity of A is in $\text{id}(A)$. Another quick but significant example is due to Amitsur.

Theorem 1.8.26. *Every algebra A of PI-degree d satisfies a power s_{2n}^k of the standard identity, for suitable $n \leq \left[\frac{d}{2}\right]$ and for suitable k.*

Proof. We can replace A by its relatively free algebra. Let N be the nilradical. A/N is a semiprime PI-algebra and thus has some PI-class $n \leq \left[\frac{d}{2}\right]$, implying that s_{2n} is a PI of A/N, i.e., $s_{2n}(\bar{x}_1, \ldots, \bar{x}_{2n}) \in N$. But this means

$$s_{2n}(\bar{x}_1, \ldots, \bar{x}_{2n})^k = 0$$

for some k, and because A is relatively free, $s_{2n}(x_1, \ldots, x_{2n})^k \in \text{id}(A)$. $\qquad\square$

An explicit bound for k in terms of d will be computed in Remark 3.4.15.

Besides providing a canonical form for a PI, Theorem 1.8.26 shows more generally that any variety of algebras (defined perhaps with extra structure), in which the semiprime members satisfy a standard identity, actually is a variety of PI-algebras satisfying the PI of Theorem 1.8.26. This argument can be used to show that classes of algebras satisfying more general sorts of identities described in Volume II are indeed PI-algebras.

This sort of result can be extended using Lewin's Theorem:

Proposition 1.8.27. *If A is a relatively free PI-algebra and its nilradical $\text{Nil}(A)$ is nilpotent, then $\mathcal{M}_{n,F} \subseteq \text{id}(A)$ for some n.*

Proof. $\mathcal{M}_m \subset \text{id}(A/\text{Nil}(A))$ for some m, so apply Theorem H(ii) of §1.5; take $n = mt$ where $\text{Nil}(A)^t = 0$. $\qquad\square$

For example, the hypothesis holds by the Braun-Kemer-Razmyslov theorem of Chapter 4 when A is relatively free and affine PI. Although we shall also see that such A is representable, the method here is much easier, and also leads to the result in Exercise 4.6, that any affine algebra satisfying a Capelli identity satisfies all identities of $M_n(\mathbb{Q})$ for some n.

One celebrated result (from 1943) predates most of the PI-theory.

Theorem 1.8.28 ((Dubnov-Ivanov)-Nagata-Higman Theorem). *Suppose a $\mathbb{Q}$-algebra A without 1 is nil of bounded index. Then A is nilpotent; more precisely, there is a function $\beta : \mathbb{N} \to \mathbb{N}$ such that if A satisfies the identity x^n, then A must satisfy the identity $x_1 \cdots x_{\beta(n)}$.*

A quick proof due to Higgens can be found in Jacobson [Ja56, Appendix C], which gives the bound $\beta(n) = 2^{n-1} - 1$, cf. Exercise 50. Later Razmyslov improved the bound to $\beta(n) = n^2$; in Exercise 8.7, we give a quick argument along the lines of his reasoning to show that $\beta(n) \leq 2n^2 - 3n + 2$, and give his bound in Exercise 8.8. On the other hand, Higman [Hig56] also showed that $\beta(n) > n^2/e^2$, which was improved by Kuzmin [Kuz75] to $\beta(n) \geq \frac{n(n+1)}{2}$, cf. Exercise 8.9; Brangsburg proved that indeed $\beta(4) = 10$.

To make best use of Theorem 1.8.28, we recall the symmetric polynomial $\tilde{s}_t$ from Equation 1.7, i.e., the multilinearization of x_1^t.

Corollary 1.8.29. *Any* $\mathbb{Q}$*-algebra without 1 satisfying the identity* $\tilde{s}_t$ *also satisfies* x_1^t *and thus* $x_1 \cdots x_{\beta(t)}$.

1.9 Generalized Identities

When dealing with a polynomial, we might not want to substitute all of the indeterminates at once into an algebra A, but rather pause after having made some substitutions. This leads us to polynomials where the coefficients are not necessarily taken from the base algebra C, but rather from A and interspersed together with the indeterminates. This leads us to another generalization of PI, to be put later into the framework of Volume II, but is treated because it seems to be necessary for the proofs of the main theorems of PI theory. We start by reviewing a standard construction of ring theory.

1.9.1 Free products

Definition 1.9.1. *The free product* $A *_C B$ *of* C*-algebras* A *and* B *is their coproduct in the category of algebras (with 1).*

This can be viewed as arising from the tensor algebra construction of $A \otimes_C B$, as described in Definition 1.3.4, but is somewhat more complicated.

The general description is a direct limit construction given in [Ro88a, Theorem 1.9.22 and Example 1.9.23]. In brief, one takes the tensor algebra

$$C \oplus (A \otimes B) \oplus (A \otimes B)^{\otimes 2} \oplus \cdots$$

of $A \otimes_C B$ (with multiplication of simple tensors given by juxtaposition) modulo the C-submodule of relations spanned by elements of the form

$$\cdots (a_i \otimes 1) \otimes (a_{i+1} \otimes b_{i+1}) \cdots \quad - \quad \cdots (a_i a_{i+1} \otimes b_{i+1}) \cdots$$

and

$$\cdots (a_i \otimes b_i) \otimes (1 \otimes b_{i+1}) \cdots \quad - \quad \cdots (a_i \otimes b_i b_{i+1}) \otimes \cdots$$

where the term on the right has been shrunk by one tensor factor. We write the elements as sums of words alternating with elements of A and B, without the symbol $\otimes$. However, the same element could be represented in various ways.

It is easier to visualize the situation when B is free as a C-module with a base containing 1, in which case we have an reasonably explicit description of multiplication in $A * B$:

Example 1.9.2. *Fix a base* $\mathcal{B}_B = \{1\} \cup \mathcal{B}_{B,0}$ *of* B *over* C. $A * B$ *is comprised* of linear combinations of elements of the form $b_0a_1b_1a_2b_2\cdots a_mb_m$, where $m \geq 0$, $b_0, b_m \in \mathcal{B}_B$, $b_1, \ldots, b_{m-1} \in \mathcal{B}_A \setminus \{1\}$, and the $a_i \in A$. The free product $A * B$ becomes an algebra via juxtaposition of terms, extended via distributivity. In other words, given

$$g_j = b_{j,0}a_{j,1}b_{j,1}a_{j,2}b_{j,2}\cdots a_{j,m_j}b_{j,m_j}$$

for $j = 1, 2$, we write $b_{1,m_1}b_{2,0} = \alpha_1 + \sum_k \alpha_k b_k$ for $\alpha_k \in CS$ and b_k ranging over $\mathcal{B}_A \setminus \{1\}$, and have

$$
\begin{aligned}
g_1 g_2 =& \alpha_1 b_{1,0}a_{1,1}b_{1,1}a_{1,2}b_{1,2}\cdots(a_{1,m_1}a_{2,1})b_{2,1}a_{2,2}b_{2,2}\cdots a_{2,m_2} \\
&+ \sum_k \alpha_k b_{1,0}a_{1,1}b_{1,1}a_{1,2}b_{1,2}\cdots a_{1,m_1}b_k a_{2,1}b_{2,1}a_{2,2}b_{2,2}\cdots a_{2,m_2}.
\end{aligned}
\tag{1.26}
$$

For example, if $b_1 b_2 = 1 + b_3 + b_4$, then

$$(b_2 a_{1,1}b_1)(b_2 a_{2,1}b_2) = b_2(a_{1,1}a_{2,1})b_2 + b_2 a_{1,1}b_3 a_{2,1}b_2 + b_2 a_{1,1}b_4 a_{2,1}b_2.$$

When A and B are both torsion-free over C, there is more elegant variant put forward by Vishne, given in Exercise 51.

1.9.1.1   The algebra of generalized polynomials

Definition 1.9.3. *In the special case that $B = C\{X\}$, we write $A\langle X\rangle$ for the free product $A * C\{X\}$, and call its elements **generalized polynomials**.*

Generalized polynomials are like polynomials, but with the coefficients interspersed together with the indeterminates. Note that here the base of B is just the monoid of words in X, so there is no collapsing from multiplication of its base elements. Hence, $A\langle X\rangle$ is graded in terms of the degrees in the indeterminates x_i.

We may take $\mathcal{B}_{B,0}$ to be the nontrivial words in the indeterminates x_i. This base is multiplicative, and noting that $C\{X\}$ is graded by its words, we can grade A according to the words in X, denoted as h, by saying that $b_0a_1h_1a_2h_2\cdots a_mh_m$ is in the $h_0h_1h_2\cdots h_m$-component. This already gives us a handle on $A\langle X\rangle$, although we find it difficult to describe addition. For example,

$$(a_1 + a_2)x_1a_3 - a_1x_1(a_3 + a_4) = a_2x_1a_3 - a_1x_1a_4.$$

Example 1.9.4. *In case C is a field F, we can describe $A\langle X\rangle$ explicitly. Fix a base $\mathcal{B}_A = \{1\}\cup\mathcal{B}_{A,0}$ of A over F, as well as the above base $\mathcal{B}_B = \{1\}\cup\mathcal{B}_{B,0}$ of $F\{X\}$. $A\langle X\rangle$ is the vector space having base comprised of all elements of the form $a_0h_1a_1h_2a_2\cdots h_ma_m$ where $m \geq 0$, $a_0, a_m \in \mathcal{B}_A$, and $h_1, \ldots, h_{m-1}$ are words. Multiplication in the free product $A*B$ is achieved via juxtaposition of terms and cancelation in A where appropriate.*

If A is faithful over C, we have the natural embedding $C\{X\} \to A\langle X\rangle$. For $g \in C\{X\}$, we write $\bar{g}$ for its natural image in $A\langle X\rangle$.

1.9.2 The relatively free product modulo a T-ideal

We introduce a rather sophisticated concept, the relatively free product modulo a T-ideal. Although difficult to describe explicitly, this construction is needed implicitly in all known proofs of the Braun-Kemer-Razmyslov Theorem (as well as Kemer's theorems) in the literature, so we must develop its properties.

Definition 1.9.5. *Suppose $\mathcal{I}$ is a T-ideal of $C\{X\}$, for which $\mathcal{I} \subseteq \mathrm{id}(A)$. The **relatively free product** $A\langle X \rangle_\mathcal{I}$ of A and $C\{X\}$ **modulo $\mathcal{I}$** is defined as $A\langle X \rangle / \bar{\mathcal{I}}$, where $\bar{\mathcal{I}}$ is the two-sided ideal $\mathcal{I}(A\langle X \rangle)$ consisting of all evaluations on $A\langle X \rangle$ of polynomials from $\mathcal{I}$.*

We can consider $A\langle X \rangle_\mathcal{I}$ as the ring of (noncommutative) polynomials but with coefficients from A interspersed throughout, taken modulo the relations in $\mathcal{I}$.

This construction is universal in the following sense: Any homomorphic image of $A\langle X \rangle$ satisfying these identities (from $\mathcal{I}$) is naturally a homomorphic image of $A\langle X \rangle_\mathcal{I}$. Thus, we have:

Lemma 1.9.6. *(i) For any $g_1, \ldots, g_k$ in $A\langle X \rangle$, there is a natural endomorphism $A\langle X \rangle \to A\langle X \rangle$ which fixes A and sends $x_i \mapsto g_i$ and fixes all other x_i.*
(ii) For any $g_1, \ldots, g_k$ in $A\langle X \rangle_\mathcal{I}$, there is a natural endomorphism

$$A\langle X \rangle_\mathcal{I} \to A\langle X \rangle_\mathcal{I},$$

which fixes A and sends $x_i \mapsto g_i$ and fixes all other x_i.

Proof. The first assertion is clear. Hence we have the composition

$$A\langle X \rangle \to A\langle X \rangle \to A\langle X \rangle_\mathcal{I}$$

whose kernel includes $\mathcal{I}$, thereby yielding (ii). □

1.9.2.1 The grading on the free product and relatively free product

Lemma 1.9.7. *We can grade $A\langle X \rangle_\mathcal{I}$ according to the degrees in the indeterminates x. When $\mathcal{I}$ is generated by homogeneous generalized polynomials, we can also grade $A\langle X \rangle_\mathcal{I}$ in this way.*

Proof. Since the formula (1.26) preserves degrees, the $\mathbb{N}$-grading is preserved under free products. When $\mathcal{I}$ is generated by homogeneous polynomials in X, it is homogeneous in X, i.e., $\mathcal{I} = \oplus_m \mathcal{I} \cap C\{X\}_m$, and the nonzero elements of $C\{X\}_m/(\mathcal{I} \cap C\{X\}_m)$ have degree m, thereby inducing the natural grading given by $\deg(f + \mathcal{I}) = \deg(f)$ for all $f \notin \mathcal{I}$. □

This enables us to define maps more easily from relatively free products.

Exercises for Chapter 1

1. (Peirce decomposition for algebras without 1). Given any idempotent e define formally $e'a$ as $a - ea$, ae' as $a - ae$, and $e'ae'$ as $(e'a)e'$. Now

$$A = eAe \oplus eAe' \oplus e'Ae \oplus e'Ae'.$$

Given a set $e_1, \ldots e_t$ of orthogonal idempotents, take $e = \sum_{i=1}^{t} e_i$, and further refine this decomposition.

2. If C is a Jacobson ring, then the polynomial ring $A = C[\lambda_1, \ldots, \lambda_t]$ is also Jacobson. (Hint: One may assume that A is prime. If P is a maximal ideal of C, then $P[\lambda_1, \ldots, \lambda_t]$ is a prime ideal of A, and the quotient algebra $(C/P)[\lambda_1, \ldots, \lambda_t]$ is affine and thus Jacobson. Thus $\mathrm{Jac}(A) \subseteq \mathrm{Jac}(C)[\lambda_1, \ldots, \lambda_t] = 0$.)

ACC

3. (Eakin-Formanek) A C-module M satisfies C-extended ACC if it satisfies the ACC on submodules of the form IM for $I \lhd C$.

   Suppose M a finite, faithful C-module, satisfying the C-extended ACC. Then M is Noetherian, and C itself is a Noetherian ring. In particular, this holds for $M = A$ when A is finite over its center. (Hint: Write $M = \sum Ca_i$. First note that if M is Noetherian then C also is Noetherian, being a subdirect product of the Noetherian rings

   By Noetherian induction, one may assume for each $0 \neq I \lhd C$ that M/IM is Noetherian. Take $N < M$ maximal with respect to M/N being faithful. For any $N = IM < N' \leq M$ one has $cM \subseteq N'$ for $c \in C$, implying that M/N', an image of M/cM, is Noetherian; hence M/N is Noetherian, implying as above that C is Noetherian.)

4. Any subalgebra of an algebra with ACC(Ann) satisfies ACC(Ann).

5. Any semiprime algebra with ACC(Ann) has a finite number of minimal prime ideals, and their intersection is 0. (Hint: These are the maximal annihilator ideals.)

Multilinearization

6. If $f(x_1, \ldots, x_m)$ is homogeneous of degree d_i in each indeterminate x_i, then the multilinearization of f is the multilinear part of

$$f(x_{1,1} + \cdots + x_{1,d_1}, \ldots, x_{m,1} + \cdots + x_{m,d_m}), \qquad (1.27)$$

where $x_{i,j}$ denote distinct indeterminates. Obtain a suitable generalization for arbitrary polynomials (but take care which degrees appear in which monomials).

7. Although the multilinear part of (1.27) is an identity whenever f is an identity, find an example for which a homogeneous component of (1.27) need not be an identity even when f is an identity. (Hint: Take a finite field.)

Homogeneous identities

8. Prove Proposition 1.2.17: If F is an infinite field, then every identity of an algebra A is a sum of homogeneous identities. (Hint: Write $f = \sum f_u \in \mathrm{id}(A)$ with each f_u homogeneous of degree u in x_i. Then for all $a_i \in A$, we have

$$
\begin{aligned}
0 &= f(a_1, \ldots, \alpha a_i, \ldots, a_m) \\
&= \sum f_u(a_1, \ldots, \alpha a_i, \ldots, a_m) = \sum \alpha^u f_u(a_1, \ldots, a_m);
\end{aligned}
\tag{1.28}
$$

view this as a system of linear equations in the $f_u(a_1, \ldots, a_m)$ whose co-efficients are α^u; the matrix of the coefficients (taken over $\deg f$ distinct values for α) is the Vandermonde matrix, which is nonsingular, so the system of linear equations has the unique solution $f_u(a_1, \ldots, a_m) = 0$ for each u. Thus, one may assume that f is homogeneous in the i inde-terminate, and repeat the procedure for each i.)

9. Here is a refinement of the linearization procedure which is useful in characteristic p. Suppose $\mathrm{char}(F) = p$. If f is homogeneous in x_i, of degree n_i which is *not* a p-power, then f can be recovered from a homo-geneous component of a partial linearization at x_i. (Hint: Take k such that $\binom{n_i}{k}$ is not divisible by p, and take the $(k, n_i - k)$ homogeneous part of the linearization.)

10. Suppose $\mathrm{char}(F) = p$. If $n_i = \deg_i f$ is a p-power, then f cannot be recovered from a homogeneous component of a partial linearization at x_i. (Hint: $\binom{n_i}{k}$ is divisible by p for each $0 < k < p_i$, so the procedure only yields 0.)

Graded algebras

In these exercises, F is a field, and $A = M_n(F)$ is G-graded by a finite group G, which is not necessarily Abelian.

11. For any fine grading, $\mathrm{supp}(A)$ is a subgroup of G. and all homogenous el-ements of A are invertible. (Hint: $\mathrm{supp}(A)$ is closed under multiplication since F is a field.)

12. A grading is fine iff all homogenous elements are invertible.

13. A group homomorphism $\phi : G \to \mathrm{PGL}_n$ is called a **projective repre-sentation**. This can be viewed as a function $\phi : G \to \mathrm{GL}_n$ such that $\phi(g)\phi(h) \in F^\times \phi(gh)$ for all $g, h \in G$. Show that any fine grading gives

rise to an irreducible projective representation, and conversely in case F is algebraically closed. (Hint: ($\Rightarrow$) is clear, defining $\phi(g)$ to act as left multiplication by g on the 1-dimensional subspaces A_h. The converse requires Hilbert's Nullstellensatz as formulated in [Ro05, Corollary 5.16'].)

14. If $A = M_n(F)$ has an elementary G-grading and B is any G-graded algebra, then $A \otimes B$ has a G-grading spanned by the homogeneous elements $\{e_{ij} \otimes b : b \in A_h,\ g_i^{-1}hg_j = g.\}$, for $g \in G$.

15. Show that Example 1.3.23(ii) is a special case of (i), by defining an appropriate function $d : \{1,\ldots,n\} \to \mathbb{Z}$, and define

$$A_m = \{Ce_{ij} : d(i) - d(j) = m\}.$$

16. Prove Theorem 1.3.25. (Hint: The identity component is semisimple, and has the form $R_1 \oplus \cdots \oplus R_q$ where each $R_k \cong M_{n_k}(F)$ is a matrix algebra, with unit element e_k. The centralizer W_k of R_k in e_kAe_k is simple, with fine grading, with some support $H_k \leq G$. Using the Peirce decomposition with respect to the idempotents e_k, and writing $e_{i,j}^k$ for the i,j matrix unit in R_k, find homogeneous elements $y_{k,\ell} \in e_{1,1}^k Ae_{1,1}^\ell$ such that

$$y_{q,q-1} \cdots y_{2,1}y_{1,2} \cdots y_{q-1,q} \neq 0.$$

Taking

$$y_{k,\ell} = y_{k,k+1} \cdots y_{\ell-1,\ell} \qquad y_{\ell,k} = y_{\ell,\ell-1} \cdots y_{k+1,k}$$

for $k < \ell$, and multiplying by suitable scalars, one may assume that

$$y_{k,\ell}y_{u,v} = \delta_{\ell,u}y_{k,v},$$

which thus span a matrix algebra with elementary grading.

$$n_k^2, n_\ell^2 \leq \dim W_k y_{k,ell} W_\ell \leq n_k n_\ell,$$

since every homogeneous element in W_k is invertible, implying all n_k are equal, and one gets an elementary grading. One gets an onto map from the tensor product, which is an isomorphism in view of dimension considerations.)

The Grassmann algebra

17. Over an arbitrary commutative ring C, define the Grassmann algebra $G(V) = T(V)/I$ where $T(V)$ is the tensor algebra and $I = \langle a^2 : a \in V \rangle$. Show that this specializes to Definition 1.3.26.

18. If $V_m = \sum_{i=1}^m Ce_i$, then any product of $m + 1$ elements of V_m is 0 in $G(V)$. (Hint: This can be checked on the generators e_i; in any product $w = e_{j_1} \cdots e_{j_{m+1}}$ some e_i repeats, so $w = 0$, by Remark 1.3.29(ii).) It follows that every element of V is nilpotent in $G(V)$. The same argument

shows that every element of $G(V)$ with constant term 0 is nilpotent, and thus, the set of such elements comprises the unique maximal ideal of $G(V)$.)

Identities of matrices

19. Prove Wagner's identity (Example 1.2.7(iii).) (Hint: $\operatorname{tr}[a,b]=0$, for any $a,b \in M_2(F)$, so apply the characteristic polynomial of $[a,b]$.)

20. (Amitsur) $s_{n+1}(y, xy, x^2 y, \ldots, x^n y) \in \mathcal{M}_{n,F}$. (Hint: Same argument as in Lemma 1.4.14.) One can cancel the y on the right and get a two-variable identity of degree $\frac{n^2 + 3n}{2}$.

21. $M_n(F)$ is PI-irreducible. (Hint: Suppose there were f.d.-algebras $A_1, \ldots, A_m$, with $\cap_{1 \le j \le m} \operatorname{id}(A_j) = \mathcal{M}_{n,F}$. Then each $A_j / \operatorname{Jac}(A_j)$ is semisimple of PI-class $< n$ and thus satisfies the standard identity $s_{2n-2}^{q_k}$ for some k. Taking $u = \max\{q_1, \ldots, q_m\}$ shows that $s_{2n-2}^u \in \mathcal{M}_{n,F}$, a contradiction.)

22. Show that if $f(x_1, \ldots, x_{2n})$ is a multilinear PI of $M_n(\mathbb{Z})$, then $f = \pm s_{2n}$. (Hint: Substitutions of matrix units as in Remark 1.4.2.)

23. Suppose $n = n_1 + \cdots + n_t$. Given $\pi \in S_n$ with

$$\pi(1) < \pi(2) < \cdots < \pi(n_1), \quad \pi(n_1+1) < \pi(n_1+2) < \cdots < \pi(n_1+n_2),$$

and so on. Write

$$c_{n_1;\pi} = c_{n_1}(x_{\pi(1)}, \ldots, x_{\pi(n_1)}; y_1, \ldots, y_{n_1});$$
$$c_{n_j;\pi} = c_{n_j}(x_{\pi(n_1+\cdots+n_{j-1}+1)}, \ldots, x_{\pi(n_1+\cdots+n_j)};$$
$$y'_{n_1+\cdots+n_{j-1}+1}, \ldots, y'_{n_1+\cdots+n_j})$$

for each $j > 1$, where at each stage y_i has been replaced by

$$y'_i := y_{\pi(i)} x_{\pi(i)+1} y_{\pi(i)+1} \cdots y_{\pi(i+1)-1},$$

the filler between $x_{\pi(i)}$ and $x_{\pi(i+1)}$. Noting that the alternator of

$$c_{n_1;\pi} c_{n_2;\pi} \cdots c_{n_t;\pi}$$

is $\operatorname{sgn}(\pi) n_1! \cdots n_t! c_n(x_{\pi(1)}, \ldots, x_{\pi(n)}; y_1, \ldots, y_n)$; show that

$$\sum_{\pi \in S_n} \operatorname{sgn}(\pi) c_{n_1;\pi} c_{n_2;\pi} \cdots c_{n_t;\pi} = k c_n(x_1, \ldots, x_n; y_1, \ldots, y_n)$$

where $k > 0$. This observation was critical to Razmyslov's proof of the Amitsur-Levitzki theorem (cf. Exercise 8.2) and also is needed in our verification of Capelli identities for affine algebras (although the notation in Chapter 5 is more user-friendly).

24. Verify the assertion in the paragraph after Lemma 1.4.14, by checking matrix units.

25. Suppose $T \in A = M_n(F)$, and let $\alpha = \text{tr}(T)$ and $\beta = \det(T)$. Then, viewing T in $\text{End}_F A$, show now that the trace is $n\alpha$ and determinant is β^2. How does this affect Equation (1.23)?

26. (Razmyslov-Bergman-Amitsur) Verify that

$$\text{tr}(w_1)\,\text{tr}(w_2)f(a_1,\ldots,a_t,r_1,\ldots,r_m)$$
$$= \sum_{k=1}^{t} f(a_1,\ldots,a_{k-1},w_1 a_k w_2, a_{k+1},\ldots,a_t,r_1,\ldots,r_m) \tag{1.29}$$

holds identically in $M_n(C)$. (Hint: Consider $a \mapsto \text{tr}(w_1)a\text{tr}(w_2)$.) This observation arises naturally when we consider $\text{End}_F A \cong A \otimes A^{\text{op}}$, and also leads to the construction of a central polynomial.

Identities of other specific rings

27. Any finite ring satisfies an identity $x^n - x^m$ for suitable $n, m \in \mathbb{N}$. (Hint: The powers of any element a eventually repeat.) Multilinearizing, show that any finite ring satisfies a symmetric identity.

28. In this exercise, "algebraic" means over $\mathbb{Z}$. Define the **algebraic radical** of an algebra A to be $\{r \in A : ra$ is algebraic, for all $a \in A.\}$ This contains the nilradical $\text{Nil}(A)$, equality holding when $\text{Cent}(A)$ contains a non-algebraic subfield. (Hint: Passing to $A/\text{Nil}(A)$, one may assume that $\text{Nil}(A) = 0$; in particular, A is semiprime. Then, passing to prime homomorphic images, one may assume that A is prime; taking central fractions, one may assume that A is simple. But any ideal must contain the center, which is not algebraic.)

29. Suppose A is a PI-algebra which is affine over a finite field. Definitions as in Exercise 28, show that the algebraic radical of A satisfies the identity $(x^m - x^n)^k$ for suitable k and $m > n$ in $\mathbb{N}$. (Hint: Since $\text{Nil}(A)$ is nilpotent, one may follow the steps of the previous exercise to assume that A is prime; then either A has algebraic radical 0 or A is a matrix algebra over a field; appeal to Exercise 27.)

30. If the algebraic radical of an algebra A is 0, then the homogeneous components of any identity of A are identities of A.

31. Amitsur's original proof of Theorem 1.8.26: All values of s_n are nilpotent. Index copies of A according to the n-tuples of its elements, and take the direct product, which is PI-equivalent to A, and use a diagonalization argument.

Representability

32. (Bergman) Suppose that the commutative ring C contains a field, and M is a Noetherian module over C. Then $\mathrm{End}_C M$ is representable. (Hint: Writing $M = \sum_{i=1}^{q} C x_i$, we have a C-module embedding

$$End_C(M) \to M^{(n)},$$

taking $f \in End_C(M)$ to $(f(x_1), ..., f(x_n))$. Thus, being a submodule of a Noetherian C-module, $End_C(M)$ is a Noetherian C-module, and thus a left Noetherian C-algebra, and is representable by Theorem 1.6.22 and Exercise 3.)

33. Suppose M is generated by n elements as an H-module, for H commutative. Then $\mathrm{End}_H M$ is a homomorphic image of a subalgebra of $M_n(H)$, and thus satisfies all the identities of $M_n(H)$. Compare with Digression 1.6.3.

34. Any semiprime PI-algebra over a field, which also satisfies ACC(Ann), is representable.

35. (Strengthened version of Proposition 1.6.8.) If C is Noetherian and $A = C\{a_1, \ldots, a_\ell\}$ is affine and representable in the weak sense of Digression 1.6.3, then $A \subseteq M_n(K)$ for a suitable commutative affine (thus Noetherian) C-algebra K.

36. Any representable algebra over an infinite field is PI-equivalent to a f.d. algebra over a (perhaps larger) field. Contrast with the example

$$\begin{pmatrix} \mathbb{Z}_p & \mathbb{Z}_p[\lambda_1, \lambda_2] \\ 0 & \mathbb{Z}_p[\lambda_1] \end{pmatrix}, \tag{1.30}$$

which satisfies the identity $(x^p - x)[y, z]$.

37. Let R be as in Example 1.6.18. Although, in the particular case $C = \mathbb{Z}/p^2\mathbb{Z}$, any ring injection from R is a C-algebra injection, in general, many ring injections are not C-algebra injections. For $C = F[\lambda]/\langle \lambda^2 \rangle$, where F is a field, for any subring $C' \supset F$ of C, there is no C'-injection into matrices over a local ring. (Hint: C' will contain an element of cR annihilating e_1, by Luroth's Theorem.)

38. Suppose C is algebraic over a field F. If C satisfies the ACC on principal annihilators, then C is a finite direct product of local rings C_i, each algebraic over some field extension K_i of F, and every element of each of these component rings is either invertible or nilpotent. (Hint: Take K to be a maximal integral domain contained in C. K is a field since it is algebraic over F. If $a \in C \setminus K$ and $K[a]$ contains a nontrivial idempotent e, then $C = Ce \oplus C(1 - e)$. Continuing to decompose in this way, we would reach a contradiction to ACC on principal annihilators unless the process terminates, at which stage we are done by Lemma 1.6.27.)

39. If an algebra A over a field F is finitely presented over its center Z, and if A satisfies the ACC (ascending chain condition) on annihilators of elements of Z, then A is representable as an F-algebra. (Hint: Suppose $R = \sum_{i=1}^{q} Z b_i$. Writing $b_i b_j = \sum_{k=1}^{q} c_{ijk} b_k$ for $c_{ijk} \in Z$, let

$$Z_1 = F_0[c_{ijk} : 1 \le i, j, k \le q, \ 1 \le \ell \le m].$$

Note that for any F_0-affine Z-subalgebra Z' of Z containing Z_1 that $R' = \sum_{i=1}^{q} Z' b_i$ is an F_0-subalgebra of R for which $R'Z = R$. Take such R' for which Z' contains all the coefficients of the finitely many relations of R as Z-module which are given to generate all the relations of R. The canonical surjection $\Phi : R' \otimes_{Z'} Z \to R$ is an isomorphism. But now embedding R' in $M_n(F')$, embed R in $M_n(F') \otimes_{Z'} Z \cong M_n(F' \otimes_{Z'} Z)$, as desired.)

40. If a weakly Noetherian PI-algebra A is affine over its center Z, and Z contains a field F, then A is representable as an F-algebra. (Hint: This proof relies on Anan'in's theorem [An92]. Assume that $R = Z[r_1, \ldots, r_t]$ is irreducible, in which case all non-nilpotent elements of Z are regular in R. As in the proof of Proposition 1.6.28, localizing, assume that all non-nilpotent elements of Z are invertible in R, and Z has a nilpotent ideal N such that Z/N is a field K. Let $N' = NR$, which is a nilpotent ideal of R and a finite R^e-module since R is a finite R^e-module. Show that R is finite over $K[r_1, \ldots, r_t] \otimes K[r_1, \ldots, r_t]$; it is enough to show that $(N')^i/(N')^{i+1}$ is finite over $K[r_1, \ldots, r_t] \otimes K[r_1, \ldots, r_t]$ for each $i \ge 0$. But $(N')^i/(N')^{i+1}$ is finite over $R/N' \otimes R/N'$.)

41. The hypotheses of Exercise 40 do not imply that Z is Noetherian. (Hint: Take the commutative polynomial algebra $A = F[\lambda_1, \lambda_2]$ and $Z = F + \lambda_1 F[\lambda_1, \lambda_2]$.)

42. (Small) A PI-ring R is called **semiprimary** if R/N is semisimple, for some nilpotent ideal N. Show that any semiprimary algebra A affine over a commutative Noetherian Jacobson base ring C is finite over C.

    One needs some hypothesis on C, since there exist fields affine over an integral domain C which are not finite over C, such as the field of rational fractions $C(\lambda^{-1})$, where C is the localization of the polynomial algebra $C[\lambda]$ at the polynomials having nonzero constant term.

Varieties of algebras

43. A class $\mathcal{V}$ of algebras is a variety iff it is closed under direct products, subalgebras, and homomorphic images. (Hint: $\mathcal{I}$ is the intersection of the kernels of homomorphisms from $C\{X\}$ to algebras in $\mathcal{V}$.)

44. Suppose $\mathcal{V}$ is a variety (perhaps with additional structure) that has the property that if $A \in \mathcal{V}$, then $A/\operatorname{Nil}(A)$ is in $\mathcal{V}$ and satisfies the identity s_n. Then every algebra in $\mathcal{V}$ satisfies s_n^k for suitable k. (Hint: Consider

the direct product $\tilde{A}$ of copies of A, indexed over all n-tuples of elements of A. $\tilde{A}/\text{Nil}(\tilde{A})$ satisfies s_n, and $\text{Nil}(\tilde{A})$ is nil of bounded index, so A satisfies s_n^k for some k. Conclude as in Exercise 31.)

45. (Lewin) Suppose M is a free A_1, A_2-bimodule, where $A_j = F\{X\}/\mathcal{I}$ are relatively free for $j = 1, 2$, and write $\bar{x}_{i,j}$ for the image of x_i in A_j. Then

$$\begin{pmatrix} A/I_1 & M \\ 0 & A/I_2 \end{pmatrix} \tag{1.31}$$

contains a relatively free algebra generated by elements of the form

$$\begin{pmatrix} \bar{x}_{i,1} & w_i \\ 0 & \bar{x}_{i,2} \end{pmatrix} \tag{1.32}$$

46. (Giambruno-Zaicev) $\text{id}(\text{UT}(n_1, \ldots, , n_q)) = \text{id}(\text{UT}(n_1)) \cdots \text{id}(\text{UT}(n_q))$, over any infinite field F. (Hint: Pass to $R := \text{UT}(n_1, \ldots, , n_q)[\Lambda]$ where Λ denotes an infinite set of commuting indeterminates. By induction the result holds for $A := \text{UT}(n_1, \ldots, , n_{q-1})$. Viewing everything in $M_n(F)$ where $n = \sum_{\ell=1}^q n_\ell$, take generic elements $\sum \lambda_{i,j}^k e_{i,j}$ in A and also in the bottom matrix block, which we call A'. Likewise define generic elements over the diagonal, and show these form a base by specializing to A and A'.)

Generic matrices over the integers

47. Define the **algebra of generic matrices** $C\{Y\}_n$ over an arbitrary commutative ring C, in analogy to Definition 1.8.11. Show that if $f(y_1, \ldots, y_m) = 0$ in $\mathbb{Z}\{Y\}_n$, then $f(x_1, \ldots, x_m) \in \text{id}(M_n(C))$ for every commutative algebra C.

48. $\text{id}(M_n(\mathbb{Z})) \subseteq \text{id}(M_n(C))$ for any commutative ring C. (Hint:

$$M_n(\mathbb{Z}[\Lambda]) \cong M_n(\mathbb{Z}) \otimes_{\mathbb{Z}} \mathbb{Z}[\Lambda],$$

so $\text{id}(M_n(\mathbb{Z})) = \text{id}(M_n(\mathbb{Z}[\Lambda]))$. Conclude with Exercise 47.)

49. If $\text{id}(M_{n_i}(\mathbb{Z})) \subseteq \text{id}(A_i)$ for $i = 1, 2$, then $\text{id}(M_{n_1 n_2}(\mathbb{Z})) \subseteq \text{id}(A_1 \otimes A_2)$. (Hint: Reduce to the ring of generic matrices via Exercise 48, and embed these into matrix algebras.)

50. Prove the Nagata-Higman Theorem (1.8.28). (Hint: Partial linearization gives the identity $\sum_{i=0}^{n-1} x^i y x^{n-i-1}$, and thus,

$$\sum_{i,j} x^i (z y^j) x^{n-i-1} y^{n-j-1},$$

which is $\sum_i x^i z \sum_j y^j x^{n-i-1} y^{n-j-1}$, and thus, $\sum x^{n-1} z y^{n-1}$. If N is the ideal of A generated by all a^{n-1}, then A/N is nilpotent by induction, but $NAN = 0$, so conclude that A is nilpotent.) Also see Exercise 8.7.

The free product

51. (Vishne's variant of the free product) Take the tensor algebra

$$C \oplus (A \oplus B) \oplus (A \oplus B)^{\otimes 2} \oplus \ldots$$

of $A \oplus B$ (with multiplication of simple tensors given by juxtaposition) modulo the C-submodule of relations spanned by elements of the form

$$\cdots (a_i \otimes a_{i+1}) \otimes \cdots \quad - \quad \cdots (a_i a_{i+1}) \otimes \cdots$$

and

$$\cdots (b_i \otimes b_{i+1}) \otimes \cdots \quad - \quad \cdots (b_i b_{i+1}) \otimes \cdots$$

where the term on the right has been shrunk by one tensor factor. Write the elements as sums of words alternating with elements of A and B, without the symbol $\otimes$. When A and B are both torsion-free over C, prove that this is isomorphic to the usual free product construction.

Chapter 2

A Few Words Concerning Affine PI-Algebras: Shirshov's Theorem

Affine algebras were defined in Definition 1.1.8. Perhaps the first major triumph of PI-theory was Kaplansky's proof [Kap50] of A.G. Kurosh's conjecture for PI-algebras, that every algebraic affine PI-algebra over a field is finite dimensional. (Later a non-PI counterexample to Kurosh's conjecture was discovered by Golod and Shafarevich [Gol64].) Kaplansky's proof was structural, based on Jacobson's radical and Levitzki's locally finite radical.

Although the structural approach to affine PI-algebras has been well understood for some time and exposed in [KrLe00] and [Ro88b, Chapter 6.3], many of these results can be streamlined and improved when viewed combinatorically in terms of words. The first combinatorial proof of Kurosh's conjecture for PI-algebras was obtained by A.I. Shirshov, later seen as a consequence of **Shirshov's Height Theorem**, which does not require any assumption on algebraicity, and shows that any affine PI-algebra bears a certain resemblance to a commutative polynomial algebra, thereby yielding an immediate proof of Kurosh's conjecture for PI-algebras. Another consequence of Shirshov's Height Theorem, proved by Berele, is the finiteness of the Gel'fand-Kirillov dimension of any affine PI-algebra. A major dividend of Shirshov's method is that it generalizes at once to an affine algebra over an arbitrary commutative ring:

If $A = C\{a_1, \ldots, a_\ell\}$ is an affine PI-algebra over a commutative ring C and a certain finite set of elements of A (e.g., the words in the a_i of degree up to some given number to be specified) is integral over C, then A is finite (finitely generated) as a C-module.

Our main goal in this chapter is to present the proof and the combinatoric theory behind Shirshov's Height Theorem, which is critical for the theory developed in the remainder of this book.

In the process, we actually give four proofs of the key Shirshov Lemma 2.3.1. We start with a quick, nonconstructive proof that suffices for the most important applications, to be described presently. Later, we present second and third proofs of Shirshov's Lemma; the second being a modification of Shirshov's original proof, which gives a recursive bound on the height. Our third proof yields a much better bound, following Belov, leading to an independence theorem on hyperwords that implies Ufnarovski'i's independence theorem and also solves a conjecture of Amitsur and Shestakov.

Although the function $\beta(\ell, k, d)$ is the focus of Shirshov's Theorem, and provides a bound on the Shirshov height function $\mu(\ell, d)$ as will be shown in the proof of Theorem 2.2.2(i), we shall see in Theorem 11.2.15 that $\mu(\ell, d)$ bounds such important invariants as the Gel'fand-Kirillov dimension, thereby motivating us to look for a better direct bound. A subexponential bound is given in Theorem 2.8.5, its proof far more intricate than the others, relying on a theorem of Dilworth to refine the words further.

2.1 Words Applied to Affine Algebras

We work throughout with a finite **alphabet** $X = \{x_1, \ldots, x_\ell\}$ of **letters** and the free monoid $\mathcal{M}\{X\}$ described in §1.1, whose elements are now called **words**.

Definition 2.1.1. *If $w = w_1 w_2 w_3$ in $\mathcal{M}\{X\}$, we call the w_i **subwords** of w; w_1 is **initial** and w_3 is a **tail**. We write $v \subseteq_{\text{init}} w$ when v is an initial subword of w.*

A subword v of w has **multiplicity** $\geq n$ if there are n nonoverlapping occurrences of v in w. If the occurrences are consecutive, we write v^k to denote $vv \ldots v$, i.e., v repeating k times. It is useful to introduce the **blank word**, denoted $\emptyset$, which is the unit element in $\mathcal{M}\{X\}$. A word $w = x_{i_1} \ldots x_{i_d}$ will be said to have **length** d, written $|w| = d$. By convention, $|\emptyset| = 0$.

Remark 2.1.2. *If $|u| \leqslant |v| < |w|$, with $u \subseteq_{\text{init}} w$ and $v \subseteq_{\text{init}} w$ for words $u, v,$ and w, then $u \subseteq_{\text{init}} v$.*

One of our main tools is the **lexicographic (partial) order** on words, in which we give each letter x_i the value i, and order words by the first letter in which the values differ. In other words, for two words $w_1 = x_{i_1} v_1$ and $w_2 = x_{i_2} v_2$, we have $w_1 \succ w_2$ if $i_1 > i_2$, or if $i_1 = i_2$ and inductively $v_1 \succ v_2$.

Note that we do not compare a word with the blank word $\emptyset$. Thus, two words $v \neq w$ with $|v| \leq |w|$ are **(lexicographically) comparable** iff v is *not* an initial subword of w. (In particular, any two words of the same length are comparable.)

Our basic set-up: $A = C\{a_1, \dots, a_\ell\}$ is an affine algebra over a commutative ring C. The homomorphism $\Phi : C\{X\} \to A$, given by $x_i \mapsto a_i \in A$ is fixed throughout this discussion. Given a word $w \in \mathcal{M}\{X\}$, we write $\bar{w}$ for its image under Φ, and shall call $\bar{w}$ a **word in A**.

For a set W of words, we write $\bar{W}$ for $\{\bar{w} : w \in W\}$. A subset S of A is **spanned** by $\bar{W}$ if each element of S can be written in the form $\{\sum_{w \in W} c_w \bar{w} : c_w \in C\}$.

We say that $\bar{w}$ is **A-minimal** if $\bar{w}$ is not spanned by the images of words in W less than w (under the lexicographic order).

Remark 2.1.3. $\{\bar{w} : w \text{ is } A\text{-minimal}\}$ *spans A. On the other hand, when C is a field, we cannot have a relation among A-minimal words, so they constitute a base of A.*

The role of the PI is to provide a reduction procedure whereby we can replace words by linear combinations of words of lower lexicographic order. In this section we write d (instead of n) for the degree of the PI, in order to reserve the symbol n for other uses. Also, we use $Y = \{y_1, y_2 \dots\}$ for an infinite set of indeterminates, in order to reserve the letters x_i for our word arguments.

The point is that if

$$f = y_1 \dots y_d + \sum_{\pi \neq 1} \alpha_\pi y_{\pi(1)} \cdots y_{\pi(d)},$$

then given any words $w_1 \succ w_2 \succ \cdots \succ w_d$, we can replace $\bar{w}_1 \dots \bar{w}_d$ by $-\sum_{\pi \neq 1} \alpha_\pi \bar{w}_{\pi(1)} \cdots \bar{w}_{\pi(d)}$, a linear combination of words in A of lower order.

2.2   Shirshov's Height Theorem

Definition 2.2.1.

(i) *A word w is a **Shirshov word** of height μ over W if $w = u_{i_1}^{k_1} \dots u_{i_\mu}^{k_\mu}$ where each $u_i \in W$. Given a set of words W, we write $\widehat{W}_\mu$ for the Shirshov words of height $\leq \mu$ over W.*

(ii) *A has **height** μ over a set of words W if A is spanned by the set*

$\{\bar{w} : w \in \widehat{W_\mu}\}$. *In this case we say that* $\bar{W}$ *is a* **Shirshov base** *of* A. *(This is a misnomer, since it is the evaluations of* $\widehat{W_\mu}$ *that span* A, *and not* $\bar{W}$, *but the terminology has become standard.)*

Note: In (i), we could have $i_j = i_{j'}$ for some $j \neq j'$. For example, the word $x_1^7 x_2^5 x_1^3$ has height 3 over $\{x_1, x_2\}$. We are ready to formulate the main result of this chapter.

Theorem 2.2.2 (Shirshov's Height Theorem).

(i) *Suppose* $A = C\{a_1, \ldots, a_\ell\}$ *is an affine PI-algebra of PI-degree* d, *and let* W *be the set of words of length* $\leq d$. *Then* $\bar{W}$ *is a Shirshov base for* A; A *has height* $\leq \mu$ *over* W, *where* $\mu = \mu(\ell, d)$ *is a function of* d *and* ℓ *(to be determined as we develop the proof). In other words,* $\{\bar{w}_1^{k_1} \ldots \bar{w}_\mu^{k_\mu} : w_i \in W, \ k_i \geq 0\}$ *span* A.

(ii) *If in addition* $\bar{w}$ *is integral over* C *for each word* w *in* W, *then* A *is finite as a* C-*module.*

Remark 2.2.3. *A trivial but crucial observation, which we often need, is that the number of words in* $X = \{x_1, \ldots, x_\ell\}$ *of length* m *is* ℓ^m, *since we can choose among* ℓ *letters in each of the* m *positions. Thus, the number of words of length* $\leq m$ *is* $\ell^m + \ell^{m-1} + \cdots + 1 = (\ell^{m+1} - 1)/(\ell - 1)$.

Remark 2.2.4. *It is easy to conclude (ii) of Theorem 2.2.2 as a consequence of (i). Indeed, if* $\bar{w}$ *satisfies a monic equation of degree* k_w, *then we can replace each* $\bar{w}^{k_w}$ *by smaller powers, thereby seeing that* A *is spanned by finitely many terms. Of course, this number is rather large, for*

$$|W| = \frac{\ell^{d+1} - 1}{\ell - 1},$$

and the number of words of height μ *over* W, *with each exponent less than or equal to* k, *is* $\big((k+1)|W|\big)^\mu$.

This motivates us to prove (ii) separately with a better bound because of its importance in its own right, and its proof also turns out to be more direct than that of (i).

Our first objective is to prove Shirshov's Height Theorem and give its host of applications to affine PI-algebras. Afterwards, since Shirshov's theorems are so important combinatorially, we shall take a hard look at the Shirshov height function μ, and also use our techniques to answer various problems of Burnside type.

Definition 2.2.5 (Shirshov). *A word* w *is called* d-**decomposable** *(with respect to* $\succ$*) if it contains a subword* $w_1 w_2 \ldots w_d$ *such that for any permutation* $1 \neq \sigma \in S_d$,

$$w_1 w_2 \ldots w_d \succ w_{\sigma(1)} w_{\sigma(2)} \cdots w_{\sigma(d)}.$$

Otherwise w *is called* d-**indecomposable.**

In other words, any A-minimal word is d-indecomposable. The following proposition relates the PI to d-decomposability.

Proposition 2.2.6 (Shirshov). *Suppose A satisfies a PI of degree d. Then any d-decomposable word w can be written in A as a linear combination of permuted words (i.e., the letters are rearranged) that are lexicographically smaller than w.*

Proof. A satisfies a multilinear identity of degree d, which can be written as

$$x_1 \cdots x_d - \sum_{\eta \neq 1,\ \eta \in S_d} \alpha_\eta x_{\eta(1)} \cdots x_{\eta(d)}.$$

But then for all $a_i \in A$,

$$a_1 \cdots a_d - \sum_{\eta \neq 1,\ \eta \in S_d} \alpha_\eta a_{\eta(1)} \cdots a_{\eta(d)},$$

so we can replace $\bar{w}_1 \cdots \bar{w}_d$ by

$$\sum_{\eta \neq 1,\ \eta \in S_d} \alpha_\eta \bar{w}_{\eta(1)} \cdots \bar{w}_{\eta(d)}$$

in A. By hypothesis, all the words on the right are smaller than $w_1 \ldots w_d$, and the proof follows. $\qquad\square$

Remark 2.2.7. *The use of a multilinear polynomial identity to lower the lexicographic order of a d-decomposable word does not change the letters used in the word, but merely rearranges them. This somewhat trivial remark is crucial both in these discussions and in considerations of growth in Chapter 11.*

2.3   Shirshov's Lemma

Shirshov's strategy to prove (ii) of Theorem 2.2.2 was to show:

Lemma 2.3.1 (Shirshov's Lemma). *For any ℓ, k, d, there is $\beta = \beta(\ell, k, d)$ such that any d-indecomposable word w of length $\geq \beta$ in ℓ letters must contain a nonempty subword of the form u^k, with $|u| \leq d$.*

This formulation does not mention any PI, but is a purely combinatorial fact about words. To prove it, we introduce some standard techniques from combinatorics. Our way of proving d-decomposability is via the following stronger version.

Definition 2.3.2. *A word w is **strongly d-decomposable** if it can be written as $w = w_0 w_1 \cdots w_n$, where the subwords $w_1, \ldots, w_n$ are in decreasing lexicographical order.*

Clearly strongly d-decomposable words are d-decomposable, and, conversely, any two words of the same length are comparable, but there is the following example of d-decomposability which is not obviously strongly d-decomposable with respect to the same factorization:

$$w = (x_1)(x_1 x_2)(x_1 x_2^2).$$

Remark 2.3.3.

(i) *The word w is strongly d-decomposable if it contains a (lexicographically) descending chain of d disjoint subwords, i.e., is of the form*

$$s_0 v_1 s_1 v_2 \cdots s_{m-1} v_d s_d,$$

where $v_1 \succ v_2 \succ \cdots \succ v_d$.

(ii) *Since comparability of words depends on the initial subwords, we could put $w_i = v_i s_i$ and note that (i) is equivalent to w containing a subword comprised of d decreasing subwords*

$$w_1 w_2 \ldots w_d, \quad \text{where} \quad w_1 \succ w_2 \succ \cdots \succ w_d.$$

Here is a variant.

Lemma 2.3.4. *If w has $(2d-1)$ pairwise incomparable words*

$$u^{k_1} v_1', \ldots, u^{k_{2d-1}} v_{2d-1}',$$

where $k_1, \ldots, k_{2d-1} > d$, and $v_1', \ldots, v_{2d-1}'$ are words of length m comparable with u, then w is strongly d-decomposable.

Proof. Among the words $v_1', \ldots, v_{2d-1}'$ either there are d words lexicographically greater than u or there are d words lexicographically smaller than u. We may assume that $v_1', \ldots, v_d' \succ u$. Then w contains the subwords $v_1', u v_2', \ldots, u^{d-1} v_d'$, which lexicographically decrease from left to right. $\square$

2.3.1　Hyperwords, and the hyperword u^∞

Although words have finite length, by definition, we can keep track of repetitions of words by broadening our horizon.

Definition 2.3.5. *A **hyperword** is a right infinite sequence of letters.*

The juxtaposition of a word followed by a hyperword produces a hyperword in the natural way; likewise, for any n, we can factor any hyperword h as

$h = wh'$ where $|w| = n$ and h' is the remaining hyperword; w is called the **initial subword** of length n.

A word is called **periodic**, with **period** u, if it has the form u^k for $k > 1$. Otherwise it is called **non-periodic**. $|u|$ is called the **periodicity** of w. A hyperword h is called **periodic with period** u, if it has the form u^∞, i.e., u repeats indefinitely. (We always choose u such that $|u|$ is minimal possible; in other words, u itself is not periodic.) A hyperword h is **preperiodic** if $h = vh'$ where v is finite and h' is periodic. Such v with $|v|$ minimal possible is called the **preperiod** of h.

Remark 2.3.6. *One could say that a hyperword h' is initial in a hyperword h if every initial subword of h' is initial in h. But then $h' = h$.*

Proposition 2.3.7. *If h is initial in uh, where h is a word or a hyperword, and u is a word, then h is initial in u^∞. (In particular, if h is a hyperword, then $h = u^\infty$.)*

Proof. For all k, $u^k h$ is initial in $u^{k+1}h$, implying h is initial in $u^{k+1}h$, by transitivity; now let $k \to \infty$. $\qquad\square$

We write $\nu_n(h)$ for the number of distinct subwords of length n in a given hyperword h. For example, $\nu_1(h)$ is the number of distinct letters appearing in h. Of course $\nu_n(h) \le \ell^n$ is finite, for any n. Let us see what happens if this number is restricted. The following trivial observation is the key to the discussion.

Remark 2.3.8.   *(i) Given any subword v of h of length n, we can get a subword of length $n + 1$ by adding on the next letter. Hence $\nu_{n+1}(h) \ge \nu_n(h)$ for any n.*

(ii) If $\nu_{n+1}(h) = \nu_n(h)$, then given a subword v of h of length n, h has a unique subword of length $n+1$ starting with v. (Indeed, suppose vx_i and vx_j both appeared as subwords of h. By (i), each other subword of length n starts a subword of length $n + 1$, implying $\nu_{n+1}(h) > \nu_n(h)$.)

Proposition 2.3.9 (Basic combinatoric lemma). *If $\nu_n(h) = \nu_{n+1}(h) = m$ for suitable m, n, then h is preperiodic of periodicity $\le m$.*

Proof. Write $h = x_{i_1} x_{i_2} x_{i_3} \dots$, and, for $1 \le j \le m + 1$, let

$$v_j = x_{i_j} x_{i_{j+1}} \dots x_{i_{j+n-1}}$$

denote the subword of h of length n starting in the j position. Then $v_1, \dots, v_{m+1}$ are $m + 1$ words of length n so, by hypothesis, two of these are equal, say $v_j = v_k$ with $j < k \le m + 1$, i.e., $j \le m$.

Let h_j denote the hyperword starting in the j position. By Remark 2.3.8(ii), any sequence of n letters of h determines the next letter uniquely, so h_j is determined by its initial subword v_j. Hence

$$h_j = h_k = x_{i_j} \dots x_{i_{k-1}} h_j,$$

so Proposition 2.3.7 implies that $h_j = u^\infty$ where $u = x_{i_j} \ldots x_{i_{k-1}}$. Thus $|u| = k - j \leq m$; furthermore,

$$h = x_{i_1} \ldots x_{i_{j-1}} h_j$$

is preperiodic.   □

Definition 2.3.10. *In analogy with Definition 2.2.5, a hyperword h is called* **strongly** *d-decomposable if it contains a (lexicographically) descending chain of d disjoint subwords. Otherwise h is called d-***indecomposable***.*

We are ready for our main result about hyperwords.

Theorem 2.3.11. *Any hyperword h is either d-strongly decomposable or is preperiodic of periodicity $< d$.*

Proof. Say a subword of length d is *unconfined* in h if it occurs infinitely many times as a subword, and is *confined* in h if it occurs only finitely many times.

First, assume that h has d unconfined subwords $v_1, \ldots, v_d$ of length d, which we write in decreasing lexicographic order. (This is possible since they have the same length.) Let us take the first occurrence of v_1, and then take the first occurrence of v_2 that starts after this occurrence of v_1, and proceeding inductively, we obtain a subword of h in which the subwords $v_1, \ldots, v_d$ appear disjointly in that order. Hence, by definition h is d-strongly decomposable.

Thus, we may assume that h has $< d$ unconfined subwords of length d. Obviously, h has a (finite) initial subword w containing all the confined subwords of length d. Writing $h = wh'$, we see by assumption that the only subwords of length d that survive in h' are the unconfined subwords. Thus h' has fewer than d words of length d, i.e., $\nu_d(h') < d$. But obviously $\nu_1(h') \geq 1$, so in view of Remark 2.3.8 (i), there is some $n \leq d$ for which $\nu_n(h') = \nu_{n+1}(h') < d$, and we conclude by Proposition 2.3.9 that h', and thus h, is preperiodic of periodicity $< d$.   □

This dichotomy of d-decomposability versus periodicity lies at the heart of the Shirshov theory.

Given a d-indecomposable hyperword h, we write $\beta(\ell, k, d, h)$ for the smallest number β for which h contains an initial subword of the form zu^k with $|z| \leq \beta$ and $|u| \leq d$. In view of Theorem 2.3.11, such $\beta(\ell, k, d, h)$ exists, being at most the length of the preperiod of h. We want to show that this number is bounded by a function independent of the choice of h.

Proposition 2.3.12. *For any numbers ℓ, k, d there is a number $\beta(\ell, k, d)$ such that $\beta(\ell, k, d, h) \leq \beta(\ell, k, d)$ for all d-indecomposable hyperwords h.*

Proof. A variant of the König graph theorem. Otherwise, for any given number β let $P_{j\beta} = \{z : z$ is the initial subword of length j of an indecomposable hyperword h with $\beta(\ell, k, d, h) > \beta\}$, and let $Q_j = \bigcap\{P_{j\beta} : \beta \in \mathbb{N}\}$.

Since each $P_{j\beta}$ is finite nonempty, this means $Q_j \neq \emptyset$. If $z_j \in Q_j$, then we can find z_{j+1} in Q_{j+1} starting with z_j, so we continue this process by induction to obtain an indecomposable hyperword h. But the initial subword of h having length $\beta(\ell, k, d, h) + kd$ already contains a subword u^k, with $|u| \leq d$, contrary to hypothesis on the z_j. $\qquad\square$

Proposition 2.3.12 yields Shirshov's Lemma, as seen by extending a word w of length $\geq \beta(\ell, k, d)$ to a hyperword.

This proof is quite nice conceptually, relying on well-known combinatoric techniques and illustrating the power of the hyperword. On the other hand, it says *nothing* about the function $\beta(\ell, k, d)$, although one could bound $\beta(\ell, k, d)$ by an induction argument based on $\nu_n(h)$. Presently all we need for any of the qualitative applications is the existence of $\beta(\ell, k, d)$, but it turns out that $\beta(\ell, k, d)$ provides bounds for important invariants of an affine PI-algebra A, such as the Gel'fand-Kirillov dimension and the degree of its minimal Capelli identity. So we shall give the applications first, and turn later to finding a reasonable bound for the Shirshov function. Toward the end of this chapter we give a second proof (based on Shirshov's original proof) that constructs an explicit bound, but this bound is so astronomical as to be useless in computations. Then we give a third proof that provides a reasonable handle on the Shirshov function. Exercise 3 is another good tool for lowering bounds in Shirshov's Lemma.

We could already prove the second part of Shirshov's Height Theorem, but to get the first part, we need another connection between decomposability and periodicity.

Proposition 2.3.13. *If a word v has multiplicity $\geq d$ in w, and v contains d lexicographically comparable subwords (not necessarily disjoint), then w is d-strongly decomposable.*

Proof. Write $w = s_0 v s_1 v \ldots s_{m-1} v s_d$. We choose the highest subword in the first occurrence of v, the second highest in the second occurrence of v, and so on. Clearly this provides a decreasing chain of non-overlapping subwords of w, so we use Remark 2.3.3. $\qquad\square$

Our goal is then to find d pairwise comparable subwords in a given word. Here is one way.

Remark 2.3.14. *Let u, v be comparable words (i.e., neither is initial in the other). Then $u^{d-1}v$ contains d different comparable subwords:*

(i) *If $v \prec u$, then $v \prec uv \prec u^2 v \prec \cdots \prec u^{d-1} v$.*

(ii) *If $v \succ u$, then $v \succ uv \succ u^2 v \succ \cdots \succ u^{d-1} v$.*

So any word in which $u^{d-1}v$ has multiplicity $\geq d$ is strongly d-decomposable, by Proposition 2.3.13.

Proof of Theorem 2.2.2(ii). Induction on the lexicographic order. Let k be the maximum of d and the maximum degree of integrality of the finite set $\{\bar{w} : w \in W\}$. By Proposition 2.2.6, we may replace d-decomposable words by linear combinations of smaller words, so we may assume that w is d-indecomposable. Since we can use integrality to reduce the power, we see in Lemma 2.3.1 that $\bar{w}$ is A-minimal only when $|w| \leq \beta(\ell, k, d)$.   □

Proof of Shirshov's Height Theorem 2.2.2(i). Let $\beta = \beta(\ell, d, d)$ as in Shirshov's Lemma 2.3.1, let β' be the smallest integer greater than $\frac{\beta}{d}$, and take

$$\mu = \mu(\ell, d) = (\beta' + 2)d^2\ell^{2d} + \beta'.$$

Again, we may assume that w is d-indecomposable, and we shall prove that the height of w over the words of length $\leq d$ is at most μ.

We describe a process for partitioning w, which will yield the result. If $|w| < \beta$, stop, since we have fewer than β' words in all. Assume that $|w| \geq \beta$. By Shirshov's Lemma 2.3.1, the initial subword of w of length β contains a subword of the form $u_1^{k_1}$ with $|u_1| \leq d$ and $k_1 \geq d$. Write $w = v_0 u_1^{k_1} w_1$ where $k_1 \geq d$ is maximal possible. Clearly $|v_0| < \beta$.

Repeat the process with w_1: If $|w_1| < \beta + |u_1|$, stop. If $|w_1| \geq \beta + |u_1|$, continue the same way: Using Shirshov's Lemma, write $w_1 = v_1 v_1' u_2^{k_2} w_2$ with k_2 maximal possible such that $|v_1| = |u_1|$, $u_j \neq v_j$, $|v_1'| < \beta$, and $|u_2| \leq d$. This implies v_1 is not comparable to u_1, since otherwise $v_1 = u_1$, contrary to the maximality of k_1.

Next apply the same process to w_2. Continue until we get

$$w = v_0 u_1^{k_1} v_1 v_1' u_2^{k_2} v_2 v_2' u_3^{k_3} v_3 \cdots u_t^{k_t} v_t w_t.$$

Here, all $|u_j| = |v_j| < \beta$, $k_j \geq d$, and $u_j \neq v_j$, implying u_j and v_j are comparable.

We want to view this as a string of subwords of length $\leq d$, so we partition each v_i' into at most β' such subwords. In bounding the height of w, we can disregard the powers k_i and thus count β' for each v_i, and 1 for each u_i. Thus, w has height at most

$$t\beta' + t + (t - 1) + \beta' + 1 = (\beta' + 2)t + \beta'.$$

On the other hand, we see that $u_i^{k_i} v_i$ is determined by $|u_i| + |v_i| \leq 2d$ letters (since they determine u_i and v_i), so there are fewer than ℓ^{2d} possibilities for $u_i^d v_i$, for any given length of $|u_i|$, or, counting all d possible lengths, fewer than $d\ell^{2d}$ possibilities in all for $u_i^d v_i$.

If $t \geq d^2\ell^{2d} = d(d\ell^{2d})$, some $u^d v$ must occur d times in w, contrary to w being d-indecomposable (by Remark 2.3.14). Hence $t < d^2\ell^{2d}$, which implies that w has height $< (\beta' + 2)d^2\ell^{2d} + \beta' = \mu$.   □

Remark 2.3.15. *In view of Remark 2.2.3, the number of elements needed to span A over C in Theorem 2.2.2 (ii) is $(\ell^{m+1} - 1)/(\ell - 1)$ where $m = \beta(\ell, k, d)$.*

Remark 2.3.16. *Since Shirshov's Theorem is combinatoric, it is applicable to Schelter's more general situations of integrality, as to be indicated in Volume II.*

We shall see many applications of Shirshov's Theorem. Here is an immediate one, to be improved later.

Corollary 2.3.17. *If A is affine PI and integral over a Noetherian Jacobson ring C, then $\mathrm{Jac}(A)$ is nilpotent, and every nil subalgebra of A is nilpotent.*

Proof. A is integral and finite over C, by Theorem 2.2.2. Hence, we apply Propositions 1.6.33 and 1.6.34. $\qquad\qquad\square$

Corollary 2.3.18. *If $A = F\{a_1, \ldots, a_\ell\}$ is an affine PI-algebra without 1 and the words in the Shirshov base are nilpotent, then A is nilpotent.*

Proof. Suppose m is the maximum nilpotence degree of the words of W, and A has PI-degree d. Checking lengths of words, we see that $A^{\ell m \beta(\ell, 2d, d)}$ is a sum of words each of which contains some $\bar{w}^m$ for a suitable $\bar{w}$ in the Shirshov base and thus is 0. $\qquad\qquad\square$

Corollary 2.3.19. *Any nil subalgebra of a PI-algebra is locally (i.e., every finite subset generates a nilpotent subalgebra).*

Proof. Any finite subset generates an affine PI-algebra without 1, so Corollary 2.3.18 is applicable. $\qquad\qquad\square$

We cannot yet say that nil subalgebras of an affine algebra are nilpotent, since they need not be affine (as algebras without 1). This assertion is true, but only as a consequence of the Braun-Kemer-Razmyslov Theorem.

Definition 2.3.20. *A subset $W \subseteq A$ is called an **essential Shirshov base** for the algebra A, if there exist a number ν and a finite set $W' \subset A$, such that A is spanned by words in A of the form*

$$v_0 w_1^{k_1} v_1 \ldots w_\nu^{k_\nu} v_\nu, \qquad\qquad (2.1)$$

where each $w_i \in W$, $v_i \in W'$, and $k_i \geq 0$.
*The minimal number $\nu(A)$ which satisfies the above conditions is called the **essential height** of A over W.*

Of course, if $\dim_F A < \infty$, then we could take W' to be a base of A, and we get $\nu = 0$.

Intuitively, having an essential Shirshov base means we can span a subspace by products of powers of given subwords (evaluated on the generators), together with an extra finite set. The essential height is the minimal such bound. In Shirshov's Theorem, for example, W is the set of words of length $\leq d$ in the generators, and $W' = \{1\}$. However, in Shirshov's original proof, cf. Section 2.6.1, it was convenient to take W to be the words in $a_1, \ldots, a_{\ell-1}$ of length $\leq d$, and W' to be various powers of a_ℓ.

2.4 The Shirshov Program

In order to appreciate how Shirshov's Lemma fits into PI-theory as a whole, let us outline a program for proving theorems about affine PI-algebras. It was really the subsequent generation (including Razmyslov, Schelter, and Kemer) who realized the full implications of Shirshov's approach. The program is:

1. Prove a special case of the theorem for representable algebras.

2. Given an affine algebra $A = F\{a_1, \ldots, a_\ell\}$, adjoin to F the characteristic coefficients of all words of length $\leq d$ in the a_i, thereby making the algebra finite over a commutative affine algebra, in view of Shirshov's Theorem.

3. Reduce to an assertion in commutative algebra.

In Chapter 4 we shall apply the Shirshov program to prove the Braun-Kemer-Razmyslov Theorem, that the radical of any affine PI-algebra is nilpotent. We shall go through the same sequence of ideas in Chapter 6, in solving Specht's problem for affine PI-algebras.

2.5 The Trace Ring

To proceed further, we need to be able to reduce to the integral (and thus finite) case, by Step 2 of Shirshov's program. Although the technique has been described in detail in several expositions, we review the description briefly here, both for its important role in studying prime affine algebras and also as preparation for the more general situation described below in §4.2. The main idea is encapsulated in the next definition.

Definition 2.5.1. *Suppose* $A = C\{a_1, \ldots, a_\ell\}$ *is an affine subalgebra of* $M_n(K)$, *where* K *is some commutative* C-algebra. *Take a finite Shirshov base* $\bar{W}$ *for* A *over which* A *has height* μ, *and form* $\hat{C}$ *by adjoining to* C *the characteristic coefficients of* $\bar{w}$ *for every* $\bar{w} \in \bar{W}$. *Define*

$$\hat{A} = \hat{C}A = \hat{C}\{a_1, \ldots, a_\ell\}.$$

Remark 2.5.2. *Over a field of characteristic 0, the construction of Definition 2.5.1 is equivalent to adjoining* $\operatorname{tr}(\bar{w}^k)$ *for all* $k \leq n$. *This observation is important since the trace is an additive function.*

Remark 2.5.3. *The Cayley-Hamilton Theorem shows that each* $\bar{w}$ *is integral over* $\hat{C}$ *in* $\hat{A}$, *implying by Theorem 2.2.2(ii) that* $\hat{A}$ *is finite as a* $\hat{C}$-module.

The point here is that $\hat{C}$ is affine as a C-algebra, generated explicitly by at most $n|W|$ elements. Hence $\hat{A}$ can be described more readily than A.

Having constructed this nice algebra $\hat{A}$, our task is to pass information back from $\hat{A}$ to A. Thus, we want to relate A to $\hat{A}$, i.e., what can we say about the characteristic coefficients of these $\bar{w}$?

2.5.1 The trace ring of a prime algebra

We first consider the special case of a prime algebra. Strictly speaking, we do not need to consider prime algebras at all, since the more technical theory to follow encompasses it. Nevertheless, the prime case provides valuable intuition, as well as a quick application, and the trace ring is one of the most important tools in the study of prime and semiprime affine PI-algebras.

Definition 2.5.4. *The* **trace ring**, *or* **characteristic closure**, *of a prime affine PI-algebra $A = C\{a_1, \ldots, a_\ell\}$ is defined as follows:*

Let $Z = \operatorname{Cent} A$ and $Q = S^{-1}A$ and $K = S^{-1}Z$, where $S = Z \setminus \{0\}$. Recall by Corollary 1.5.3 that $[Q : K] = n^2$, so any element a of A, viewed as a linear transformation of Q (given by $q \mapsto aq$), satisfies a characteristic polynomial

$$a^{n^2} + \sum_{i=0}^{n^2-1} \alpha_i a^i = 0, \qquad (2.2)$$

where the $\alpha_i \in K$. We work inside Q, which satisfies the standard identity s_{2n} since Q has PI-class n. For each word of length $\leq 2n$ in $a_1, \ldots, a_\ell$, adjoin the appropriate coefficients $\alpha_0, \ldots, \alpha_{n^2-1}$ to C. Thus, all in all, we need to adjoin at most $n^2(\ell^{2n+1} - 1)/(\ell - 1)$ elements to C to get a new algebra $\hat{C} \subseteq K$ over which $\hat{A}$ becomes finite. Although $\hat{C}$ is not a field, when C is Noetherian, $\hat{C}$ is commutative affine and thus Noetherian. Let $\hat{Z} = Z\hat{C}$, and let $\hat{A} = \hat{C}A \subseteq Q$, called the **characteristic closure** *of A.*

Note: In the above construction, since Q is central simple, we could have used instead the reduced characteristic polynomial of degree n, and then Equation (2.2) could be replaced by an equation of degree n. This alternate choice does not affect the development of the theory.

Remark 2.5.5. *In view of Corollary 1.1.18, $\hat{A} = \hat{C}\{a_1, \ldots, a_\ell\}$ is a prime algebra, which is finite over $\hat{C}$. In particular, if C is Noetherian, $\hat{A}$ is Noetherian as well as C-affine. (This is a special case of Remark 2.5.3.)*

We need some device to transmit our knowledge back to A. This is provided by the following fact, which is another immediate consequence of Theorem J of §1.5.

Remark 2.5.6. *$\hat{Z}$ and Z have a nonzero ideal $\mathcal{I}$ in common. In other words, $\mathcal{I} \subseteq Z$ but $\mathcal{I} \triangleleft \hat{Z}$. Consequently, $\hat{A}$ and A have the ideal $A\mathcal{I}$ in common.*

One application of the trace ring is the following famous result (which we do not need for our proof of the general theorem, but include to show how easily this method works):

Theorem 2.5.7 (Amitsur). *If A is affine PI over a field, then its Jacobson radical* $\mathrm{Jac}(A)$ *is locally nilpotent.*

Proof. First note that the nilradical N is locally nilpotent, in view of Corollary 2.3.19. Modding out N, we may assume that $N = 0$, and we want to prove $\mathrm{Jac}(A) = 0$. But now A is a subdirect product of prime algebras, so passing to the prime images, we may assume that A is prime. In the notation of Remark 2.5.6, $\mathrm{Jac}(A)\mathcal{I}$ is quasi-invertible, but also an ideal of $\hat{A}$, so $\mathrm{Jac}(A)\mathcal{I} \subseteq \mathrm{Jac}(\hat{A})$. Since $\mathcal{I} \neq 0$ and A is prime, it suffices to prove that $J := \mathrm{Jac}(\hat{A}) = 0$.

$\hat{C}$ is an affine domain so $\mathrm{Jac}(\hat{C}) = 0$. In view of Corollary 1.5.1 and "going up" for maximal ideals of an integral extension of a central subalgebra, it follows that $J \cap \hat{C} = 0$. But any element $a \in J$ satisfies an equation $\sum_{i=0}^{t} c_i a^i = 0$ for $c_i \in \hat{C}$; take such an equation with t minimal. Then $c_0 \in J \cap \hat{C} = 0$. If $0 \neq a \in \mathrm{Cent}(\hat{A})$, we can cancel out a and lower t, contradiction; hence $\mathrm{Cent}(\hat{A}) \cap J = 0$. But then $J = 0$ by Theorem C of §1.5.   □

Note that Theorem 2.5.7 does not generalize to affine PI-algebras over commutative Noetherian rings, since any local Noetherian integral domain has non-nilpotent Jacobson radical. It can be generalized to affine PI-algebras over commutative Noetherian Jacobson rings.

2.5.2   The trace ring of a representable algebra

We would like to backtrack, to show that the hypothesis in Definition 2.5.4 that A is prime is superfluous, and that we only use the fact that the affine F-algebra A is representable, i.e., $A \subseteq M_n(K)$ for a field K; since K is included in its algebraic closure, we may assume that K is algebraically closed. Anticipating Chapter 6, we rely on a famous classical theorem of Wedderburn:

Theorem 2.5.8 (Wedderburn's Principal Theorem and Mal'cev's Inertia Theorem [Ro88b, p.193]). *For any f.d. algebra $\tilde{A}$ over an algebraically closed field, there is a vector space isomorphism*

$$\tilde{A} = \bar{A} \oplus \tilde{J}$$

where $\tilde{J} = \mathrm{Jac}(\tilde{A})$ *is nilpotent and $\bar{A}$ is a semisimple subalgebra of $\tilde{A}$ isomorphic to* $\tilde{A}/\tilde{J}$.

Furthermore, if there is another decomposition $\tilde{A} = \bar{A}' \oplus \tilde{J}$, *then there is some invertible $a \in \tilde{A}$ such that* $\bar{A}' = a\bar{A}a^{-1}$.

Remark 2.5.9. *Let $\tilde{A} = KA$, a f.d. subalgebra of $M_n(K)$. Since we have little control over the embedding into $M_n(K)$, we look for a more effective way of computing the traces. Take the decomposition into simple algebras*

$$\bar{A} = \bar{R}_1 \times \cdots \times \bar{R}_q,$$

where $\bar{R}_k = M_{n_k}(K)$ (since K is algebraically closed), and we arrange the $\bar{R}_k$ so that $n_1 \geq n_2 \geq \cdots \geq n_q$.

For convenience, we assume first that $\mathrm{char}(F) = 0$ in order to be able to appeal to Newton's formulas (Remark 1.4.5), which show that all characteristic coefficients rely on traces.

We want to define traces in $\tilde{A}$ in an effective way. Since any nilpotent element should have trace 0, we ignore $\tilde{J}$ and focus on $\bar{A}$. Since $\bar{A}$ is a direct product of matrix algebras $\bar{A}_1 \times \cdots \times \bar{A}_q$, we can define tr_j to be the trace in the j component and then for any $r = (r_1, \ldots, r_q) \in \bar{A}$, define

$$\mathrm{tr}(r) = \sum_{j=1}^{q} n_j \, \mathrm{tr}_j(r_j) \in K. \tag{2.3}$$

The reason we weight the traces in the sum is that this gives the trace of r in the regular representation of $\bar{A}$.

Remark 2.5.10. *Notation as in Remark 2.5.9, let $t = n_1^2$, and take j' to be the maximal j such that $n_j^2 = t_1$. In other words, $n_1 = \cdots = n_{j'} > n_{j'+1} \geq \ldots$. Suppose $\mathcal{S}$ is any set of t-alternating polynomials, and let $\bar{\mathcal{I}}$ be the ideal of A generated by evaluations of polynomials of $\mathcal{S}$. Thus, any element of $\bar{\mathcal{I}}$ is a sum of specializations that we typically write as*

$$f(\bar{x}_1, \ldots, \bar{x}_t; \bar{y}_1, \ldots, \bar{y}_m).$$

Let $\pi_j : \bar{A} \to \bar{A}_j$ be the natural projection. If $j > j'$, then c_t is an identity of $\bar{A}_j$, so $\pi_j(\bar{\mathcal{I}}) = 0$.

If $j = j'$, then Theorem J of §1.5 implies for any $r \in A$ that $\pi_j(\bar{\mathcal{I}}) \, \mathrm{tr}(r_j) \subseteq \pi_j(\bar{\mathcal{I}})$, and in fact, Equation (1.23) shows for any $\bar{x}_i, \bar{y}_i \in \bar{A}_j$ that

$$n_j \, \mathrm{tr}_j(r_j) f(\bar{x}_1, \ldots, \bar{x}_t, \bar{y}_1, \ldots, \bar{y}_m) = \sum_{i=1}^{t} f(\bar{x}_1, \ldots, r_j \bar{x}_i, \bar{x}_{i+1}, \ldots, \bar{x}_t, \bar{y}_1, \ldots, \bar{y}_m).$$
$$\tag{2.4}$$

(We had to multiply by n_1, since left multiplication by r is viewed in (1.23) as a $t \times t$ matrix, whose trace is n_1 times the trace of r as an $n_1 \times n_1$ matrix.) Hence, for any $\bar{x}_i, \bar{y}_i$ in $\tilde{A}$, in view of (2.3), we have

$$\mathrm{tr}(r) f(\bar{x}_1, \ldots, \bar{x}_t, \bar{y}_1, \ldots, \bar{y}_m)$$
$$= \sum_{j=1}^{t} f(\bar{x}_1, \ldots, r_j \bar{x}_i, \bar{x}_{i+1}, \ldots, \bar{x}_t, \bar{y}_1, \ldots, \bar{y}_m) \in \tilde{J}. \tag{2.5}$$

We conclude that traces are compatible in $\bar{A}$ with t-alternating polynomials, where t is the size of the largest matrix component in $\bar{A}$.

In characteristic 0, we now define $\hat{A}$ as in Definition 2.5.1 and Remark 2.5.2. Let $\tilde{J} = \mathrm{Jac}(A)$. We would like to say that $\overline{\mathcal{I}} \triangleleft \hat{A}$, but unfortunately, we do not know much about how t-alternating polynomials interact with $\tilde{J}$. (This is because Wedderburn's principal theorem does not relate the multiplicative structures of $\bar{A}$ and $\tilde{J}$.) Thus, the best we can say is that $\overline{\mathcal{I}}\hat{A} \subseteq A + \tilde{J}$. This is a serious obstacle, and to circumvent it in solving Specht's problem, we need the rather sophisticated analysis of Kemer polynomials in Chapter 6. Of course, since $\mathrm{Jac}(A)$ is locally nilpotent, $\mathrm{Jac}(A)K$ is nil so $\mathrm{Jac}(A) \subseteq \tilde{J}$. But $\tilde{J}$ is nilpotent by Corollary 2.3.17, so its subset $\mathrm{Jac}(A)$ is nilpotent. (This also is seen at once from the Braun-Kemer-Razmyslov Theorem below.)

When $\tilde{J}^s = 0$, we can say that $\tilde{J}^{s-1}\overline{\mathcal{I}}\hat{A} \subseteq \tilde{J}^{s-1}\overline{\mathcal{I}}$, so $\tilde{J}^{s-1}\overline{\mathcal{I}} \triangleleft \hat{A}$. $\tilde{J}^{s-1}\overline{\mathcal{I}}$ could inconveniently be 0, but this in turn could be valuable information, enabling us to induct on s when passing to $A/\overline{\mathcal{I}}$, as we shall see in Chapter 11.

With all of these difficulties in characteristic 0, one might fear that the characteristic $p > 0$ situation is altogether unwieldy, since we do not have control on the characteristic coefficients (other than trace) of elements of J. But there is a cute little trick to take care of this, which leads to a better final result. We give the philosophy, leaving most applications for Volume II.

Remark 2.5.11. *If a and b are commuting matrices over an algebraically closed field K of characteristic p, then $(a+b)^{p^k} = a^{p^k} + b^{p^k}$. Hence, if we take the Jordan decomposition $T = T_s + T_r$ of a linear transformation T into its semisimple and radical parts, and if p^k is greater than the nilpotence degree of T_r, then $T_r^{p^k} = 0$ and thus $T^{p^k} = T_s^{p^k}$ is semisimple.*

In other words, taking high enough powers of an element destroys the radical part. Here is an abstract formulation.

Lemma 2.5.12. *Suppose C is a commutative affine algebra of characteristic p, and $q = p^k$ is at least the nilpotence index of the radical of C. Then the map $\varphi : c \mapsto c^q$ is an endomorphism of C, and the algebra $\varphi(C)$ is semiprime and affine.*

Proof. Suppose $a = c^q$ is nilpotent in $\varphi(C)$. Then c is nilpotent, so $0 = c^q = a$. $\qquad\square$

Definition 2.5.13. *Suppose F is a field, and the affine algebra $A = F\{a_1,\ldots,a_\ell\}$ is representable.*

If $\mathrm{char}(F) = 0$, define $\hat{A}$ as in Definition 2.5.1 (taking $C = F$), using the trace as defined above.

If $\mathrm{char}(F) = p > 0$, take q greater than the nilpotence index of $\mathrm{Jac}(A)$ (cf. Theorem 1.6.34). Now take a finite Shirshov base $\bar{W}$ for A over which A has height μ, and for every $\bar{w} \in \bar{W}$, form $\hat{F}$ by adjoining to F the characteristic coefficients of $\bar{w}^q$ to A, and let

$$\hat{A} = A\hat{F} = \hat{F}\{a_1,\ldots,a_\ell\}.$$

*In either case, $\hat{A} = \hat{F}A$ is called the **characteristic closure** of A.*

Let us summarize these results.

Proposition 2.5.14. *Notation as above, suppose that* char$(F) = 0$, *A is F-affine and representable, and $\tilde{J}^s = 0$. Let $\mathcal{I}$ be any T-ideal generated by t-alternating polynomials, and let $\overline{\mathcal{I}}$ be their specializations in A. Then*

  (i) $J^{s-1}\overline{\mathcal{I}}$ is a common ideal of $\hat{A}$ and A.

  (ii) $\mathcal{I} \subseteq \mathrm{id}(A/\overline{\mathcal{I}})$.

  (iii) $\hat{A}$ is algebraic of bounded degree and thus $\hat{A}$ is finite over its center $\hat{Z}$; in particular, $\hat{Z}$ and $\hat{A}$ are Noetherian algebras.

Proof. $J^{s-1}\overline{\mathcal{I}} \subseteq A$ by definition, but is closed under multiplication by the traces, so is a common ideal with $\hat{A}$, proving (i); furthermore, (ii) is immediate by definition of $\overline{\mathcal{I}}$.

  (iii) By Shirshov's Height Theorem, the projection of $\hat{A}$ into $\bar{A}$ is algebraic of degree t. But this means for any $a \in \hat{A}$, there is a polynomial f of degree t for which $f(a) \in \mathrm{Jac}(\hat{A})$. But $\mathrm{Jac}(\hat{A})^s$ is nil, implying $\mathrm{Jac}(\hat{A})^s = 0$ (viewed in $\tilde{A}$), so $f(a)^s = 0$ and a is algebraic of degree $\leq ts$. Thus, we can apply Shirshov's Theorem to all of $\hat{A}$. The rest is standard structure theory: $\hat{Z}$ is commutative affine by Proposition 1.1.21 and so is Noetherian; $\hat{A}$, being finite over $\hat{Z}$, is Noetherian.   □

To show that these results subsume the previous case when A is prime, note there that $\tilde{A}$ is simple so $\tilde{J} = 0$ and $s = 1$. The situation is easier in nonzero characteristic, since we can bypass J and $\tilde{J}$ at the outset.

Proposition 2.5.15. *Notation as above, suppose* char$(F) > 0$ *and A is F-affine and representable. Then*

  (i) $\overline{\mathcal{I}}$ is a common ideal of $\hat{A}$ and A.

  (ii) $\mathcal{I} \subseteq \mathrm{id}(A/\overline{\mathcal{I}})$.

  (iii) $\hat{A}$ is algebraic of bounded degree and thus $\hat{A}$ is finite over its center $\hat{Z}$; in particular, $\hat{Z}$ and $\hat{A}$ are Noetherian algebras.

Proof. We need a few modifications in the previous proof.

  (i) We need to show that $\overline{\mathcal{I}}$ absorbs the characteristic coefficients of $\bar{w}^q$, where q is as in Definition 2.5.13. This is true since $\bar{w}^q \in \bar{A}$ by Remark 2.5.11.

  (iii) Each $\bar{w}^q$ is algebraic, implying $\bar{w}$ is algebraic, so Shirshov's Theorem is applicable again.   □

2.6   Shirshov's Lemma Revisited

Although the main applications of Shirshov's Theorem are qualitative, insofar as we do not need the precise function $\beta(\ell, k, d)$, we would like methods of computing his function $\beta(\ell, k, d)$. Thus, as promised earlier, we shall now give two other proofs of Shirshov's Lemma 2.3.1, and a better bound for the Shirshov height, which could be skipped if the reader wants to focus on the Braun-Kemer-Razmyslov Theorem and Specht's problem.

2.6.1   Second proof of Shirshov's Lemma

Here we use Shirshov's original idea, which although rather straightforward in principle, is rather tricky to execute. Although this proof is of mostly historical interest, we present it in full. We start with a stronger version of Lemma 2.3.1, which does not have any stipulation on the length of the repeated word u:

Lemma 2.6.1 (Shirshov's Lemma strengthened). *For any ℓ, k, d, there is $\beta = \bar{\beta}(\ell, k, d)$ such that any d-indecomposable word w must contain a nonempty subword of the form u^k, with $|u| \leq d$.*

Proof. By double induction, on d and ℓ. If $d = 1$, then the assertion is tautological. If $\ell = 1$, then just take $\beta = k$, and $v = x_1$.

Suppose we have $\bar{\beta}(\ell - 1, k, q)$ and $\bar{\beta}(p, k, d - 1)$ for any p, q.

Assume that w is any word which does not contain any subword of the form v^k. We shall calculate $\beta = \bar{\beta}(\ell, k, d)$ large enough such that w must be d-decomposable whenever $|w| \geq \beta$.

We list each occurrence of x_ℓ separately in w, i.e.,

$$w = v_0 x_\ell^{t(1)} v_1 x_\ell^{t(2)} v_2 \cdots x_\ell^{t(m)} v_m,$$

where v_i are words in $x_1, \ldots, x_{\ell-1}$ and v_0, v_m could be blank, but the other v_i are not blank. (We wrote the power as $t(i)$ instead of the more customary t_i, in order to make the notation more readable.) By induction on ℓ, we are done unless each $|v_i| < \bar{\beta}(\ell - 1, k, d)$. Also, by hypothesis on w, $t(i) < k$ for each i.

We form a new alphabet

$$X' := \{v_1 x_\ell^{t(2)}, v_2 x_\ell^{t(3)}, \ldots, v_{m-1} x_\ell^{t(m)}\}.$$

Thus, every new letter $x' \in X'$ starts with x_j with $1 \leq j \leq \ell - 1$. To estimate $|X'|$, we noted above that $|v_i| < \bar{\beta}(\ell - 1, k, d)$, and each letter in v_i has $\ell - 1$ possibilities, so there are fewer than $\ell^{\bar{\beta}(\ell-1,k,d)}$ choices for v_i. Multiplying this by the $k - 1$ possible values for $t(i)$, we see that the number of letters in X' is bounded by "only" $(k - 1)\ell^{\bar{\beta}(\ell-1,k,d)}$.

Consider the following new order $\succ'$ on these letters: $v_{i-1} x_\ell^{t(i)} \succ' v_{j-1} x_\ell^{t(j)}$, iff $v_{i-1}x_\ell^{t(i)} \succ v_{j-1}x_\ell^{t(j)}$ in the old lexicographic order $\succ$, or $v_{j-1}x_\ell^{t(j)}$ is an initial subword of $v_{i-1}x_\ell^{t(i)}$ (in which case, $v_{i-1} = v_{j-1}$ and $t(j) < t(i)$). This is a total order on X' which induces the corresponding lexicographic order on the submonoid $\mathcal{M}(X')$ generated by X', also denoted by $\succ'$.

Claim 1. Let $f, g \in \mathcal{M}(X')$. If $f \succ' g$, then either $f \succ g$ or g is an initial subword of f with respect to the original alphabet X.

Indeed, assume that g is not an initial subword of f. Then $f = uxr$ and $g = uys$, with $u, r, s \in \mathcal{M}(X')$, $x, y \in X'$ and $x \succ' y$. Let $x = v_{i-1}x_\ell^{t(i)}$ and $y = v_{j-1}x_\ell^{t(j)}$.

Case 1. $v_{i-1} \succ v_{j-1}$. Then clearly $f \succ g$.

Case 2. $v_{i-1} = v_{j-1}$. Then $t(i) \geq t(j)$. Rewrite $f = uv_{i-1}x_\ell^{t(i)}r$ and $g = uv_{j-1}x_\ell^{t(j)}s$. Since g is not an initial subword of f, $s \neq 1$. Now s starts with some letter from $\{x_1, \ldots, x_{\ell-1}\}$; hence $x_\ell \succ s$, which implies that $f \succ g$.

Claim 2. Let $w = vw'$ where $v \in \mathcal{M}(X)$ and $w' \in \mathcal{M}(X')$, and assume for some integer q that w' is q-decomposable in the $\succ'$ order. Then w', and therefore also w, is q-decomposable in the original ($\succ$) order.

Indeed, by assumption, $w' = w_1 \cdots w_q$ satisfies $w_1 \cdots w_q \succ' w_{\pi(1)} \cdots w_{\pi(q)}$ for any $1 \neq \pi \in S_q$. Since both words have the same length but are different, one cannot be an initial subword of the other. By Claim 1 it follows that $w_1 \cdots w_q \succ w_{\pi(1)} \cdots w_{\pi(q)}$, as desired.

We can now complete the proof of Shirshov's Lemma. Write

$$w = v_0 x_\ell^{t(1)} v_1 x_\ell^{t(2)} v_2 \cdots x_\ell^{t(m)} v_m = vw',$$

where $v = v_0 x_\ell^{t(1)}$ and $w' = v_1 x_\ell^{t(2)} v_2 \cdots x_\ell^{t(m)} v_m$. Let

$$\bar{\beta}(\ell, k, d) = \bar{\beta}(\ell-1, k, d) + k + \bar{\beta}\big(k\ell^{\bar{\beta}(\ell-1,k,d)}, k, d-1\big).$$

Then we must have at least one of three possibilities:
  (i) $|v_0| \geq \bar{\beta}(\ell-1, k, d)$ or
  (ii) $t(1) \geq k$ or
  (iii) $|w'| \geq \bar{\beta}\big(k\ell^{\beta(\ell-1,k,d)}, k, d-1\big)$.

In case (i), we are done, since by induction (on ℓ), v_0, and hence w, contains a subword that is d-decomposable.

In case (ii), we are done since x_ℓ^k appears in w.

So assume that case (iii) holds. By induction (on d), w' is $(d-1)$-decomposable with respect to $\succ'$ and therefore also with respect to $\succ$. Thus,

$$w' = w_0 w_1 \cdots w_{d-1}$$

and $w_0 w_1 \cdots w_{d-1} \succ w_0 w_{\pi(1)} \cdots w_{\pi(d-1)}$ for any $1 \neq \pi \in S_{d-1}$. Thus, w contains the subword

$$(x_\ell^{t(1)} w_0) w_1 \cdots w_{d-1}$$

which is ($\succ$) d-decomposable, since $w_1, \ldots, w_{d-1}$ all start with letters $\prec x_\ell$.

$\square$

It remains to bound the size of the repeating subword u. We need a basic fact about periodicity, which could be seen via Proposition 2.3.7, but for which we give a direct proof:

Lemma 2.6.2. *If $w = uv = vu$ for u, v nonempty, then w is periodic of periodicity dividing $\gcd(|u|, |v|)$, and u, v are periodic with the same period as w.*

Proof. We may assume that $|u| \leq |v|$. Then u is an initial subword of v, so we write $v = uv'$. Now $uuv' = w = vu = uv'u$, so $uv' = v'u$ and by induction on $|w|$, we have uv' periodic with period $\hat{u}$ whose length divides $\gcd(|u|, |v'|)$. Writing $u = \hat{u}^i$ and $v' = \hat{u}^j$ we see that $v = \hat{u}^{i+j}$ and $w = \hat{u}^{2i+j}$. $\square$

We formalize Lemma 2.6.2 with a definition.

Definition 2.6.3. *Words w and w' are called* **cyclically conjugate** *if $w = uv$ and $w' = vu$ for suitable subwords v, u of w. Note that w' is a subword of w^2.*

The **cyclic shift** *δ is defined by $\delta(vx) = xv$, for any word v and letter x.*

Remark 2.6.4. *$\{\delta^k(u) : 0 \leq k < |u|\}$ is the set of words which are cyclically conjugate to the word u. By Lemma 2.6.2, $\delta^k(u) = u$ for some $0 < k < |u|$, iff u is periodic with periodicity dividing k. Thus, we have the following dichotomy:*

  (i) *u is periodic of periodicity $< d$ (in case $\delta^k(u) = u$ for some $0 < k < d$), or*

  (ii) *$\{\delta^k(u) : 0 \leq k < d\}$ are d distinct comparable subwords of u^2. In this case, u^{2d} is d-decomposable by Proposition 2.3.13.*

Applying this dichotomy to the weak version of Shirshov's Lemma yields the stronger result:

Proposition 2.6.5. *If w is d-indecomposable and $|w| \geq \beta(\ell, k, d)$ with $k \geq 2d$, then w contains a word of the form u^k with $|u| \leq d$.*

Proof. We already know that w contains a subword u^k. Take such u with $|u|$ minimal; in particular, u is non-periodic. We are done unless $|u| > d$, then (i) of Remark 2.6.4 fails, so u^{2d} is d-decomposable, implying w is d-decomposable. $\square$

This completes the second proof of Shirshov's Lemma, when we replace k by the maximum of k and $2d$.

2.6.2   Third proof of Shirshov's Lemma: Quasi-periodicity

Our third proof of Shirshov's Lemma yields a much more reasonable bound than what we gave above. This study will also have other applications, to Burnside-type problems and, later, to Gel'fand-Kirillov dimension. We fix ℓ and d throughout. Disregarding possibility (ii) of Lemma 2.6.8 below (working with $\beta > kd$), we search for β such that every word of greater length is d-decomposable. Then β would be an upper bound for $\beta(\ell, k, d)$.

Our key is a more careful analysis of periodic words, so let us record some related conditions.

Definition 2.6.6. *A word z is* **quasi-periodic** *with* **period** u *if* $z = u^k r$ *with* $r \subseteq_{\text{init}} u$.

A word w is **pre-periodic** *if* $w = vz$ *where z is quasi-periodic.*

The **normal form** *of a word w is vz, where the word $z = u^k r$ is quasi-periodic and $|v| + |u|$ is minimal possible. In particular, r is not an initial subword of u. Then v is called the* **pre-period** *of w, u is the* **period**, $|u|$ *is the* **periodicity**, k *is the* **exponent**, *and $|v| + |u|$ is the* **pre-periodicity** *of w.*

Quasiperiodicity is left-right symmetric. Explicitly, if $w = u^k v$, where $u = vv'$, then $w = v(v'v)^k$. In the same way, we see that words with cyclically conjugate periods can have arbitrarily large common subwords.

We turn again to hyperwords to simplify a proof.

Proposition 2.6.7. *If $w = uw'$ is a word with $0 < |u| < |w|$, and $w' \subseteq_{\text{init}} w$, then w has pre-period an initial subword of v, and w' is quasi-periodic with u a power of its period.*

Proof. By Proposition 2.3.7. $\square$

We want the pre-periodicity to be close to d, and we establish a dichotomy which underpins our estimate.

Lemma 2.6.8. *Assume that $|w| \geq kd$, for $k \geq 1$, and write*

$$w = v_1 u_1 w' = v_2 u_2 w' = \cdots = v_d u_d w',$$

where $|u_i| = i$. Then one of the following two statements holds:

(i) the words $u_i w'$ all are lexicographically comparable and comprise a chain;

(ii) $k \geq d - 1$, w' is quasi-periodic with some period u of periodicity $< d$ and exponent $\geq k$, and w is pre-periodic of pre-periodicity $< d$.

Proof. We are done unless $u_j w'$ is initial in $u_i w'$ for some pair $i > j$. Write $u_i = u_j u$. Then w' is initial in uw', so Proposition 2.6.7(ii) shows that w' is quasi-periodic with period u. But

$$v_j u_j = v_i u_i = v_i u_j u,$$

so we see that the pre-period is contained in $v_i u_j$, and the pre-periodicity $\leq |v_i u_j| + |u| < d$. $\square$

Definition 2.6.9. *We call a word w (**Shirshov**) (k, d)-reducible if w has either a subword u^k with $|u| \leq d$ or is d-decomposable.*

Remark 2.6.10. *Let us focus on two ways for a word to be (k, d)-reducible. We call a word w (**Shirshov**) (k, d)-**admissible of type (1), (2)** respectively if either*

1. *w has a subword of the form u^k for $1 \leq |u| \leq d$ or*

2. *w contains d comparable subwords.*

By Lemma 2.6.8, any word of length $\geq kd$ is (k, d)-admissible. But then by Proposition 2.3.13, if w has a (k, d)-admissible subword of multiplicity d, then w is (k, d)-reducible. So we look for the smallest β to guarantee this condition.

The easiest way to do this is to take all words of length kd, namely ℓ^{kd} such words. If $|w|$ can be partitioned into $d\ell^{kd}$ subwords of length kd, then at least one of these has multiplicity d and is (k, d)-admissible, since it has length kd; we then conclude that w is d-decomposable. So we could take

$$\beta = d^2 k \ell^{kd}.$$

To lower this bound, we need to exclude many of these words. The underlying question is, "What conditions are satisfied by a word, such that condition (ii) of Lemma 2.6.8 fails?"

Definition 2.6.11. *A word w is called d-**critical**, if w has length and preperiodicity d.*

Remark 2.6.12. *Every d-critical word contains d lexicographically comparable subwords.*

Lemma 2.6.13.

(i) *Writing $z = u^k r$ we cannot have rx an initial subword of u.*

(ii) *For any d-critical word w, the tail obtained by erasing the first letter has pre-periodicity $d - 1$. Thus, $|w| \leq d - 1 + (d-1)(k-1) + 1 = kd - k + 1$.*

Proof. (i) Otherwise it must be absorbed into z.
(ii) Follows from 2.6.8.   □

Let us use this observation to tighten our estimates on β. We consider the "extreme case" for a word w. w' denotes the subword obtained by omitting the last letter.

Lemma 2.6.14. *Any word w of length $\geq kd$ is (k, d)-admissible and if not of type (1), it contains a d-critical initial subword of length $< kd$.*

Proof. If w has pre-periodicity $< d$, then its period u has length $\leq d - 1$, and its pre-period v has length $\leq d - 1 - |u|$, implying vu^k appears in w' (as defined above), so w is (k, d)-admissible of type (1).

Thus we may assume that w has pre-periodicity $\geq d$, so its initial subword of length d is d-critical.   □

Let us bound the number of such subwords.

Remark 2.6.15. *For $|X| = \ell$, the number of distinct d-critical words of length t is $\leq \ell^d$.*

Theorem 2.6.16 (Improved bound for Shirshov's function).

$$\beta(\ell, k, d) \leq d^2 k \ell^d.$$

Proof. Partitioning any word w of length $> d^2 k \ell^d$ into subwords each of length kd as above, we proved that either one of them is (k, d)-admissible of type (1) or each of these subwords contains a d-critical subword, so we have at least $d\ell^d$ disjoint d-critical subwords. By Remark 2.6.15(iii), at least d of them are the same. Thus, w is d-decomposable, as desired.   □

Corollary 2.6.17. *Suppose $A = C\{a_1, \dots, a_\ell\}$ is an affine PI-algebra (not necessarily with 1), of PI-degree d. If all words of length $\leq d$ in the a_i are nilpotent of index k, then A is nilpotent of index $\leq n = d^2 k \ell^d$.*

Proof. In view of Proposition 2.2.6, A^n is spanned by those $\bar{w}$ for which w has length n and is not d-decomposable and thus contains a k-th power of a word of length $\leq d$, so by hypothesis each $\bar{w} = 0$.   □

2.7   Appendix A: The Independence Theorem for Hyperwords

Having utilized hyperwords to good effect, we pause for one more application — a version of a theorem of Ufnarovski'i — which also settles a conjecture of Amitsur and Shestakov.

Clearly, if two hyperwords h, h' are unequal, then they must differ at some position. It follows that hyperwords (unlike finite words) are totally ordered with respect to the lexicographic ordering. Also, each subset of right hyperwords has an infimum and a supremum (obtained by starting with the smallest or largest possible letter, and continuing inductively).

Although we cannot define the substitution $\bar{h}$ of a right hyperword h, since we do not have infinite products in A, we can define when $\bar{h}$ is 0:

Definition 2.7.1. $\bar{h} = 0$ *iff* $\bar{w} = 0$ *for some (finite) initial subword w of h. We say that h is **nonzero** in A if $\bar{h} \neq 0$.*

Let w be a word which is minimal in the set of all words of length $\leq n$ which are nonzero in A. It is conceivable that w cannot be extended to a longer nonzero word in A. Hence, to use hyperwords we apply the following formalism.

Example 2.7.2. *Let $A = F\{a_1, \ldots, a_\ell\}$. Take an indeterminate $x_{\ell+1}$ and formally declare $x_{\ell+1} \succ x_i$ for all $1 \leq i \leq \ell$. Consider the free product $A' = A * F\{x_{\ell+1}\}$.*

Each word w in A is initial in some hyperword in A'. If w is the minimal (nonzero) word in A in the set of all words of length $\leq |w|$, then clearly w is initial in certain hyperwords of A' (for example, $wx_{\ell+1}^\infty$), among which the hyperwords in A are minimal (if they exist).

Theorem 2.7.3 (Independence of hyperwords). *Let h be a minimal non-zero hyperword on A, and w_i denote its initial subword of length i. Then, for any n, one of the following two conditions holds:*

(i) *$\bar{w}_1, \ldots, \bar{w}_n$ are linearly independent.*

(ii) *h is pre-periodic of pre-periodicity $\leq n$, and its period is a tail of w_n.*

Proof. Let us define the hyperwords h_i by $h = w_i h_i$. As in Lemma 2.6.8, if (ii) does not hold, then the set $\{h_i\}$ constitutes a chain. Suppose (i) does not hold, so $\sum \alpha_i \bar{w}_i = 0$ with $\alpha_i \in F$ not all 0. Take a minimal hyperword h_j in the set $\{h_i : \alpha_i \neq 0\}$. Then $h = w_i h_i \succ w_i h_j$, for all $i \neq j$ where $\alpha_i \neq 0$. We can truncate the hyperwords $\{h_i\}$ to words $\{\hat{h}_i\}$ which comprise a chain under the same order, and then $h \succ w_i \hat{h}_i$ for each i.

Now $\bar{w}_j$ is a linear combination of the $\bar{w}_i$, so $\overline{\hat{h}} = \overline{w_j \hat{h}_j}$ also is spanned by a linear combination of the $\overline{w_i \hat{h}_i}$, each of which is smaller. If all of the $\overline{w_i \hat{h}_i}$ were 0 on A, then $\overline{\hat{h}}$, being a linear combination, would also be 0 on A. Hence, some $\overline{w_i \hat{h}_i}$ is nonzero on A, contradicting the minimality of h. ◻

Minimality can be replaced by maximality in the formulation of Theorem 2.7.3.

Corollary 2.7.4. *Let $A \subset M_n(F)$. Then any minimal nonzero (hyper)word h is pre-periodic of pre-periodicity $\leq n^2 + 1$, since (i) cannot hold.*

By the independence theorem for hyperwords and Example 2.7.2, we also have

Corollary 2.7.5 (Ufnarovskiĭ's independence theorem). *Suppose:*

(i) *$|w| = d$, and w is minimal with respect to the left lexicographic ordering in the set of all nonzero products of length $\leq d$.*

(ii) The tails of w are all nilpotent.

   Then the initial subwords of w are linearly independent.

The next result was proved by Ufnarovski'i and Chekanu [Che94b].

Corollary 2.7.6 (Amitsur and Shestakov's conjecture). *If A is an affine sub-algebra without* 1 *of* $M_n(F)$, *such that all words of degree* $\leq n$ *in the generators are nilpotent, then* $A^n = 0$.

Proof. Corollary 2.7.5 shows that any word of length $> n^2$ is 0, so A is nil. Then $A^n = 0$, by Remark 1.6.32.   □

2.8   Appendix B: A Subexponential Bound for the Shirshov Height

We turn now to a direct subexponential bound for the Shirshov height function $\mu(\ell, d)$. In particular, we consider the essential height, since among other things it bounds the Gel'fand-Kirillov dimension, as we shall see in Theorem 11.2.15. Shirshov's original proof, presented above, yielded only crude recursive estimates. Later Kolotov [Kol81] obtained the estimate $\mu(\ell, d) \leqslant \ell^{\ell^d}$. Belov in [Bel92] showed that $\mu(d, \ell) < 2d\ell^{\ell^{d+1}}$. An exponential bound is given for $\beta(\ell, k, d)$ and thus for $\mu(\ell, d)$, in Theorem 2.6.16, cf. also [Dr00, Kh2015]. Sharper estimates for nilpotence had already been obtained by Klein [Kl85, Kl00].

Zelmanov raised the question in the 1993 *Dniester Notebook* of whether the nilpotence index of any 2-generated associative algebra satisfying the identity $x^m = 0$ has exponential growth. The connection to Shirshov's Lemma is that any associative ℓ-generated algebra A without 1 satisfying $x^d = 0$ must have nilpotence index $< \beta(k, k, \ell)$.

Lopatin [Lop11] showed that if $\frac{n}{2}!$ is invertible, then the nilpotence degree of the free ℓ-generated algebra satisfying the identity $x^n = 0$ is $< 4 \cdot 2^{n/2}\ell$. Our subexponential bound for Shirshov's height, following [BelK12], answers Zelmanov's question positively. It is obtained by applying a technique of Latyshev suggested by G. R. Chelnokov in 1996.

The idea is to refine the techniques of Theorem 2.6.16, to partition a word w into d-decomposable pre-periodic subwords, and then reduce further by applying the following theorem of Dilworth to piece new subwords from the letters of w (where the letters are not necessarily taken consecutively, but the order of the letters is preserved). Dilworth's Theorem is formulated for an arbitrary finite partially ordered set $M = \{x_1, \ldots, x_\ell\}$ and a sequence $w = \{y_1, \ldots, y_n\}$ where $y_j = x_{i_j}$ for $1 \leq j \leq n$.

Definition 2.8.1. *A **chain** is a series of indices $i_1 < i_2 < \cdots < i_t$ (not necessarily consecutive) for which $y_{i_1} < y_{i_2} < \cdots < y_{i_t}$. An **antichain** is a series of indices $i_1 < i_2 < \cdots < i_t$ for which $y_{i_1} > y_{i_2} > \cdots > y_{i_t}$.*

For example, the two antichains in the word $x_1 x_4 x_2 x_3 x_5$ are $4, 2$ and $4, 3$.

Theorem 2.8.2 (Dilworth's Theorem). *Let d be the maximal number of elements of an antichain in the sequence w. Then w can be subdivided into d disjoint chains.*

Proof. (due to Amitsur, cf. [Ro80, Theorem 6.1.8]). We construct two tables $T = (t_{i,j})$ and $T' = (t'_{i,j})$, as follows: $t_{1,1} = 1$ and $t'_{1,1} = i_1$. If possible, given $t_{1,j}$ and $t'_{1,j}$, we define $t_{1,j+1}$ to be the smallest $k > t_{1,j}$ such that $x_k > t'_{1,j}$, and we put $t'_{1,j+1} = i_k$. When we cannot continue extending the first row, we start the second row taking $t_{2,1}$ to be the smallest k not contained in the first row and $t'_{2,1} = i_k$, and at each stage, given $t_{i,j}$ and $t'_{i,j}$, we define $t_{i,j+1}$ to be the smallest $k > t_{i,j}$ (not yet having appeared) such that $y_k > y_{t'_{ij}}$, and we put $t'_{i,j+1} = i_k$.

By construction, each row of T' is a chain, so it suffices to prove that T (and thus T') have precisely d rows. Suppose T has r rows. Clearly the entries of any antichain are in distinct rows, so $r \geq d$. Conversely, define $m_r = t_{r,1}$ and by reverse induction, given m_i from row i, take $m_{i-1} = t_{i,k}$, with k maximal such that $t_{i-1,k} < m_i$. Then $t'_{i-1,k} > m_i$, since otherwise m_i would have appeared in the $i-1$ row of T. Thus we have built an antichain of length r, so $r \leq d$, proving $r = d$. ☐

Using the usual partial order on $M = \{x_i, \ldots, x_\ell\}$, we see that an antichain of length d would make w d-decomposable. Thus Dilworth's Theorem permits us to piece out $d - 1$ chains from w, and examine the ensuing tails and their inits. This is the new ingredient used here. (Also at one key stage we need to apply Dilworth's Theorem to sets of subwords with a certain partial order.)

In related work Kharitonov [Kh2011a, Kh2011b, Kh2015] obtained lower and upper estimates for "piecewise periodicity," described in Exercises 33ff., which also could aid in obtaining a polynomial bound (still an open problem).

2.8.1   Statements of the main results

We aim for the following theorems:

Theorem 2.8.3. *The function $\beta(d, k, \ell)$ of Shirshov's Lemma 2.3.1 is bounded by*

$$2^{27} \ell (kd)^{3 \log_3 (kd) + 9 \log_3 \log_3 (kd) + 36}.$$

Note as $kd \to \infty$, that

$$\beta(d, k, \ell) \leq (kd)^{3(1+o(1)) \log_3 (kd)} \ell,$$

which is linear in ℓ for fixed d.

For any real number γ we define $\ulcorner\gamma\urcorner$ to be $-[-\gamma]$. This replaces any non-integer by the next larger integer. En route to Theorem 2.8.5 we also prove the following theorem estimating essential height:

Theorem 2.8.4. *The essential height of an ℓ-generated PI-algebra satisfying a polynomial identity of degree d over the set of words of length $\leq d$ is less than $2d^{3\ulcorner\log_3 d\urcorner+4}\ell$.*

This yields our main theorem:

Theorem 2.8.5 ([BelK12]). $\mu(\ell,d) \leq 2^{96}d^{12\log_3 d+36\log_3\log_3 d+91}\ell$.

For fixed ℓ and $d \to \infty$ we obtain the asymptotic bound $d^{12(1+o(1))\log_3 d}\ell$.

Corollary 2.8.6. *The height of an ℓ-generated PI-algebra of PI-degree d over the set of words of length $\leq d$ is bounded as in Theorem 2.8.5.*

Corollary 2.8.7. *There is a subexponential bound for the nilpotence index of ℓ-generated nil-algebras of bounded degree d, in arbitrary characteristic.*

We extend these results in Exercises 37ff.
To get started, we elaborate on Definition 2.1.1.

Definition 2.8.8.

(i) *The j-**tail** of a word w is the tail following the first j letters of w.*

(ii) *The j-**init** of a tail is the first j letters of the tail.*

(iii) *A subword u is **to the left** of a subword v if u starts before v.*

For example, the 3-tail of the word $x_1x_4x_2x_3x_5$ is $x_2x_3x_5$, and its 2-init is x_2x_3.

In proving Theorem 2.8.5 we deal first with each pre-periodic subword separately (cf. Definition 2.6.6), improving on our previous proof of Shirshov's Lemma by a closer examination of the tails (instead of general arguments about hyperwords). We also reduce to the case that each $m_i < d$. Specifically, Lemmas 2.8.10, and 2.8.11 provide sufficient conditions for a d-indecomposable word w to contain a period of length $\geq d$. Lemma 2.8.14 connects d-decomposability of a word w with its tails; d-indecomposability of w implies the absence of an antichain of d elements.

By Dilworth's Theorem, the word w partitions into $(d-1)$ chains of pre-periodic words, whose periods we color distinctly (in w) according to their location in the chains obtained from Dilworth's Theorem. The number of these colorings does not exceed $(n-1)^d(n-1)^d = (n-1)^{2d}$.

Moreover we generalize our arguments, viewing tails as having the form vu^kr, taking k maximal possible. We call vu^k the **fragment** of the tail, and say that the tail **breaks** before u^k. Thus the break determines where fragments may differ. Of course, writing $u = u'u''$, we could apply cyclic shifts to u

and replace it by $u''u'$, which starts the tail $v'u'(u''u')^k$. In this way, we may identify periods of fragments. Exercise 13 connects the "frequencies" (after the break) of p-inits and kp-inits for $k = 3$.

To finish the proof, we construct a hierarchical structure based on Exercise 13, that is, we consecutively consider fragments of d-tails, fragments of words obtained by omitting these fragments, and so on. Furthermore we consider the greatest possible number of tails in the subset to which Dilworth's Theorem is applicable, and from this we estimate from above the total number of tails and hence of the letters in the word w.

2.8.2   More properties of d-decomposability

The proof uses the following sufficient conditions for presence of a period:

Lemma 2.8.9. *If a word w is incomparable with a d-tail v where $j|v| < |w|$, then w contains a period of length $\leq d$ and exponent j.*

Proof. Write $w = uv$. Then v starts with u, by hypothesis, so $v = uv_1$ for some word v_1. Hence $w = u^2 v_1$, and hence $v_1 = u^2 v^2$ for some word v_2. Continuing in this way we can write $v_{j-1} = u^j v_j$ for each $j < \frac{|w|}{|v|}$.   □

Lemma 2.8.10. *If $|w| = m$, then either the j-tails are pairwise comparable for $1 \leq j \leq \lceil m/t \rceil$ or w contains a period of length t.*

Proof. We can apply Lemma 2.8.9 to any pair of incomparable j-tails.   □

Lemma 2.8.11. *If a word w of length $\geq jt$ contains at most t different subwords of length t, then w contains a period of length t.*

Proof. Induction on t. The case $t = 1$ is obvious. If there are at most $t-1$ different subwords of length $t-1$, then we are done by induction. If there exist precisely k different subwords of length $t-1$, then every subword of length t is uniquely determined by its first $t-1$ letters. But then the t-tail of w must start with the initial subword u of w, and then continue with the first letter of u (seen by disregarding the first letter of w), and so forth, implying w contains u^j.   □

Definition 2.8.12. *A d-**tail-decomposition** of a word w is a chain of tails $w_1 \succ w_2 \succ \ldots \succ w_d$ such that for any $i = 1, 2, \ldots, d-1$, w_i is to the left of w_{i+1}.*

Now we describe a sufficient condition for (k, d)-reducibility (Definition 2.6.9), and its connection with d-decomposability, in the same spirit as Lemma 2.6.8.

Lemma 2.8.13. *If a d-indecomposable word w contains d identical disjoint subwords v of length kd, then w is (k, d)-reducible.*

Proof. Take the j-tails $v_j : 1 \leq j \leq d$ of the word v, arranged such that

$$v_1 \succ \ldots \succ v_d.$$

By lemma 2.8.10, we are done unless the tails are comparable. Take the sub-word v_1 in the left-most copy of v, the subword v_2 in the second copy from the left, $\ldots$, v_d in the d-th copy from the left. We get a d-decomposition of w, contrary to hypothesis.   □

We are ready for one of the main computations. Given some k, define
$p_{d,k} := [\frac{3}{2}(d-1)d(\log_3(kd) + 2)] - 1$.

Lemma 2.8.14. *Any d-indecomposable word w with a $(p_{d,k} + 1)$-tail-decomposition is (k,d)-reducible.*

Proof. Suppose that the word w has a $(p_{d,k} + 1)$-tail-decomposition

$$w_1 \succ w_2 \succ \cdots \succ w_{p_{d,k}+1}.$$

Write $w_i = u_i w_{i+1}$ for $1 \leq i \leq u_{p_{d,k}}$, and let w_0 be the initial subword of w before w_1, and $u_{p_{d,k}+1} = w_{p_{d,k}+1}$. Thus

$$w = w_0 u_1 u_2 \cdots u_{p_{d,k}+1}.$$

Define $u_{i,j} = u_i u_{i+1} \cdots u_j$.
    First suppose that for each $1 \leq i \leq p_{d,k} + 1$ there exist integers

$$0 \leq j_i \leq k_i < m_i \leq q_i < [3d(\log_3(kd) + 2)]$$

such that

$$u_{j_i+i,k_i+i} \succ u_{m_i+i,q_i+i}.$$

Let $i_s = \frac{2s}{d}(p_{d,k} + 1) + 1$, where $0 \leq s \leq [\frac{d-1}{2}]$. Then the sequence

$$u_{j_{i_0}+i_0,k_{i_0}+i_0} \succ u_{m_{i_0}+i_0,q_{i_0}+i_0} \succ u_{j_{i_1}+i_1,k_{i_1}+i_1} \succ u_{m_{i_1}+i_1,q_{i_1}+i_1} \succ \cdots$$

$$\cdots \succ u_{j_{i_{[\frac{d-1}{2}]}}+i_{[\frac{d-1}{2}]},k_{i_{[\frac{d-1}{2}]}}+i_{[\frac{d-1}{2}]}} \succ u_{m_{i_{[\frac{d-1}{2}]}}+i_{[\frac{d-1}{2}]},q_{i_{[\frac{d-1}{2}]}}+i_{[\frac{d-1}{2}]}}$$
$$(2.6)$$

is a $2[\frac{d+1}{2}]$-decomposition of w, and w is (k,d)-reducible.
    Hence we may assume for some integer $1 \leq i \leq p_{d,k} + 1$ that

$$u_{j+i,k+i} \preceq u_{m+i,q+i}$$

for each $0 \leq j \leq k < m \leq q < [3d(\log_3(kd) + 2)]$. We may assume that $i = 1$.
    For a given integer t we take a sequence $\{k_i\}_{i=1}^t$ such that $k_1 = 3$ and
$\prod_{i=2}^t k_i > kd$. Note that $\inf_{0<j\leq k_1 d} |u_j| \leq kd$, by Lemma 2.8.13. Suppose that $\inf_{0<j\leq k_1 d} |u_j|$ is attained at u_{j_1}, where $0 < j_1 \leq k_1 d$. Write w_1' and w_1'' respectively for the $d|u_{j_1}|$-inits of the respective tails w_{j_1}, w_{j_1+1}.

- If $j_1 \leqslant d$, then both w_1' and w_1'' do not overlap the word $u_{2d+1,3d}$, and are incomparable with it. Furthermore, $|w_1'|, |w_1''| \leqslant \left| \prod\limits_{j=2d+1}^{3d} u_j \right|$. Hence, w_1' and w_1'' are incomparable by Remark 2.1.2, and repeating this argument implies that $u_{j_1}^d$ is a subword of w.

- If $d < j_1 \leqslant 2d$, then both w_1' and w_1'' do not overlap the word u_{3d+1}, and are incomparable with $u_{1,d}$. Furthermore, $|w_1'|, |w_1''| = |u_{1,d}|$. Hence $u_{j_1}^d$ is a subword of the word w.

- Likewise, if $2d < j_1 \leqslant 3d$ and w_1'' does not overlap $u_{[3d(\log_3 (kd)+2)]}$, then $u_{j_1}^d$ is a subword of w.

In each case, we conclude with Lemma 2.8.13. So we may assume that w_1' overlaps $u_{[3d(\log_3 (kd)+2)]}$. It is easy to show that there exists t and a sequence of integers $\{k_j\}_{j=2}^t$ such that $\prod\limits_{j=2}^t k_j \geqslant kd$ and $\sum\limits_{j=2}^t k_j \leqslant (\log_3 (kd) + 1)$. Then the word $u_{k_1 d+1, (k_1+k_2)d}$ is a subword of w_1''.

Write w_2' and w_2'' respectively for the $d|u_{j_2}|$-inits of the respective tails w_{j_2}, w_{j_2+1}. Consider the rightmost word u_{j_2} whose length is minimal among the words u_j, $k_1 d < j \leqslant (k_1 + k_2)d$. Then $|u_{j_2}| \leqslant \left\lceil \frac{|u_{j_1}|}{k_2} \right\rceil$. If w_2'' does not overlap the word $u_{[3d(\log_3 (kd)+2)]}$, then $u_{j_2}^d$ is a subword of w. So we may assume that the word $u_{(k_1+k_2)d+1, (k_1+k_2+k_3)d}$ is a subword of w_2''.

Suppose for each $2 \leqslant i < t$ that there exists an integer j_i such that
$$|u_{j_i}| \leqslant \left\lceil \frac{|u_{j_1}|}{\prod\limits_{s=2}^i k_s} \right\rceil, \quad \text{where } d \sum_{s=1}^{i-1} k_s < j \leqslant d \sum_{s=1}^{i} k_s.$$
Write w_i' and w_i'' respectively for the $d|u_{j_i}|$-inits of the respective tails w_{j_i}, w_{j_i+1}. If w_i'' does not overlap $u_{[3d(\log_3 (kd)+2)]}$, then $u_{j_i}^d$ is a subword of w. So we may assume that $u_{d \sum\limits_{s=1}^{i} k_s+1, d \sum\limits_{s=1}^{i+1} k_s}$ is a subword of w_i''. Consider the word $u_{j_{i+1}}$ whose length is minimal among the words u_j, where $d \sum\limits_{s=1}^{i} k_s < j \leqslant d \sum\limits_{s=1}^{i+1} k_s$. Now
$$|u_{j_{i+1}}| \leqslant \left\lceil \frac{|u_{j_i}|}{k_{i+1}} \right\rceil \leqslant \left\lceil |u_{j_1}| / \prod_{s=2}^{i+1} k_s \right\rceil. \quad \text{Thus}$$

$$|u_t| \leqslant \left\lceil \frac{|u_{j_1}|}{\prod\limits_{s=2}^{t} k_s} \right\rceil < 1,$$

a contradiction which yields the lemma.

□

Let us record the contrapositive.

Corollary 2.8.15. *If a word w is not (k, d)-reducible, then w is $8kd\log_3(kd)$-tail-indecomposable.*

The proofs of Theorem 2.8.3 and Theorem 2.8.5 are completed in Exercises 10ff. The proofs of Theorem 2.8.3 and Theorem 2.8.5 both are obtained by means of companion sets of words, which we respectively call $B^p(i)$ and $C^p(i)$. The first bound, given in Theorem 2.8.3, is somewhat easier to obtain, so we start with it. Note the similarity to our earlier proof of Shirshov's Height Theorem, but now we examine the tails and their inits more closely.

Our technique estimating the height remains somewhat crude and possibly could be improved by means of more sophisticated combinatorial arguments, but so far it remains subexponential. New ideas and methods concerning relations between subwords are needed to obtain or refute a polynomial estimate. In Exercises 37ff., we obtain specific lower and upper bounds using pre-periodic subwords. If we only use the colors of the initial positions of the subwords then we only obtain an exponential bound. But on the other hand, if we utilize the colors of all the positions in a single subword, then we still only obtain an exponential estimate.

Exercises for Chapter 2

Periodic words

1. If all subwords of w of length $|u|$ are cyclically conjugate to u, then w is periodic with periodicity $|u|$. (Hint: Start with the initial subword of w of length $|u|$; we may take this subword to be u. Then the subword of length $|u|$ starting with the second position must be $\delta(u)$; continue letter by letter.)

2. If $w = vu^k r$ with $|v| < |u|$ and u is not periodic, then any appearance of u as a subword of w must be in one of the k obvious positions. In other words, if $w = v'ur'$, then $|v'| = |v| + i|u|$ for some integer i. (Hint: Otherwise $u = u'u''$ where one of the obvious occurrences of u ends in u' or begins with u''. But then apply Lemma 2.6.2 to conclude that u is periodic, a contradiction.)

3. If $|v| > |u|$ and v^2 is a subword of a quasiperiodic word w of period u, then v is cyclically conjugate to a power of u. (Hint: Cyclically conjugating v if necessary, write $v = u^j v'$ with j maximal possible. Then $v^2 = u^j v' u^j v'$. Since w is quasiperiodic, some

$$u^i u' = w = w' v^2 w'' = w' u^j v' u^j v' w''.$$

Then $w' = u^k u''$ where u'' is an initial subword of u, and matching parts shows $u''u = uu''$.)

4. If quasiperiodic words w_1, w_2 with respective periods u_1, u_2 of lengths d_1, d_2 have a common subword v of length $d_1 + d_2$, then u_1 is cyclically conjugate to u_2. (Hint: Assume $d_1 \geq d_2$. Replacing u_1 by a cyclic conjugate, write $v = u_1 u_1'$, where $|u_1'| = d_2$. Since v starts with a cyclic conjugate $\tilde{u}_2$ of u_2, one can write $u_1 = \tilde{u}_2^k u_1''$, for $|u_1''| < d_2$, so v starts with $\tilde{u}_2^k u_1''$; show that $u_1'' = \emptyset$.)

5. (Ambiguity in the periodicity.) The periodic word 12111211 can also be viewed as the initial subword of a periodic word of periodicity 7.

6. (i) If two subwords v_1 and v_2 of u^∞ have the same initial subword of length $|u|$, then one of them is a subword of the other. Put another way, if $|c| \geq |u|$ and d_1 and d_2 are lexicographically comparable words, then either cd_1 or cd_2 is not a subword of u^∞.

(ii) Occurrences of any subword v of length $\geq |u|$ in u^∞ differ by a multiple of $|u|$.

(iii) If v^2 is a subword of u^∞ with $|v| \geq |u|$, then v is cyclically conjugate to a power of u.

(iv) Suppose $h = wvw'$ is a subword of u^∞, with v cyclically conjugate to u. Then, for all $k \geq 0$, wv^k is a subword of u^∞. In particular, $ww' \subset u^\infty$ and wv^2w' is a subword of u^∞.

Shirshov's Height Theorem

7. (Counterexample to a naive converse of Shirshov's Height Theorem.) The algebra $A = \mathbb{C}[\lambda, \lambda^{-1}]$ has the property that for any homomorphism sending λ to an algebraic element, the image of A is f.d., but nevertheless $\{\lambda\}$ is not a Shirshov base.

8. As in Definition 2.7.1, given an affine algebra A without 1, a right A-module M and a hyperword h, define when $Mh(A) = 0$. Prove that if $MA^k \neq 0$, $\forall k$, for some right module M, then the set of all right hyperwords h, such that $Mh \neq 0$, has maximal and minimal hyperwords.

9. If $A = F\{a_1, \ldots, a_\ell\}$ is semiprime without 1, of PI-class n, and if all words of degree $\leq n$ in the a_i are nilpotent, then $A = 0$. (Hint: First suppose that A is prime. Then $A \subset M_n(K)$ for some field K so $A^n = 0$ by Corollary 2.7.6, implying $A = 0$..)

Proof of Theorem 2.8.3

The idea is to observe the "evolution" of tails in a chain, with respect to the sequence of positions in w.

10. Let Ω be the set of tails of w which begin within the $[|w|/d]$-tail of w. Any two elements of Ω are comparable. There is a natural bijection between Ω, the letters of v, and positive integers from 1 to $|\Omega| = |v|$. (Hint: Lemma 2.8.10.)

11. Adjoin a formal word $\emptyset$ which is lexicographically smaller than any other word. For tails v and v', define the partial order $v < v'$ if $v \prec v'$ and v is to the left of v'. By Dilworth's Theorem, Ω can be partitioned into $p_{d,k}$ chains such that in each chain $v \prec v'$ when v is to the left of v'. Color the initial position of each tail (and continue with the same color until we reach the next tail) into one of $p_{d,k}$ colors, according to the chain to which it belongs.

Take a positive number p. For each $1 \leq i \leq |\Omega|$, define an ordered set $B^p(i)$ of $p_{d,k}$ words $\{b(i,1), \cdots, b(i,p_{d,k})\}$, as follows:

Let $b(i,j) = \{\max b \leqslant i : b$ is colored $j\}$. If there is no such b, then the word from $B^p(i)$ at place j is taken to be $\emptyset$; otherwise, we take it to be the p-init of the $b(i,j)$-tail.

Suppose a sequence S of length $|S|$ consists of words of length $j - 1$, where each word consists of $j - 2$ symbols "0" and a single symbol "1." Let S satisfy the following condition:

If for some $0 < s, t \leqslant j - 1$ there are $p_{d,j}$ words with "1" in the t-th position, then among these words there exists a word with "1" in position $< s$. Put $L(j - 1) = \sup |S|$ (ranging over such S).

Prove that $L(k - 1) \leqslant p_{d,k}^{k-1} - 1$. (Hint: $L(1) \leqslant p_{d,k} - 1$. Inductively, assume that $L(j - 1) \leqslant p_{d,k}^{j-1} - 1$. To show that $L(j) \leqslant p_{d,k}^{j} - 1$, note that there are at most $p_{d,k} - 1$ words in which "1" occupies the first position. Between any two of them as well as before the first one and after the last one, the number of words is at most $L(j - 1) \leqslant p_{d,k}^{j-1} - 1$. Hence

$$L(k) \leqslant (p_{d,k} - 1) + p_{d,k}\left((p_{d,k})^{j-1} - 1\right) = (p_{d,k})^{j} - 1,$$

as required.)

12. Set $\psi(p) := \{\max j : B^p(i) = B^p(i + j - 1)\}$. Then $\psi(p_{d,k}) \leqslant p_{d,k}d$.

13. $\psi(q) \leqslant q_{d,k}^k \psi(kq) + kq$. (Extensive hint: For any given q, divide the sequence of the first $|\Omega|$ positions i of w into equivalence classes with respect to the equivalence: $i \sim_q j$ if $B^q(i) = B^q(j)$. If $p < q$, then $\psi(p) \leqslant \psi(q)$.

Taking the least representative in each class of $\sim_{kq}$ yields a sequence of positions $\{i_1, i_2, \ldots\}$. Take the interval $[m_1, m_2]$ such that $B^{kq}(i_j)$ belongs to the equivalence class of $\sim_q$ for each $i_j \in [m_1, m_2]$. Let $\{i_j\}'$ denote the subsequence $\{i_j\}$, for which $i_j \in [m_1, m_2 - kq]$.

Fix some $1 \leqslant r \leqslant q_{d,k}$. All kp-inits of color r beginning with positions $\{i_j\}'$ will be called **representatives of type** r. All representatives of type r are pairwise distinct because they begin from the least positions in equivalence classes of $\sim_{kq}$. Divide each representative of type r into k segments of length q. Enumerate them from left to right by the integers 0 to $(k-1)$. If there exist $(p_{d,k}+1)$ representatives of type r with the same first $(t-1)$ segments but with pairwise different t-th segments where $1 \leqslant t \leqslant k-1$, then there are two t-th segments whose first letters are of the same color. The initial positions of these segments belong to different equivalence classes of $\sim_q$.

In all representatives of type r except the rightmost, mark a segment as a *1-segment* if it contains the least position where this representative of type r differs from the preceding one. All other segments are marked as *0-segments*.

By Exercise 11, the sequence $\{i_j\}'$ contains at most $p_{d,k}^{k-1}$ representatives of type r. Then the sequence $\{i_j\}'$ contains at most $p_{d,k}^k$ terms, and
$$m_2 - m_1 \leqslant p_{d,k}^k \psi(kq) + kq.))$$

14. Prove Theorem 2.8.3. (Hint: By Lemma 2.8.10, $|w| \leqslant d\,|\Omega| + d$. Hence $|w| \leqslant d(1+p_{d,k}\ell)\psi(1)+d$, since no more than $(1+p_{d,k}\ell)$ different values are possible for the set $B^1(i)$.

Iterate Exercise 13, taking the $\lceil \log_3 p_{d,k} \rceil + 1$ values $3^{\lceil \log_3 p_{d,k} \rceil}$, $3^{\lceil \log_3 p_{d,k} \rceil - 1}, \ldots, 1$, to get

$$\psi(1) < (p_{d,k}^3 + p_{d,k})\psi(3) < \cdots < (p_{d,k}^3 + p_{d,k})^{\lceil \log_3 p_{d,k} \rceil}\psi(p_{d,k}) \leqslant$$

$$\leqslant (p_{d,k}^3 + p_{d,k})^{\lceil \log_3 p_{d,k} \rceil} p_{d,k} d$$

Take $p_{d,k} = 8kd\log_3(kd) - 1$, to get

$$|w| < 2^{27}\ell(kd)^{3\log_3(kd)+9\log_3\log_3(kd)+36}.)$$

Proof of Theorem 2.8.4

The underlying idea is to apply Dilworth's Theorem to "pump" sets of tails (starting where the period ends) of big quasi-period subwords that are d-indecomposable.

15. Define a **word cycle** u to be the set consisting of the word u and all its cyclic shifts. **A circular word** u is the word u arranged in a circle, i.e., the first letter of u is also viewed after the last letter of u.

Any circular word u gives rise to a word cycle constituted of words of length $|u|$, obtained by starting at any position of u, and conversely a word cycle can be glued into a single circular word.

16. Let s denote the maximal number of periodic subwords $u_1^{2d}, u_2^{2d}, \ldots, u_s^{2d}$ in w separated by subwords of length $> d$, each of periodicity $< d$ and exponent $\geq 2d$. Thus $w = v_0 u_1^{2d} v_1 u_2^{2d} \cdots u_s^{2d} v_s$. Then s is at least the essential height of w over the set of words of length less than d.

17. Words u and u' are called **strongly incomparable** if no pair of cyclic shifts of u and u' are comparable. **Strong comparability** of word cycles and circular words is defined analogously. Show that strong incomparability yields an equivalence relation on the words of a given length $m < d$. (Hint: Two words of the same length are strongly incomparable iff they belong to the same word cycle.)

18. The word u_i^{2d} is strongly d-decomposable.

19. Partition all u_i of length m into s_m equivalence classes with respect to cyclic shifts, and choose a single representative $u_{i_1}, \ldots, u_{i_{s_m}}$, from each class. To ease notation, re-index them as $u_1, \ldots, u_{s_m}$, the inits of tails $w_1, \ldots, w_{s_m}$. Since the subwords u_i are periods and thus repeat, they can be viewed as circular words.

    For each $1 \le i \le m$, let $u(k,i) = \delta^k(u_i)$, cf. Definition 2.6.3. Thus $\{u(k,i) : 1 \le k \le m\}$ is the word cycle of u_i. For any $1 \le i_1 < i_2 \le p$, $1 \le j_1, j_2 \le m$, the word $u(i_1, j_1)$ is strongly incomparable with $u(i_2, j_2)$.

20. Define $\Omega' = \{u(i,j) : 1 \le i \le p, 1 \le j \le m.\}$ Define the partial order $\succ'$ by writing $u(i_1, j_1) \succ' u(i_2, j_2)$ if $u(i_1, j_1) > u(i_2, j_2)$ and $i_1 > i_2$.

    Show that if Ω' has an antichain of length d with respect to $\succ'$, then w is strongly d-decomposable. (Hint: Suppose there exists an antichain $u(i_1, j_1), u(i_2, j_2), \ldots, u(i_d, j_d)$ of length d; here $i_1 \le i_2 \le \cdots \le i_d$. If all of these inequalities are strict, then one is done.

    On the other hand, suppose that there exist $i_r < i_{r+1} = \cdots = i_{r+k} < i_{r+k+1}$ (where the inequalities at the end are not required for $r = 0$ or $r + k = d$ respectively).

    By hypothesis, $u_{i_{r+1}}^{k_{i_{r+1}}}$ contains $(u_{i_{r+1}}^2)^d$. But $u_{i_{r+1}}^2$ contains a word cycle of $u_{i_{r+1}}$. Hence one obtains disjoint subwords placed in lexicographically decreasing order and equal respectively to $u(i_{r+1}, j_{r+1}), \ldots, u(i_{r+k}, j_{r+k})$. Dealing in this manner with all sets of equal indices in the sequence $\{i_r\}_{r=1}^d$ yields the result.)

21. Ω' can be divided into $d - 1$ chains. (Hint: Apply Dilworth's Theorem to Exercise 20.)

22. Color each period from Ω' according to the chain in which it belongs (thus, $d - 1$ colors are used). Also color the integers from 1 to $|\Omega'|$ with the colors of their corresponding chains. Fix a positive integer $q \le m$. To each integer i from 1 to $|\Omega'|$, attach the following ordered set $C^q(i)$ of $d - 1$ words:

    For each $j = 1, 2, \ldots, d - 1$, define $c(i,j)$ to be the maximum of those $c \le i$: for which there exists k such that $u(c, k)$ is colored j and the q-init of $u(c, k)$ is an init of a word from Ω' (if such c exist).

$C^q(i)$ is the set of q-inits of $c(i,j)$, $1 \leq j \leq d$ (for those j for which $c(i,j)$ exists).

Define $\phi(q) = \{\max \; k : C^q(i) = C^q(i+k-1) \text{ for some } i\}$.

Then $\phi(q) \leq \phi(q')$, for any $q < q'$.

Partition Ω' via their indices, by defining $i \sim_q j$ if $C^q(i) = C^q(j)$.

Show that $\phi(m) \leq \frac{d-1}{m}$. (Hint: This construction is analogous to that used in Exercise 13. Likewise, $C^q(i)$ is analogous to $B^q(i)$, and $\phi(q)$ to $\psi(q)$. Consider the word cycles in Ω' indexed by $i, i+1, \ldots, i+[\frac{d-1}{m}]$. The first letter in each word cycle has some position in w, and the total number of such positions is at most d. Hence at least two of these positions are colored the same, so strong incomparability of the word cycles implies our assertion.)

23. $\phi(q) \leq p_{d,k}^k \phi(kq)$, whenever $kq \leq m$. Hint: Taking the least representative in each class of $\sim_{kq}$ yields a sequence of positions $\{i_1, i_2, \ldots\}$. Now take the interval $[m_1, m_2]$ such that $C^{kq}(i_j)$ belongs to the equivalence class of $\sim_q$ for each $i_j \in [m_1, m_2]$. Let $\{i_j\}'$ denote the subsequence $\{i_j\}$ for which $i_j \in [m_1, m_2 - kq)$.

Fix $1 \leq r \leq p_{d,k}$. All kq-inits colored r of tails beginning with positions in $\{i_j\}'$ will be called *representatives of type* r. All representatives of type r are distinct because they begin at the lowest positions in equivalence classes of $\sim_{k \cdot q}$. Divide each representative of type r into k segments of length q. Enumerate the segments of each representative of type r from left to right by integers from 0 to $k-1$. If there exist d representatives of type r having the same first $t-1$ segments but pairwise different t-th segments where $1 \leq t \leq k-1$, then there are two t-th segments whose first letters are of the same color. Then the initial positions of these segments belong to different equivalence classes of $\sim_q$.

Apply Exercise 11 as before, terminology as in Exercise 13: The sequence $\{i_j\}'$ contains no more than $(d-1)^{k-1}$ representatives of type r. Hence, the sequence $\{i_j\}'$ contains no more than $(d-1)^k$ terms, and $m_2 - m_1 \leq (d-1)^k \phi(kq)$.)

24. Prove Theorem 2.8.4. (Hint: Taking q respectively to have the $\lceil \log_3 p_{d,k} \rceil$ values

$$3^{\lceil \log_3 p_{d,k} \rceil}, \quad 3^{\lceil \log_3 p_{d,k} \rceil - 1}, \quad \ldots, \quad 1.$$

Iterating Exercise 22 and then applying Exercise 23, yields

$$\phi(1) \leq (d-1)^3 \phi(3) \leq (d-1)^9 \phi(9) \leq \cdots$$
$$\cdots \leq (d-1)^{3^{\lceil \log_3 m \rceil}} \phi(m) \leq (d-1)^{3^{\lceil \log_3 m \rceil + 1}}. \tag{2.7}$$

Since C_i^1 takes no more than $1 + (d-1)\ell$ distinct values, one has

$$|\Omega'| < (d-1)^{3^{\lceil \log_3 m \rceil}+1}(1+(d-1)\ell) < d^{3^{\lceil \log_3 d \rceil}+2}\ell.$$

By virtue of Lemma 2.3.4, the number of periods u_i of given length m is at most $2d^{3^{\lceil \log_3 d \rceil}+3}\ell$. Summing over all $m \leq d$ yields $2d^{3^{\lceil \log_3 d \rceil}+4}\ell$.)

Proof of Theorem 2.8.5

Outline of the proof: First find the necessary number of fragments in w of periodicity at most $2d$. For this, it suffices to apply Theorem 2.8.3 to subwords of w of large length. However the estimate can be improved. One finds a fragment $v_1 u_1^{k_1}$ in w having periodicity $|u_1| \leq 4d$. Temporarily discarding this fragment, one obtains a fragment $v_2 u_2^{k_2}$ having periodicity at most $\leq 4d$, and discard it to get a word w_2. Continue in this way, each time considering only the letters of w that have not been discarded, as is described in more detail in the algorithm of Exercise 25 below. The reason that we need $4d$ instead of $2d$ is that the repetitions of the period could pass over a discarded part, but a continuous repetition of the period at least $2d$ times in the original word is required. Then restore the original word w bringing back the discarded fragments. Furthermore, show that a subword of w usually is not a product of a large number of non-neighboring subwords. In Exercise 26, the algorithm enables one to find the number of subwords of w of periodicity $\leq 4d$ that must be discarded.

Define $\Phi(d,\ell) = 2^{27}\ell(kd)^{3\log_3(kd)+9\log_3\log_3(kd)+36}$ (from Theorem 2.8.3).

25. Here is the algorithm to be applied to a word w of Shirshov height $\mu(w) > \Phi(d,\ell)$ (cf. Definition 2.2.1).

Step 1. By Theorem 2.8.3 the word w contains a periodic subword z_1 of exponent $\geq 4d$, which is taken maximal possible. Write $w = w_1' v_1 z_1 w_1''$, where $|v_1| = d$. A position is **inessential of type 0** if it is contained in z_1. The j-position of v_1 is called **inessential of type j**, for $1 \leq j \leq d$. The words w_1' and w_1'' are called **significant**.

Step k. Disregard all inessential positions of w of type $\leq kq$, determined in the first $k-1$ steps. What remains is a new word which we call w_k. If $|w_k| \geqslant \Phi(d,\ell)$, apply Theorem 2.8.3 to it.

Thus w_{k-1} contains a non-periodic subword z_k with period of exponent $\geq 4d$ such that

$$w_{k-1} = w_k' v_k z_k w_k''',$$

taken maximal possible. At least half of the exponent occurs in some essential part, i.e., has a period in the original word w of exponent at least $2d$. Again, the positions in z_k or v_k are called **inessential**, and the j-position in v_k is called **inessential of type j**. If a position has already been declared inessential of some type, then we assign it the lower type.

Pick a number $t \leq \log_3 p_{d,k} + 1$. After $4t + 1$ steps of the algorithm, for each integer $1 \leq m \leq t$, one has

$$w = w_0 v_1 z_{m,1} w_1 v_2 z_{m,2} w_2 \cdots v_{t_m} z_{m,t_m} w_{t_m}$$

for various subwords v_j, with $z_m = z_{m,1} \cdots z_{m,t_i}$. Assume that the subword w_j is nonempty for $1 \leqslant j \leqslant t_m - 1$. Let $\varepsilon(k)$ be the number of indices $i \in [1, 4t]$ such that $t_i = k$.

One needs as many long periodic fragments as possible. For this, use the next exercise, taking $\varepsilon(j)$ to be the number of m for which $t_m = j$, and $\varepsilon = \varepsilon(1) + \varepsilon(2)$.

26. $\varepsilon \geqslant 2t$, and

$$\sum_{j=1}^{\infty} j \cdot \varepsilon(j) \leqslant 10t \leqslant 5\varepsilon. \tag{2.8}$$

(Hint: A subword v of w will be called *pivotal* if v is a product of words of the form $z_{i,j}$, but v is not a proper subword of another subword which satisfies (2.8). Suppose that after the $(i-1)$th step of the algorithm, the word w contains j_{i-1} pivotal subwords. Note that $j_i \leqslant j_{i-1} - t_i + 2$.

If $t_i \leqslant 2$, then $j_i \leqslant j_{i-1}$. Furthermore $j_1 = 1$ and $j_2 \geqslant j_1 = 1$.

If $t_i \geqslant 3$, then $j_i \leqslant j_{i-1} - 1$. Hence $\sum\limits_{t_i \geqslant 3} (t_i - 2) \leqslant 2\varepsilon$.

By definition $\sum\limits_{j=1}^{\infty} \varepsilon(j) = 4t$, i.e., $\sum\limits_{j=1}^{\infty} 2\varepsilon(j) = 8t$.

Obtain the desired inequality by summing these two inequalities and applying (2.8).)

27. $\mu(w) \leq \beta(d, 4d, \ell) + \sum\limits_{j=1}^{\infty} j\varepsilon(j) \leqslant \beta(d, 4d, \ell) + 5\varepsilon$.

28. Consider only z_i with $t_i \leqslant 2$.

If $t_i = 1$ then we put $z_i' = z_i$. If $t_i = 2$, then take z_i' to be the word of maximal length between $z_{i,1}$ and $z_{i,2}$.

Ordering the words z_i' according to their distance from the beginning of w, yields a sequence $z_{m_1}', \ldots, z_{m_\varepsilon}'$. Put $z_i'' := z_{m_i}'$. Write $z_i'' = w_i' u_i''^{p_i''} w_i''$ where at least one of the words w_i', w_i'' is empty.

Then one may assume that at the earlier steps of the algorithm, the z_i have been chosen such that $t_i = 1$.

29. Any word w of height $\geq \beta(d, 4d, \ell) + 5\varepsilon$ contains at least ε disjoint periodic subwords of periodicity $\geq 2d$. Furthermore between each pair of these periodic subwords, there is a subword having the same exponent

as the leftmost of this pair. (Hint: Consider the subwords (of w) of the form

$$z'_j = u_{(2j-1)''}^{p_{(2j-1)''} + q} v_j$$

where $q \geqslant 0$, $|v_j| = |u_{(2j-1)''}| \neq u_{(2j-1)''}$, and the period of z'_j is initial in some cyclic shift of the period of z''_{2j-1}.

The z'_j are disjoint. Indeed, if $z''_{2j-1} = z_{m_{2j-1}}$ then $z'_j = z_{m_{2j-1}} v_j$. If $z''_{2j-1} = z^{(k)}_{m_{2j-1}}$ for $k = 1, 2$, and z'_j overlaps z'_{j+1}, then

$$z''_{2j} \subset z'_i.$$

Since $u_{(2j)''}$ and $u_{(2j-1)''}$ are not cyclic, $|u_{(2j)''}| = |u_{(2j-1)''}|$. But then the periodicity of z'_j is at most $\leq 4d$, contradicting Exercise 28.)

30. Obtain the assertion of Theorem 2.8.5, by repeating the argument of Exercise 24, now using ε from Exercise 29 instead of s, and thereby obtaining

$$\mu(w) \leq \beta(d, 4d, \ell) + 5\varepsilon < 2^{96}\ell \cdot d^{12 \log_3 d + 36 \log_3 \log_3 d + 91}.$$

Estimates using quasiperiodic subwords

The following exercises yield estimates of the number of m-quasiperiodic subwords of a d-indecomposable word w, for $m \in \{2, 3, \ldots, (d-1)\}$, especially in the cases 2 and 3, and provide a lower bound for the number of 2-quasiperiodic subwords.

Let Ω'' be the set of word cycles of non-overlapping words of the form u^k, where u is a non-periodic word of length m; these are called m-*patterns*. Thus, there are exactly m different m-letter subwords in any m-pattern of length $2m$. Partially order these words as follows:

$u \prec v$, if

- $u < v$ lexicographically, and
- the m-pattern of u is to the left of the 2-pattern of v.

31. The word w can be partitioned into $d - 1$ chains of the appropriate m-letter words. Color the letters according to the chains to which they belong.

32. There are $1 : 1$ correspondences between the following:

- Positive integers from 1 to $|\Omega''|$,
- equivalence classes under strong comparability,
- two-letter word cycles from the equivalence classes of strong comparability,
- colorings of these word cycles.

33. [Kh2011a] Let M be the set of d-indecomposable words of finite essential height over the words of degree 2. Then there are at most $\frac{(2\ell-1)(d-1)(d-2)}{2}$ different lexicographically comparable 2-preperiodic subwords in any word from M. (Hint: Take $m = 2$. Consider the graph Γ having vertices of the form (k, i), where $0 < k < d$ and $0 < i \leqslant \ell$. The first coordinate corresponds to the color and the second to the letter. Two vertices $(k_1, i_1), (k_2, i_2)$ are connected by an edge of **weight** j if the j-th 2-pattern has letters i_1, i_2, respectively colored k_1, k_2.

    Fixing k_1, k_2, consider two edges ℓ_1 and ℓ_2 of respective weights $j_1 < j_2$ connecting vertices $A = (k_1, i_1), B = (k_2, i_2)$ and $C = (k_1, i_1'), D = (k_2, i_2')$ respectively, with $i_1 \leqslant i_1'$ and $i_2 \leqslant i_2'$. Then some $i_1 + i_2 < i_1' + i_2'$. Hence, $|E_{k_1,k_2}| < 2\ell$. Since each vertex has first coordinate $< d$, the number of edges in the graph is at most $\frac{(2\ell-1)(d-1)(d-2)}{2}$.)

34. Let $\mathcal{M} = \{d$-indecomposable words of finite essential height over the words of degree 3$\}$. Then there are at most $(2\ell - 1)(d - 1)(d - 2)$ different lexicographically comparable 3-preperiodic subwords in any word from $\mathcal{M}$.

    (Hint: Take $m = 3$. Proceed as in Exercise 33, but now let Ω'' be the set of 3-patterns, and introduce an oriented analog G of the graph Γ. Its vertices are of the form (k, i), where $0 < k < d$ and $0 < i \leqslant \ell$. The first coordinate corresponds to a color and the second to a letter. Two vertices $(k_1, i_1), (k_2, i_2)$ are connected by an edge of **weight** j if for some i_3, k_3 the word $i_1 i_2 i_3$ is in the j-th 3-pattern, and the letters i_1, i_2, i_3 are colored k_1, k_2, k_3, respectively.

    Thus, the graph G consists of oriented triangles with edges of the same weight, but now may have multiple edges.

    Show that if A, B and C are vertices of the graph G, such that the edges of the oriented triangle $A \to B \to C \to A$ all have weight j, and if there exist other edges $A \to B$, $B \to C$, $C \to A$ of respective weight a, b, c, then one of the numbers a, b, c is bigger than j.

    Now consider the graph G_1 obtained from G by replacing any multiple edge by the edge of smallest weight. Conclude as in Exercise 33, where the number of edges in the graph now is at most $(2\ell - 1)(d - 1)(d - 2)$.)

35. [Kh2011b] Let $\mathcal{M}$ be the set of d-indecomposable words of finite essential height over the words of length $(d - 1)$. Then the number of different lexicographically comparable $(d - 1)$-preperiodic subwords in any word of $\mathcal{M}$ is at most $(\ell - 2)(d - 1)$. (Hint: Here Ω'' is the set of $(d-1)$-patterns. The oriented graph G is as in Exercise 34. Two vertices $(k_1, i_1), (k_2, i_2)$ are connected by an edge of **weight** j if there is a cycle of words $i_1 i_2 \cdots i_{d-1}$ for some $i_3, i_4, \ldots, i_{d-1}$ in the j-th $(d - 1)$-pattern, and the letters i_1, i_2 are colored k_1, k_2 respectively.

    An index is needed that grows monotonically with the appearance of new $(d-1)$-patterns as one moves from the beginning to the end of the word. In the proof of Exercise 34, this index was the number of pairs of connected vertices of the graph G. Now it is the sum of the second coordinates of the non-isolated vertices of the graph G.

Suppose that $A_1, A_2, \ldots, A_{d-1}$ are vertices of the graph G, such that the edges of the oriented cycle $A_1 \to A_2 \to \cdots \to A_{d-1} \to A_1$ all have weight j. If there exist other edges $A \to B$, $B \to C$, $C \to A$ of respective weight a, b, c, then there is no other cycle of the same weight between the vertices $A_1, A_2, \ldots, A_{d-1}$.

There exist k, i such that the vertex (k, i) belongs to a cycle C of weight $(j+1)$ but does not belong to any cycle of weight j. Define

$$\pi(j) = \sum_{k=1}^{d-1} i_{(j,k)},$$

where the cycle C has vertices $(k, i_{(j,k)})$, $k = 1, 2, \ldots, d-1$. Then $\pi(j+1) \geqslant \pi(j) + 1$. Hence there exists k such that $i_{(1,k)} > 1$, implying $\pi(1) > d-1$. $\pi(j) \leqslant (\ell-1)(d-1)$ because $\forall j : i_{(j,k)} \leqslant \ell - 1$. Thus, $j \leqslant (\ell-2)(d-1)$, yielding $t \leqslant (\ell-2)(d-1)$.)

36. The essential height of the set of non-strongly d-decomposable words over the set W of 2-preperiodic subwords is at least

$$\gamma(2, d, \ell) = (\ell - 2^{d-1})(d-2)(d-3)/2,$$

which is approximately $\frac{d^2 \ell}{2}(1 - o(\ell))$. (Hint: One may assume that $\ell > 2^{d-1}$. In the notation of Exercise 33, in order to construct an example, one may build edges in a graph with ℓ vertices. This process is done via a series of "big" steps. For each $2 \leq i \leq \ell - 2^{d-1} + 1$, connect pairs of vertices in the following order during the i-th big step:

- $(i, 2^{d-2} + i)$,
- $(i, 2^{d-2} + 2^{d-3} + i)$, $(2^{d-2} + i, 2^{d-2} + 2^{d-3} + i)$,
- $(i, 2^{d-2} + 2^{d-3} + 2^{d-4} + i)$, $(2^{d-2} + i, 2^{d-2} + 2^{d-3} + 2^{d-4} + i)$, $(2^{d-2} + 2^{d-3} + i, 2^{d-2} + 2^{d-3} + 2^{d-4} + i), \ldots$,
- $(i, 2^{d-2} + \ldots + 2 + 1 + i), \ldots, (2^{d-2} + \ldots + 2 + i, 2^{d-2} + \ldots + 2 + 1 + i)$.

No edge is counted twice. Hence, a vertex is connected only to such vertices that differ by non-repeating sums of powers of two. We say that a vertex A is has type (k, i) if it is connected to k vertices of smaller value during i-th big step. For all i there exist vertices of the form $(0, i), (1, i) \ldots, (d-2, i)$.

If a word w beginning with a letter of the vertex of type (k, i) also ends with a vertex connected to A as described in the i-th "big" step, then color it i. One has a $(d-1)$-coloring, so w is d-decomposable.

Build $\frac{(d-2)(d-3)}{2}$ edges during the i-th big step. Then the number of edges in Γ is $q = (\ell - 2^{d-1})(d-2)(d-3)/2$.)

Bounds on the essential height

Continuing Exercise 36, one can obtain an explicit exponential bound for the essential height, by considering words of degree 2. This estimate has polynomial growth when the degree d of the PI-identity is fixed, and has exponential growth when the number ℓ of generators is fixed. The bijection between word cycles and circular words is utilized.

Take the alphabet $\tilde{A} = \{x_1, x_2, \ldots, x_\ell\}$, under the lexicographical order. We call a set W of words d-*inadequate* if it does not contain any antichain of length d. W is called d-*adequate* if it contains an antichain of length d. Consider the set $W(t, \ell)$ of non-periodic, pairwise strongly comparable word cycles $\{u_1, \ldots, u_m\}$ of distinct periods in w of the same length t and exponent $\geq d$, partially ordered as follows: $u_i \prec u_j$ if

- $u_i < u_j$ as words;

- u_i appears to the left of u_j in w.

Let $w(i, j) \in W(t, \ell)$ denote the word of length t beginning at the j-th position in the i-th word cycle.

Denote the largest possible number of elements in a d-inadequate class $W(t, \ell)$ as $\gamma(t, \ell, d)$, where $t < d$.

37. $\gamma(t, \ell^2, d) \geqslant \gamma(2t, \ell, d)$. (Hint: Partition the positions in all the word cycles of a d-inadequate set $W(2t, \ell)$ into pairs. Define a new alphabet $B = \{b_{i,j}\}_{i,j=1}^{\ell}$, where $b_{i_1, j_1} > b_{i_2, j_2}$, if $i_1 \ell + j_1 > i_2 \ell + j_2$. Denote the pair of letters (x_i, x_j) as $b_{i,j}$, thereby yielding a new set $W(t, \ell^2)$. Suppose that $W(t, \ell^2)$ has an antichain $w(i_1, j_1), w(i_2, j_2), \ldots, w(i_d, j_d)$ of length d. Taking preimages in $W(2t, \ell)$ yields an antichain of length d in $W(2t, \ell)$, contradiction.)

38. $\gamma(t, \ell^2, d) \leqslant \gamma(2t, \ell, 2d-1)$ (Hint: Analogous to Exercise 37. This time, consider a set $W(2t, \ell)$ containing an antichain of length $2d - 1$.

One may assume that d words $w(i_1, j_1), w(i_2, j_2), \ldots, w(i_d, j_d)$ in the antichain begin from odd positions of word cycles. Divide the positions in the word cycles of $W(2t, \ell)$ into pairs so that every position is exactly in one pair and the first position in each pair is odd. Then take the alphabet $\tilde{B} = \{b_{i,j}\}_{i,j=1}^{\ell}$, ordered as in Exercise 37. Again, replacing (x_i, x_j) by $b_{i,j}$ yields a new set $W(t, \ell^2)$, and conclude as in Exercise 37.)

39. $\gamma(t, \ell, d) \leqslant \gamma(2^s, \ell+1, 2^s(d-1)+1)$, where $s = \lceil \log_2(t) \rceil$. (Hint: Given a d-inadequate set $W(t, \ell)$, form the alphabet $\tilde{A}' = \tilde{A} \cup \{x_0\}$, where x_0 is less than any other letter from $\tilde{A}$. We expand any word cycle from the set $W(t, \ell)$ to length $2s$, assigning the letter x_0 to the $(t+1), (t+2), \ldots, 2^s$

positions, thereby yielding a new set $W(2^s, \ell + 1)$. Suppose that it is not $(2^s(d-1)+1)$-inadequate. Then for some j there exist a sequence of words $w(i_1, j), w(i_2, j), \ldots, w(i_d, j)$, which is an antichain in the set $W(2^s, \ell + 1)$.

(a) If $j > t$, then $w(i_1, 1), w(i_2, 1), \ldots, w(i_d, 1)$ is an antichain in the set $W(t, \ell)$.

(b) If $j \leqslant t$, then $w(i_1, j), w(i_2, j), \ldots, w(i_d, j)$ is an antichain in the set $W(t, \ell)$.

In each case one has a contradiction.)

40. $\gamma(t, \ell, d) \leqslant \gamma(t, \ell, d+1)$ (Hint: By Exercise 39,

$$\gamma(t, \ell, d) \leqslant \gamma(2^s, \ell + 1, 2^s(d-1)+1).$$

But $t < d$, so $2^s < 2d$, and

$$\gamma(2^s, \ell+1, 2^s(d-1)+1) \leqslant \gamma(2^s, \ell+1, 2d^2).$$

Exercise 37 yields the inequalities

$$\gamma(2^s, \ell+1, 2d^2) \leqslant \gamma(2^{s-1}, (\ell+1)^2, 2d^2) \leqslant \cdots \leqslant \gamma(2, (\ell+1)^{2^{s-1}}, 2d^2).)$$

41. $\gamma(t, \ell, d) < 4(\ell+1)^d d^4$. (Hint: Exercise 33 yields

$$\gamma(2, (\ell+1)^{2^{s-1}}, 2d^2) < (\ell+1)^{2^{s-1}} \cdot 4d^4 < 4(\ell+1)^d d^4.)$$

42. If a word w has $(2d-1)$ subwords such that some period u repeats more than d times in each subword, then their periods are pairwise non-strongly comparable, and the word w is d-decomposable. (Hint: There exist comparable, non-overlapping subwords $u^{k_1} v'_1, \ldots, u^{k_{2d-1}} v'_{2d-1}$ that are with the word u, where $k_1, \ldots, k_{2d-1}$ are positive integers bigger than d and the words $v'_1, \ldots, v'_{2d-1}$ are of length $|u|$. Then there exist d words from the set $\{v'_1, \ldots, v'_{2d-1}\}$ which all are either lexicographically greater or lexicographically less than u, say lexicographically greater. Then $v'_1, uv'_2, \ldots, u^{d-1}v'_d$, are in lexicographically decreasing order.)

43. Let W be the set of d-indecomposable words with finite essential height over the words of periodicity m. Then the number of lexicographically comparable m-preperiodic subwords in any word from W is less than $8(\ell+1)^d d^5 (2d-1)$, where $\widehat{\gamma}(m, d, \ell)$ is the maximal number of different lexicographically comparable m-preperiodic subwords in any word from W. The essential height over the set of words degree $< d$ of the relatively free algebra generated by ℓ elements and satisfying a polynomial identity degree d is $8(\ell+1)^d d^5 (d-1)$. (Hint: Suppose that a d-indecomposable word w contains a subword with some period u of

length $\geqslant d$ that repeats more than $2d$ times. Then the tails of u^2 which begin from the first, second, $\ldots, d$-th positions are pairwise comparable, contradicting the d-decomposability of the word w. Conclude by Exercises 41 and 42.)

Words on semirings

The following set of exercises, following I. Bogdanov [Bog01] and Belov, enables us to put Shirshov's theory in a general context, applicable to automata. A **generalized semiring** is an algebra $(S, +, \cdot, 0, 1)$ for which $(S, +, 0)$ and $(S, \cdot, 1)$ are semigroups and multiplication is distributive on both sides over addition, and $0S = S0 = 0$. A generalized semiring is called a **semiring** if addition is commutative. A **(generalized) semialgebra** over a commutative associative semiring Ψ is defined as usual, but not assumed to have a multiplicative unit.

44. The axiom $0S = S0 = 0$ does not follow from the other axioms, even if addition is commutative; however, it does follow for any nilpotent element. (Hint: If $x^n = 0$, then $0 = x \cdot x^{n-1} = x(x^{n-1} + 0) = x^n + x0 = x0$.)

45. (The band) There exists a generalized semiring S with non-commutative addition: Given any semigroup $(S, \cdot)$; define addition by $x + y = y$, and adjoin 0.)

46. If $(S, +)$ is a group, then $(S^2, +)$ is commutative.

The following exercises show that long words in a nil semiring have the same behavior as in a nil ring. By **negative**, we mean "additive inverse."

47. If S is a semiring satisfying the identity $x^n = 0$, then S^n is a ring (without 1). (Hint: Expanding the brackets in $(s_1 + \cdots + s_n)^n = 0$ yields a negative for $s_1 \cdots s_n$.)

48. If $(S, +)$ is a (not necessary commutative) monoid in which

$$s_1 + s_2 + \cdots + s_t = s_t + s_{t-1} + \cdots + s_1 = 0,$$

then all the s_is have negatives. (Hint: $s_1 + \cdots + s_{t-1} = s_{t-1} + \cdots + s_1$, since right and left negatives coincide if both exist. Conclude by induction.)

49. If S is a generalized semiring satisfying the identity $x^n = 0$, then any monomial of the form $s_1 \cdots s_n$ has a negative. (Hint: Expand all brackets in $(s_1 + \cdots + s_n)^n = (s_n + \cdots + s_1)^n = 0$, and apply Exercise 48.)

50. Under the same assumptions, S^n is a ring with commutative addition. (Hint: If $a, b \in S$ and $a', b' \in S^{n-1}$, then

$$ab' + aa' + bb' + ba' = (a + b)(b' + a') = ab' + bb' + aa' + ba'.$$

Add the negatives of ab' from the left and of ba' from the right.)

51. Suppose $\mathcal{F}_C$ and $\mathcal{F}_\Psi$ are the free C-algebra and free Ψ-semialgebra introduced above.

    Denote by $\mathcal{I}_n$ and $\mathcal{I}_n^+$ the ideals in $\mathcal{F}_C$ generated by $\{r^n : r \in \mathcal{F}_C\}$ and $\{r^n : r \in \mathcal{F}_\Psi\}$, respectively. Then $\mathcal{I} = \mathcal{I}^+$. (Hint: Denote by h the determinant of the Vandermonde matrix generated by powers of $0, 1, \ldots, n$. If $(x + y)^n = f_0(x, y) + \cdots + f_n(x, y)$ is the expansion of $(x + y)^n$ into a sum of homogeneous components, then $h f_i(x, y) \in \mathcal{I}_n^+$, implying $(x - hy)^n \in \mathcal{I}_n^+$.)

52. Suppose d_n is a natural number such that each (ℓ-generated) C-algebra satisfying the identity $x^n = 0$ is nilpotent of degree d_n. Then each (ℓ-generated) generalized Ψ-semialgebra satisfying the identity $x^n = 0$ is also nilpotent of degree d_n. (Hint: Suppose S is such a generalized semialgebra. There exists an obvious morphism of semialgebras $\psi : \mathcal{F}_\Psi^n \to S^n$ induced by $x_i \to a_i$, and $\mathcal{I}_n \cap \mathcal{F}_\Psi = \mathcal{I}_n^+ \cap \mathcal{F}_\Psi \subseteq \psi^{-1}(0)$.)

Chapter 3

Representations of S_n and Their Applications

In this chapter we turn to a sublime connection, first exploited by Regev, of PI-theory with classical representation theory (via the group algebra $F[S_n]$), leading to many interesting results (including Regev's Exponential Bound Theorem). Throughout, we assume that C is a field F, and we take an algebra A over F. Perhaps the key result of this chapter, discovered independently by Amitsur-Regev and Kemer, shows in characteristic 0 that any PI-algebra satisfies a "sparse" identity. This is the key hypothesis in translating identities to affine superidentities in Chapter 7, leading to Kemer's solution of Specht's problem.

3.1 Permutations and Identities

As usual S_n denotes the symmetric group on $\{1, \ldots, n\}$. Since any multilinear polynomial in $x_1, \ldots, x_n$ is a C-linear combination of the monomials $x_{\sigma(1)} x_{\sigma(2)} \cdots x_{\sigma(n)}$, $\sigma \in S_n$, we are led to introduce the following notation. Assume $X = \{x_1, x_2, \ldots\}$ is infinite.

Definition 3.1.1. *For each n, $V_n = V_n(x_1, \ldots, x_n)$ denotes the C-module of multilinear polynomials in $x_1, \ldots, x_n$, i.e.,*

$$V_n = \mathrm{span}_C \{ x_{\sigma(1)} x_{\sigma(2)} \cdots x_{\sigma(n)} \mid \sigma \in S_n \}.$$

We identify V_n with the group algebra $C[S_n]$, by identifying a permutation $\sigma \in S_n$ with its corresponding monomial (in $x_1, x_2, \ldots, x_n$):

$$\sigma \leftrightarrow M_\sigma(x_1, \ldots, x_n) = x_{\sigma(1)} \cdots x_{\sigma(n)}.$$

Any polynomial $\sum \alpha_\sigma x_{\sigma(1)} \cdots x_{\sigma(n)}$ corresponds to an element $\sum \alpha_\sigma \sigma \in C[S_n]$, and conversely, $\sum \alpha_\sigma \sigma$ corresponds to the polynomial

$$\left(\sum \alpha_\sigma \sigma \right) x_1 \cdots x_n = \sum \alpha_\sigma x_{\sigma(1)} \cdots x_{\sigma(n)}.$$

Definition 3.1.1 identifies V_n with the group algebra $C[S_n]$. Let us study some basic properties of this identification.

Given $\sigma, \pi \in S_n$, the product $\sigma\pi$ which corresponds to the monomial

$$M_{\sigma\pi} = x_{\sigma\pi(1)} \cdots x_{\sigma\pi(n)}$$

can be viewed in two ways, corresponding to **left** and **right** actions of S_n on V_n, described respectively as follows:

(i) $\sigma M_\pi(x_1 \ldots, x_n) = M_{\sigma\pi}$,   and

(ii) $M_\sigma(x_1 \ldots, x_n)\pi = M_{\sigma\pi}$.

Lemma 3.1.2. *Let $\sigma, \pi \in S_n$, and let $y_i = x_{\sigma(i)}$. Then*

(i) $\sigma M_\pi(x_1 \ldots, x_n) = M_\pi(x_{\sigma(1)}, \ldots, x_{\sigma(n)})$;

(ii) $M_\sigma(x_1 \ldots, x_n)\pi = (y_1 \cdots y_n)\pi = y_{\pi(1)} \cdots y_{\pi(n)}.$

Thus, the effect of the right action of π on a monomial is to permute the places of the variables according to π.

Proof. $y_{\pi(j)} = x_{\sigma\pi(j)}$. Then

$$M_{\sigma\pi} = x_{\sigma\pi(1)} \cdots x_{\sigma\pi(n)} = y_{\pi(1)} \cdots y_{\pi(n)} =$$

(which already proves (ii))

$$= M_\pi(y_1, \ldots, y_n) = M_\pi(x_{\sigma(1)}, \ldots, x_{\sigma(n)}),$$

which proves (i).   □

Extending by linearity, we obtain

Corollary 3.1.3. *If $p = p(x_1, \ldots, x_n) \in V_n$ and $\sigma, \pi \in S_n$, then*

(i) $\sigma p(x_1, \ldots, x_n) = p(x_{\sigma(1)}, \ldots, x_{\sigma(n)})$;

(ii) $p(x_1, \ldots, x_n)\pi = q(x_1, \ldots, x_n)$, *where $q(x_1, \ldots, x_n)$ is obtained from $p(x_1, \ldots, x_n)$ by place-permuting all the monomials of p according to the permutation π.*

For any finite group G and field F, there is a well-known correspondence between the $F[G]$-modules and the representations of G over the algebraically closure of F. The simple modules correspond to the irreducible representations, which in turn correspond to the irreducible characters; in fact, in characteristic 0 there is a 1:1 correspondence between the isomorphic classes of simple modules and the irreducible characters, which we shall use freely.

Definition 3.1.4. *Define $\Gamma_n = \operatorname{id}(A) \cap V_n \subseteq V_n$.*

We shall study the multilinear identities $\operatorname{id}(A) \cap V_n$, $n = 1, 2, \ldots$. In characteristic zero, this sequence of spaces determines $\operatorname{id}(A)$, by Proposition 1.2.18.

Remark 3.1.5. *If $p \in \operatorname{id}(A)$, then $\sigma p \in \operatorname{id}(A)$ since the left action is just a change of variables.*

Hence, for any PI-algebra A, the spaces

$$\Gamma_n \subseteq V_n$$

are in fact left ideals of $F[S_n]$, thereby affording certain S_n representations. We shall study the quotient modules

$$V_n/\Gamma_n$$

and their corresponding S_n-characters. Of course, these characters are independent of the particular choice of $x_1, \ldots, x_n$. The above obviously leads us into the theory of group representations, which we shall review in the next section.

However, $p\sigma$ need not be an identity. Let us consider some examples.

Example 3.1.6. *Applying the transposition (12) on the left to the Grassmann identity*

$$f = [[x_1, x_2], x_3] = x_1 x_2 x_3 - x_2 x_1 x_3 - x_3 x_1 x_2 + x_3 x_2 x_1$$

yields $-f$; applying (23) yields $[[x_1, x_3], x_2]$, which also is a PI of the Grassmann algebra G.

However, applying (12) to f on the right yields

$$x_2 x_1 x_3 - x_1 x_2 x_3 - x_1 x_3 x_2 + x_2 x_3 x_1 = -[x_2, x_1 x_3] - [x_1, x_2 x_3],$$

which clearly is not an identity of G (seen by specializing $x_2 \mapsto 1$).

Here is a slightly more complicated example, which is much more to the point.

Example 3.1.7. *The Capelli polynomial c_n can be written as*

$$(s_n(x_1, \ldots, x_n)x_{n+1} \cdots x_{2n})\pi,$$

where

$$\pi = \begin{pmatrix} 1 & 2 & 3 & 4 & \ldots & 2n-1 & 2n \\ 1 & n+1 & 2 & n+2 & \ldots & n & 2n \end{pmatrix}.$$

This example is so crucial that we formalize it.

Remark 3.1.8. *(The embedding $V_n \subseteq V_m$) Let $n \leq m$. The natural embedding $S_n \subseteq S_m$ (namely, fixing all indices $> n$) induces an embedding $V_n \subseteq V_m$, given by $f(x_1, \ldots, x_n) \mapsto f(x_1, \ldots, x_n)x_{n+1} \cdots x_m$.*

Definition 3.1.9. *Given $f(x_1, \ldots, x_n) = \sum_{\sigma \in S_n} a_\sigma x_{\sigma(1)} \cdots x_{\sigma(n)} \in V_n$ we form the **Capelli-type** polynomial in V_{2n}:*

$$f^*(x_1, \ldots, x_n; x_{n+1}, \ldots, x_{2n})$$

$$= \sum_{\sigma \in S_n} a_\sigma x_{\sigma(1)} x_{n+1} x_{\sigma(2)} x_{n+2} \cdots x_{\sigma(n-1)} x_{2n-1} x_{\sigma(n)} x_{2n}.$$

Thus, $c_n = s_n^*$. Note the last indeterminate x_{2n} is superfluous, since we could specialize it to 1, and so strictly speaking we need only pass to V_{2n-1}. However, the notation is more symmetric (with n permuted variables and n nonpermuted variables), so we keep x_{2n}. The following observation is a straightforward application of Corollary 3.1.3.

Remark 3.1.10. *For any polynomial $f(x_1, \ldots, x_n) \in V_n$,*

$$f^* = (f(x_1, \ldots, x_n)x_{n+1} \cdots x_{2n})\pi = \sum_{\sigma \in S_n} a_\sigma x_{\sigma(1)} x_{n+1} x_{\sigma(2)} x_{n+2} \cdots x_{\sigma(n)} x_{2n},$$

(3.1)

where the permutation π is given in Example 3.1.7.

Note that if $f^* \in \mathrm{id}(A)$, then $f \in \mathrm{id}(A)$, seen by specializing $x_j \mapsto 1$ for all $j > n$.

3.2 Review of the Representation Theory of S_n

In this section we describe, without proof, the classical theory of Frobenius and Young about the representations of S_n. Although formulated here in characteristic 0, there is a characteristic p version given in Green [Gre80], building from Remark 3.2.5 below.

3.2.1   The structure of group algebras

The structure of the group algebra $F[G]$ is particularly nice in characteristic 0; for the reader's convenience, we shall review some of the main structural results, including those involving idempotents.

Note that an algebra is a left module over itself, in which case a submodule is the same as a left ideal, and a simple submodule is the same as a minimal left ideal. We start with Maschke's Theorem and its consequences.

Theorem 3.2.1 (Maschke). *Let $A = F[G]$, where G is a finite group and F is a field of characteristic 0. Then A is semisimple, implying that any module M over A is semisimple; in particular, any submodule N of M has a* **complement** *N' such that $N \oplus N' = M$. Furthermore,*

$$A = \bigoplus_{i=1}^{r} I_i$$

where the $\mathcal{I}_i$'s are the minimal two-sided ideals of A. When F is algebraically closed, r equals the number of conjugacy classes of G,

$$\mathcal{I}_i \cong M_{n_i}(F),$$

the $n_i \times n_i$ matrices over F, and the numbers n_i satisfy

$$n_1^2 + \cdots + n_r^2 = |G|.$$

Maschke's Theorem thereby leads us to a closer study of direct sums of matrix rings.

Remark 3.2.2. *Notation as above, let $J_i \subseteq \mathcal{I}_i$ be minimal left ideals, so for $i = 1, \ldots r$, $\dim J_i = n_i$ and $\operatorname{End}_{\mathcal{I}_i} J_i = F$. Then any module M finite over A can be decomposed as*

$$M = \bigoplus_{i=1}^{r} M_i \quad \text{where} \quad M_i = \mathcal{I}_i M \quad \text{and} \quad M_i \cong J_i^{\oplus m_i}$$

for suitable multiplicities m_i, $i = 1, \ldots, r$. This is called the **isotypic decomposition** *of M as an A-module.*

Proposition 3.2.3. *Suppose A is a semisimple ring. For any left ideal $J \subseteq A$, there exists an idempotent $e \in J$ such that $J = Ae$. Moreover, the idempotent e is primitive if and only if J is a minimal left ideal.*

Proof. Take a complement J' of J in A and write $1 = e + e'$ where $e \in J$, $e' \in J'$. Then $e = e1 = e^2 + ee'$; matching components in $J \oplus J'$ shows $e = e^2$ and $ee' = 0$. Likewise, e' is idempotent. Clearly, $J \supseteq Ae$. But if $a \in J$, then $a = a1 = ae + ae'$, implying $a = ae$ (and $ae' = 0$); hence, $J = Ae$. The last assertion is now clear. $\qquad\square$

Remark 3.2.4. *In practice, it is more convenient to work with* **semi-idempotents,** *i.e., elements e such that $0 \neq e^2 \in Fe$. Of course, if $e^2 = \alpha e$ for $\alpha \neq 0$, then $\frac{e}{\alpha}$ is idempotent and generates the same left ideal as e.*

Remark 3.2.5. *Even in characteristic $p > 0$, enough structure theory carries over to enable us to obtain useful results. For G finite, $F[G]$ is finite dimensional and thus as a left module over itself has a composition series*

$$L_0 = F[G] \supset L_1 \supset L_2 \supset \cdots \supset L_r = 0, \tag{3.2}$$

whose simple factors L_i/L_{i+1} comprise the set of irreducible representations of G. (Indeed, any irreducible representation is a simple $F[G]$-module, thus of the form $F[G]/L$ for L a suitable maximal left ideal; conversely, for any maximal left ideal L, the chain $F[G] \supset L$ can be refined to a composition series $F[G] \supset L \supset \ldots$, and by the Jordan-Holder theorem, the factors $F[G]/L$ and L_i/L_{i+1} are isomorphic for some i.)

Lemma 3.2.6. *Suppose L is a minimal left ideal of an algebra A. Then the minimal two-sided ideal of A containing L is a sum of minimal left ideals of A isomorphic to L as modules.*

Proof. For any $a \in A$, there is a module homomorphism $L \to La$ given by $r \mapsto ra$, so La is either 0 or isomorphic to L. But $\sum_{a \in A} La$ is clearly the minimal ideal of A containing L. $\qquad\square$

From now on, we take $G = S_n$, and I_λ denotes the minimal two-sided ideal of $F[G]$ corresponding to the partition λ of n.

Lemma 3.2.7. *Let A be an F-algebra, and let λ be a partition of n. If $\dim J_\lambda > c_n(A)$, then $I_\lambda \subseteq \mathrm{Id}(A) \cap V_n$.*

Proof. By Lemma 3.2.6, I_λ is a sum of minimal left ideals, with each such minimal left ideal J isomorphic to J_λ. Thus, $\dim J = \dim J_\lambda > c_n(A)$. Since J is minimal, either $J \subseteq \mathrm{Id}(A) \cap V_n$ or $J \cap (\mathrm{Id}(A) \cap V_n) = 0$. If $J \cap (\mathrm{Id}(A) \cap V_n) = 0$ then it follows that

$$c_n(A) = \dim V_n/(\mathrm{Id}(A) \cap V_n) \geq \dim J > c_n(A),$$

a contradiction. Therefore each $J \subseteq \mathrm{Id}(A) \cap V_n$. Hence $I_\lambda \subseteq \mathrm{Id}(A) \cap V_n$ since I_λ equals the sum of these minimal left ideals. $\qquad\square$

3.2.2 Young's theory

Throughout this section (since it is customary in the representation theory of S_n), $\lambda = (\lambda_1, \ldots \lambda_k)$ denotes a partition of n, denoted $\lambda \vdash n$, if $\lambda_1, \ldots, \lambda_k$ are integers satisfying

(i) $\lambda_1 \geq \lambda_2 \geq \cdots \geq \lambda_k \geq 0$, and

(ii) $\lambda_1 + \cdots + \lambda_k = n$.

Let us apply Theorem 3.2.1 to S_n, where F is a field of characteristic zero, for example, $F = \mathbf{Q}$. The conjugacy classes of S_n, and therefore the minimal two-sided ideals of $F[S_n]$, are indexed by the partitions of n.

3.2.2.1 Diagrams

The **Young diagram** of $\lambda = (\lambda_1, \ldots \lambda_k)$ is a left-justified array of empty boxes such that the i-th row contains λ_i boxes. For example,

is the Young diagram of $\lambda = (3,1)$. We identify a partition with its Young diagram. For partitions $\mu = (\mu_1, \mu_2, \ldots)$ and $\lambda = (\lambda_1, \lambda_2, \ldots)$, we say that $\mu \geq \lambda$ if the Young diagram of μ contains that of λ, i.e., if $\mu_i \geq \lambda_i$ for all i.

3.2.2.2 Tableaux

Given $\lambda \vdash n$, we fill its diagram with the numbers $1, \ldots, n$ without repetition, to obtain a **Young tableau** T_λ of **shape** λ. Such a Young tableau is called a **standard tableau** if its entries increase in each row from left to right and in each column from top to bottom. For example,

$$\begin{array}{|c|c|} \hline 1 & 2 \\ \hline \end{array} \quad \begin{array}{|c|c|} \hline 1 & 3 \\ \hline \end{array}$$

are the only standard tableaux of shape $\lambda = (2,1)$.

The number of standard tableaux of shape λ is denoted as s^λ. For example, $s^{(2,1)} = 2$. The Young-Frobenius formula and the hook formula, given below, are convenient formulas for calculating s^λ. (In the literature, one writes f^λ instead of s^λ, but here we have used f throughout for polynomials.)

3.2.2.3 Left ideals

The tableau T_λ defines two subgroups of S_n, where $\lambda \vdash n$:

(i) The row permutations R_{T_λ}, i.e., those permutations that leave each row invariant.

(ii) The column permutations C_{T_λ} i.e., those permutations that leave each column invariant.

Define

$$R^+_{T_\lambda} = \sum_{\pi \in R_{T_\lambda}} \pi, \qquad C^-_{T_\lambda} = \sum_{\tau \in C_{T_\lambda}} \mathrm{sgn}(\tau)\tau.$$

Then T_λ defines the semi-idempotent

$$e_{T_\lambda} := \sum_{\tau \in C_{T_\lambda}} \sum_{\pi \in R_{T_\lambda}} \mathrm{sgn}(\tau)\tau\pi = C^-_{T_\lambda} R^+_{T_\lambda}.$$

The left ideal $J_\lambda := F[S_n]e_{T_\lambda}$ is minimal in $F[S_n]$. For the tableaux T_λ of shape λ, these left ideals are all isomorphic as left $F[S_n]$ modules. In particular, they give rise to equivalent representations, and thus afford the same character, which we denote as χ^λ. However, if $\lambda \neq \mu \vdash n$, with corresponding tableaux T_λ and T_μ, then

$$F[S_n]e_{T_\lambda} \not\cong F[S_n]e_{T_\mu}.$$

Details can be found in [Jac80, pp. 266ff].

Remark 3.2.8. *One can always reverse directions by working with π^{-1} instead of π, and thereby replace e_{T_λ} by the semi-idempotent $e'_{T_\lambda} = R^+_{T_\lambda} C^-_{T_\lambda}$.*

Example 3.2.9. *Let us write out explicitly the polynomial corresponding to e'_{T_λ}, where we take T_λ to be*

$$\begin{pmatrix} 1 & u+1 & 2u+1 & \dots & (v-1)u+1 \\ 2 & u+2 & 2u+2 & \dots & (v-1)u+2 \\ \dots & & & & \\ u & 2u & 3u & \dots & vu \end{pmatrix}.$$

$$C^-_{T_\lambda} x_1 \cdots x_{uv} = \sum_{\pi \in C_{T_\lambda}} \mathrm{sgn}(\pi) x_{\pi(1)} \cdots x_{\pi(u)} \cdots x_{\pi((v-1)u+1)} \cdots x_{\pi(vu)}$$

$$= s_u(x_1, \dots, x_u) \cdots s_u(x_{(v-1)u+1}, \dots, x_{vu}).$$

Thus, $e'_{T_\lambda} = R^+_{T_\lambda} C^-_{T_\lambda}$ is the multilinearization of s^v_u. The element e_{T_λ} is more difficult to describe but also plays an important role in the sequel.

In the literature (see for example [Sag01]), there is a construction of an $F[S_n]$-module $S^\lambda \cong F[S_n]e_{T_\lambda}$, called the **Specht module** (not directly connected to Specht's problem).

Given $\lambda \vdash n$, the sum of the minimal left ideals corresponding to the T_λ is a minimal two-sided ideal, denoted as

$$I_\lambda := \sum_{T_\lambda} F[S_n]e_{T_\lambda}. \tag{3.3}$$

It is well known ([Jac80, §5.4] or [Ro08, Theorem 19.61] that

$$I_\lambda = \bigoplus_{T_\lambda \text{ is standard}} F[S_n]e_{T_\lambda},$$

and as an F-algebra, I_λ is isomorphic to $M_{s^\lambda}(F)$. This leads to the following extended S_n-version of Theorem 3.2.1.

Remark 3.2.10.

(i) *There is a $1:1$ correspondence between the minimal ideals I_λ (of $F[S_n]$) and the irreducible S_n-characters χ^λ.*

(ii) *By general representation theory, the element*

$$E_\lambda = \sum_{\sigma \in S_n} \chi^\lambda(\sigma)\sigma$$

satisfies $E_\lambda^2 = n!E_\lambda$, and thereby is a central semi-idempotent in $F[S_n]$.

(iii) *For example, let $\lambda = (1^n)$. Then $\chi = \chi^{(1^n)}$ is the sign character given by $\chi(\sigma) = \operatorname{sgn}(\sigma)$, and $E_{(1^n)}$ is the n-th standard polynomial s_n.*

In summary, we have:

Theorem 3.2.11. *We have the following decompositions, notation as above:*

(i)

$$F[S_n] = \bigoplus_{\lambda \vdash n} I_\lambda = \bigoplus_{\lambda \vdash n} \left(\bigoplus_{T_\lambda \text{ is standard}} F[S_n]e_{T_\lambda} \right).$$

(ii)

$$I_\lambda \cong (S^\lambda)^{\oplus s^\lambda},$$

and hence,

$$F[S_n] \cong \bigoplus_{\lambda \vdash n} (J_\lambda)^{\oplus s^\lambda}$$

as S_n-modules.

Remark 3.2.12. *Since F can be any field of characteristic 0, it follows that the decomposition of an S_n-module (resp. character) into simple submodules (resp. irreducible characters) is the same for any field extension of $\mathbb{Q}$.*

Example 3.2.13. *We can view the standard identity*

$$s_n = \sum_{\sigma \in S_n} \operatorname{sgn}(\sigma)\sigma$$

as the (central) unit in the two-sided ideal I_λ, where $\lambda = (1^n)$.

3.2.2.4   The Branching Theorem

Let $\lambda = (\lambda_1, \lambda_2, \ldots)$ be a partition of n, and let λ^+ denote those partitions of $n+1$ that contain λ, i.e., of the form $(\lambda_1, \ldots, \lambda_{i-1}, \lambda_i + 1, \lambda_{i+1}, \ldots)$. For example, if $\lambda = (3, 3, 1, 1)$, then

$$(3, 3, 1, 1)^+ = \{(4, 3, 1, 1), (3, 3, 2, 1), (3, 3, 1, 1, 1)\}.$$

Similarly, if $\mu = (\mu_1, \mu_2, \ldots)$ is a partition of $n+1$, we let μ^- denote the partitions of n of the form $(\mu_1, \ldots, \mu_i - 1, \ldots)$. For example,

$$(3, 3, 1, 1, 1)^- = \{3, 2, 1, 1, 1), (3, 3, 1, 1)\}.$$

We have the following important theorem. Recall that if χ^λ is a character corresponding to an $F[S_n]$-module M, then viewing $F[S_n] \hookrightarrow F[S_{n+1}]$ naturally, we have the $F[S_{n+1}]$-module $F[S_{n+1}] \otimes_{F[S_n]} M$, whose character is called the **induced character**, denoted $\chi^\lambda \uparrow^{S_{n+1}}$. In the other direction, if χ^μ is a character corresponding to an $F[S_{n+1}]$-module M, then we can view M a fortiori as an $F[S_n]$-module, and call the corresponding character the **restricted character** denoted as $\chi^\mu \downarrow_{S_n}$.

Theorem 3.2.14 (Branching).

(i) *(Branching-up) If $\lambda \vdash n$, then*

$$\chi^\lambda \uparrow^{S_{n+1}} = \sum_{\mu \in \lambda^+} \chi^\mu.$$

(ii) *(Branching-down) If $\mu \vdash n+1$, then*

$$\chi^\mu \downarrow_{S_n} = \sum_{\lambda \in \mu^-} \chi^\lambda.$$

Note by Frobenius reciprocity that the two parts of Theorem 3.2.14 are equivalent. The following are obvious consequences of the Branching Theorem.

Corollary 3.2.15.

(i) *Let $\lambda \vdash n$ with I_λ the corresponding minimal two-sided ideal in $F[S_n]$, and let $m \geq n$. Then*

$$F[S_m] I_\lambda F[S_m] = \bigoplus_{\mu \vdash m; \, \mu \geq \lambda} I_\mu.$$

(ii) *Let $\lambda \vdash n$, $\mu \vdash m$ with $m \geq n$, and assume $\mu \geq \lambda$. Then $s^\mu \geq s^\lambda$; furthermore, if $E_\lambda \in \mathrm{id}(A)$, then $E_\mu \in \mathrm{id}(A)$.*

3.2.3   The RSK correspondence

The Robinson-Schensted-Knuth (RSK) correspondence is an algorithm that corresponds a permutation $\sigma \in S_n$ with a pair of standard tableaux (P_λ, Q_λ), both of the same shape $\lambda \vdash n$. See Knuth [Kn73] or Stanley [St99] for a description of the RSK correspondence.

This algorithm and its generalizations have become important tools in combinatorics. For example, it instantly yields the formula

$$n! = \sum_{\lambda \vdash n} (s^\lambda)^2.$$

This formula can be obtained by calculating the dimensions on both sides of Theorem 3.2.11. However, in the development of the S_n-representation theory, this formula is usually proved earlier, as a first step toward proving Theorem 3.2.11.

Among other properties, the RSK correspondence $\sigma \mapsto (P_\lambda, Q_\lambda)$ has the following property. Recall that the **length** $\ell(\lambda) = \ell(\lambda_1, \lambda_2, \ldots)$ is d if $\lambda_d > 0$ and $\lambda_{d+1} = 0$.

Theorem 3.2.16. *Let $\sigma \in S_n$ and*

$$\sigma \mapsto (P_\lambda, Q_\lambda)$$

in the RSK correspondence, so $\lambda \vdash n$, and let $d = \ell(\lambda)$. Then d is the length of a maximal chain $1 \leq i_1 < \cdots < i_d \leq n$ such that $\sigma(i_1) > \cdots > \sigma(i_d)$. In other words, $\ell(\lambda)$ is the length of a maximal descending subsequence in σ.

To understand this condition, recall Shirshov's notion of d-decomposability and Remark 2.3.3(ii).

Remark 3.2.17. *A multilinear monomial $x_{\sigma(1)} \cdots x_{\sigma(n)}$ is d-decomposable iff the sequence $\{\sigma(1), \ldots, \sigma(n)\}$ contains a descending subsequence of length d, i.e., iff there are $i_1 < \cdots < i_d$ such that $\sigma(i_1) > \cdots > \sigma(i_d)$.*

This led Latyshev to the following definition.

Definition 3.2.18. *A permutation $\sigma \in S_n$ is d-good if its monomial $x_{\sigma(1)} \cdots x_{\sigma(n)}$ is d-decomposable. We denote the number of d-good permutations in S_n as $g_d(n)$.*

(For $d = 3$, $g_3(n)$ can be shown to be the so-called "Catalan number" of n.) Theorem 3.2.16 clearly gives a bijective proof of the following fact:

Lemma 3.2.19. *Notation as above,*

$$g_d(n) = \sum_{\lambda \vdash n, \ \ell(\lambda) \leq d-1} (s^\lambda)^2.$$

3.2.4   Dimension formulas

Here are some formulas for calculating s^λ. First, for commuting variables $\xi_1, \ldots, \xi_k$ consider the discriminant

$$\text{Disc}_k(\xi) = \text{Disc}_k(\xi_1, \ldots, \xi_k) = \prod_{1 \le i < j \le k} (\xi_i - \xi_j).$$

Theorem 3.2.20 (The Young-Frobenius formula). *Given*

$$\lambda = (\lambda_1, \ldots, \lambda_k) \vdash n,$$

let $t_i = \lambda_i + k - i, \ 1 \le i \le k$. *Then*

$$s^\lambda = \frac{n!}{t_1! \cdots t_k!} \text{Disc}_k(t_1, \ldots, t_k).$$

A very convenient formula for calculating s^λ is the hook formula, due to J. S. Frame, G. de B. Robinson, and R. M. Thrall.

Definition 3.2.21. *Let* $\lambda = (\lambda_1, \lambda_2, \ldots)$ *be a partition and let* $\lambda' = (\lambda'_1, \lambda'_2, \ldots)$ *denote its* **conjugate partition**, *where* λ'_j *is the length of the* j *column of the Young diagram of* λ. *For* $(i,j) \in \lambda$, *the* (i,j)-**hook number** n_{ij} *in* λ *is the length of the "hook" from* (i,j) *to the right border and to the bottom of the Young diagram of* λ, *i.e.,*

$$h_{(i,j)} = h_{(i,j)}(\lambda) = \lambda_i - j + \lambda'_j - i + 1.$$

Theorem 3.2.22 (The S_n hook formula, [Sag01] or [JamK80]). *If* $\lambda \vdash n$, *then*

$$s^\lambda = \frac{n!}{\prod_{(i,j) \in \lambda} h_{(i,j)}(\lambda)}.$$

We need one more formula concerning s^λ, for whose preparation we introduce some more structure.

3.3 S_n-Actions on $T^n(V)$

Each of our two actions on multilinear polynomials extends naturally to the space of homogeneous polynomials — the left action yields more generally the polynomial representations of the general linear group $GL(V)$ (to be described in Chapter 12), and the right action has some remarkable applications, especially for superalgebras. It is this second action on which we focus here.

We recall the **tensor algebra** $T(V)$ of V, from Definition 1.3.4. We denote $\text{End}_F(T^n(V))$ by $E_{n,k}$.

There is a natural action of S_n on $T^n(V)$ by permuting coordinates — or places — in "monomials":

Definition 3.3.1. *Let $\sigma \in S_n$ and define $\hat{\sigma} \in E_{n,k}$ via*

$$\hat{\sigma}(v_1 \otimes \cdots \otimes v_n) = v_{\sigma(1)} \otimes \cdots \otimes v_{\sigma(n)},$$

extended by linearity to all of $T^n(V)$. It is routine to verify that $\hat{\sigma}$ is well-defined. Denote by $\varphi_{n,k} : S_n \to E_{n,k}$ the map given by $\varphi_{n,k}(\sigma) = \hat{\sigma}$, and extend $\varphi_{n,k}$ by linearity to $\varphi_{n,k} : F[S_n] \to E_{n,k}$. Define

$$A(n,k) = \varphi_{n,k}(F[S_n]) \subseteq E_{n,k}.$$

Let $a \in F[S_n]$ and let $w \in T^n(V)$; by abuse of notation we write $a \cdot w$ for $\hat{a}(w)$.

Although this action extends the right action defined in Definition 3.1.2(ii), we write it now on the left, to conform with the literature.

Remark 3.3.2.

(i) *$\widehat{\sigma\pi} = \hat{\pi}\hat{\sigma}$ for $\sigma, \pi \in S_n$; namely, $\varphi_{n,k}$ is an anti-homomorphism. Thus, if $a, b \in F[S_n]$ and $w \in T^n(V)$ then*

$$(ab) \cdot w = b \cdot (a \cdot w).$$

Therefore, $T^n(V)$ is a left module over the opposite algebra $(F[S_n])^{op}$.

(ii) *If $f = f(x_1, \ldots, x_n) \in V_n = F[S_n]$ is written as a multilinear polynomial and $y_1 \otimes \cdots \otimes y_n \in T^n(V)$, then identifying $T(V)$ with the free algebra $F\{x\}$ gives us the identification*

$$f \cdot (y_1 \otimes \cdots \otimes y_n) = f(y_1, \ldots, y_n).$$

(iii) *In the literature, the action of S_n on $W_{n,k}$ is usually defined via*

$$\sigma'(v_1 \otimes \cdots \otimes v_n) = v_{\sigma^{-1}(1)} \otimes \cdots \otimes v_{\sigma^{-1}(n)},$$

with corresponding $\varphi'_{n,k} : \sigma \mapsto \sigma'$ which is a homomorphism, since $\sigma'\pi' = (\sigma\pi)'$. However, since $\varphi_{n,k}(F[S_n]) = \varphi'_{n,k}(F[S_n]) = A(n,k)$, the two approaches lead to the same results. The reason we chose the action $\hat{\sigma}$ is because of the formula (ii) above.

Recall that $F[S_n] = \oplus_{\lambda \vdash n} I_\lambda$ with each I_λ isomorphic to a matrix algebra, and any matrix algebra is anti-isomorphic with itself, via the transpose map. It follows that each I_λ is mapped by $\varphi_{n,k}$ either to zero or to $\varphi_{n,k}(I_\lambda) \cong I_\lambda$. The following theorem, due to Schur and Weyl, describes $A(n,k) = \varphi_{n,k}(F[S_n])$, and in particular, it describes those λ's satisfying $\varphi_{n,k}(I_\lambda) \cong I_\lambda$. We first introduce another notion.

Definition 3.3.3. *$H(k, 0; n)$ denotes those partitions of n contained in the infinite k-strip, namely with at most k rows:*

$$H(k, 0; n) = \{\lambda \vdash n \mid \lambda_{k+1} = 0\}.$$

Theorem 3.3.4. *Notation as above,*

(i) *If λ is a partition of n, then $\varphi_{n,k}(I_\lambda) \cong I_\lambda$ if $\lambda \in H(k,0;n)$, and $\varphi_{n,k}(I_\lambda) = 0$ otherwise.*

(ii)
$$A(n,k) = \varphi_{n,k}(F[S_n]) = \bigoplus_{\lambda \in H(k,0;n)} A_\lambda \cong \bigoplus_{\lambda \in H(k,0;n)} I_\lambda$$

where for each $\lambda \in H(k,0;n)$, $A_\lambda = \varphi_{n,k}(I_\lambda) \cong I_\lambda$, a matrix algebra.

(iii) *By Remark 3.2.2,*
$$T^n(V) = \bigoplus_{\lambda \in H(k,0;n)} W_\lambda,$$

where $W_\lambda = I_\lambda T^n(V) = A_\lambda T^n(V)$, and this is the isotypic decomposition of $T^n(V)$ as an $A(n,k)$-module, hence, as an $F[S_n]$-module.

Corollary 3.3.5. *Notation as above,*

(i)
$$\dim A(n,k) = \sum_{\lambda \in H(k,0;n)} (s^\lambda)^2.$$

(ii) *In particular, since $A(n,k) \subseteq E_{n,k} = \operatorname{End}_F(T^n(V))$, and $\dim E_{n,k} = k^{2n}$,*
$$\sum_{\lambda \in H(k,0;n)} (s^\lambda)^2 \le k^{2n}. \tag{3.4}$$

Note that the inequality (3.4) is a combinatoric fact about Young diagrams, and does not refer directly to $E_{n,k}$ or $A(n,k)$.

3.4 Codimensions and Regev's Theorem

Having brought in some basic tools from representation theory, we already can apply them to obtain some deep results in PI-theory. Recall that $\Gamma_n = \operatorname{id}(A) \cap V_n$, cf. Remark 3.1.4. We shall now bring in a technique that applies in all characteristics, although we suppose for the time being that $\operatorname{char}(F) = 0$. The sequence $\{\Gamma_n : n \in \mathbb{N}\}$ determines $\operatorname{id}(A)$. Being a left ideal of the semisimple ring $F[S_n]$, Γ_n has a direct sum complement (isomorphic as an $F[S_n]$-module to V_n/Γ_n), which is a direct sum of minimal left ideals, each of which can be described as $F[S_n]e_{T_\lambda}$, for a suitable Young tableau T_λ. This already constitutes a major step, providing a fast algorithm for determining the multilinear identities of degree n. The V_n/Γ_n have other remarkable properties to be described below.

Definition 3.4.1. *We call*

$$c_n(A) := \dim_F (V_n/\Gamma_n)$$

the n-th **codimension** *of the algebra A. The character of the representation corresponding to the module V_n/Γ_n is called the n-th* **cocharacter**.

Thus, the algebra·A defines a sequence of codimensions $c_n(A)$, for $n = 1, 2, \ldots$ (Later we shall study the cocharacters themselves in Chapter 12.)

Codimensions have played a key role in PI-theory since Regev used them in his dissertation to prove his Tensor Product Theorem (Theorem 3.4.7). Our notation is a bit unfortunate, since we already have used c_n to denote the Capelli polynomial. However, the contexts are completely different, since here $c_n(A)$ is a number, so there should not be ambiguity.

Observation 3.4.2. *A satisfies a PI of degree $\leq n$ iff $\Gamma_n \neq 0$, which holds iff $c_n(A) < n!$.*

The surprising basic fact Regev discovered about codimensions of a PI-algebra, which lies at the foundation of quantitative PI-theory, is their exponential bound, which presently we shall prove.

Theorem 3.4.3. *If A satisfies a PI of degree d, then*

$$c_n(A) \leq (d-1)^{2n}.$$

An important step toward proving this theorem is the next lemma. Recall from Remark 3.2.17 and Definition 3.2.18 that $g_d(n)$ is the number of d-indecomposable monomials in S_n.

Lemma 3.4.4 (Latyshev). *Let A be an algebra satisfying a polynomial identity of degree d. Then*

$$c_n(A) \leq g_d(n).$$

(This holds in any characteristic.)

Proof. By Proposition 2.2.6, V_n is spanned modulo $\mathrm{id}(A) \cap V_n$ by the d-indecomposable monomials, so

$$c_n(A) = [V_n : \mathrm{id}(A) \cap V_n] \leq g_d(n).$$

$\square$

Proof of Theorem 3.4.3 when $\mathrm{char}(F) = 0$. By Lemma 3.2.19,

$$g_d(n) = \sum_{\lambda \vdash n,\, \ell(\lambda) \leq d-1} (s^\lambda)^2.$$

Note that $\{\lambda \vdash n : \ell(\lambda) \leq d - 1\} = \{\lambda \in H(d - 1, 0; n)\}$. Thus, by Corollary 3.3.5(ii), $g_d(n) \leq (d - 1)^{2n}$, so we conclude with Lemma 3.4.4. $\square$

This proof is an immediate application of standard results about combinatorics. In [Ro80], the use of Corollary 3.3.5(ii) was bypassed by means of Amitsur's version of Dilworth's Theorem, which works in all characteristics, and even over arbitrary commutative rings. Thus, Theorem 3.4.3 is true over any field.)

3.4.1 Regev's Tensor Product Theorem

As a first application, we prove (Theorem 3.4.7). The key is

Remark 3.4.5. *Let $a_1, \ldots, a_n \in A$ be ml-generic elements (cf. Definition 1.8.9). Denote*

$$V_n(a) = V_n(a_1, \ldots, a_n) = \{f(a_1, \ldots, a_n) \mid f \in V_n(x_1, \ldots, x_n)\}.$$

Then $V_n(a)$ is a subspace of A. Also,

$$V_n(a) \cong \frac{V_n(x)}{\Gamma_n}$$

via the canonical map $f(x_1, \ldots, x_n) \mapsto f(a_1, \ldots, a_n)$, whose kernel is Γ_n, and therefore $c_n(A) = \dim V_n(a)$. In particular, if $W \subseteq A$ is a subspace such that $V_n(a) \subseteq W$, then $c_n(A) \leq \dim W$.

We can now prove a key result.

Proposition 3.4.6. $c_n(A \otimes B) \leq c_n(A) \cdot c_n(B)$.

Proof. Let $U_A = F\{\bar{x}_1, \bar{x}_2, \ldots\}$, $U_B = F\{\bar{x}'_1, \bar{x}'_2, \ldots\}$ be the respective relatively free algebras, and write $\bar{x}$ for $(\bar{x}_1, \ldots, \bar{x}_n)$ and $\bar{x}'$ for $(\bar{x}'_1, \ldots, \bar{x}'_n)$, as well as $\bar{x} \otimes \bar{x}'$ for $(\bar{x}_1 \otimes \bar{x}'_1, \ldots, \bar{x}_n \otimes \bar{x}'_n)$. Then $c_n(A) = \dim V_n(\bar{x})$ and $c_n(B) = \dim V_n(\bar{x}')$. It easily follows that

$$V_n(\bar{x} \otimes \bar{x}') \subseteq V_n(\bar{x}) \otimes V_n(\bar{x}').$$

Since $\bar{x}_1 \otimes \bar{x}'_1$, $\bar{x}_2 \otimes \bar{x}'_2$, ... are ml-generic,

$$c_n(A \otimes B) = c_n(U_A \otimes U_B) = \dim V_n(\bar{x} \otimes \bar{x}')$$
$$\leq \dim V_n(\bar{x}) \dim V_n(\bar{x}').$$

$\square$

As an immediate application, we have Regev's landmark theorem.

Theorem 3.4.7. *If A and B are two PI-algebras over a field F, then $A \otimes B$ is also PI, where a bound for the degree of the PI is indicated in the proof.*

Proof. By Theorem 3.4.3 the codimension sequences $c_n(A)$, $c_n(B)$ are bounded exponentially, and therefore by Proposition 3.4.6, so are the codimensions $c_n(A \otimes B)$. Thus, $c_n(A \otimes B) < n!$ for large enough n, and we conclude with Observation 3.4.2. $\square$

From now on, $e = 2.718281828\ldots$ is the base of the natural logarithms.

Remark 3.4.8. *Stirling's formula says $n! > (n/e)^n$.*

Remark 3.4.9. *Combining the bound of Theorem 3.4.3 with Proposition 3.4.6 and the fact that $n! > (n/e)^n$ yields a bound*

$$\text{PI-}\deg(A \otimes B) \le (\text{PI-}\deg(A) - 1)^2 (\text{PI-}\deg(B) - 1)^2 e.$$

In certain cases, a much better bound can be given for the degree. For example, if $A_i \sim_{\text{PI}} M_{n_i}(F)$ for $i = 1, 2$, then $\text{PI-}\deg(A_i) = 2n_i$ by the Amitsur-Levitzki theorem, whereas $\text{PI-}\deg(A_1 \otimes A_2) \le 2n_1 n_2$ since $A_1 \otimes A_2$ satisfies the identities of $n_1 n_2 \times n_1 n_2$ matrices, cf. Exercise 1.49.

Remark 3.4.10. *The implications of Theorem 3.4.3, even asymptotically, are so far-reaching that one is led to ask exactly what happens at the asymptotic level. Giambruno and Zaicev [GiZa98] proved that Theorem 3.4.3 implies $\sqrt[n]{c_n(A)} \le (d-1)^2$, leading us to define the **PI-exponent** of a PI-algebra A as $\varlimsup_{n \to \infty} \sqrt[n]{c_n(A)}$. In the affine case, Kemer's PI-representability Theorem 6.3.1 enables us to assume that A is finite dimensional (and, in fact, PI-basic), and Giambruno and Zaicev succeeded in proving the exponent is an integer which also equals $\varliminf_{n \to \infty} \sqrt[n]{c_n(A)}$. Using the techniques of Chapter 7, they extended this result to arbitrary PI-algebras and obtained other asymptotic results about $c_n(A)$, cf. [GiZa03a]. Unfortunately, since the proof of Theorem 6.3.1 is nonconstructive, it may be difficult to study the exponent of a given PI-algebra. From a different point of view, in Chapter 12, we discuss methods for obtaining finer information about specific T-ideals.*

3.4.2 The Kemer-Regev-Amitsur trick

Applying the same methods with care provides a powerful computational technique discovered independently by Kemer [Kem80] and Amitsur-Regev [AmRe82] following Regev [Reg78], which was used to great effect by Kemer. Recall $\Gamma_n = \text{id}(A) \cap V_n$, cf. Remark 3.1.4. The basic idea is contained in:

Lemma 3.4.11. *If $s^\lambda > c_n(A)$, then $I_\lambda \subseteq \Gamma_n$.*

Proof. Take any minimal left ideal $J = F[S_n] e_{T_\lambda} \subseteq I_\lambda$ of $F[S_n]$. $\dim J = s^\lambda$. If $J \not\subseteq \Gamma_n$, then $J \cap \Gamma_n = 0$, which implies that $s^\lambda \le c_n(A)$, a contradiction. Hence, $J \subseteq \Gamma_n$, and the result follows since I_λ is the sum of such J's. □

Next, we confront the codimension c_n with the computation in the next lemma.

Example 3.4.12. *Consider the "rectangle" of u rows and v columns, so $n = uv$. By [MacD95, p. 11], the hook numbers of the partition $\mu = (u^v)$ satisfy*

$$\sum_{x \in \mu} h_x = uv(u+v)/2 = n\frac{u+v}{2}.$$

Let us review the proof, for comparison in Chapter 5. For any box x in the $(1, j)$ position, the hook has length $u + v - j$, so the sum of all hook numbers in the first row is

$$\sum_{j=1}^{v}(u + v - j) = uv + \frac{v(v-1)}{2} = v\left(u + \frac{v-1}{2}\right).$$

Summing this over all rows yields

$$v\frac{u(u+1)}{2} + uv\frac{v-1}{2} = uv\left(\frac{u+1}{2} + \frac{v-1}{2}\right) = uv\frac{u+v}{2},$$

as desired.

Lemma 3.4.13. *Let u, v be integers, $n = uv$ and let $\lambda = (v^u) \vdash n$ be the $u \times v$ rectangle. Then*

$$s^\lambda > \left(\frac{uv}{u+v}\right)^n \cdot \left(\frac{2}{e}\right)^n. \tag{3.5}$$

In particular, if $\frac{uv}{u+v} > \frac{e}{2}\alpha$, then $s^\lambda > \alpha^n$.

Proof. Notation as in Definition 3.2.21, we estimate

$$s^\lambda = \frac{n!}{\prod_{x \in \lambda} h_{(i,j)}}.$$

Since the arithmetic mean is greater than the geometric mean, we deduce that

$$\left(\prod_{x \in \lambda} h_{(i,j)}\right)^{1/n} \leq \frac{1}{n}\sum_{x \in \lambda} h_{(i,j)} = \frac{u+v}{2},$$

by Example 3.4.12, and therefore $(\prod_{x \in \lambda} h_{(i,j)})^{-1} \geq (2/(u+v))^n$. Thus, in view of Remark 3.4.8,

$$s^\lambda > \left(\frac{n}{e}\right)^n \cdot \left(\frac{2}{u+v}\right)^n = \left(\frac{uv}{u+v}\right)^n \left(\frac{2}{e}\right)^n,$$

and the formula (3.5) follows. $\square$

Now we bring in the exponential bound of $c_n(A)$.

Proposition 3.4.14. *Suppose A has PI-degree d. Take $u \geq (d-1)^4e$, $v \geq \frac{(d-1)^4eu}{2u-(d-1)^4e}$, $n = uv$, and any number m between n and $2n$. Let λ_0 denote the rectangular shape (v^u), i.e., u rows of v columns each. For any $\lambda \vdash m$ containing λ_0 and any Young tableau T_λ, the semi-idempotents $e_{T_\lambda}, e'_{T_\lambda} \in \Gamma_m$.*

Proof. Noting $\frac{uv}{u+v} \geq \frac{1}{2}(d-1)^4 e$ and using Lemma 3.4.13 and branching, we have

$$s^\lambda \geq s^{\lambda_0} > \left(\frac{uv}{u+v}\right)^n \left(\frac{2}{e}\right)^n \geq \left(\frac{1}{2}(d-1)^4 e\frac{2}{e}\right)^n$$

$$= (d-1)^{4n} \geq c_{2n}(A) \geq c_m(A),$$

implying $I_\lambda \subseteq \Gamma_m$ by Lemma 3.4.11; in particular $e_{T_\lambda}, e'_{T_\lambda} \in \Gamma_m$. $\qquad\square$

Remark 3.4.15 (Regev). *In view of Example 3.2.9, this provides a second proof of Theorem 1.8.26, that every PI-algebra satisfies a power of the standard identity, but now we have the explicit identity s_u^v (estimated in terms of the PI-degree of A).*

So far Proposition 3.4.14 provides a way of inducing from certain identities of A of degree n to identities of degree m where $m \leq 2n$. Our next theorem will remove the restriction on m. This is an amazing result. It gives us an ideal I (generated by $e_{T_{\lambda_0}}$) which, viewed inside *any* $F[S_m]$, will generate a two-sided ideal of multilinear identities of A. Clearly, we can then replace I by any element f in such an ideal, so f will remain an identity after left or right actions by permutations. The key to this improvement is the $*$ construction from Definition 3.1.9.

Theorem 3.4.16. *Let A be a PI-algebra and let $I \subseteq F[S_n]$ be a two-sided ideal in $F[S_n] = V_n$.*

(i) Let $f = f(x) \in I$ and assume for $n \leq m \leq 2n - 1$ that the elements of the two-sided ideal generated by f in $F[S_m]$ are identities of A. Then $f^* \in \text{id}(A)$.

(ii) Assume $I \subseteq \text{id}(A)$ and that for each $f \in I$, $f^* \in \text{id}(A)$.

Then for all $m \geq n$, all the elements of the two-sided ideal generated by I in $F[S_m]$ are identities of A: $F[S_m]IF[S_m] \subseteq \text{id}(A)$.

Proof. (i) follows easily from Remark 3.1.10.

(ii) Since $\text{id}(A) \cap V_m$ is a left ideal, it suffices to prove $f\pi \in \text{id}(A)$ for all $f \in I$ and $\pi \in S_m$. Write

$$(x_1 \ldots x_n \ldots x_m)\pi = p_0 x_{\eta(1)} p_1 x_{\eta(2)} p_2 \cdots p_{n-1} x_{\eta(n)} p_n,$$

where $\eta \in S_n$ and p_j are either 1 or are monomials in $x_{n+1}, \ldots, x_m$. Then

$$(x_1 \ldots x_m)\eta^{-1}\pi = p_0 x_1 p_1 x_2 p_2 \cdots p_{n-1} x_n p_n. \qquad (3.6)$$

Since $I \lhd F[S_r]$, we see $f\eta \in I$; writing $f\eta = \sum_{\sigma \in S_n} a_\sigma \sigma$, we have

$$f\pi = (f\eta)(\eta^{-1}\pi) = \sum_{\sigma \in S_n} a_\sigma p_0 x_{\sigma(1)} p_1 x_{\sigma(2)} \cdots p_{n-1} x_{\sigma(n)} p_n$$

$$= p_0 (f\eta)^*(x_1, \ldots, x_n; p_1, \ldots, p_n).$$

By assumption $(f\eta)^* \in \mathrm{id}(A)$. Substituting $x_{n+j} \mapsto p_j$ and multiplying by p_0 and p_n, we conclude that $f\eta \in \mathrm{id}(A)$.   □

Finally we bring in the branching theorem, Corollary 3.2.15, to get a powerful result.

Theorem 3.4.17. *Suppose A has PI-degree d, and m, n, u, v are as in Proposition 3.4.14. Then:*

   (i) $F[S_m]e_{T_{\lambda_0}}F[S_m] \subseteq \Gamma_m$, *for any tableau T_{λ_0} of $\lambda_0 = (v^u)$.*

   (ii) Γ_m *contains every two-sided ideal of S_m corresponding to a shape containing the rectangle (v^u).*

Proof. In view of Corollary 3.2.15, it suffices to prove (ii). By Theorem 3.4.16, we need to show that if

$$f \in I = F[S_n]e_{T_{\lambda_0}}F[S_n],$$

then $f^* \in \mathrm{id}(A)$. This is true by Proposition 3.4.14, taking $m = 2n$.   □

Corollary 3.4.18. *Hypotheses as in the theorem, $e_{T_{\lambda_0}}\pi^* \in \mathrm{id}(A)$, for any $\pi \in S_{uv}$.*

Proof. We just displayed $e_{T_{\lambda_0}}\pi^*$ in the right ideal generated by $e_{T_{\lambda_0}}$, so we apply Theorem 3.4.17(ii).   □

3.4.3 Hooks

In Theorem 3.4.17 we proved that if A has PI-degree d, then any multilinear polynomial whose Young tableau has a shape containing the rectangle v^u is an identity of A. Let us restate the contrapositive, starting with a definition generalizing Definition 3.3.3.

Definition 3.4.19. $H(k, \ell; n) = \{\lambda \vdash n \mid \lambda_{k+1} \leq \ell\}$ *is the collection of all shapes such that the $k + 1$ row (as well as all subsequent rows) has length $\leq \ell$.*

Equivalently, the $\ell + 1$ column (as well as all subsequent columns) has length $\leq k$.

Proposition 3.4.20. *Suppose A has PI-degree d. Then there are k and ℓ such that $I_\lambda \subseteq \mathrm{id}(A)$, for any shape λ not lying in $H(k, \ell; n)$.*

Proof. Take $k = u - 1$ and $\ell = v - 1$ for u, v as in Proposition 3.4.14. If the shape does not lie in $H(k, \ell; n)$, then it must contain an entry in the $(k+1, \ell+1)$-position, and thus contain the $u \times v$ rectangle (which we denoted as v^u). But then we are done by Theorem 3.4.17.   □

3.5   Multilinearization

In the key Theorem 3.3.4, one moves back and forth between multilinear polynomials (corresponding to V_n) and homogeneous polynomials (corresponding to $T^n(V)$). Our objective here is to correlate the two theories, explaining why they coincide in characteristic 0. To do this, we must first translate the multilinearization process (Definition 1.2.12) to the S_n-actions we have been studying.

Let A be a PI-algebra over an infinite field. By Proposition 1.2.17, every homogeneous part of an identity is an identity, so the study of the identities of A is reduced to the study of its homogeneous identities. Given a homogeneous polynomial p, we now describe its multilinearization $L(p)$ in terms of our S_n-action on $T^n(V)$. Furthermore, in characteristic zero, we reprove that the multilinear identities of A determine all other identities. In Chapter 1 we defined the multilinear process inductively, in terms of a sequence of substitutions. Here we define it in terms of a single operator L. We write y_i for the indeterminates used in the polynomial before the application of L, and x_i for the indeterminates after L is applied.

We recall the action of Definition 3.3.1, also cf. Remark 3.3.2, which is a "place-permutation." Throughout this section, $\underline{d} = (d_1, \ldots, d_k)$ with

$$d_1 + \cdots + d_k = n.$$

Remark 3.5.1. *If $f = f(x_1, \ldots, x_n) \in V_n$, then*

$$f \cdot (y_1^{d_1} \ldots y_k^{d_k}) = f(y_1, \ldots, y_1, \ldots, y_k, \ldots, y_k),$$

where each y_j is repeated d_j times. It follows that if $f \in \mathrm{id}(A)$, then $f \cdot y_1^{d_1} \ldots y_k^{d_k} \in \mathrm{id}(A)$. Working backwards, if $p = p(x_1, \ldots, x_k) \in T^n(V)$ is homogeneous of multidegree $(d_1, \ldots, d_k)$, then there exists an element $a \in F[S_n]$ such that $p(x_1, \ldots, x_k) = a \cdot (x_1^{d_1} \cdots x_k^{d_k})$.

Given i, i' write $S_{i,\ldots,i'}$ for the subgroup of S_n fixing all j other than for $i \leq j \leq i'$. Thus, $S_{1,\ldots,n} = S_n$.

Definition 3.5.2. *We define the subgroup*

$$Y_{\underline{d}} = S_{1,\ldots,d_1} \times S_{d_1+1,\ldots,d_1+d_2} \times \cdots \times S_{n-d_k+1,\ldots,n} \subseteq S_n.$$

Note that

$$\left(\sum_{\sigma \in Y_{\underline{d}}} \sigma \right) \cdot (x_1 \cdots x_n) = \left(\sum_{\sigma \in Y_{d_1,\ldots,d_k}} \sigma \right) \cdot (x_1 \cdots x_n) =$$

$$\left(\sum_{\sigma \in S_{1,\ldots,d_1}} x_{\sigma(1)} \cdots x_{\sigma(d_1)} \right) \left(\sum_{\sigma \in S_{d_1+1,\ldots,d_1+d_2}} x_{\sigma(d_1+1)} \cdots x_{\sigma(d_1+d_2)} \right) \cdots$$

Definition 3.5.3. *Let p be a homogeneous polynomial of multidegree $\underline{d}$, so that $p(y) = a \cdot (y_1^{d_1} \cdots y_k^{d_k})$ where $a \in F[S_n]$. Define the* **multilinearization operator** L *as follows:*

$$L(y_1^{d_1} \cdots y_k^{d_k}) = \left(\sum_{\sigma \in Y_{\underline{d}}} \sigma \right) \in F[S_n],$$

and, in general,

$$L(p) = L(a \cdot (y_1^{d_1} \cdots y_k^{d_k})) = L(y_1^{d_1} \cdots y_k^{d_k})a$$

$$= \left(\left(\sum_{\sigma \in Y_{d_1, \dots, d_k}} \sigma \right) \cdot (x_1 \cdots x_n) \right) a \in F[S_n].$$

Lemma 3.5.4. *Let $\underline{d} = (d_1, \dots, d_k)$ with $d_1 + \cdots + d_k = n$, and let p be a homogeneous polynomial of multidegree $\underline{d}$. Then*

(i) $L(p) \cdot (y_1^{d_1} \cdots y_k^{d_k}) = d_1! \cdots d_k! p(y_1, \dots, y_k).$

(ii) *Under the specializations*

$$y_1 \mapsto \bar{y}_1 = x_1 + \cdots + x_{d_1}, \quad y_2 \mapsto \bar{y}_2 = x_{d_1+1} + \cdots + x_{d_1+d_2}, \quad \dots,$$

we have $p(\bar{y}_1, \dots, \bar{y}_k) = L(p) + h$, where h is a linear combination of non-multilinear monomials.

Proof. (i) Write $\bar{y}$ for $y_1^{d_1} \cdots y_k^{d_k}$, and b for $\sum_{\sigma \in Y_{d_1, \dots, d_k}} \sigma$. Writing $p = a \cdot bary$, we have $L(p) = ba$. Hence,

$$L(p) \cdot \bar{y} = (ba) \cdot \bar{y} = a \cdot (b \cdot y),$$

and the proof follows from the obvious fact that

$$\left(\sum_{\sigma \in S_d} \sigma \right) \cdot y^d = d! y^d.$$

(ii) First, $(x_1 + \cdots + x_d)^d = \sum_{\sigma \in S_d} x_{\sigma(1)} \cdots x_{\sigma(d)} + g(x_1, \dots, x_d)$, where $g(x_1, \dots, x_d)$ is a linear combination of non-multilinear monomials. By a similar argument,

$$\bar{y}_1^{d_1} \cdots \bar{y}_k^{d_k} = \ell(x_1, \dots, x_n) + H(x_1, \dots, x_n)$$

where $\ell(x_1, \dots, x_n) = \left(\sum_{\sigma \in Y_{\underline{d}}} \sigma \right) \cdot (x_1 \cdots x_n)$ and $H(x_1, \dots, x_d)$ is a linear combination of non-multilinear monomials. Thus,

$$\bar{y}_1^{d_1} \cdots \bar{y}_k^{d_k} = L(y_1^{d_1} \cdots y_k^{d_k}) + H(x_1, \dots, x_n).$$

Let $p(y_1, \dots, y_k) = a \cdot (y_1^{d_1} \cdots y_k^{d_k})$ where $a \in V_n = F[S_n]$, then $p(\bar{y}_1, \dots, \bar{y}_k) = a \cdot (\bar{y}_1^{d_1} \cdots \bar{y}_k^{d_k}) = L(y_1^{d_1} \cdots y_k^{d_k})a + a \cdot H(x_1, \dots, x_n)$. By definition, $L(y_1^{d_1} \cdots y_k^{d_k})a = L(p)$ and also, $a \cdot H(x_1, \dots, x_n) = h(x_1, \dots, x_n)$ is a linear combination of non-multilinear monomials. $\square$

Let us apply this result to identities.

Proposition 3.5.5. *Let A be a PI-algebra over F, and let p be a homogeneous polynomial of multidegree $\underline{d} = (d_1, \ldots, d_k)$, with multilinearization $L(p)$.*

(i) *Assume the field F is infinite, of arbitrary characteristic. If $p \in \mathrm{id}(A)$, then $L(p) \in \mathrm{id}(A)$.*

(ii) *Assume in addition that $\mathrm{char}(F) = 0$ (or $\mathrm{char}(F) > \max\{d_1, \ldots, d_k\}$). If $L(p) \in id(A)$, then $p \in id(A)$.*

Proof. (i) Recall that $p(\bar{y}_1, \ldots, \bar{y}_k) = L(p)(x_1, \ldots, x_n) + H(x_1, \ldots, x_n)$ as above, with $\bar{y}_1 = x_1 + \cdots + x_{d_1}$, etc. Since $p(\bar{y}_1, \ldots, \bar{y}_k)$ is an identity of A, the right-hand side is also an identity of A; hence, $L(p)(x_1, \ldots, x_n)$, its homogeneous part of degree 1 in each i, is an identity of A.

(ii) This part clearly follows from Lemma 3.5.4(i): If $L(p) \in \mathrm{id}(A)$, then by Remark 3.5.1

$$L(p) \cdot (y_1^{d_1} \cdots y_k^{d_k}) = d_1! \cdots d_k! p(y_1, \ldots, y_k) \in \mathrm{id}(A),$$

implying $p(y_1, \ldots, y_k) \in \mathrm{id}(A)$ since $d_1! \cdots d_k!$ is invertible in F. $\qquad\square$

Proposition 3.5.6. *If $\mathrm{char}(F) = 0$, then*

$$\Gamma_n T^n(V) = \Gamma_n.$$

Thus, by homogeneity,

$$\mathrm{id}(A) = \bigoplus_n (\Gamma_n T^n(V)).$$

Proof. ($\subseteq$) If $f \in \Gamma_n$ and $w \in T^n(V)$ is a monomial, then by Remark 3.3.2(ii), $fw \in \mathrm{id}(A) \cap T^n(V)$, which implies the inclusion $\subseteq$.

($\supseteq$) Suppose $q(y_1, \ldots, y_k) \in \mathrm{id}(A) \cap T^n(V)$. By Proposition 1.2.17, we may assume that q is homogeneous, say of multidegree $\underline{d}$. By Proposition 3.5.5(i), its multilinearization $L(q) \in \Gamma_n$, and by Lemma 3.5.4(i), $L(q) \cdot (y_1^{d_1} \cdots y_k^{d_k}) = d_1! \cdots d_k! q(y_1, \ldots, y_k)$. Since $\mathrm{char}(F) = 0$, it follows that $q \in \Gamma_n T^n(V)$, which completes the proof. $\qquad\square$

Exercises for Chapter 3

1. Fill in the $u \times v$ rectangular tableau λ with the numbers

$$
\begin{array}{cccc}
1 & 2 & \cdots & v \\
v+1 & v+2 & \cdots & 2v \\
\vdots & \vdots & \ddots & \vdots \\
(u-1)v+1 & (u-1)v+2 & \cdots & uv
\end{array}
$$

and compute its semi-idempotents e_{T_λ} and e'_{T_λ}. In case $u = v = 2$, one of them is the multilinearization of S_2^2, but the other corresponds to the multilinearization of

$$
x_1^2 x_3^2 - x_3 x_1^2 x_3 - x_1 x_3^2 x_1 + x_3^2 x_1^2,
$$

which can be rewritten as

$$
x_1[x_1, x_3^2] - x_3[x_1^2, x_3].
$$

2. Let $\lambda \vdash n$ and let $E_\lambda \in I_\lambda$ be the corresponding central idempotent. Then $I_\lambda \subseteq \mathrm{id}(A)$ if and only if $E_\lambda \in \mathrm{id}(A)$. (Hint: $I_\lambda = F[S_n]E_\lambda$.)

3. In Theorem 1.8.26, if $\mathrm{char}(F) = 0$, then k is bounded by a function of d. (Hint: One considers only a finite number of semi-idempotents of $\mathbb{Q}[S_d]$, and takes the maximal k corresponding to them all.)

Amitsur-Capelli polynomials

4. Recall the polynomials p^* from Definition 3.1.9. Taking E_λ as in Exercise 2, define the **Amitsur-Capelli polynomial**

$$
E_\lambda^* = \sum_{\sigma \in S_n} \chi^\lambda(\sigma) x_{\sigma(1)} y_1 x_{\sigma(2)} y_2 \cdots x_{\sigma(n)} y_n,
$$

a natural generalization of the Capelli polynomial. If $E_\lambda^* \in \mathrm{id}(A)$, then $f^* \in \mathrm{id}(A)$ for all $f \in I_\lambda$. (Hint: Since $f = aE_\lambda$, $a \in F[S_n]$, it suffices to show that for $\sigma \in S_n$, $(\sigma E_\lambda)^* \in \mathrm{id}(A)$. Let $\pi \in S_{2n}$ be the permutation of Lemma 3.1.10. Then

$$
\begin{aligned}
(\sigma E_\lambda)^* ((\sigma E_\lambda)(x_1, \ldots, x_n, x_{n+1}, \ldots, x_{2n}))\pi \\
= \sigma(E_\lambda(x_1, \ldots, x_{2n})\pi) = \sigma(E_\lambda^*) \in \mathrm{id}(A)).
\end{aligned}
\tag{3.7}
$$

5. Let E_λ^* be the Amitsur-Capelli polynomial of a partition $\lambda \vdash n$. Then $E_\lambda^* \in \mathrm{id}(A)$, if and only if the two-sided ideal $I_\mu \subseteq \mathrm{id}(A)$ for every partition μ containing λ. (Hint: Exercise 4, Corollary 3.2.15, and Theorem 3.4.16(ii).)

Part II

Affine PI-Algebras

Chapter 4

The Braun-Kemer-Razmyslov Theorem

We turn now to one of the major theorems in the structure of affine PI-algebras. Recall Definition 1.1.14 of a Jacobson ring.

Theorem 4.0.1 (The Braun-Kemer-Razmyslov Theorem). *The Jacobson radical* $\mathrm{Jac}(A)$ *of any affine PI-algebra* A *over a commutative Jacobson ring is nilpotent.*

The Razmyslov-Kemer-Braun Theorem has an interesting history, having been proved in several stages. The nilpotence of $\mathrm{Jac}(A)$ is well known for finite dimensional algebras, and is used in the proof of Wedderburn's Principle Theorem, quoted in Theorem 2.5.8. Amitsur [Am57, Theorem 5], generalizing the weak Nullstellensatz, proved that if A is affine over a commutative Jacobson ring, then $\mathrm{Jac}(A)$ is nil. In particular, A is a Jacobson ring. (Later, Amitsur and Procesi [AmPr66, Corollary 1.3] proved that $\mathrm{Jac}(A)$ is locally nilpotent.) Thus, it remained to prove that every nil ideal of A is nilpotent. It was soon proved that this does hold for weakly representable algebras, see Proposition 1.6.34. However, Small [Sm71] showed the existence of an affine

149

PI-algebra which cannot be embedded into any matrix algebra, and proving Jac(A) nilpotent for affine PI-algebras over a field turned out to be very challenging. Thus, the theorem by Razmyslov [Raz74a] was a major breakthrough in this area.

Razmyslov [Raz74a] (also cf. Schelter [Sch78]) proved that Jac(A) is nilpotent for any affine PI-algebra that satisfies the identities of $n \times n$ matrices for some n. Schelter's proof, given in Exercises 3,4, is an elegant application of Shirshov's Height Theorem. Also using Shirshov's Height Theorem, Razmyslov [Raz74a] proved:

Theorem 4.0.2 (Razmyslov). *If an affine algebra A over a field satisfies a Capelli identity, then its Jacobson radical Jac(A) is nilpotent.*

Although Razmyslov's theorem was given originally in characteristic zero, he later found a proof that works in any characteristic. As we shall see, the same ideas yield the parallel result:

Theorem 4.0.3. *Let A be an affine algebra over a commutative Noetherian ring C. If A satisfies a Capelli identity, then any nil subring (without 1) of A is nilpotent.*

Following Razmyslov's theorem, Kemer [Kem80] then proved

Theorem 4.0.4 ([Kem80]). *In characteristic zero, any affine PI-algebra satisfies some Capelli identity.*

Thus, Kemer completed the proof of the following theorem:

Theorem 4.0.5 (Kemer-Razmyslov). *If A is an affine PI-algebra over a field F of characteristic zero, then its Jacobson radical Jac(A) is nilpotent.*

Then Braun [Br82], using a clever series of reduction to Azumaya algebras, proved the following result, which together with the Amitsur-Procesi Theorem immediately yields Theorem 4.0.1:

Theorem 4.0.6. *Any nil ideal of an affine PI-algebra over an arbitrary commutative Noetherian ring is nilpotent.*

Braun's qualitative proof was also presented in [Ro88b, Theorem 6.3.39], and a detailed exposition, by L'vov is available (unpublished) in Russian. A sketch of Braun's proof is also given in [AmSm93, Theorem 3.1.1].

Meanwhile, Kemer [Kem95] proved:

Theorem 4.0.7. *If A is a PI-algebra (not necessarily affine) over a field F of characteristic $p > 0$, then A satisfies some Capelli identity.*

Together with a characteristic-free proof of Razmyslov's theorem 4.0.2 due to Zubrilin [Zub95b], Kemer's Theorems 4.0.4 and 4.0.7 yield another proof of the Braun-Kemer-Razmyslov Theorem 4.0.1. The paper [Zub95b] is given in rather general circumstances, with some non-standard terminology. Zubrilin's method was given in the first edition, although it glosses over a key point (given here as Lemma 4.3.4), so a complete combinatoric proof had not yet appeared in print with all the details. Furthermore, full combinatoric details were provided in the first edition only in characteristic 0 because the conclusion of the proof required Kemer's difficult Theorem 4.0.7. We need the special case, which we call "Kemer's Capelli Theorem," that every affine PI-algebra A over an arbitrary field satisfies some Capelli identity, proved in §5.1. It should be noted that every proof that we have cited of the Braun-Kemer-Razmyslov Theorem ultimately utilizes an idea of Razmyslov defining a module structure on generalized polynomials with coefficients in the base ring, but we cannot locate full details of its implementation in the literature. One of our objectives is to provide these details.

In this exposition, we give a characteristic-free proof based on the approach of Razmyslov (as refined by Zubrilin) and Kemer . We prove that $\mathrm{Jac}(A)$ is nilpotent for any affine algebra A satisfying a Capelli identity, leading us to Kemer's Capelli Theorem, which we prove twice in the next chapter. First we introduce the more historic argument of Kemer, which introduces two important techniques: The use of "sparse" identities, and the "pumping procedure" developed by Belov. Then we turn to representation theory for the second argument.

Aside from the intrinsic interest in having such a proof available of this important theorem (and characteristic-free), these methods generalize easily to nonassociative algebras, and in Volume II we use this approach as a framework for the nonassociative PI-theory, as initiated by Zubrilin. (The proofs are nearly the same, but the statements are somewhat more complicated.)

In the original edition we emphasized the case where the base ring C is a field, and one can prove Theorem 4.0.1 directly by an induction argument without subdividing it into Theorem 4.0.6 and the weak Nullstellensatz, but the general case where C is a Noetherian Jacobson ring does not require much more work, once one has the easy commutative case given in Proposition 1.6.31.

Small has pointed out that Theorems 4.0.6 and 4.0.1 actually are equivalent, in view of a trick of [ReSS82], cf. Exercise 1.

4.1 Structure of the Proof

We assume that A is an affine C-algebra satisfying the $n+1$ Capelli identity c_{n+1} (but not necessarily c_n), and we induct on n: if such A satisfies c_n then we assume that $\mathrm{Jac}(A)$ is nilpotent, and we prove this for c_{n+1}.

We might as well assume that the algebra A has a unit element 1, by means of the standard construction described in Remark 1.1.1 of adjoining a unit element.

The same argument shows that any nil ideal N of an affine algebra A over a Noetherian ring is nilpotent, yielding Theorem 4.0.3. For this result we would replace $\mathrm{Jac}(A)$ by N throughout our sketch.

When working in the free associative algebra, it now is convenient to use different letters to denote different sets of indeterminates (according to their function in the proof), so now we write $C\{X, Y, Z\}$ for the free algebra over the base ring C, in finitely many noncommuting indeterminates $X = \{x_1, x_2, \dots\}$ (which we always notate), as well as finitely many noncommuting indeterminates $Y = \{y_1, \dots, y_n\}$ (which we sometimes notate), and Z containing $\{z_1, \dots\}$ (possibly infinite), which we often do not notate, and containing one extra indeterminate t for further use. The x and y indeterminates play special roles and need to be treated separately. We write $C\{Z\}$ for the free subalgebra generated by the z_i and t, omitting the x and y indeterminates.

1. The induction starts with $n = 1$. Then $n + 1 = 2$, and any algebra satisfying c_2 is commutative. This classical case was given in Remark 1.6.30 and Proposition 1.6.31.

2. Next is the *representable* case, which was known well before Razmyslov's Theorem, and was done in Proposition 1.6.34.

3. Let $\mathcal{CAP}_n = T(c_n)$ denote the T-ideal generated by the Capelli polynomial c_n, and let $\mathcal{CAP}_n(A) \subseteq A$ be the ideal generated in A by the evaluations of c_n on A, so $A/\mathcal{CAP}_n(A)$ satisfies c_n. Therefore, by induction on n, $\mathrm{Jac}(A/\mathcal{CAP}_n(A))$ is nilpotent. Hence there exists q such that
$$\mathrm{Jac}(A)^q \subseteq \mathcal{CAP}_n(A), \quad \text{so} \quad \mathrm{Jac}(A)^{2q} \subseteq (\mathcal{CAP}_n(A))^2.$$

4. Following Zubrilin [Zub95b], we work over an arbitrary commutative base ring C (which need not even be Noetherian), and obtain a Cayley-Hamilton type theorem for algebras satisfying a Capelli identity.

5. For any algebra A we introduce the ideal $\mathcal{I}_{n,A} \subset A[\xi_{n,a} : a \in A]$, for commuting indeterminates $\xi_{n,a}$, which provides "generic" integrality relations for elements of A.

6. Let $\overline{C\{X, Y, Z\}} := C\{X, Y, Z\}/\mathcal{CAP}_{n+1}$ denote the relatively free algebra of c_{n+1}. In Section 4.3 we construct the key $\overline{C\{Z\}}$-module $\overline{\mathcal{M}} \subset \overline{C\{X, Y, Z\}}$, which contains the image of the "double Capelli" polynomial
$$
f = z_{2n+1} c_n(x_1, \dots, x_n, z_1, \dots, z_n) z_{2n+2} \cdot \\
\cdot c_n(y_1, \dots, y_n, z_{n+1}, \dots, z_{2n}) z_{2n+3}.
\tag{4.1}
$$
But $\delta_{k,h}^{(n)}$ induces a well-defined map $\bar{\delta}_{k,h}^{(n)}$ on $\overline{\mathcal{M}}$ via Lemma 4.3.3. A combinatoric argument given in Proposition 4.3.9 applied to $\overline{C\{X, Y, Z\}}$ (together with substitutions) shows that $\mathcal{I}_{n, \overline{C\{X,Y,Z\}}} \cdot \overline{\mathcal{M}} = 0$.

7. We introduce the **obstruction to integrality** $\mathrm{Obst}_n(A) = A \cap \mathcal{I}_{n,A}$ and show that $A/\mathrm{Obst}_n(A)$ can be embedded into an algebra finite over an affine central F-subalgebra; hence $\mathrm{Jac}(A/\mathrm{Obst}_n(A))$ is nilpotent. This means that there exists m such that

$$\mathrm{Jac}(A)^m \subseteq \mathrm{Obst}_n(A).$$

The proof of this step applies Shirshov's Height Theorem 2.2.2.

8. We prove that $\mathrm{Obst}_n(A) \cdot (\mathcal{CAP}_n(A))^2 = 0$. This is obtained from Step 6 via a sophisticated specialization argument involving generalized polynomials and **relatively free products,** described in §1.9.2.

9. We put the pieces together. When C is Jacobson, Step 3 shows that

$$\mathrm{Jac}(A)^q \subseteq \mathcal{CAP}_n(A)$$

for some q, and Step 5 shows that $\mathrm{Jac}(A)^m \subseteq \mathrm{Obst}_n(A)$ for some m. Hence

$$\mathrm{Jac}(A)^{2q+m} \subseteq \mathrm{Obst}_n(A) \cdot (\mathcal{CAP}_n(A))^2 = 0,$$

which completes the proof of Theorem 4.0.2.

When C is Noetherian, any nil subring of C is nilpotent, so the analogous argument shows that $N^m \subseteq \mathrm{Obst}_n(A)$ for some m. Hence

$$N^{2q+m} \subseteq \mathrm{Obst}_n(A) \cdot (\mathcal{CAP}_n(A))^2 = 0.$$

At this stage, many steps involve concepts not yet defined, and so are difficult to comprehend. At the end of the proof we recapitulate the proof more explicitly, together with a discussion of its basic ideas.

As already noted, Steps 1 and 2 are standard, cf. Propositions 1.6.34 and 1.6.33. But since not every affine PI-algebra satisfies the hypotheses of Proposition 1.6.34, we must proceed further.

4.2 A Cayley-Hamilton Type Theorem

Our main tool is a powerful abstract version of the Cayley-Hamilton Theorem, which will put the characteristic closure in a much broader framework. For notational convenience, we let Y and Z denote extra sets of indeterminates, whose elements are designated as $y_1, y_2, \ldots$ and $z_1, z_2, \ldots$ but we also designate one extra indeterminate of Z as t.

We start with a polynomial $f := f(\vec{x}, \vec{y}, \vec{z}) \in C\{X, Y, Z\}$, where each vector denotes a collection of indeterminates appearing in f. Here, $\vec{x}$ denotes a fixed set of n indeterminates of f which may be $\{x_1, \ldots, x_n\}$ but not necessarily. Also $\vec{y} \subset Y$ and $\vec{z} \subset Z$, which may be empty.

Throughout, we denote the relatively free algebra $C\{X, Y, Z\}/\mathcal{CAP}_{n+1}$ by $\overline{C\{X, Y, Z\}}$, and the image of $f \in C\{X, Y, Z\}$ in $\overline{C\{X, Y, Z\}}$ by $\bar{f}$.

4.2.1   The operator $\delta_{k,t}^{(\vec{x},n)}$

The motivation for our next definition comes from (1.22) of Chapter 1.

Definition 4.2.1. *Let* $f(\vec{x}, \vec{y}, \vec{z}) \in C\{X, Y, Z\}$ *be* n-linear, *where* $\vec{x} = \{x_{i_1}, \ldots, x_{i_n}\}$. *We write*

$$\delta_{k,t}^{(\vec{x},n)}(f) = \sum_{1 \leq i_1 < \cdots < i_k \leq n} f(x_1, \ldots, x_n, \vec{y}, \vec{z})\,|_{x_{i_j} \to t x_{i_j}} =$$

$$= \sum_{1 \leq i_1 < \cdots < i_k \leq n} f(x_1, \ldots, t x_{i_1}, \ldots, t x_{i_k}, \ldots, x_n, \vec{z}) \in C\{X, Y, Z, t\},$$

where t *is an extra noncommutative indeterminate.*

Remark 4.2.2. *An equivalent formulation: Take* $0 \leq k \leq n$, *and define*

$$\check{f} = f((t+1)x_{i_1}, \ldots, (t+1)x_{i_n}, \vec{y}, \vec{z}).$$

Then $\delta_{k,t}^{(\vec{x},n)}(f) := \delta_{k,t}^{(\vec{x},n)}(f)(x_{i_1}, \ldots, x_{i_n}, \vec{y}, \vec{z}, t)$ *is the homogeneous component of* $\check{f}$ *of degree* k *in the indeterminate* t.

For example let $n = 2$ *and* $f = x_1 x_2$. *Then*

$$(t+1)x_1(t+1)x_2 = t x_1 t x_2 + t x_1 x_2 + x_1 t x_2 + x_1 x_2.$$

Hence $\delta_{0,t}^{(x,2)}(x_1 x_2) = x_1 x_2$, $\delta_{1,t}^{(x,2)}(x_1 x_2) = t x_1 x_2 + x_1 t x_2$, *and* $\delta_{2,t}^{(x,2)}(x_1 x_2) = t x_1 t x_2$.

More generally, for any $h \in C\{t\}$ *we write*

$$\delta_{k,h}^{(\vec{x},n)}(f) := \delta_{k,h}^{(\vec{x},n)}(f)(x_1, \ldots, x_n, \vec{y}, \vec{z}, h), \tag{4.2}$$

i.e., the specialization of $\delta_{k,t}^{(\vec{x},n)}(f)$ *under* $t \mapsto h$.

Remark 4.2.3.

1. *Let* $v = 1 + \varepsilon t$, *where* ε *is a central indeterminate. If*

$$f(v x_1, \ldots, v x_n, \vec{y}, \vec{z}) = \sum_{k=0}^{n} \left(\delta_{k,t}^{(\vec{x},n)}(f(x_1, \ldots, x_n, \vec{y}, \vec{z})) \right) \cdot \varepsilon^k$$

is alternating in $x_1, \ldots, x_n$, *it follows that each* $\delta_{k,t}^{(\vec{x},n)}(f(x_1, \ldots, x_n, \vec{z})$

is alternating in $x_1, \ldots, x_n$. Since $\mathcal{CAP}_n$ is generated as a T-ideal by polynomials alternating in $x_1, \ldots, x_n$, we have

$$\delta_{k,t}^{(\vec{x},n)}(\mathcal{CAP}_n) \subseteq \mathcal{CAP}_n \quad \text{and} \quad \delta_{k,t}^{(\vec{x},n)}(\mathcal{CAP}_{n+1}) \subseteq \mathcal{CAP}_{n+1}.$$

Similarly, let $\mathcal{CAP}_{n,\vec{x}}$ denote the subset of $\mathcal{CAP}_n$ of polynomials which are n-linear in $\vec{x}$. Then

$$\delta_{k,t}^{(\vec{x},n)}(\mathcal{CAP}_{n,\vec{x}}) \subseteq \mathcal{CAP}_{n,\vec{x}} \quad \text{and} \quad \delta_{k,t}^{(\vec{x},n)}(\mathcal{CAP}_{n+1,x}) \subseteq \mathcal{CAP}_{n+1,x}.$$

2. *The results proved for the indeterminate t specialize to an arbitrary polynomial h, and thus can be formulated for h.*

Lemma 4.2.4. *The $\delta_{k,t}^{(\vec{x},n)}(f)$-operator is functorial, in the sense that if $\vec{a} = (a_1, \ldots, a_m) \in A$ and $h(\vec{a}) = h'(\vec{a})$, then $\delta_{k,h}^{(\vec{x},n)}(f)(\vec{a}) = \delta_{k,h'}^{(\vec{x},n)}(f)(\vec{a})$.*

Proof. We get the same result in Definition 4.2.1 by specializing t to h and then to $\vec{a}$, as we get by specializing t to h', and then to $\vec{a}$. $\square$

This observation is needed in our later specialization arguments.

4.2.2 Zubrilin's Proposition

Our next main goal is Proposition 4.3.9. Let us prepare some terminology used there.

Throughout, the relatively free algebra $C\{X,Y,Z\}/\mathcal{CAP}_{n+1}$ is denoted as $\overline{C\{X,Y,Z\}}$, and the image of $f \in C\{X,Y,Z\}$ in $\overline{C\{X,Y,Z\}}$ as $\bar{f}$.

Remark 4.2.5. *If A satisfies c_{n+1}, then any algebra homomorphism*

$$\varphi : C\{X,Y,Z\} \to A$$

naturally induces an algebra homomorphism

$$\overline{\varphi} : \overline{C\{X,Y,Z\}} \to A$$

given by $\overline{\varphi}(\bar{f}) = \varphi(f)$, since $\mathcal{CAP}_{n+1} \subseteq \ker \varphi$.

We recall Remark G from §1.5: Given $f(x_1, \ldots, x_{n+1}, \vec{y}, \vec{z})$ multilinear in $x_1, \ldots, x_{n+1}$ and n-alternating, we define

$$\tilde{f} = \sum_{k=1}^{n} f(x_1, \ldots, x_{k-1}, x_{n+1}, x_{k+1}, \ldots, x_n, x_k, \vec{y}, \vec{t}) \tag{4.3}$$

Then $\tilde{f}$ is $(n+1)$-alternating in $x_1, \ldots, x_{n+1}$.

Proposition 4.2.6. *Suppose* $f(x_1, ..., x_{n+1}, \vec{y}, \vec{z})$ *is multilinear on* $x_1, ..., x_{n+1}$ *and also alternates on* $x_1, ..., x_n$. *Then*

$$\sum_{j=0}^{n}(-1)^j \delta_{j,t}^{(\vec{x},n)}\left(f|_{x_{n+1}\to t^{n-j}x_{n+1}}\right) \equiv 0 \text{ modulo } \mathcal{CAP}_{n+1}.$$

Proof. Throughout we work modulo $\mathcal{CAP}_{n+1}$ and delete $\vec{x}, \vec{y}$ from the notation. It easy to see that $\delta_{j,t}^{(\vec{x},n)}(f)$ is multilinear on $x_1, ..., x_{n+1}$ and alternates on $x_1, ..., x_n$. Therefore,

$$\widetilde{\delta_{j,t}^{(\vec{x},n)}}(f) = \left(\delta_{j,t}^{(\vec{x},n)}(f)\right) - \sum_{k=1}^{n}\left(\delta_{j,t}^{(\vec{x},n)}(f)\right)|_{x_k \leftrightarrow x_{n+1}}$$

is zero modulo $\mathcal{CAP}_{n+1}$.

Substituting $t^{n-j}x_{n+1}$ for x_{n+1} yields:

$$\left(\delta_{j,t}^{(\vec{x},n)}(f)\right)|_{x_{n+1}\to t^{n-j}x_{n+1}}$$

$$\equiv \sum_{k=1}^{n}\left(\left(\delta_{j,t}^{(\vec{x},n)}(f)\right)|_{x_k \leftrightarrow x_{n+1}}\right)|_{x_{n+1}\to t^{n-j}x_{n+1}} \pmod{\mathcal{CAP}_{n+1}}$$

$$= \sum_{k=1}^{n}\left(\delta_{j,t}^{(\vec{x},n)}(f)\right)|_{x_{n+1}\to x_k, \ x_k \to t^{n-j}x_{n+1}}.$$

Since $\delta_{j,t}^{(\vec{x},n)}(f)$ does not affect x_{n+1}, we get

$$\left(\delta_{j,t}^{(\vec{x},n)}(f)\right)|_{x_{n+1}\to t^{n-j}x_{n+1}} = \delta_{j,t}^{(\vec{x},n)}\left(f|_{x_{n+1}\to t^{n-j}x_{n+1}}\right),$$

yielding:

$$\delta_{j,t}^{(\vec{x},n)}\left(f|_{x_{n+1}\to t^{n-j}x_{n+1}}\right) \equiv$$

$$\sum_{k=1}^{n}\left(\delta_{j,t}^{(\vec{x},n)}(f)\right)|_{x_{n+1}\to x_k, \ x_k \to t^{n-j}x_{n+1}} \pmod{\mathcal{CAP}_{n+1}}.$$

So we need to show that: So we need to show that:

$$\sum_{j=0}^{n}\sum_{k=1}^{n}(-1)^j\left(\delta_{j,t}^{(\vec{x},n)}(f)\right)|_{x_{n+1}\to x_k, \ x_k \to t^{n-j}x_{n+1}} \in \mathcal{CAP}_{n+1}.$$

Therefore, after changing the order of summation, it is clear that it is enough to show that for every k:

$$\sum_{j=1}^{n}(-1)^j\left(\delta_{j,t}^{(\vec{x},n)}(f)\right)|_{x_{n+1}\to x_k, \ x_k \to t^{n-j}x_{n+1}} = 0.$$

For simplicity we show this only for $k = 1$: The following equality holds:

$$\left(\delta_{j,t}^{(\vec{x},n)}(f)\right)\Big|_{x_{n+1}\to x_1,\; x_1\to t^{n-j}x_{n+1}}$$

$$= \underbrace{\sum_{1<i_1<\cdots<i_j\leq n}\left(f_{i_1,\ldots,i_j}\right)\Big|_{x_1\to x_{n+1},\; x_1\to t^{n-j}x_{n+1}}}_{g_j}$$

$$+ \underbrace{\sum_{1=i_1,i_2<\cdots<i_j\leq n}\left(f_{i_1,\ldots,i_j}\right)\Big|_{x_{n+1}\to x_1,\; x_1\to t^{n-j}x_{n+1}}}_{h_j}$$

where

$$f_{i_1,\ldots,i_j} = f\Big|_{x_{i_1}\to tx_{i_1},\ldots,x_{i_j}\to tx_{i_j}}.$$

Thus,

$$g_j = \sum_{1<i_1<\cdots<i_j\leq n} f\Big|_{x_1\to t^{n-j}x_{n+1},\; x_{i_1}\to tx_{i_1},\ldots,x_{i_t}\to tx_{i_j},\; x_{n+1}\to x_1}$$

$$h_j = \sum_{1<i_2<\cdots<i_j\leq n} f\Big|_{x_1\to t^{n-j+1}x_{n+1},\; x_{i_2}\to tx_{i_2},\ldots,x_{i_t}\to tx_{i_j},\; x_{n+1}\to x_1}.$$

Observing that $g_n = h_0 = 0$ and $h_{j+1} = g_j$ for $t = 0\ldots n-1$, we conclude:

$$\sum_{j=1}^{n}(-1)^j\left(\delta_{j,t}^{(\vec{x},n)}(f)\right)\Big|_{x_1\to x_{n+1},\; x_1\to t^{n-j}x_{n+1}} = \sum_{j=0}^{n}(-1)^j(h_j + g_j) = 0.$$

$\square$

4.2.3   Commutativity of the operators $\delta_{k,h_j}^{(n)}$ modulo $\mathcal{CAP}_{n+1}$

We need a special sort of alternating polynomials.

Definition 4.2.7. *A polynomial* $f(x_1,\ldots,x_n;y_1,\ldots,y_n;\vec{z})$, *where* $\vec{z}$ *denotes other possible indeterminates, is **doubly alternating** if f is linear and alternating in* $x_1,\ldots,x_n$ *and* $y_1,\ldots,y_n$.

Our main example is the **double Capelli** polynomial

$$\mathrm{DCap}_n = z_{2n+1}c_n(x_1,\ldots,x_n;z_1,\ldots,z_n)z_{2n+2}\cdot$$
$$\cdot c_n(y_1,\ldots,y_n;z_{n+1},\ldots,z_{2n})z_{2n+3}. \tag{4.4}$$

Definition 4.2.8. *An **RZ polynomial** (for Razmyslov and Zubrilin) for an algebra A is a non-identity* $f(x_1,\ldots,x_n,y_1,\ldots,y_n\vec{z})$ *that is doubly alternating in* $x_1,\ldots,x_n$ *and in* $y_1,\ldots,y_n$, *but such that any alternator of f in $n+1$ indeterminates from* $x_1,\ldots,x_n,y_1,\ldots,y_n$ *(cf. Definition 1.2.31) is an identity of A.*

The case needed for this chapter is Definition 4.2.7. However, Zubrilin's theory can be made applicable for all affine PI-algebras, as we will see in Proposition 6.9.5.

The theory hinges on Proposition 4.2.9, which is needed in order to justify the specialization argument of §4.6. It is motivated by the following application of Theorem I of §1.5:

$$\delta_{k,t}^{(\vec{x},n)}(c_n(x_1,\ldots,x_n;\vec{y})) = \sum_{1\leq i_1<\cdots<i_k\leq n} c_n(x_1,\ldots,tx_{i_1},\ldots,tx_{i_k},\ldots,x_n;\vec{y})$$

$$= \alpha_k \cdot c_n(x_1,\ldots,x_n;\vec{y}),$$

and the coefficients α_k are independent of the particular indeterminates $x_1,\ldots,x_n$. Proposition 4.2.9 below displays a similar phenomenon.

In case $f = f(\vec{x},\vec{y},\vec{z})$ also involves indeterminates $\vec{y}$, which say are $y_1,\ldots,y_n$, we still have

$$\delta_{k,t}^{(\vec{x},n)}(f) = \sum_{1\leq i_1<\cdots<i_k\leq n} f\,|_{x_{i_j}\to tx_{i_j}},$$

indicating that the other indeterminates $y_1,\ldots,y_n$ remain fixed. But analogously, we could use the set Y instead of X in our definition, obtaining

$$\delta_{k,t}^{(\vec{y},n)}(f) = \sum_{1\leq i_1<\cdots<i_k\leq n} f\,|_{y_{i_j}\to ty_{i_j}},$$

with the indeterminates $x_1,\ldots,x_n$ fixed.

Proposition 4.2.9.

(i) *Let $f(x_1,\ldots,x_n;y_1,\ldots,y_n,\vec{z})$ be doubly alternating in $x_1,\ldots,x_n$ and in $y_1,\ldots,y_n$. Then for any polynomial h,*

$$\delta_{k,h}^{(\vec{x},n)}(f) \equiv \delta_{k,h}^{(\vec{y},n)}(f) \quad mod\,\mathcal{CAP}_{n+1}; \tag{4.5}$$

namely,

$$\sum_{1\leq i_1<\cdots<i_k\leq n} f\,|_{x_{i_j}\to hx_{i_j}} \equiv \sum_{1\leq i_1<\cdots<i_k\leq n} f\,|_{y_{i_j}\to hy_{i_j}} \quad mod\,\mathcal{CAP}_{n+1}.$$

(ii) *Let $f(x_1,\ldots,x_n;y_1,\ldots,y_n,\vec{z})$ be an RZ-polynomial for A in $x_1,\ldots,x_n$ and in $y_1,\ldots,y_n$, where $\mathcal{CAP} \in \mathrm{id}(A)$. Then for any polynomial h,*

$$\delta_{k,h}^{(\vec{x},n)}(f) \equiv \delta_{k,h}^{(\vec{y},n)}(f) \quad mod\,\mathcal{CAP}_{n+1}. \tag{4.6}$$

Remark 4.2.10. *It follows from Proposition 4.2.9 that*

$$\delta_{k,h}^{(\vec{x},n)}(f) - \delta_{k,h}^{(\vec{y},n)}(f) \in \mathcal{CAP}_{n+1}$$

whenever $f \in \overline{\mathcal{M}}$, so working modulo $\mathcal{CAP}_{n+1}$ we can suppress $\vec{x}$ and $\vec{y}$ in the notation, writing $\bar{\delta}_{k,h}^{(n)}(\bar{f})$ for $\bar{\delta}_{k,h}^{(\vec{x},n)}(\bar{f})$.

4.2.3.1   The connection to the group algebra of S_n

We are about to present the proof of the crucial Proposition 4.2.9. Since the proofs of the two parts are analogous and we only need (i) here, we only do full details for (i), but indicate the changes in (ii). We need considerable preparation for the proof, starting with the basic action of the group algebra S_n on the multilinear polynomials, given in Definition 3.1.1.

Here is a combinatorial identity of interest of its own.

Consider two disjoint sets $X \cap Y = \emptyset$, each of cardinality n, and the symmetric group $S_{2n} = S_{X \cup Y}$ acting on $X \cup Y$. For each subset $Z \subseteq X$, we define an element $P(Z) \in C[S_{2n}]$ as follows:

$$P(Z) = \sum_{\sigma(Z) \subseteq Y} \text{sgn}(\sigma) \cdot \sigma.$$

In particular

$$P(\emptyset) = \sum_{\sigma \in S_{2n}} \text{sgn}(\sigma) \cdot \sigma.$$

Proposition 4.2.11.

$$\sum_{Z \subseteq X} (-1)^{|Z|} P(Z) = \sum_{\sigma(X)=X} \text{sgn}(\sigma) \cdot \sigma. \tag{4.7}$$

Proof. Let $\sigma \in S_{2n}$ and let a_σ (resp. b_σ) be the coefficient of σ on the l.h.s. (resp. r.h.s.) of (4.7). We show that $a_\sigma = b_\sigma$.

Let $Z(\sigma) = \sigma^{-1}(Y) \cap X$ be the largest subset $Z \subseteq X$ such that $\sigma(Z) \subseteq Y$. Note that $\sigma(X) = X$ if and only if $Z(\sigma) = \emptyset$. Therefore

$$b_\sigma = \text{sgn}(\sigma) \text{ if } Z(\sigma) = \emptyset \quad \text{and} \quad b_\sigma = 0 \text{ if } Z(\sigma) \neq \emptyset, \tag{4.8}$$

We claim that

$$a_\sigma = \text{sgn}(\sigma) \cdot \sum_{Z \subset Z(\sigma)} (-1)^{|Z|}.$$

To show this, recall that

$$l.h.s = \sum_{Z \subseteq X} (-1)^{|Z|} \sum_{\sigma(Z) \subseteq Y} \text{sgn}(\sigma) \cdot \sigma.$$

In $P(Z)$ the coefficient of σ is $\text{sgn}(\sigma)$ if $Z \subseteq Z(\sigma)$ (since then $\sigma(Z) \subseteq Y$), and is zero if $Z \not\subseteq Z(\sigma)$ (since if $\sigma(Z) \subseteq Y$, then $\sigma(Z \cup Z(\sigma)) \subseteq Y$, contradicting the maximality of $Z(\sigma)$). It follows that as claimed,

$$a_\sigma = \text{sgn}(\sigma) \cdot \sum_{Z \subseteq Z(\sigma)} (-1)^{|Z|}.$$

It is well known that $\sum_{Z \subseteq Z(\sigma)} (-1)^{|Z|} = 1$ when $Z(\sigma) = \emptyset$ and $= 0$ otherwise. Therefore

$$a_\sigma = \operatorname{sgn}(\sigma) \quad \text{if} \quad Z(\sigma) = \emptyset, \quad \text{and} \quad a_\sigma = 0 \quad \text{if} \quad Z(\sigma) \neq \emptyset. \qquad (4.9)$$

The proof now follows by comparing (4.8) with (4.9).

$\square$

Lemma 4.2.12. *Let* $f(x_1, \ldots, x_n, y_1, \ldots, y_n, \vec{z})$ *be doubly alternating. Then, modulo* $\mathcal{CAP}_{n+1}$,

$$f(x_1, \ldots, x_n, y_1, \ldots, y_n, \vec{y}, \vec{z}) \equiv f(y_1, \ldots, y_n, x_1, \ldots, x_n, \vec{y}, \vec{z}).$$

Proof. Let $X = \{x_1, \ldots, x_n\}$ and $Y = \{y_1, \ldots, y_n\}$. Then $|X| = |Y| = n$ and $X \cap Y = \emptyset$, and we identify $S_{2n} = S_{X \cup Y}$. Let $M = \{x_{i_1}, \ldots, x_{i_k}\} \subseteq X$, with $1 \leq i_1 < \cdots < i_k \leq n$, and $N = \{y_{j_1}, \ldots, y_{j_k}\} \subseteq Y$, with $1 \leq j_1 < \cdots < j_k \leq n$. Thus, $|M| = |N| = k \leq n$. M will play the role of Z in Proposition 4.2.11. We consider permutations $\sigma \in S_{2n}$ with $\sigma(M) = N$. Define the permutation

$$\tau_{MN} = (x_{i_1}, y_{j_1}) \cdots (x_{i_k}, y_{j_k}).$$

Since $M \cap N = \emptyset$, τ_{MN} has order 2 in S_{2n}, and satisfies $\operatorname{sgn}(\tau_{MN}) = (-1)^k$. If $M = X$, then $N = Y$ and $\operatorname{sgn}(\tau_{MN}) = \operatorname{sgn}(\tau_{XY}) = (-1)^n$. Moreover, $\tau_{MN}(M) = N$ and $\tau_{MN}(N) = M$.

Next, we define

$$T_{MN} = \sum_{\pi(M)=N} \operatorname{sgn}(\pi) \cdot \pi \in C[S_n].$$

Let $\rho = \tau_{MN} \cdot \pi$, so that $\rho(M) = M$. Then $\pi = \tau_{MN} \cdot \rho$ and

$$T_{MN} = \operatorname{sgn}(\tau_{MN}) \cdot \tau_{MN} \cdot \left(\sum_{\rho(M)=M} \operatorname{sgn}(\rho) \cdot \rho \right).$$

But by Proposition 4.2.11,

$$\sum_{M \subseteq X} (-1)^{|M|} P(M) = \sum_{\sigma(X)=X} \operatorname{sgn}\sigma \cdot \sigma. \qquad (4.10)$$

If $M \subsetneq X$, then $P(M)$ is a sum of polynomials alternating on the same set of $2n - |M| \geq n + 1$ indeterminates, and hence is 0 modulo $\mathcal{CAP}_{n+1}$. Thus, modulo $\mathcal{CAP}_{n+1}$, the left-hand side of (4.10) equals the unique summand with $M = X$, which is

$$(-1)^n \sum_{\sigma(X)=Y} \operatorname{sgn}(\sigma) \cdot \sigma = (-1)^n \operatorname{sgn}(\tau_{XY}) \cdot \tau_{XY} \cdot \left(\sum_{\sigma(Y)=Y} \operatorname{sgn}(\sigma) \cdot \sigma \right)$$

$$= \tau_{XY} \cdot \left(\sum_{\sigma(Y)=Y} \operatorname{sgn}(\sigma) \cdot \sigma \right).$$

$$(4.11)$$

Since $\sigma(X) = X$ if and only if $\sigma(Y) = Y$, it follows that

$$\sum_{\sigma(Y)=Y} \mathrm{sgn}(\sigma) \cdot \sigma = \sum_{\sigma(X)=X} \mathrm{sgn}(\sigma) \cdot \sigma \equiv \tau_{XY} \cdot \left(\sum_{\sigma(Y)=Y} \mathrm{sgn}(\sigma) \cdot \sigma \right),$$

modulo $\mathcal{CAP}_{n+1}$.

Let us write

$$f(x_1, \ldots, x_n, y_1, \ldots, y_n; \vec{z}) := \left(\sum_{\sigma(Y)=Y} \mathrm{sgn}(\sigma) \cdot \sigma \right) h.$$

for a polynomial $h(x_1, \ldots, x_n, y_1, \ldots, y_n; \vec{z})$ multilinear in $x_1, \ldots, x_n, y_1, \ldots, y_n$. Then

$$\tau_{XY} \cdot \left(\sum_{\sigma(Y)=Y} \mathrm{sgn}(\sigma) \cdot \sigma \right) h = f(y_1, \ldots, y_n, x_1, \ldots, x_n; \vec{z}).$$

Again, since $\sigma(X) = X$ if and only if $\sigma(Y) = Y$, it follows that $f(x_1, \ldots, x_n, y_1, \ldots, y_n; \vec{z})$ is doubly alternating, proving that

$$f(x_1, \ldots, x_n, y_1, \ldots, y_n; \vec{z}) \equiv f(y_1, \ldots, y_n, x_1, \ldots, x_n; \vec{z}) \mod \mathcal{CAP}_{n+1},$$

as desired.

$\square$

Lemma 4.2.13. *Let $f(x_1, \ldots, x_n, y_1, \ldots, y_n, \vec{y}, \vec{z})$ be an RZ polynomial for A. Then, modulo $\mathrm{id}(A)$,*

$$f(x_1, \ldots, x_n, y_1, \ldots, y_n, \vec{z}) \equiv f(y_1, \ldots, y_n, x_1, \ldots, x_n, \vec{z}).$$

Proof. Same as for Lemma 4.2.12. $\square$

4.2.3.2 Proof of Proposition 4.2.9

We are ready to prove Proposition 4.2.9. We may assume that h is an indeterminate t. Recall that

$$\delta_{k,t}^{(\vec{x},n)} (f(x_1, \ldots, x_n, y_1, \ldots, y_n, \vec{z})) =$$

$$\sum_{1 \le i_1 < \cdots < i_k \le n} f(x_1, \ldots, x_n, y_1, \ldots, y_n, \vec{z})|_{x_{i_u} \mapsto z x_{i_u}}; \quad u = 1, \ldots, k,$$

$$(4.12)$$

and

$$\delta_{k,t}^{(\vec{y},n)}(f(x_1,\ldots,x_n,y_1,\ldots,y_n,\vec{z})) =$$

$$\sum_{1\le i_1<\cdots<i_k\le n} f(x_1,\ldots,x_n,y_1,\ldots,y_n,\vec{z})|_{y_{i_u}\mapsto zy_{i_u}}; \quad u=1,\ldots,k.$$

(4.13)

Let $t' = 1 + \varepsilon t$, ε being a central indeterminant. Then clearly

$$f(t'x_1,\ldots,t'x_n,y_1,\ldots,y_n,\vec{z}) =$$

(4.14)

$$\sum_{k=0}^{n} \varepsilon^k \cdot \delta_{k,t}^{(\vec{x},n)}(f(x_1,\ldots,x_n,y_1,\ldots,y_n,\vec{z}))$$

and

$$f(x_1,\ldots,x_n,t'y_1,\ldots,t'y_n,\vec{z}) =$$

$$\sum_{k=0}^{n} \varepsilon^k \cdot \delta_{k,t}^{(\vec{y},n)}(f(x_1,\ldots,x_n,y_1,\ldots,y_n,\vec{z})).$$

(4.15)

By Equations (4.14) and (4.15) it is enough to show that

$$f(t'x_1,\ldots,t'x_n,y_1,\ldots,y_n,\vec{z}) \equiv f(x_1,\ldots,x_n,t'y_1,\ldots,t'y_n,\vec{z}),$$

modulo $\mathcal{CAP}_{n+1}$.

Let

$$g_1(x_1,\ldots,x_n,y_1,\ldots,y_n,\vec{z}) = f(t'x_1,\ldots,t'x_n,y_1,\ldots,y_n,\vec{z})$$

and

$$g_2(x_1,\ldots,x_n,y_1,\ldots,y_n,\vec{z}) = f(x_1,\ldots,x_n,t'y_1,\ldots,t'y_n,\vec{z}).$$

We have to show that

$$g_1 \equiv g_2 \mod \mathcal{CAP}_{n+1}.$$

Denote $x_i' = t'x_i$, $y_i' = t'y_i$; $i = 1,\ldots,n$. Then

$$g_1(x_1,\ldots,x_n,y_1,\ldots,y_n,\vec{z}) = f(x_1',\ldots,x_n',y_1,\ldots,y_n,\vec{z})$$
$$\equiv f(y_1,\ldots,y_n,x_1',\ldots,x_n',\vec{z})$$
$$= g_2(y_1,\ldots,y_n,x_1,\ldots,x_n,\vec{z})$$
$$\equiv g_2(x_1,\ldots,x_n,y_1,\ldots,y_n,\vec{z}) \mod \mathcal{CAP}_{n+1}.$$

The congruences follow from Lemma 4.2.12 since both f and g_2 are doubly alternating. This concludes the proof of Proposition 4.2.9.

4.3 The Module $\mathcal{M}$ over the Relatively Free Algebra $\overline{C\{X,Y,Z\}}$ of c_{n+1}

At this stage, we need to use partial specializations and generalized polynomials, introduced in §1.9.

Definition 4.3.1. *Let $\mathcal{M}$ denote the $C\{Z\}$-submodule of $C\{X,Y,Z\}$ consisting of all doubly alternating polynomials (in $x_1 \ldots, x_n$, and in $y_1, \ldots, y_n$).*
$\overline{\mathcal{M}}$ denotes the image of $\mathcal{M}$ in $\overline{C\{X,Y,Z\}}$, i.e., the C-submodule of $\overline{C\{X,Y,Z\}}$ consisting of the images of all doubly alternating polynomials (in $x_1 \ldots, x_n$, and in $y_1, \ldots, y_n$).

Note that $\mathcal{M}$ is a set of RZ polynomials for $\overline{C\{X,Y,Z\}}$.

Remark 4.3.2. $\overline{\mathcal{M}}$ *is a $\overline{C\{t\}}$-submodule of $\overline{C\{X,Y,Z\}}$. Indeed, let $h \in C\{t\}$ and $f \in \mathcal{M}$. If either h or f is in $\mathcal{CAP}_{n+1}$ then $hf \in \mathcal{CAP}_{n+1}$; hence the product $\bar{h}\bar{f} = \overline{hf}$ is well-defined. Moreover, if $f = f(x_1, \ldots, x_n, y_1, \ldots, y_n, \vec{z})$ is doubly alternating in the x's and in the y's, and $h \in C\{\vec{z}\,\}$, then hf is doubly alternating in the x's and in the y's.*

We use $\overline{\mathcal{M}}$ instead of $\mathcal{M}$ because the following lemma enables us to apply $\bar{\delta}_{k,h}^{(n)}$.

Lemma 4.3.3.

(i) $\delta_{k,h}^{(n)}$ *induces a well-defined map $\bar{\delta}_{k,h}^{(n)} : \overline{\mathcal{M}} \to \overline{\mathcal{M}}$ given by $\bar{\delta}_{k,h}^{(n)}(\bar{f}) = \overline{\delta_{k,h}^{(\vec{x},n)}(f)}$.*

(ii) $\bar{\delta}_{k,h}^{(n)}$ *produces the same result using the indeterminates x or y.*

Proof. (i) If $f(x_1, \ldots, x_n, y_1, \ldots, y_n)$ and $g(x_1, \ldots, x_n, y_1, \ldots, y_n)$ are doubly alternating polynomials, with $\bar{f} = \bar{g}$, then $f - g \in \mathcal{CAP}_{n+1}$, so by Remark 4.2.3(1), $\delta_{k,h}^{(n)}(f-g) \in \mathcal{CAP}_{n+1}$ and hence $\overline{\delta_{k,h}^{(n)}(f-g)} = 0$. Therefore we have

$$0 = \overline{\delta_{k,h}^{(n)}(f-g)} = \overline{\delta_{k,h}^{(n)}(f)} - \overline{\delta_{k,h}^{(n)}(g)} = \bar{\delta}_{k,h}^{(n)}(\bar{f}) - \bar{\delta}_{k,h}^{(n)}(\bar{g}),$$

proving that $\bar{\delta}_{k,h}^{(n)}$ is well-defined.

(ii) The assertion follows from Remark 4.2.10, which shows that $\overline{\delta_{k,h}^{(\vec{x},n)}(f)} = \overline{\delta_{k,h}^{(\vec{y},n)}(f)}$. $\qquad\square$

Lemma 4.3.4. *Let* $f = f(x_1, \ldots, x_n; y_1, \ldots, y_n)$ *be doubly alternating in* $x_1, \ldots, x_n$ *and in* $y_1, \ldots, y_n$ *(and perhaps involving other indeterminates). Let* $1 \le k, \ell \le n$. *Then for any* $h_1, h_2 \in C\{t\}$,

$$\bar{\delta}_{k,h_1}^{(n)} \bar{\delta}_{\ell,h_2}^{(n)}(\bar{f}) = \bar{\delta}_{\ell,h_2}^{(n)} \bar{\delta}_{k,h_1}^{(n)}(\bar{f}). \tag{4.16}$$

Proof. Equation (4.16) claims that modulo $\mathcal{CAP}_{n+1}$,

(i)
$$\delta_{k,h_1}^{(\vec{x},n)} \delta_{\ell,h_2}^{(\vec{x},n)}(f) \equiv \delta_{\ell,h_2}^{(\vec{x},n)} \delta_{k,h_1}^{(\vec{x},n)}(f) \quad \text{and}$$

(ii)
$$\delta_{k,h_1}^{(\vec{x},n)} \delta_{\ell,h_2}^{(\vec{y},n)}(f) \equiv \delta_{\ell,h_2}^{(\vec{y},n)} \delta_{k,h_1}^{(\vec{x},n)}(f) \quad \text{and}$$

(iii)
$$\delta_{k,h_1}^{(\vec{y},n)} \delta_{\ell,h_2}^{(\vec{y},n)}(f) \equiv \delta_{\ell,h_2}^{(\vec{y},n)} \delta_{k,h_1}^{(\vec{y},n)}(f).$$

The middle equivalence (ii) is an obvious equality. The first and third equivalences are similar, and we prove the first. By Proposition 4.2.9, by (ii), and again by Proposition 4.2.9, modulo $\mathcal{CAP}_{n+1}$ we can write

$$\delta_{k,h_1}^{(\vec{x},n)} \delta_{\ell,h_2}^{(\vec{x},n)}(f) \equiv \delta_{k,h_1}^{(\vec{x},n)} \delta_{\ell,h_2}^{(\vec{y},n)}(f) \equiv \delta_{\ell,h_2}^{(\vec{y},n)} \delta_{k,h_1}^{(\vec{x},n)}(f) \equiv \delta_{\ell,h_2}^{(\vec{x},n)} \delta_{k,h_1}^{(\vec{x},n)}(f).$$

$\square$

Definition 4.3.5. *Given a subset* $W \subset A$, *for each* $a \in W$ *let* $\xi_{1,a}, \ldots, \xi_{n,a}$ *be* n *corresponding new commuting variables, and construct*

$$A[\xi_{n,W}] = A[\xi_{1,a}, \ldots, \xi_{n,a} : a \in W].$$

Let $\mathcal{I}_{n,W} \subseteq A[\xi_{n,W}]$ *be the ideal generated in* $A[\xi_{n,W}]$ *by the elements*

$$a^n + \xi_{1,a} a^{n-1} + \cdots + \xi_{n,a}, \quad a \in W,$$

namely

$$\mathcal{I}_{n,W} = \langle a^n + \xi_{1,a} a^{n-1} + \cdots + \xi_{n,a} : a \in W \rangle.$$

We start by taking $W = A$, but later want to restrict W to words of bounded length in the generators of A.

Remark 4.3.6. *In view of Proposition 4.2.9, the map* $\bar{\delta}_{k,h}^{(n)} : \overline{\mathcal{M}} \to \overline{\mathcal{M}}$ *of Lemma 4.3.3 yields an action of the C-algebra* $\overline{C\{t\}}[\xi_{n,\overline{C\{t\}}}]$ *on* $\overline{\mathcal{M}}$, *given by* $\xi_{k,h} f = \delta_{k,h}^{(n)}(f)$.

Because of its importance, we describe it explicitly:

Lemma 4.3.7. $\overline{\mathcal{M}}$ *becomes a* $\overline{C\{t\}}[\xi_{n,C\{t\}}]$-*module via the following action:*
Order the $\xi_{k,h}$ *as* $\xi_j = \xi_{k_j,h_j}$ *for* $1 \leq j < \infty$.
For a letter $\xi_j = \xi_{k_j,h_j}$, *define*

$$\xi_j \bar{f} = \bar{\delta}^{(n)}_{k_j,h_j}(\bar{f}),$$

and, inductively,

$$\xi_j^d \bar{f} = \bar{\delta}^{(n)}_{k_j,h_j}(\xi_j^{d-1}\bar{f}).$$

For a monomial $h = \xi_j^{d_j} \dots \xi_1^{d_1}$ *of degree* $d = d_1 + \dots + d_j$, *define*

$$hf = \xi_j^{d_j}(\xi_{j-1}^{d_{j-1}} \dots \xi_1^{d_1}\bar{f})$$

inductively on j.
Finally, define

$$\sum(c_i h_i)\bar{f} = \sum c_i(h_i\bar{f})$$

where $c_i \in C$ *and* h_i *are distinct monomials.*

Proof. The action is clearly well-defined, so we need to verify the associativity and commutativity of the action, which follows easily from the fact that $\xi_j(\xi_{j'}\bar{f}) = \xi_{j'}(\xi_j\bar{f})$ for any ξ_j and $\xi_{j'}$. $\qquad\square$

Working with the relatively free algebra, our next goal is to prove that $\mathcal{I}_{n,\overline{C\{t\}}} \cdot \overline{\mathcal{M}} = 0$. For that we shall need the next result.

Proposition 4.3.8 (Zubrilin). *Assume that a multilinear polynomial* $g(x_1, \dots, x_n)$ *is alternating in* $x_1, \dots, x_n$. *Then, modulo* $\mathcal{CAP}_{n+1}$,

$$\sum_{k=0}^{n}(-1)^k h^{n-k}\delta^{(n)}_{k,h}(g) \equiv 0$$

for any $h \in C\{t\}$. *In particular, this holds if* g *is doubly alternating.*

Proof. First we take h to be an indeterminate t. Let $f(x_1, \dots, x_{n+1}) = x_{n+1}g(x_1, \dots, x_n)$. By Proposition 4.2.6, $\overline{C\{X,Y,Z\}}$ satisfies the identity

$$\sum_{j=0}^{n}(-1)^j\delta^{(n)}_{j,t}(f(x_1, \dots, x_n, t^{n-j}x_{n+1})) \equiv 0.$$

Note that in computing $\delta^{(n)}_{j,t}(f(x_1, \dots, x_n, t^{n-j}x_{n+1}))$, the last indeterminate is x_{n+1} and is unchanged, cf. Remark 4.2.2, so

$$\delta^{(n)}_{j,t}(f(x_1, \dots, x_n, t^{n-j}x_{n+1})) = t^{n-j}x_{n+1}\delta^{(n)}_{j,t}g(x_1, \dots, x_n).$$

In view of Proposition 4.2.6,

$$\sum_{j=0}^{n}(-1)^j t^{n-j}x_{n+1}\delta^{(n)}_{j,t}(g(x_1, \dots, x_n)) \in \mathcal{CAP}_{n+1}.$$

The proof now follows by substituting $x_{n+1} \mapsto 1$ and $t \mapsto h \in C\{t\}$.

$\square$

As a consequence we can now prove the key result:

Proposition 4.3.9. *Let $\overline{M}$ be the module given by Definition 4.3.1. Then*
$\mathcal{I}_{n,\overline{C\{t\}}} \cdot \overline{M} = 0.$

Proof. We prove that $\mathcal{I}_{n,\overline{C\{t\}}} \cdot \overline{M} = 0$, by showing for any doubly alternating polynomial $f(x_1, \ldots, x_n, y_1, \ldots, y_n) \in \overline{M}$ and $h \in \overline{C\{t\}}$, that

$$(h^n + \xi_{1,h}h^{n-1} + \cdots + \xi_{n,h})f \equiv 0 \quad (\mathrm{mod}\ \mathcal{CAP}_{n+1}).$$

It follows from the action $\xi_{k,h}f = \delta_{k,h}^{(n)}(f)$ and from Proposition 4.3.8 that modulo $\mathcal{CAP}_{n+1}$,

$$(h^n + \xi_{1,h}h^{n-1} + \cdots + \xi_{n,h})f = \sum_{k=0}^{n}(-1)^k h^{n-k}\delta_{k,h}^{(n)}(f) \equiv 0.$$

$\square$

4.4   The Obstruction to Integrality $\mathrm{Obst}_n(A) \subseteq A$

In order to utilize these results about integrality, we need another concept. Recalling Definition 4.3.5, we define $\mathrm{Obst}_n(A) = A \cap \mathcal{I}_{n,A}$, viewing $A \subset A[\xi_{n,A}]$.

Remark 4.4.1.

1. *Let*

$$\tilde{A} = A[\xi_{n,A}]/\mathcal{I}_{n,A}, \tag{4.17}$$

   with $\varphi : A[\xi_{n,A}] \to \tilde{A}$ the natural homomorphism, and $\varphi_r : A \to \tilde{A}$ be the restriction of φ to A. Then

$$\ker(\varphi_r) = A \cap \mathcal{I}_{n,A} = \mathrm{Obst}_n(A).$$

2. *Note that for every $a \in A$, $\varphi(a)$ is n-integral (i.e., integral of degree n) over $C[\tilde{\xi}_{i,A}]$, and thus over the center of $\tilde{A}$. Indeed, apply the homomorphism φ to the element*

$$a^n + \xi_{1,a}a^{n-1} + \cdots + \xi_{n,a} \quad (\in \mathcal{I}_{n,A})$$

   to get

$$\tilde{a}^n + \tilde{\xi}_{1,a}\tilde{a}^{n-1} + \cdots + \tilde{\xi}_{n,a} = (a^n + \xi_{1,a}a^{n-1} + \cdots + \xi_{n,a}) + \mathcal{I}_{n,A} = 0.$$

Lemma 4.4.2. $\mathrm{Obst}_n(A)$ *also is the intersection of all kernels* $\ker(\psi)$ *of the following maps* ψ:

$\psi : A \to B$, *where B is a C-algebra, and $\psi : A \to B$ is a homomorphism such that for any $a \in A$, $\psi(a)$ is n-integral over the center of B.*

Proof. Denote this intersection $\cap_\psi \ker(\psi)$ as $\mathrm{Obst}_n'(A)$. Then $\mathrm{Obst}_n'(A) \subseteq \mathrm{Obst}_n(A)$ since $ker(\varphi_r)$ is among these $\ker(\psi)$. To show the opposite inclusion we prove

Claim: For such $\psi : A \to B$, $\ker(\psi) \supseteq A \cap \mathcal{I}_{n,A} = \mathrm{Obst}_n(A)$.

By assumption there exist $\beta_{1,a}, \ldots, \beta_{n,a}$ in the center of B satisfying

$$\psi(a)^n + \beta_{1,a}\psi(a)^{n-1} + \cdots + \beta_{n,a} = 0. \tag{4.18}$$

Extend ψ to $\check{\psi} : A[\xi_{n,A}] \to B$ as follows: $\check{\psi}(a) = \psi(a)$ if $a \in A$, while $\check{\psi}(\xi_{i,a}) = \beta_{i,a}$. We claim that $\check{\psi}(\mathcal{I}_{n,A}) = 0$. Indeed, for any

$$r = a^n + \xi_{1,a}a^{n-1} + \cdots + \xi_{n,a},$$

we have

$$\check{\psi}(r) = \psi(a)^n + \beta_{1,a}\psi(a)^{n-1} + \cdots + \beta_{n,a} = 0.$$

This shows that as claimed, $\check{\psi}(\mathcal{I}_{n,A}) = 0$.

Finally, if $a \in A \cap \mathcal{I}_{n,A}$ then $\psi(a) = \check{\psi}(a) = 0$. Hence $a \in \ker(\psi)$, so $\ker(\psi) \supseteq A \cap \mathcal{I}_{n,A} = \mathrm{Obst}_n(A)$. $\square$

Corollary 4.4.3.

(i) $\mathrm{Obst}_n(A/\mathrm{Obst}_n(A)) = 0$.

(ii) *If every $a \in A$ is n-integral (over the base field), then $\mathrm{Obst}_n(A)) = 0$.*

This corollary explains the notation $\mathrm{Obst}_n(A)$: it is the **obstruction** for each $a \in A$ to be n-integral. Exercise 2 helps to show how Obst behaves.

4.5   Reduction to Finite Modules

The reduction to finite modules is done using Shirshov's Theorem.

Proposition 4.5.1. *Let $A = C\{a_1, \ldots, a_\ell\}$ have PI degree d over the base ring C. Then the affine algebra $A/\mathrm{Obst}_n(A)$ can be embedded into an algebra which is finite over a central affine subalgebra.*

Proof. Replacing A by $A/\mathrm{Obst}_n(A)$, we may assume that $\mathrm{Obst}_n(A) = 0$. Let $W \subseteq A$ be the subset of the words in the alphabet $a_1, \ldots, a_\ell$ of length $\leq d$. By Shirshov's Height Theorem, there exists an integer μ such that the set

$$W = \{b_1^{k_1} \cdots b_\mu^{k_\mu} : b_i \in W, \, k_i \geq 0\}$$

spans A over the base ring C.

Denote

$$A' = A[\xi_{n,W}]/\mathcal{I}_{n,W}. \tag{4.19}$$

We show that A' is finite over an affine central subalgebra and thus is Noetherian.

Given $a \in A$, denote $a' = a + \mathcal{I}_{n,W} \in A'$, and similarly $\xi'_{i,b} = \xi_{i,b} + \mathcal{I}_{n,W}$. Then for every $b \in W$, b' is n-integral over $C[\xi'_{n,W}]$, where

$$C[\xi'_{n,W}] = C[\xi'_{1,b}, \ldots, \xi'_{n,b} : b \in W] \subseteq \mathrm{center}(A').$$

Hence the finite subset

$$\{b'^{k_1}_1 \cdots b'^{k_\mu}_\mu : b_i \in W, \quad k_i \leq n - 1\} \quad (\subseteq A') \tag{4.20}$$

spans A' over $C[\xi'_{n,W}]$. Thus A' is finite over the affine central subalgebra $C[\xi'_{n,W}] \subseteq \mathrm{center}(A')$ and thus is Noetherian.

Restricting the natural map $g : A[\xi_{n,W}] \to A[\xi_{n,W}]/\mathcal{I}_{n,W}$ to A, we have

$$g|_A : A \to A' \quad (a \mapsto a' = a + \mathcal{I}_{n,W}) \tag{4.21}$$

which satisfies

$$\ker(g|_A) = A \cap \mathcal{I}_{n,W} \subseteq A \cap \mathcal{I}_{n,A} = \mathrm{Obst}_n(A) = 0. \tag{4.22}$$

$\square$

In this case, we conclude from Step 2 and the argument outlined in Step 9 that any nil ideal is nilpotent, and $\mathrm{Jac}(A)$ is nilpotent when C is Jacobson.

4.6   Proving that $\mathrm{Obst}_n(A) \cdot (\mathcal{CAP}_n(A))^2 = 0$

It remains to show that $\mathrm{Obst}_n(A) \cdot (\mathcal{CAP}_n(A))^2 = 0$, thereby completing the proof of Razmyslov's Theorem 4.0.2. We present two proofs, one faster but more ad hoc (since we intersect with A and bypass difficulties involving the action of $\mathcal{I}_{n,W}$), and the second more structural.

Both approaches involve taking the free product of A with the free associative algebra, and then modding out the identities defining its variety.

For the remainder of this section, we assume that id(A) *contains* $\mathcal{CAP}_{n+1}$, *so that we can work with* $\overline{\mathcal{M}}$. In this case, every doubly alternating polynomial is clearly RZ for A.

4.6.1 The module $\overline{\mathcal{M}}_A$ over the relatively free product $C\{X,Y\} * A$ of c_{n+1}

Let $\overline{\mathcal{M}}_A$ denote the image of $\mathcal{M}$ under substitutions of $\vec{z}$ to A, i.e., the C-submodule of $\overline{A\langle X \rangle} = A\langle X \rangle / \text{id}(A)$ consisting of the images of all doubly alternating polynomials (in $x_1 \ldots, x_n$, and in $y_1, \ldots, y_n$). In view of Lemma 4.3.4, the natural action of Obst$_n(A)$ on $\overline{\mathcal{M}}_A$ respects multiplication by the $\delta_{k,h}^{(n)}$-operators.

Proposition 4.6.1. Obst$_n(A)\overline{\mathcal{M}}_A = 0$.

Proof. If $a \in$ Obst$_n(A)$, then (after any specialization of the y_i to elements of A), $a\mathcal{M} \in \text{id}(A)$, in view of Lemmas 4.2.4 and 4.3.3 and Proposition 4.3.9, so is 0 modulo id(A). $\qquad\square$

4.6.2 A more formal approach to Zubrilin's argument

Alternatively, rather than push immediately into A, one could perform these computations first at the level of polynomials and then specialize. This requires a bit more machinery, since it requires adjoining the commuting indeterminates $\xi_{n,A}$ to the free product, but might be clearer conceptually since it works at the level of generalized polynomials.

Remark 4.6.2. *Clearly* $A[\xi_{n,A}]*_C C\{t\} \subset A[\xi_{n,A}]*_C C\{X,Y,Z\}$ *in the natural way, and then*

$$\mathcal{I}(A[\xi_{n,A}] *_C C\{t\}) = (A[\xi_{n,A}] *_C C\{t\}) \cap \mathcal{I}(A[\xi_{n,A}] *_C C\{X;Y;Z\})$$

since we are just restricting the indeterminates $\vec{x}, \vec{y}, \vec{z}$ *to the indeterminates* $\vec{z}$.

It follows from Noether's Isomorphism Theorem that

$$(A[\xi_{n,A}] *_C C\{t\})/\mathcal{I}(A[\xi_{n,A}] *_C C\{t\}),$$

can be viewed naturally in $(A[\xi_{n,A}]*_C C\{X,Y,Z\})/\mathcal{I}(A[\xi_{n,A}] *_C C\{X,Y,Z\})$. Viewing $\mathcal{M} \subset C\{X,Y,Z\} \subset A[\xi_{n,A}] *_C C\{X,Y,Z\}$, we define

$$\widetilde{\mathcal{M}'} = (A[\xi_{n,A}] *_C C\{t\})\mathcal{M} \subset A[\xi_{n,A}] *_C C\{X,Y,Z\}, \qquad (4.23)$$

and its image in $(A[\xi_{n,A}]*_C C\{X,Y,Z\})/\mathcal{I}(A[\xi_{n,A}]*_C C\{X;Y;Z\})$, which we call $\widetilde{\mathcal{M}}$ (intuitively consisting of terms ending with images of RZ polynomials).

$\widetilde{\mathcal{M}}$ can be viewed as a $C[\xi_{n,\overline{C\{t\}}}]$-module, via the crucial Lemma 4.3.3. Then, as above, $\mathcal{M}$ is an $A * C\{t\}$-module (viewing $\mathcal{M} \subset C\{X, Y, Z\}$), implying $\overline{\mathcal{M}}$ is an $A * C\{t\}$-module annihilated by CAP_{n+1}. $\widetilde{\mathcal{M}}$ thereby becomes an $A[\xi_{n,\overline{C\{t\}}}] * \overline{C\{t\}}$-module, where we define

$$\xi_{k,h}\bar{f} = \bar{\delta}_{k,h}^{(n)}\bar{f}$$

for $h \in \overline{C\{t\}}$ and $\bar{f} \in \overline{\mathcal{M}}$, by means of the action given in Lemma 4.3.7, also cf. Remark 4.3.6. Our main task is to understand how this two action is specialized to A.

Having in hand the module $\widetilde{\mathcal{M}}$ on which $A[\xi_{n,\overline{C\{X,Y,Z\}}}]$ acts, we can specialize the assertion of Proposition 4.3.9 down to A once we succeed in matching the actions of $A[\xi_{n,\overline{C\{t\}}}]$ and $A[\xi_{n,A}]$ when specializing to A.

Remark 4.6.3. $CAP_k(A[\xi_{n,A}]) = CAP_k(A)[\xi_{n,A}]$, since c_k is multilinear.

Lemma 4.6.4. *Any specialization* $\varphi : C\{t\} \to A$ *gives rise naturally to a map*

$$\Phi : A[\xi_{n,A}]\mathcal{M} \to \widetilde{\mathcal{M}}$$

given by

$$\sum_i a_i \xi_{k,\varphi(h_{j_i})}\overline{f_i} \mapsto \sum_i a_i \overline{\varphi}(\overline{\xi_{k,h_{j_i}}f_i}) = \sum_i a_i \overline{\varphi}(\overline{\delta_{k,h_{j_i}}^{(\vec{x},n)}f_i})$$

where $\overline{f_i} \in \mathcal{M}$.

Proof. We need to show that this is well-defined, which follows from the functoriality property given in Lemma 4.2.4. Namely, if $\varphi(h_{j_i}) = \varphi(h'_{j_i})$, then $\overline{\varphi}(\overline{h_{j_i}}) = \overline{\varphi}(\overline{h'_{j_i}})$ and

$$\sum_i a_i \varphi(\overline{\delta_{k,h_{j_i}}^{(\vec{x},n)}f_i}) = \sum_i a_i \varphi(\delta_{k,h_{j_i}}^{(\vec{x},n)}f_i) = \sum_i a_i \varphi(\delta_{k,h'_{j_i}}^{(\vec{x},n)}f_i) = \sum_i a_i \varphi(\overline{\delta_{k,h'_{j_i}}^{(\vec{x},n)}f_i}).$$

$\square$

The objective of this lemma was to enable us to replace $A[\xi_{n,A}]$ by A in our considerations. Note that $\ker\Phi$ contains all $\overline{\delta_{k,h}^{(n)}f} - \xi_{k,h}\bar{f}$ (cf. Remark 4.2.10) as well as $(\bar{h}^n - \sum_{k=0}^{n-1} \bar{h}^k \xi_{k,h})\overline{f}$, where h ranges over all words and $\bar{f} \in \mathcal{M}$, so we see that the Zubrilin integrality relations are passed on.

We write $DCAP_n$ for the $C\{t\}$-submodule of $C\{X, Y, Z\}$ generated by DCap_n, cf. (4.4).

Lemma 4.6.5. *If Z is an infinite set of noncommuting indeterminates whose cardinality $\aleph$ is at least that of A, then for any given evaluation w in $DCAP_n(A *_C C\{X, Y, Z\})$, there is a map*

$$\varphi_w : C\{X, Y, Z\} \to A *_C C\{X, Y, Z\},$$

sending $\mathcal{M}$ to $\mathcal{M}(A *_C C\{X, Y, Z\})$, such that w is in the image of φ_w.

Proof. Note that $A *_C C\{X, Y, Z\}$ has cardinality $\aleph$. Setting aside indeterminates

$$\{t_g : g \in A *_C C\{X, Y, Z\}\},$$

we still have $\aleph$ indeterminates left over, to map onto our original set t of $\aleph$ indeterminates. But any evaluation w of $\mathcal{M}$ on $A *_C C\{X, Y, Z\}$ can be written as

$$w = gc_n(x_1, \ldots, x_n; g_1, \ldots, g_n)g'c_n(y_1, \ldots, y_n; h_1, \ldots, h_n)g'', \qquad (4.24)$$

for suitable $g, g', g'', g_i, h_j \in A *_C C\{X, Y, Z\}$. Defining φ_w by sending $x_i \mapsto x_i$, $y_j \mapsto y_j$, and sending the appropriate $t_g \mapsto g$, $t_{g'} \mapsto g'$, $t_{g''} \mapsto g''$, $t_{g_i} \mapsto g_i$, and $t_{h_j} \mapsto h_j$, we have an element in $\varphi_w^{-1}(w)$. $\qquad \square$

Clearly $\varphi_w(\mathcal{CAP}_{n+1}) \subseteq \mathcal{CAP}_{n+1}(A *_C C\{X, Y, Z\})$, so, when $c_{n+1} \in \mathcal{I}$, φ_w induces a map

$$\bar{\varphi}_w : \overline{C\{X, Y, Z\}} \to (A *_C C\{X, Y, Z\})_{\mathcal{I}},$$

which sends $\overline{\mathcal{M}} \to \widetilde{\mathcal{M}}$.

Although $\mathcal{CAP}_{n+1}$ need not be mapped onto $\mathcal{CAP}_{n+1}(A *_C C\{X, Y, Z\})$ by a single map, Lemma 4.6.5 says that it is "pointwise" onto, according to any chosen point, and this is enough for our purposes.

Theorem 4.6.6. $\mathrm{Obst}_n(A) \cdot (\mathcal{CAP}_n(A))^2 = 0$, *for any PI-algebra* $A = C\{a_1, \ldots, a_\ell\}$ *satisfying the Capelli identity* c_{n+1}.

Proof. We form the free algebra $C\{X, Y, Z\}$ by taking a separate indeterminate z_j for each element of $A[\xi_{n,A}]\mathcal{M}$. We work with $A[\xi_{n,A}]\mathcal{M}$, viewed in the relatively free product $\tilde{A} := (A[\xi_{n,A}] *_C C\{X, Y, Z\})_{\mathcal{I}}$, where $\mathcal{I} = \mathcal{CAP}_{n+1}(A[\xi_{n,A}] *_C C\{X, Y, Z\})$. In view of Lemma 4.6.4, the relation

$$I_{n, \overline{C\{X, Y, Z\}}} \cdot \overline{\mathcal{M}} \equiv 0 \quad (\mathrm{mod}\ \mathcal{CAP}_{n+1}(A[\xi_{n,A}] *_C C\{X, Y, Z\}))$$

restricts to the relation

$$I_{n, \overline{C\{X, Y, Z\}}} \mathcal{M} \equiv 0 \quad (\mathrm{mod}\ \mathcal{CAP}_{n+1}(A[\xi_{n,A}] *_C C\{X, Y, Z\})).$$

But the various specializations of Lemma 4.6.5 cover all of $\mathcal{DCAP}_n(A)$. Hence Lemma 4.2.4 applied to Proposition 4.3.9 and Lemma 4.6.4 implies $I_{n,A}\mathcal{DCAP}_n(A) = 0$, and thus

$$\mathrm{Obst}_n(A) \cdot (\mathcal{CAP}_n(A))^2 \subseteq I_{n,A}\mathcal{DCAP}_n(A) = 0,$$

as desired. $\qquad \square$

Remark 4.6.7.

(i) $\Phi(A)M = 0$ *(since the image of $\Phi(A)$ in $\tilde{A} = A/\Phi(A)$ is 0). Recall this was for any $M_{\psi;f}$ as well as any ψ, so in particular, $\Phi(A)$ annihilates every substitution of every n-alternating polynomial.*

(ii) *By construction, $\tilde{A}$ has a central extension $\tilde{A}' = C'\{\tilde{a}_1, \ldots, \tilde{a}_\ell\}$ for which every word of length $\leq d$ is integral; thus Shirshov's Height Theorem 2.2.2 implies $\tilde{A}'$ is finite over C'. It follows by Proposition 1.6.33 that the image in $\tilde{A}$ of $\mathrm{Jac}(A)$ is nilpotent, and the image of any nil subalgebra N is nilpotent.*

Remark 4.6.8. *Let us quickly recapitulate this proof of Braun's Theorem 4.0.6. The proof was by induction on the degree of the Capelli identity c_n, the case $n = 2$ being the well-known commutative case. By induction on n, the image of N in $A/\mathcal{CAP}_n(A)$ is nilpotent, so some power of N lies in $\mathcal{CAP}_n(A)$. On the other hand, for some m, Shirshov's Theorem 2.2.2 implies that every integral affine extension is representable, so the representable case (Proposition 1.6.34) implies $N^m \subseteq \mathrm{Obst}_n(A)$ (the obstruction to integrality), and it remains to prove that $\mathrm{Obst}_n(A)$ annihilates a power of $\mathcal{CAP}_n(A)$. In fact we proved that $\mathrm{Obst}_n(A) \cdot (\mathcal{CAP}_n(A))^2 = 0$, and this is the hardest part of the proof. There is a formal combinatorial argument for annihilation of $\mathcal{CAP}_n$ at the level of polynomials and abstractions of the characteristic coefficients, given in Proposition 4.2.6, using the operators $\delta_{j,t}^{(\vec{x},n)}(f)$. But this has to be lowered to the level of A by means of a specialization argument, which requires "enough" commutativity in these operators. This is proved for doubly alternating polynomials using the sophisticated argument in the proof, Proposition 4.3.9. Even so, applying the specialization requires maintaining the doubly alternating property of the vecx-indeterminates, which requires us first to specialize the other indeterminates and work over the generalized polynomial algebra $A\langle X \rangle$, using the module generated by the doubly alternating generalized polynomials.*

To obtain Theorem 4.0.1 directly (without appealing to the Amitsur-Procesi theorem), one notes that the commutative base for the induction works because we have Jacobson rings (Proposition 1.6.31), and the inductive step works formally, using $\mathrm{Jac}(A)$ instead of N.

4.7   The Shirshov Closure and Shirshov Closed Ideals

The obstruction to integrality used above gives rise to a generalization of the characteristic closure, which we will need for answering Specht's conjecture. Suppose A is any affine PI-algebra. Let us fix a Shirshov base $\bar{W}$ of A, which by Shirshov's Theorem is finite. Generalizing (4.17), we obtain an algebra homomorphism $\varphi_W : A \to \bar{A} = A\bar{F}$, where $\bar{F}$ is a central affine subalgebra of $\bar{A}$, satisfying the property that $\varphi_W(\bar{w})$ is integral of degree κ in $\bar{A}$ over $\bar{F}$, for some rather large number κ, and $\bar{F}$ is generated by the coefficients of the equations of integrality of the elements of $\bar{W}$; furthermore, if $\psi : A \to R$ is any homomorphism of A into an algebra R for which the image of each element of $\bar{W}$ is integral of degree κ, then $\ker \psi \subseteq \ker \varphi_W$.

Definition 4.7.1. *We call* $\ker \varphi_W$ *the* **Shirshov obstruction to integrality for** A *(with respect to W), and* $\bar{A}$ *the* **Shirshov closure** *of A (with respect to W). (Although $\bar{A}$ formally depends on the choice of W, this is not important.)*

Remark 4.7.2.

(i) *Shirshov's Height Theorem implies that $\bar{A}$ is finite over $\bar{F}$, of some very large dimension κ' and in particular, is Noetherian. Thus, $A/\ker \varphi_W$ is a representable algebra. Thus, the Shirshov obstruction to integrality of A (with respect to W) is 0, iff A can be centrally embedded into an algebra $\bar{A}$ integral of some bounded degree κ' over its center.*

(ii) *The Shirshov obstruction to integrality is precisely* $\mathrm{Obst}_\kappa(A)$, *by Lemma 4.4.2.*

This leads to a closure operation:

Definition 4.7.3. *Given an ideal $\mathcal{I} \lhd A$, define $\mathcal{I}^{\mathrm{cl}}$ to be the preimage in A of the obstruction to integrality of degree κ of $A/\mathcal{I}$. $\mathcal{I}$ is* **Shirshov closed** *if $\mathcal{I} = \mathcal{I}^{\mathrm{cl}}$, i.e., the obstruction to integrality of degree κ of $A/\mathcal{I}$ is 0.*

This gives us a chain condition for affine PI-algebras.

Proposition 4.7.4. *Any affine PI-algebra satisfies ACC on Shirshov closed ideals.*

Proof. Let $\bar{A}$ be the Shirshov closure of A. Let $\bar{\mathcal{I}} = \mathcal{I}\bar{F}$. $\mathcal{I}$ is closed in $\bar{A}$ iff $\mathcal{I} = A \cap \bar{\mathcal{I}}$. But the $\bar{\mathcal{I}}$, being ideals of the Noetherian algebra $\bar{A}$, satisfy the ACC, so their contractions to A also satisfy the ACC. $\square$

Exercises for Chapter 4

1. Theorem 4.0.1 implies Theorem 4.0.6. (Hint: Modding out the nilradical, and then passing to prime images, assume that A is prime. Then embed A into the polynomial algebra $A[\lambda]$ over the Noetherian ring $C[\lambda]$, and localize at the monic polynomials over $C[\lambda]$, yielding a Jacobson base ring by [ReSS82, Theorem 2.8].)

2. $\mathrm{Obst}_{n-1}(A) \supseteq \mathrm{Obst}_n(A)$. (Hint: Any element integral of degree $n-1$ is also integral of degree n.)

Nil implies nilpotent results

3. (Schelter) Suppose that A is any prime affine PI-algebra over a commutative Noetherian ring, and $\mathcal{I} \lhd A$. $A/\mathcal{I}$ has a finite set of prime ideals whose intersection is nilpotent, and any nil subalgebra $N/\mathcal{I}$ of $A/\mathcal{I}$ is nilpotent. (Hint: Otherwise, by Noetherian induction, assume that the assertion holds in A/P for every nonzero prime ideal P of A. Pass to the characteristic closure $\hat{A}$. Take $\mathcal{I}_A$ as in Proposition 2.5.6. $\mathcal{I}_A I$ is also an ideal of $\hat{A}$, which is Noetherian, so has a finite set of prime ideals $\hat{P}_1, \ldots, \hat{P}_m$ minimal over $\mathcal{I}_A I$, such that $\hat{P}_1 \cdots \hat{P}_m \subseteq \mathcal{I}_A I$. Let $P_i = A \cap \hat{P}_i$, a prime ideal of A. Apply the assertion to A/P_i and its ideal $(\mathcal{I} + P_i)/P_i$, and thus, there are a finite set of prime ideals $Q_{i1}/(\mathcal{I}+P_i), \ldots, Q_{it}/(\mathcal{I}+P_i)$ in $A/(\mathcal{I}+P_i)$ whose intersection is nilpotent, i.e.,
$$(Q_{i1} \cap \cdots \cap Q_{it})^k \subseteq \mathcal{I} + P_i.$$
But then, $\prod_{i=1}^m (Q_{i1} \cap \cdots \cap Q_{it})^k \subseteq \mathcal{I} + P_1 \ldots P_m \subseteq \mathcal{I}$.)

4. If A satisfies the identities of $n \times n$ matrices and is affine over a Noetherian commutative ring, then any nil subalgebra N is nilpotent. (Hint: The relatively free algebra of $\mathrm{id}(M_n(A))$ is the algebra of generic matrices, which is a prime algebra. Apply Exercise 3.)

Internal characteristic coefficients

5. If A is affine and satisfies c_{n+1}, then the obstruction to integrality of order n has nilpotence index not greater than $1 + 2^2 + \cdots + n^2 = \frac{n(n+1)(2n+1)}{6}$. (For characteristic p, one needs the Donkin-Zubkov theory.)

6. Any affine algebra satisfying a Capelli identity also satisfies the identities of $n \times n$ matrices for some n. (Hint: Using Exercise 5, show that if $\mathcal{M}_n(A)$ is the set of evaluations of identities of $n \times n$ matrices on A, then $\mathcal{M}_n(A)^k = 0$, for some k. Using Lewin's embedding Theorem 1.6.13, conclude that A satisfies all identities of $m \times m$ matrices for some m depending only on k.)

Chapter 5

Kemer's Capelli Theorem

So far we know that any affine algebra satisfying a Capelli identity has nilpotent radical. In order to conclude the proof of the Razmyslov-Kemer-Braun Theorem, we need to prove what we have called Kemer's Capelli Theorem. In fact, in characteristic p Kemer obtained the following stronger result, even for non-affine algebras:

Theorem 5.0.1 ([Kem95]). *Any PI-algebra over a field F of characteristic $p > 0$ satisfies a Capelli identity c_n, for large enough n.*

This fails in characteristic 0, since the Grassmann algebra does not satisfy a Capelli identity. Since the proof of Theorem 5.0.1 given in [Kem95] is quite complicated, here we only give the affine case since it is more straightforward. A slightly weaker result, that every PI-algebra A (not necessarily affine) of characteristic p satisfies a standard identity is given in Exercises 6–10. Some key ideas of the proof of the general case of Theorem 5.0.1 are given in Exercises 11–13.

We shall prove this assertion twice. The first proof works in all characteristics, but is tricky, relying on the "identity of algebraicity," and gives a huge estimate of the degree.

The second proof, closer to Kemer's original proof in characteristic 0, can be accomplished in two steps: First, that A satisfies a "sparse" system, by means of the first author's "pumping procedure" which he developed to answer Specht's question in characteristic p, and then a formal argument that every sparse system implies a Capelli identity. The latter is attractive since it utilizes the representation theory developed in Chapter 3, which is classical in characteristic 0. In characteristic $p > 0$, we have the opportunity to draw on recent results in the modular representation theory. This proof also provides a reasonable quartic bound $((p-1)p\binom{u+1}{2} = \mathrm{O}(p^4 d^4))$, where $u = \frac{2pe(d-1)^2}{3}$, for the degree of the sparse system of A in terms of the degree d of the given PI of A.

5.1   First Proof (Combinatoric)

In this section, we provide the tricky but concise combinatoric proof of Theorem 5.0.1 that works over an arbitrary commutative base ring C.

5.1.1   The identity of algebraicity

We start the proof with a general version of algebraicity. Take the free associative algebra $C\{Y, t\}$ in the indeterminates Y, and an extra noncommuting indeterminate t. Let V_n be the subspace of $C\{Y, t\}$ spanned by all monomials in $y_1, \ldots, y_n, t$ which are linear in $y_1, \ldots, y_n$, and for any permutation π of $\{1, \ldots, n\}$, let $V_{n,\pi}$ denote the subspace of monomials in which the variables $y_1, \ldots, y_n$ occur in the order $y_{\pi(1)}, \ldots, y_{\pi(n)}$. Then $V_n = \bigoplus_{\pi \in S_n} V_{n,\pi}$.

For any polynomial $f = f(y_1, \ldots, y_n, t)$ we define $\mathrm{Ad}_{\ell k}^t : V_n \to V_n$ by

$$\mathrm{Ad}_{\ell k}^t(f) = f|_{y_k \to t y_k} - f|_{y_\ell \to y_\ell t},$$

for our extra indeterminate t. (These are the D_{ij} of [BelRo05, Definition 1.56].)

Let $I = (i_0, \ldots, i_n) \in \mathbb{N}^{(n+1)}$, and consider the monomial

$$h_I = h_I(y_1, \ldots, y_n, t) = t^{i_0} y_1 t^{i_1} y_2 \cdots t^{i_{n-1}} y_n t^{i_n}.$$

We call I the **power vector** of h_I, and write $h_{I,\pi}$ for $\pi(h_I)$. Then

$$\begin{aligned}
\mathrm{Ad}_{\ell-1,k}^t(h_{I;\pi}) = {} & t^{i_0} y_1 t^{i_1} y_1 t^{i_2} y_2 \cdots y_{k-1} t^{i_k+1} \cdots t^{i_{n-1}} y_n t^{i_n} \\
& - t^{i_0} y_0 t^{i_1} y_1 t^{i_2} y_2 \cdots y_{\ell-1} t^{i_\ell+1} \cdots t^{i_{n-1}} y_n t^{i_n},
\end{aligned} \tag{5.1}$$

which is 0 if $k = \ell + 1$ but nonzero if $k \neq \ell + 1$.

It is easy to see that the $\mathrm{Ad}_{\ell k}^t$ commute as operators, so it makes sense to define the linear transformation,

$$D^t = \prod_{k \neq \ell+1} \mathrm{Ad}_{\ell k}^t,$$

and more generally, for any $\pi \in S_n$, we define

$$D_\pi^t = \prod_{\pi(k) \neq \pi(\ell)+1} \mathrm{Ad}_{\ell k}^t,$$

Remark 5.1.1. $D^t(h_I) \neq 0$, but any permutation $\pi \neq (1)$ of $\{1, \ldots, n\}$ satisfies $\pi^{-1}(\ell+1) + 1 \neq \pi^{-1}(\ell)$ for some ℓ, and letting $\ell' = \pi^{-1}(k+1)$ and $k' = \pi^{-1}(k)$, we get

$$\mathrm{Ad}_{\ell',k'}^t(h_{I;\pi}) = \mathrm{Ad}_{\ell+1,\ell}^t(h_I) = 0.$$

Lemma 5.1.2. *Suppose A is a PI-algebra satisfying a multilinear PI $p = p(y_1, \ldots, y_n)$. Write $D^t(y_1 \cdots y_n) \in id(A)$ as the sum*

$$\sum_I \alpha_I y_1 t^{i_1} \cdots y_n t^{i_n}.$$

Then for every multilinear polynomial $f = f(y_1, \ldots, y_n, \vec{x})$ and all polynomials $h_1, h_1', \ldots, h_n, h_n'$ the polynomial

$$g = \sum_I \alpha_I f(h_1 t^{i_0} h_1', \ldots, h_n t^{i_n} h_n', \vec{x})$$

is in $id(A)$.

Proof. It is enough to prove this for $f = x_0 y_{\tau(1)} x_1 \cdots x_{n-1} y_{\tau(n)} x_n$, where $\tau \in S_n$. Since

$$D_\tau^t(y_{\tau(1)} \cdots y_{\tau(n)}) = \sum_I \alpha_I y_{\tau(1)} t^{i_{\tau(1)}} \cdots y_{\tau(n)} t^{i_{\tau(n)}},$$

we deduce that

$$g = x_0 \cdot D_\tau^t(y_{\tau(1)} \cdots y_{\tau(n)})|_{y_{\tau(i)} \to y_{\tau(i)} x_i}.$$

The Lemma follows because for every $\tau \in S_n$ the polynomial

$$D_\tau^t(p(y_{\tau(1)}, \ldots, y_{\tau(n)})) = D_\tau^t(y_{\tau(1)} \cdots y_{\tau(n)})$$

f is an identity of A assuming that the monomial $y_1 \cdots y_n$ appears with nonzero coefficient in p. $\qquad \square$

We extend the definition of our operator $\mathrm{Ad}_{\ell k}^t$ to all of the free associative algebra $C\{\vec{x}, \vec{y}, t\}$, as follows: Take an arbitrary monomial

$h = c_0 t^{i_1} c_1 t^{i_2} c_2 \cdots t^{i_n} c_n \in C\{\vec{x}, \vec{y}, t\}$, where each $c_j \in C\{\vec{x}, \vec{y}\}$, and now define

$$\text{Ad}^t_{\ell k}(h) = \text{Ad}^t_{\ell k}(t^{i_0} y_0 t^{i_1} y_1 t^{i_2} y_2 \cdots t^{i_n - 1} y_n t^{i_n})_{y_i \mapsto c_i, 1 \le i \le n};$$

we extend this action to all polynomials $f = \sum_\pi \alpha_\pi h_\pi$ via

$$\text{Ad}^t_{\ell k}(f) = \sum_\pi \alpha_\pi \, \text{Ad}^t_{\ell k}(h_\pi).$$

Proposition 5.1.3. *If A satisfies an n-linear identity*

$$f = \sum_{\pi \in S_d} \alpha_\pi \pi(h),$$

where $h \in V_{n,(1)}$, then $0 \ne D^t(h) \in \text{id}(A)$.

Proof. Clearly $D^t(f) \in \text{id}(A)$, since it sums different evaluations of f. But $D^t(\pi(h)) = 0$ for all $\pi \ne (1)$, by Remark 5.1.1. Hence, by Lemma 5.1.2,

$$D^t(h) = D^t(f) \in \text{id}(A).$$

$\square$

Corollary 5.1.4. *If A satisfies a multilinear PI of degree n, then $0 \ne D^t(y_1 \cdots y_n) \in \text{id}(A)$.*

Ironically, the only property we used about the identity f was its degree.

Lemma 5.1.5. *When $f \in \text{id}(A)$ is homogeneous and n-linear, $D^t(f)$ is a nontrivial identity which is homogeneous of degree $n^2 - n + 1$ in t and total degree $n^2 - n + 1 + \deg f$.*

Proof. There are $n^2 - n + 1$ ways of choosing $k < \ell$ with $k \ne \ell + 1$. From each application of Ad, we may choose ℓ or k, so $D^t(y_1 \cdots y_d)$ is a sum of monomials, precisely one of which is

$$t^n y_1 \cdots t^{n-1} y_2 \cdots \cdots t^{n-1} y_{n-1} \cdots t^{n-1} y_n.$$

(This is the unique monomial in which each occurrence of t is as far to the left as possible.) $\square$

Thus, $D^t(y_1 \cdots y_n)$ is a nontrivial polynomial which is a sum of monomials of the form

$$t^{i_0} y_1 t^{i_1} y_2 \ldots y_n t^{i_n},$$

where $i_0 + \cdots + i_n = n^2 - n + 1$. We call $(i_0, \ldots, i_n)$ the **power vectors** and order them as follows:

First we define the degree $\deg(i_0, \ldots, i_n) = \max\{i_0, \ldots, i_n\}$. Then we say $\mathbf{i} > \mathbf{j}$ if either:

- $\deg(\mathbf{i}) > \deg(\mathbf{j})$;

- $\deg(\mathbf{i}) = \deg(\mathbf{j})$ but the number of maximal entries of $\mathbf{i}$ (i.e., of degree equal to $\deg(\mathbf{i})$) is greater than the number of maximal entries of $\mathbf{j}$;

- $\mathbf{i}$ and $\mathbf{j}$ have the same degree and the same number of maximal entries, but $\mathbf{i} > \mathbf{j}$ in the lexicographical order.

(This differs from the usual lexicographic order.)

Remark 5.1.6. *Writing $D^t(h) = g + \sum g_i$, where g is the monomial with leading power vector and the g_i are the other power vectors, we can replace g by $-\sum g_i$, and thereby lower the power vectors, in a way reminiscent of algebraicity. In fact, we have lowered the lexicographic order of the vector of leading components.*

Accordingly, $D^t(y_1 \cdots y_n)$ is called the **identity of algebraicity**.

Remark 5.1.7. *For any monomial $h_0 t^{i_0} h_1 t^{i_1} \cdots h_n t^{i_n} h_{n+1}$ such that $i_k \geq n$ and $h_k \in C\{X,Y\}$ for each k, we could then use $D^t(y_1 \cdots y_n)$ to "reduce" the power vector $\mathbf{i}$ of t until one of these components becomes less than n.*

5.1.2 Conclusion of the first proof of Kemer's Capelli Theorem

We are ready for an explicit version of Kemer's Capelli Theorem.

Theorem 5.1.8. *Let $A = C[a_1, \ldots, a_\ell]$ be an affine PI-algebra satisfying a multilinear PI f of degree n. Let W be the set of words in a_i of degree $< n$, and $D^t(f)$ be the identity of algebraicity (Lemma 5.1.5) holding in A.*
Then A satisfies the Capelli identity c_q for q given in the proof.

Proof. First we pass to the relatively free algebra of A, since this is a result about identities (and we thereby avoid the any difficulties arising from degenerate specializations). The a_i now are generic images of the indeterminates x_i. By linearity, we need only to evaluate c_q on words $w_1, \ldots, w_q$ in the a_i. Note that $|W| \leq \sum_{k=1}^{n-1} \ell^k$.

We consider an arbitrary evaluation $c_q(w_1, \ldots, w_q)$, where the v^{t_j} appear inside w_{i_j}, i.e., $w_{i_j} = h_{i_j} v^{t_j} h'_{i_j}$.

The main tools at our disposal are Shirshov's Height Theorem, this new identity of algebraicity to reduce the power vectors, and then induction on the lexicographical order.

Suppose that each w_k has length $\mu_k > \mu = \mu(\ell, d)$ of Shirshov's Theorem 2.2.2. Thus $w_k = u_{k_1}^{i_{1,k}} \ldots u_{k_{\mu_k}}^{i_{\mu,k}}$ where each $u_{k_j} \in W$. We call an exponent $i_{j,k}$ "modest" if $i_{j,k} < n$, and otherwise it is "oversized." We need to work with some word built of elements of W where "most" of the exponents are modest. We see that whenever we have n oversized powers, we can use Remark 5.1.7 to reduce one of them to modest size. Doing this for each word in W in turn, we have the exponents of all words from W modest except in at most $m_1 = (n-1)|W|\mu$ places.

Disregarding these oversized positions, we now have entries each of Shirshov height $\leq \mu$ and exponent $< n$. The total number of these "modest" possibilities are at most $m_2 = |W|^{\mu n}$, so taking $q = m_1 + m_2 + 1$, we see that some w_k must repeat in the entries of c_q, so the value of c_q must be 0. □

Ironically, we have proved combinatorially that Kemer's Capelli Theorem holds for an affine PI-algebra over any commutative ring C (not necessarily Noetherian), although the Braun-Kemer-Rasmylov theorem required C to be Noetherian, and the representation-theoretic proof to be given also requires some extra assumption on C.

5.2 Second Proof (Pumping Plus Representation Theory)

The downside of our first proof of Kemer's Capelli Theorem is that it gives a huge bound for the degree of the Capelli identity, with little hope for improvement. Our second proof is more intricate, but gives a better bound. It starts with a technical definition that encapsulates a basic property of the Capelli identity.

5.2.1 Sparse systems and d-identities

Definition 5.2.1. *A set $\{\alpha_\sigma : \sigma \in S_m\}, \alpha_\sigma \in C$ with $\alpha_{(1)} = 1$ is a **sparse system** of an algebra A (of degree m) if, for any polynomial $f(x_1, \ldots, x_m; \vec{y})$ linear in $x_1, \ldots, x_m$, we have*

$$\sum \alpha_\sigma f(x_{\sigma(1)}, \ldots, x_{\sigma(m)}; y) \in \mathrm{id}(A).$$

Instead of sparse systems, writers often deal with what they call the "strongly sparse" polynomial $\sum_{\sigma \in S_m} \alpha_\sigma x_{\sigma(1)}, \ldots, x_{\sigma(m)} \in \mathrm{id}(A)$, and weaken our running assumption that $\alpha_{(1)} = 1$. We delete the word "strongly" to keep the terminology more concise, and focus on the coefficients α_σ. Note that in checking strong sparseness, one may assume that f is a monomial $y_0 x_{\tau(1)} y_1 \cdots y_{m-1} x_{\tau(m)} y_m$.

Our motivating example of a sparse system comes from the Capelli identity $c_{n^2+1} \in \mathrm{id}(M_n(F))$, cf. Exercise 1. Here $m = n^2 + 1$ and $\alpha_\sigma = \mathrm{sgn}(\sigma)$; we call this the **Capelli system**. Although there are PI-algebras (such as the Grassmann algebra in characteristic 0) which do not satisfy a Capelli identity, a key feature of PI-theory, to be proved below, is that every PI-algebra over a field satisfies a suitable sparse system.

Sparse systems play a key role in the representation theory in characteristic 0, since they generate two-sided ideals in the group algebra of S_n, in a way to be clarified shortly. At this stage, we have a combinatoric reason why sparse systems are important.

Any sparse system can be viewed as a powerful **sparse reduction procedure**. Namely, given $a_1, \ldots, a_m$ in A, we can replace any term $f(a_1, \ldots, a_m)$ by

$$-\sum_{\sigma \neq 1} \alpha_\sigma f(a_{\sigma(1)} \cdots a_{\sigma(m)}, a_{m+1}, \ldots, a_n).$$

The way to obtain a Capelli identity from a sparse system (Theorem 5.2.8) is through **pumping**, which we discuss presently. Before pumping in full force, let us introduce an interesting variant of PI and its relationship to sparseness.

Definition 5.2.2. *A polynomial $f(x_1, \ldots, x_m; \bar{y})$ is a d-**identity** of $A = F\{a_1, \ldots, a_\ell\}$ (with respect to designated indeterminates $x_1, \ldots, x_m$) if f vanishes for all specializations of $x_1, \ldots, x_m$ to words in the a_i of length $\leq d$.*

Remark 5.2.3. *Let $m = (\ell + 1)^d$. The Capelli polynomial c_m is always a d-identity of $A = F\{a_1, \ldots, a_\ell\}$ (with respect to the m alternating variables $x_1, \ldots, x_m$), since there are fewer than m words of length $\leq d$ in $a_1, \ldots, a_\ell$.*

Thus, when m is large with respect to d, the verification of d-identities can be immediate.

5.2.2 Pumping

In using Dilworth's Theorem for our subexponential bound of Shirshov height in Theorem 2.8.5, we introduced the powerful technique of changing substitutions in a polynomial so that occurrences of the largest letter are moved closer together. We formalize the basic technique here, called **pumping**, and apply it to Capelli polynomials and Kemer index, transforming the Kemer index as tool for induction in PI-theory and later leading to Theorem 10.1.2 as a major example.

Nevertheless, d-identities fit in well with pumping. First a diversion.

Remark 5.2.4. *Pumping game: There are t piles of counters, denoted $v_1, \ldots, v_t$, where pile v_i contains m_i counters. Player 1 can select d piles $v_{i_1}, \ldots, v_{i_t}$, and from the pile v_{i_j} he selects a certain number $k_{i_j} < m_{i_j}$ of counters. Player 2 permutes these selected sub-piles (with a nonidentical permutation). For example, if $t = 5$, suppose the 5 piles respectively have 11, 3, 5, 7, and 4 counters, and Player 1 selects sub-piles of 4, 2, and 3 counters from piles 1, 2, and 5. Then Player 2 could perform the cyclic permutation in the sub-piles to get 3, 4, and 2, and the new piles have respective sizes*

$$11 - 4 + 3 = 10, \quad 3 - 2 + 4 = 5, \quad 5, \quad 7, \quad 4 - 3 + 2 = 3.$$

   Problem: Find a strategy such that Player 1 can eventually force $t - d$ of the piles to have at most d counters.

   Solution: Each time, Player 1 takes d counters from the largest pile, $d - 1$ from the second largest pile, and so on. Player 2 must then reduce the d-tuple of the d largest piles lexicographically, so, arguing by induction, we see that Player 1 can force the desired outcome.

Proposition 5.2.5 (Sparse Pumping Lemma). *Let A be a PI-algebra satisfying a d-sparse system $\{\alpha_\sigma : \sigma \in S_d\}$. Then for each homogeneous d-linear polynomial $f(x_1, \ldots, x_m; \bar{y})$, any specializations $\bar{x}_i, \bar{y}$ of the x_i and $\bar{y}$, and any words v_i in the $\bar{x}_i$,*

$$f(v_1, \ldots, v_m; \bar{y})$$

is a linear combination of elements of the form $f(v_1', \ldots, v_m'; \bar{y})$ where at most $d - 1$ of the v_i' have length $> d$.

Proof. By induction on the number of v_i' of length $> d$. Assume the lemma is false. Then we have some specialization with at least d of the v_i of length $> d$, which cannot be written as a linear combination of specializations with a smaller number of v_i of length $> d$. Renumbering the indices if necessary, we may assume that $|v_1| \geq |v_2| \geq \cdots \geq |v_d| > d$.

Claim. *We can rearrange the letters of these v_i (without altering $v_{d+1}, \ldots, v_m$) to reduce v_d to a word of length $< d$, contrary to hypotheses.*

   The proof of the claim is by induction on $|v_1|, |v_2|, \ldots, |v_d|$. (In other words, we induct first on $|v_1|$, then on $|v_2|$, and so forth.) We want to use the sparse system, which we now write as

$$g = \sum_{\sigma \in S_d} \alpha_\sigma z_{\sigma(1)} \cdots z_{\sigma(d)}$$

for indeterminates $z_1, \ldots, z_d$, to decrease the length of the first component that it alters, and thereby conclude by the induction hypothesis. Toward this end, we are done unless $v_1, \ldots, v_d$ all have length $> d$. Write $v_i = u_i' u_i$ where $|u_i| = d - i$, for $1 \leq i \leq d$. Define

$$h(x_1, \ldots, x_m, \bar{y}, z_1, \ldots, z_d) = f(x_1 z_1, x_2 z_2, \ldots, x_d z_d, x_{d+1}, \ldots, x_m, \bar{y})$$

where we have expanded our set of indeterminates to include the $z_1, \ldots, z_d$. The sparse reduction procedure (for $g(z_1, \ldots, z_d)$) says that we can replace h by

$$-\sum_{\sigma \neq 1} \alpha_\sigma h(x_1, \ldots, x_m, \bar{y}, z_{\sigma(1)}, \ldots, z_{\sigma(d)}) =$$

$$-\sum_{\sigma \neq 1} \alpha_\sigma f(x_1 z_{\sigma(1)}, x_2 z_{\sigma(2)}, \ldots, x_d z_{\sigma(d)}, x_{d+1}, \ldots, x_m, \bar{y}).$$

But then substituting u'_i for x_i and u_i for z_i shows that we could replace $f(v_1, \ldots, v_d, v_{d+1}, \ldots, v_m; \bar{y})$ by linear combinations of

$$f(u'_1 u_{\sigma(1)}, u'_2 u_{\sigma(2)}, \ldots, u'_d u_{\sigma(d)}, v_{d+1}, \ldots, v_m; \bar{y}). \tag{5.2}$$

Letting $v_{i,\sigma} = u'_i u_{\sigma(i)}$, we see that

$$|v_{1,\sigma}| = |v_1| - (d-1) + (d - \sigma(1)) = |v_1| + 1 - \sigma(1) \le |v_1|,$$

with inequality holding iff $\sigma(1) = 1$. Hence, by induction, each term in (5.2) with $\sigma(1) \ne 1$ can be reduced. This leaves the terms in (5.2) with $\sigma(1) = 1$ but $\sigma(2) \ne 2$, which by induction can be reduced, continuing down the line. □

Corollary 5.2.6. *The conclusion of Proposition 5.2.5 holds for any PI-algebra which satisfies either:*

(i) some Capelli d-identity c_d (see Definition 5.2.2);

(ii) some sparse system of degree d.

Proof. Both are special cases of the Proposition. □

The proof worked because we had $d - 1$ "extra" alternating indeterminates x_i which could absorb the excess length of the substitutions and then be ignored (i.e., considered as part of the indeterminates y).

This process of "pumping" all words of length $> d$ into a small number of indeterminates is delicate, but produces surprising results. Here is a taste.

Theorem 5.2.7. *If c_d is a d-identity of A (with respect to the d alternating variables $x_1, \ldots, x_d$), then $c_{2d-1} \in \mathrm{id}(A)$.*

Proof. Since any d-alternating polynomial is a consequence of c_d, the hypothesis implies that any d-alternating polynomial is a d-identity of A with respect to those d variables. We want to show that

$$c_{2d-1}(\bar{x}_1, \ldots, \bar{x}_{2d-1}, \bar{y}_1, \ldots, \bar{y}_{2d-1}) = 0$$

evaluated on any words $\bar{x}_i, \bar{y}_i$ in A. If at least d of the $\bar{x}_i$ have length $\le d$, then we view c_{2d-1} as alternating in these particular $\bar{x}_i$, and are done by the first sentence. Thus, we may assume that at most $d - 1$ of the $\bar{x}_i$ have length $\le d$, and thus, at least d of the $\bar{x}_i$ have length $> d$; for notational convenience, we assume that $\bar{x}_1, \ldots, \bar{x}_d$ have length $> d$. Assume we have some

$$a = c_{2d-1}(\bar{x}_1, \ldots, \bar{x}_{2d-1}, \bar{y}_1, \ldots, \bar{y}_{2d-1}) \ne 0,$$

with the smallest possible number of $\bar{x}_i$ of length $> d$.

Given these restrictions, we choose a with $\bar{x}_1$ of minimal length; given this $\bar{x}_1$, we choose a with $|\bar{x}_2|$ minimal, and so forth. We induct on the d-tuple $(|\bar{x}_1|, \ldots, |\bar{x}_d|)$. Viewing c_{2d-1} as a d-alternating polynomial in $x_{i_1}, \ldots, x_{i_d}$, we apply Proposition 5.2.5 to show that a is a linear combination of smaller terms, each of which is 0 by induction, contradiction. □

5.2.3   Affine algebras satisfying a sparse system

We are ready to utilize sparse systems. This result, like the previous one, holds over an arbitrary commutative ring (assuming that $\alpha_{(1)} = 1$).

Theorem 5.2.8. *Every affine algebra $A = C\{a_1, \ldots, a_\ell\}$ satisfying a sparse system of degree d satisfies the Capelli identity c_n, where $n = \ell^d + d$.*

Proof. We claim that $c_n(A) = 0$. By linearity, we can specialize all of the x_i to words v_i in $a_1, \ldots, a_\ell$, and need to show that $c_n(v_1, \ldots, v_n; \bar{y})$ always vanishes. Applying Proposition 5.2.5 repeatedly, it suffices to show that $f(v_1', \ldots, v_n'; \bar{y})$ vanishes under this substitution when at most $d-1$ of the v_i' have length $> d$. But this means that there are at least

$$n - (d-1) = \ell^d + 1$$

positions for which $|v_i'| < d$. The number of such words is

$$\ell^{d-1} + \ell^{d-2} + \cdots + 1 < \ell^d.$$

Thus, two of the words must be the same, so by the alternating hypothesis, $f(v_1', \ldots, v_n', \bar{y})$ vanishes.   □

Although the proof here is rather easy, the value for n still is huge. By a clever substitution and careful analysis of Young tableaux, Kemer [Kem80] actually proved that $n \leq d$. Parts of his argument are given in Exercise 11.

5.2.4   The Representation Theoretic Approach

To conclude the second proof of Kemer's Capelli Theorem, we must show that the hypothesis in Theorem 5.2.8 is superfluous. Surprisingly, this is done by means of representation theory, and does not require the algebra to be affine.

Theorem 5.2.9. *Every PI-algebra satisfies a sparse system (of suitable degree).*

We prove this theorem in three stages:

1. characteristic 0, using the classical representation theory of the symmetric group;

2. characteristic $p > 0$, using the more recent modular representation theory of the symmetric group;

3. ring-theoretic reduction to algebras over arbitrary commutative rings.

5.2.4.1 The characteristic 0 case

The characteristic 0 case is treated separately here, since it can be handled via the classical representation theory of the symmetric group. By Maschke's Theorem, the group algebra FS_n now is a finite direct product of matrix algebras over F. We have the decomposition $FS_n = \bigoplus_{\lambda \vdash n} I_\lambda$.

(Here I_λ is the sum of those $F[S_n]e_T$ for which T is a standard tableau with partition λ. These I_λ are minimal two-sided ideals, each a sum of s^λ minimal left ideals isomorphic to J_λ.)

Definition 5.2.10. *Let A be a PI-algebra. The multilinear polynomial $g \in V_n$ is a strong identity of A if for every $m \geq n$ we have $FS_m \cdot g \cdot FS_m \subseteq \mathrm{Id}(A)$.*

Remark 5.2.11. *The set of coefficients of a strong identity is a sparse system, since right multiplication by permutations inserts arbitrary monomials.*

To obtain strong identities, we utilize the following construction, due to Amitsur, recalling Remark 1.9.6 and Definition 3.1.9 (of f^*).

Let $L \subseteq \{x_{n+1}, \ldots, x_{2n-1}\}$ and denote by f_L^* the polynomial obtained from f^* by substituting $x_j \to 1$ for all $x_j \in L$. Rename the indeterminates in $\{x_{n+1}, \ldots, x_{2n-1}\} \setminus L$ as $\{x_{n+1}, \ldots, x_{n+q}\}$ (where $q = n - 1 - |L|$) and denote the resulting polynomial as f_L^*. Then similarly to (3.1) of Remark 3.1.10, there exists a permutation $\rho \in S_{n+q}$ such that $f_L^* = (f x_{n+1} \cdots x_{n+q})\rho$.

Note that if $1 \in A$ and $f^* \in \mathrm{Id}(A)$, then also $f_L^* \in \mathrm{Id}(A)$ for any such L, and in particular $f \in \mathrm{Id}(A)$. The converse is not true: it is possible that $f \in \mathrm{Id}(A)$ but $f^* \notin \mathrm{Id}(A)$.

Remark 5.2.12. *To apply this, we need Regev's estimate $c_m(A) \leq (d-1)^{2m}$ of Theorem 3.4.3.*

Remark 5.2.13. *Hypotheses as in Proposition 3.4.14, for $n \leq m \leq 2n$, if $f \in I_\lambda$, then $f^* \in \mathrm{Id}(A) \cap V_{2n-1}$ (see (3.1)). Also, $f_L^* \in \mathrm{Id}(A)$ for any subset $L \subseteq \{n+1, \ldots, 2n-1\}$, and in particular $f \in \mathrm{Id}(A)$. (This follows from Theorem 3.4.17. If $f \in I_\lambda$ and $\rho \in S_m$, then $f\rho \in \mathrm{Id}(A)$.)*

In view of Proposition 3.4.14 and Theorem 3.4.16(ii), we have

Theorem 5.2.14. *Every PI-algebra over a field of characteristic 0 satisfies non-trivial strong identities.*

Explicitly, let $\mathrm{char}(F) = 0$ and let A satisfy an identity of degree d. Let u, v be as in Proposition 3.4.14, and let $\lambda = (u^v)$ be the $u \times v$ rectangle. Then every $g \in I_\lambda$ is a strong identity of A. The degree of such a strong identity g is uv. We can choose for example $u = v = \lceil e \cdot (d-1)^4 \rceil$, so $\deg(g) = \lceil e \cdot (d-1)^4 \rceil^2 = e^2(d-1)^8$.

We summarize:

Theorem 5.2.15. *Every affine PI-algebra over a field of characteristic 0 satisfies some Capelli identity. Explicitly, we have the following:*

(a) *Suppose the F-algebra A satisfies an identity of degree d. Then A satisfies a strong identity of degree*

$$d' = \lceil e(d-1)^4 \rceil^2 = e^2(d-1)^8.$$

(b) *Suppose $A = F\{a_1, \ldots, a_\ell\}$, and A satisfies an identity of degree d and take d' as in (a). Let $n = r^{d'} + d' \cong r^{e^2(d-1)^8}$. Then A satisfies the Capelli identity c_n.*

Proof. (a) is by Theorem 5.2.14, and then (b) follows from Theorem 5.2.8, since every strong identity has its sparse system.   □

5.2.4.2   Actions of the group algebra on sparse systems

Although the method of §5.2.4.1 developed by Amitsur and Regev is the one customarily used in the literature, it does rely on branching and thus only is effective in characteristic 0. A slight modification enables us to avoid branching. The main idea, inspired from the Capelli system, is that any sparse system of degree n arises from an identity of the form

$$f = \sum_{\sigma \in S_n} \alpha_\sigma x_{\sigma(1)} x_{n+1} \cdots x_{\sigma(n)} x_{2n},$$

since we can specialize $x_{n+1}, \ldots, x_{2n}$ to whatever we want.

S_n acts on $F[S_{2n}]$ by fixing $x_{n+1}, \ldots, x_{2n}$, thereby making $F[S_{2n}]$ an $F[S_n]$-module. Thus, letting V_n' denote the subspace of V_{2n} generated by the words $x_{\sigma(1)} x_{n+1} \cdots x_{\sigma(n)} x_{2n}$, we can identify the sparse systems of degree n with $F[S_n]$-bisubmodules of V_n' inside V_{2n}. But there is an as $F[S_n]$-bimodule isomorphism $\varphi : V_n \to V_n'$, given by $x_{\sigma(1)} \cdots x_{\sigma(n)} \mapsto x_{\sigma(1)} x_{n+1} \cdots x_{\sigma(n)} x_{2n}$. In particular V_n' has the same simple $F[S_n]$-bisubmodule structure as V_n and can be studied with the same representation theory, although now we only utilize the left action of permutations.

Thus, for any PI-algebra A, the spaces

$$\mathrm{Id}(A) \cap V_n' \subseteq V_n'$$

are $F[S_n]$-bisubmodules of V_n'.

Remark 5.2.16. *For any partition $\lambda \vdash n$, any tableau $T := T_\lambda$ of λ gives rise to an element*

$$a_T = \varphi \left(\sum_{q \in \mathcal{C}_T,\ p \in \mathcal{R}_T} \mathrm{sgn}(q) qp \right) \in F[S_{2n}],$$

where $\mathcal{C}_T$ (resp. $\mathcal{R}_T$) denotes the set of column (resp. row) permutations of the tableau T.

Thus, $F[S_{2n}]a_T$ (if nonzero) is an $F[S_n]$-submodule, which we call J_T. If J_T contains an element corresponding to a nontrivial PI of A, a_T itself must correspond to a PI of A.

We let I_λ denote the minimal $F[S_n]$-bisubmodule of $F[S_{2n}]$ containing J_T.

Lemma 5.2.17. *Let A be an F-algebra, with $\lambda \vdash n$. If $\dim J_T > c_n(A)$ and J_T is a simple $F[S_n]$-module, then $I_\lambda \subseteq \mathrm{Id}(A) \cap V_n'$.*

Proof. I_λ is a sum of $F[S_n]$-submodules $J_T a$ each isomorphic to J_T. Thus, taking such J, one has

$$c_n(A) = \dim \left(\frac{V_n'}{\mathrm{Id}(A) \cap V_n'} \right) \geq \dim J > c_n(A),$$

a contradiction. Therefore each $J \subseteq \mathrm{Id}(A) \cap V_n'$, implying $I_\lambda \subseteq \mathrm{Id}(A) \cap V_n$. $\quad\square$

This all can be viewed directly in terms of partitions of $2n$ and their Young tableaux, cf. Exercise 5.

5.2.4.3 Simple Specht modules in characteristic $p > 0$

Although we cannot prove Theorem 5.2.14 in characteristic $p > 0$ through branching, the ideas of the previous section still apply, replacing the application of Lemma 3.2.7 by Lemma 5.2.17.

Our main difficulty for $\mathrm{char}(F) = p > 0$, is to find some J_λ which is simple. James and Mathas [JamMa99, Main Theorem] determined when J_λ is simple for $p = 2$.

One such example is when λ is the **staircase**, which we define to be the Young tableau T_u whose u rows have length $u, u - 1, \ldots, 1$. This gave rise to the James-Mathas conjecture [JamMa99] of conditions on λ characterizing when J_λ is simple in characteristic $p > 2$, which was solved by Fayers [Fay05].

In order to obtain a p-version of Proposition 3.4.14 in characteristic $p > 2$, we need to find a class of partitions satisfying Fayer's criterion.

For a positive integer m, define v_p to be the p-adic valuation, i.e., $v_p(m)$ is the largest power of p dividing m. Also, temporarily write $h_{(i;j)}$ for h_x where x is the box in the i, j position. The James-Mathas conjecture for $p \neq 2$, proved in [Fay05], is that J_λ is simple if and only if there do not exist i, j, i', j' for which $v_p(h_{(i;j)}) > 0$ with $v_p(h_{(i;j)}), v_p(h_{(i';j)}), v_p(h_{(i;j')})$ all distinct. Of course this is automatic when each hook number is prime to p, since then every $v_p(h_{(i;j)}) = 0$.

Example 5.2.18. *A **wide staircase** is a Young tableau T_u whose u rows all have lengths different multiples of $p - 1$, the first row of length $(p - 1)u$, the second of length $(p - 1)(u - 1)$, and so forth until the last of length $p - 1$. The number of boxes is*

$$n = \sum_{j=1}^{u} (p - 1)j = (p - 1)\binom{u + 1}{2}.$$

When $p = 2$, the wide staircase just becomes the staircase described above. In analogy to Example 3.4.12, the dimension of the "wide staircase" T_u

can be estimated as follows: We write $j = (p-1)j' + j''$ for $1 \le j'' \le p-1$. The hook of a box in the (i, j) position has length $(u+1-i)(p-1) + 1 - j$, and depth $u + 1 - j' - i$, so the hook number is

$$(u+1-i)(p-1)+1-j+u-j'-i = (u+1-i)p-j-j' = (u+1+j'-i)p-j'',$$

which is prime to p. Thus each wide staircase satisfies a stronger condition than Fayer's criterion.

The dimension can again be calculated by means of the hook formula. The first $p-1$ boxes in the first row have hook numbers

$$pu - 1, \ pu - 2, \ \ldots, \ pu - (p-1),$$

whose sum is $(p-1)pu - \binom{p}{2} = \binom{p}{2}(2u-1)$.

The next $p-1$ boxes in the first row have hook numbers

$$p(u-1) - 1, \ p(u-1) - 2, \ \ldots, \ p(u-1) - (p-1),$$

whose sum is $(p-1)p(u-1) - \binom{p}{2} = \binom{p}{2}(2u-3)$, and so forth. Thus the sum of the hook numbers in the first row is

$$\binom{p}{2} \left((2u-1) + (2u-3) + \cdots + 1 \right) = \binom{p}{2} u^2.$$

Summing over all rows yields

$$\sum h_x = \binom{p}{2} \sum_{k=1}^{u} k^2 = \binom{p}{2} \frac{u(u+1)(2u+1)}{6} = \frac{(2u+1)np}{6}.$$

Lemma 5.2.19. *For any integer u, let μ be the wide staircase T_u of u rows. Let $n = (p-1)\binom{u}{2}$. Then*

$$\left(\frac{6n}{p(2u+1)} \right)^n \cdot \left(\frac{1}{e} \right)^n < s^\mu \qquad (\text{where} \quad e = 2.718281828\ldots).$$

In particular, if $\alpha \le \frac{6n}{p(2u+1)e}$, then $\alpha^n \le s^\mu$.

Proof. We imitate the proof of Lemma 3.4.13. Since the geometric mean is bounded by the arithmetic mean,

$$\left(\prod_{x \in \mu} h_x \right)^{1/n} \le \frac{1}{n} \sum_{x \in \mu} h_x \le \frac{p(2u+1)}{6} = \frac{p(2u+1)}{6},$$

which, in view of Example 5.2.18, together with Stirling's formula (Remark 3.4.8), implies that

$$\left(\frac{6n}{p(2u+1)} \right)^n \cdot \left(\frac{1}{e} \right)^n = \left(\frac{n}{e} \right)^n \cdot \left(\frac{6}{p(p-1)(2u+1)} \right)^n < \frac{n!}{\prod_{x \in \mu} h_x} = s^\mu.$$

$\square$

Lemma 5.2.20. *Let A be a PI algebra over a field of characteristic p, that satisfies an identity of degree d. Choose a natural number u such that, for $n = (p-1)\binom{u+1}{2}$,*

$$\frac{6n}{p(2u+1)} \cdot \frac{1}{e} \geq (d-1)^4.$$

Let $\lambda \vdash n$ be any partition of n corresponding to the "wide staircase" T_u. Then the elements of the corresponding $F[S_n]$-bimodule $I_\lambda \subseteq V'_n$ correspond to sparse systems of A of degree n.

Proof. By Remark 5.2.12,

$$c_{2n}(A) \leq (d-1)^{4n} \leq \left(\frac{6n}{p(2u+1)} \cdot \frac{1}{e}\right)^n < s^\lambda,$$

and we conclude from Lemma 5.2.17.   □

5.2.4.4   Capelli identities in characteristic p

We are ready for the characteristic p version of Theorem 5.2.14.

Theorem 5.2.21. *Any algebra A over a field F of characteristic $p > 0$ satisfying an identity of degree d, satisfies a sparse system of degree $n = (p-1)\binom{u+1}{2}$, where*

$$\frac{3(p-1)u(u+1)}{p(2u+1)} \geq (d-1)^4 e.$$

Proof. By Lemma 5.2.20.   □

Corollary 5.2.22. *Suppose $A = F\{a_1, \ldots, a_\ell\}$, and A satisfies an identity of degree d and take n as in Theorem 5.2.21. Then A satisfies the Capelli identity c_n.*

Proof. Since $\frac{u+1}{2u+1} \geq \frac{1}{2}$, we could take $u \geq \frac{2pe(d-1)^2}{3}$.   □

This concludes the second proof of Theorem 5.0.1 in the affine case over a field.

5.2.5   Kemer's Capelli Theorem over Noetherian base rings

We turn to the case where C is a commutative Noetherian ring. One could handle this directly using Young diagrams, but there also is a ring-theoretic reduction.

Theorem 5.2.23. *Any affine PI-algebra over a commutative Noetherian base ring C satisfies some Capelli identity.*

Proof. By Noetherian induction, we may assume that the theorem holds for every affine PI-algebra over a proper homomorphic image of C.

First we do the case where C is an integral domain, and $A = C\{a_1, \ldots, a_\ell\}$ satisfies some multilinear PI f. It is enough to assume that A is the relatively free algebra $C\{x_1, \ldots, x_n\}/\mathcal{I}$ (where $\mathcal{I}$ is the T-ideal generated by f). Let F be the field of fractions of C. Then $A_F := A \otimes_C F$ is also a PI-algebra, and thus, by Theorem 5.0.1 satisfies some Capelli identity $f_1 = c_n$. Thus the image $\bar{f}_1$ of f_1 in A becomes 0 when we tensor by F, which means that there is some $s \in C$ for which $sf_1 = 0$. Letting $\mathcal{I}'$ denote the T-ideal of A generated by the image of f_1, we see that $s\mathcal{I}' = 0$. If $s = 1$, then we are done, so we may assume that $s \in C$ is not invertible. Then A/sA is an affine PI-algebra over the proper homomorphic image C/sC of C, and by Noetherian induction, satisfies some Capelli identity c_m, so $A/(sA \cap \mathcal{I}')$ satisfies $c_{\max\{m,n\}}$. But $sA \cap \mathcal{I}'$ is nilpotent modulo $sA\mathcal{I}' = As\mathcal{I}' = 0$, implying by Lemma 1.2.27(iii) that A satisfies some Capelli identity.

In general, the nilpotent radical of C is a finite intersection

$$N = P_1 \cap \cdots \cap P_t$$

of prime ideals. By the previous paragraph, $A/P_j A$, being an affine PI-algebra over the integral domain C/P_j, satisfies a suitable Capelli identity c_{n_j}, for $1 \le j \le t$, so $A/\cap(P_j A)$ satisfies c_n, where $n = \max\{n_1, \ldots, n_t\}$. But $\cap(P_j A)$ is nilpotent modulo NA, so, by Lemma 1.2.27(iii), A/NA satisfies a suitable Capelli identity $c_{n'}$. Furthermore, $N^m = 0$ for some m, implying again by Lemma 1.2.27(iii) that A satisfies $c_{mn'}$. $\square$

Exercises for Chapter 5

1. The Capelli identity c_d satisfies the following stronger property than being a sparse system:

   Given any polynomial $f(x_1, \ldots, x_d; y_1, \ldots, y_m)$ and any substitutions $\bar{x}_1, \ldots, \bar{x}_d, \bar{y}_1, \ldots, \bar{y}_m$ in A, one can rewrite

   $$\sum_{\sigma \in S_d} \text{sgn}(\sigma) f(\bar{x}_{\sigma(1)}, \ldots, \bar{x}_{\sigma(d)}, \bar{y}_1 \ldots, \bar{y}_m)$$

   as

   $$\sum_i \bar{w}_{i0} c_d(\bar{x}_{\sigma(1)}, \ldots, \bar{x}_{\sigma(1)}, \bar{w}_{i1} \ldots, \bar{w}_{im}), \qquad (5.3)$$

   for suitable words $\bar{w}_j$ in the $\bar{x}$ and $\bar{y}$. (Hint: First, assume that f is a monomial $h = w_0 x_1 w_1 \ldots x_d w_d$ for suitable words w_j. But then

$$\sum_{\sigma \in S_d} \mathrm{sgn}(\sigma) h(\bar{x}_{\sigma(1)}, \ldots, \bar{x}_{\sigma(d)}, \bar{y}_1 \ldots, \bar{y}_m) = \tag{5.4}$$

$$\bar{w}_0 c_d(\bar{x}_{\sigma(1)}, \ldots, \bar{x}_{\sigma(d)}, \bar{w}_1 \ldots, \bar{w}_d),$$

as desired.)

Pumping words

2. Suppose that c_d is a d-identity of A with respect to d alternating variables.

   For each homogeneous m-linear polynomial $f(x_1, \ldots, x_n; y)$, any specializations $\bar{x}_i, \bar{y}$ of the x_i and y, and any words v_i in the $\bar{x}_1, \ldots, \bar{x}_n$ such that $m_i = |v_i| \geq d$, one can rewrite $f(v_1, \ldots, v_n; \bar{y})$ as a linear combination

   $$\sum_{1 \neq \sigma \in S_d} \alpha_\sigma f(v_{1,\sigma}, \ldots, v_{d,\sigma}, v_{d+1}, \ldots, v_n; \bar{y}),$$

   where $v_{1,\sigma}, \ldots v_{d,\sigma}$ are words formed by rearranging the letters of $v_1, \ldots, v_d$ and lexicographically the d-tuple

   $$(|v_{1,\sigma}|, |v_{2,\sigma}|, \ldots, |v_{d,\sigma}|) < (m_1, m_2, \ldots, m_d).$$

   (Hint: Identical, word for word, to the proof in the induction procedure of Proposition 5.2.5.)

3. Let

   $$w = v_0 c_1 v_1 c_2 v_2 \ldots c_m v_m$$

   where the c_i are powers of x_ℓ, and x_ℓ does not appear in the v_j. If d of the v_j have length $> d$, then w is d-decomposable with respect to the lexicographic order assigning value 0 to each letter of v_j. (Hint: By induction on the order of w. Discarding the initial segment v_0, assume that w starts with c_1. If $v_{j_1}, \ldots, v_{j_d}$ have length $\geq d$, write $w = c'_1 w_1 c'_2 w_2 \ldots w_d c'_{d+1} \cdots$ where each w_u ends with v_{j_u}, and c'_i starts with c_i. Write $w_j = w'_j w''_j$, where w''_j has length j. Letting $a_j = w''_j c'_{i_j} w'_{j+1}$, note that $a_j > a_{j+1}$ for $1 \leq j \leq d-1$. Thus, w is d-decomposable.)

4. Apply Exercise 3 to an identity of degree d, to rewrite any word as a linear combination of words in which all but at most $d-1$ of the words separating the occurrences of x_ℓ have length $< d$.

5. Any Young diagram can be **doubled** by repeating each column. Then one doubles a Young tableau by sending $i \mapsto 2i - 1$ in the first column and $2i-2$ in the second column. Now embedding S_n into S_{2n} by sending

each permutation to a permutation of the odd numbers, replace $\mathcal{C}_T$ by the column permutations fixing each even column, and $\mathcal{R}_T$ by the row permutations only acting on the entries in the odd columns. Show that the corresponding semi-idempotent is a_T of Remark 5.2.16.

Kemer's s_n Theorem [Kem93]

6. Let $F[\varepsilon]_0 = F[\varepsilon_i : i \in \mathbb{N}]_0$ denote the commutative polynomial algebra without 1 modulo the relations $\varepsilon_i^2 = 0$. Then A satisfies the symmetric identity $\tilde{s}_n$, iff $A \otimes F[\varepsilon]_0$ satisfies x_1^n. Conclude that A satisfies $\tilde{s}_n$ iff A satisfies the same multilinear identities of a suitable nil algebra (without 1) of bounded index. (Hint: Recall that $\tilde{s}_n$ is the multilinearization of x_1^n.)

7. For any t, an algebra A satisfies $\tilde{s}_n$ iff $M_t(A)$ satisfies $\tilde{s}_m$ for some m. (Hint: Using Exercise 6, translate to nil of bounded index, which is locally nilpotent; do this argument generically.)

8. Define the k-**Engel polynomial** f_k inductively, as $[f_{k-1}, x_2]$, where $f_0 = x_1$. (Thus, $f_2 = [[x_1, x_2], x_2]$.) If $\operatorname{char}(F) = p$ and $k = p^j$, then f_k is the homogeneous component of the partial linearization of x_1^k. Conclude in this case that any (associative) algebra satisfying an Engel identity satisfies some $\tilde{s}_n$.

9. (Difficult) Prove by induction on the Kemer index that every associative algebra in characteristic p satisfies a suitable Engel identity. This is a tricky computational argument that takes about five pages in [Kem93]. In view of Exercise 8, this concludes the proof of Kemer's Capelli Theorem in characteristic p.

10. Every PI-algebra A of characteristic p satisfies a standard identity. (Hint: If $A \otimes G$ satisfies $\tilde{s}_n$, then A satisfies s_n, so put together the previous exercises.)

Kemer's General Capelli Theorem [Kem95]

11. If $n = uv$ is as in Example 3.2.9, then

$$c_n(x_1, x_1^2, \ldots, x_1^n, y_1, \ldots, y_n) \in \operatorname{id}(A).$$

(Hint: Take $p = e'_{T_\lambda}$, where λ is the rectangular tableau v^u of Example 3.2.9 which can be written explicitly as

$$p(x_1, \ldots, x_n) = \sum_{\sigma \in C} \sum_{\tau \in R} \operatorname{sgn}(\sigma) x_{\sigma\tau(1)} \cdots x_{\sigma\tau(n)},$$

where $C = C_{T_\lambda}$ and $R = R_{T_\lambda}$. In view of Corollary 3.4.18 and Remark 1.2.32, one wants to take the alternator of p^*. However, this could be 0. So look for a clever substitution of p^* that will preserve the column permutations (which create alternating terms) while neutralizing the row transformations.

Toward this end, for each $1 \leq k \leq n$, write $k = i + (j-1)u$, i.e., k is in the (i, j) position in the tableau T_λ, and substitute $x_k \mapsto x_1^i$ and $y_k \mapsto x_1^{(j-1)u} y_k$. Since x_1^i now repeats throughout the i row, this neutralizes the effect of row permutations, and

$$p^*(x_1, \ldots, x_1, x_1^2, \ldots, x_1^2, \ldots; y_1, \ldots, y_u, x_1 y_{u+1}, \ldots, x_1^{(v-1)u} y_n) =$$

$$v! \sum_{\sigma \in C} \mathrm{sgn}(\sigma) x_1^{\sigma(1)} y_1 x_2^{\sigma(2)} y_2 \ldots x_k^{\sigma(k)} y_k \ldots x_n^{\sigma(n)} y_n.$$

Since these terms respect the sign of the permutation, one can apply the alternator (but treating these power of x_1 as separate entities) to get the identity

$$u! v! \sum_{\sigma \in \pi} \mathrm{sgn}(\pi) x_1^{\pi(1)} y_1 x_2^{\pi(2)} y_2 \ldots x_k^{\pi(k)} y_k \ldots x_n^{\pi(n)} y_n,$$

which is $u! v! c_{2n}(x_1, x_1^2, \ldots, x_1^n, y_1, \ldots, y_n).$)

12. If $n = uv$ are as in Theorem 3.4.14, then for any n-alternating polynomial $f(x_1, \ldots, x_n; y)$,

$$f(w_1 x_1 w_2, w_1 x_1^2 w_2, \ldots, w_1 x_1^n w_2, y_1, \ldots, y_n) \in \mathrm{id}(A),$$

for any words $w_1, w_2 \in C\{X\}$. (Hint: Apply Corollary 1.2.30 to Exercise 11.)

13. Using Exercise 11, prove that $c_n(x_1^{k_1}, x_1^{k_2}, \ldots, x_1^{k_n}; y) \in \mathrm{id}(A)$. (Hint: An induction argument.)

Kemer then uses an induction argument on the tableaux to prove that if A satisfies c_n for n as in Proposition 3.4.14. then A satisfies s_n.

Part III

Specht's Conjecture

Chapter 6

Specht's Problem and Its Solution in the Affine Case (Characteristic 0)

In this part, we turn to Specht's Problem, perhaps the question most discussed in PI-theory. Its solution remains one of the most difficult in the theory, and also, as we shall see, different versions of the question have varying answers: There is a positive solution for (associative) algebras of characteristic 0, but in positive characteristic the answer is positive only for affine algebras, as shown in Volume II (where various results also are given for nonassociative algebras).

6.1   Specht's Problem Posed

Perhaps the most interesting question about T-ideals is in generating them as efficiently as possible, in the sense of Definition 1.1.27, where $\mathcal{I}$ is the class of T-ideals. This is because if f is in the T-ideal generated by S, then f is a consequence of S.

The first basic question about generation of T-ideals was asked by Specht. Since we know by Hilbert's Basis Theorem that every algebraic variety can be defined by a finite number of commutative polynomials, we might ask whether any T-ideal can be generated (as a T-ideal) by a finite number of noncommutative polynomials. In order to avoid confusion with f.g. as a module, a finitely generated T-ideal is usually called **finitely based**, so we have:

Problem 6.1.1 (Specht's problem [Sp50]). *Is every T-ideal of $F\{X\}$ finitely based, for a given field F? (This property is called the **finite basis property** for T-ideals for T-ideals.*

If so, then the set of identities of any PI-algebra A is generated as a T-ideal by a finite set of identities $\{f_1, \ldots, f_n\}$, and so every identity of A is a consequence of $\{f_1, \ldots, f_n\}$.

Ironically, Specht's problem can be reformulated in a slightly different context of "trace identities," in which one proves (Corollary 8.2.3) that every PI of matrices is a consequence of the Cayley-Hamilton polynomial, and in fact the proof given in Exercise 8.2 of the Amitsur-Levitzki theorem was based on it. Nevertheless, the solution of Specht's problem is considered one of the major achievements in PI-theory for several reasons:

First, one wants to know about the polynomial identities themselves. For example, for computational purposes, one can use central polynomials and Theorem J to invert matrices, leading to the question (still open) of the minimal central polynomial of $n \times n$ matrices. Although in principle, one can derive this from Theorem 8.2.3, nobody knows how to do it yet.

Secondly, Kemer's theorem can be viewed more structurally — namely, $\mathbb{Q}\{X\}$ satisfies the ACC on T-ideals, cf. §1.1.7.1. In this way, parts of the theory of Noetherian rings become applicable, and we shall describe these in Chapter 7. Let us note here that the applications of Specht's problem should contain an analysis of the "prime" T-ideals in the sense of Definition 1.1.29.

Thirdly, the solution as it stands is a tour de force, requiring virtually every major theorem and technique in PI-theory developed over 35 years (from 1950 until 1985). Unlike many of the other major theorems discussed in Chapter 1, whose original proofs were intricate but then were reduced to half a page each, the optimal proof of Kemer's Theorem still takes several chapters of this volume, and it is hard to see where it can be shortened significantly.

Specht's Problem generated a large literature, especially in the Russian school, which at first dealt with various special cases. In a striking breakthrough, Kemer [Kem87] solved Specht's problem affirmatively for char(F) = 0, although Belov later found counterexamples in positive characteristic. Nevertheless, Specht's problem has an affirmative answer for all affine algebras over a commutative Noetherian ring (proved by Kemer for algebras over an infinite field, and by Belov in the general case), which will be presented in Volume II.

As beautiful as Kemer's proof is, it still does not answer some of the most basic questions, such as "What is a base for the T-ideal id($M_2(\mathbb{Q})$)?" Razmyslov [Raz74c] proved that id($M_2(\mathbb{Q})$) is based by s_4 and Wagner's identity (Example 1.2.8). His proof is difficult and computational, and requires passing to the theory of Lie identities, so is deferred until Volume II, also cf. [Raz89] and [Ba87]. This makes it all the more important to have a convincing exposition available in the literature.

The solution to Specht's problem in characteristic 0 requires techniques to be developed in the next two chapters, and only will be completed in Chapter 7. We start by describing the T-ideal of identities of the Grassmann algebra as a motivating example, showing that it is generated by a single identity.

Then we provide the details of Kemer's proof of the finite basis property in characteristic 0. We work initially over an infinite field F of arbitrary characteristic, although soon Kemer's lemmas require char(F) = 0 in order for us to pass to multilinear polynomials. By Theorems 7.1.15 and 7.2.1, it is enough to prove the finite basis property for affine superalgebras. Since the proof for the super-case differs from the ungraded case only with regard to some technical difficulties, we prove the finite basis property first for affine algebras of characteristic 0, and then indicate (in Chapter 7) how to modify the proof in the super-case to get the full solution of Specht's problem in characteristic 0. Along the way, we prove the important PI-representability theorem of Kemer (Theorem 6.3.1). In Chapter 9, we shall see that there are non-affine counterexamples to Specht's problem in characteristic p, although Capelli identities hold for arbitrary PI-algebras of characteristic p ([Kem93], cf. Exercises 5.6–5.10. In Volume II, we show that Specht's problem has a positive solution for affine algebras of characteristic p.

Although Kemer's paper also handled the important case of algebras without 1, we continue to assume for convenience that our algebras contain 1. For the reader interested in algebras without 1, Exercise 9 contains the essential observation needed for the modification of the proof.

6.2   Early Results on Specht's Problem

The naive approach to answering Specht's Problem is to try to search for a set of candidates for the generators, usually polynomials having "nice" properties so that we can work with them.

Specht contributed the next result. We define **higher (Lie) commutators** inductively:

$[f, g]$ is a higher commutator if f is a letter or a higher commutator, and g is a letter or a higher commutator.

For example, $[[x_1, x_2], [x_5, [x_3, x_4]]]$ is a higher commutator. We call a polynomial **Spechtian** if it is a sum of products of higher commutators.

Proposition 6.2.1. *Every identity of an algebra A can be written*

$$f(x_1, \ldots, x_m) = \sum_{(u_1, \ldots, u_m) \in \mathbb{N}} f_{(u_1, \ldots, u_m)} x_1^{u_1} \ldots x_m^{u_m}$$

where each $f_{(u_1, \ldots, u_m)}$ is a Spechtian polynomial or a constant.

Proof. Note that $x_{i_{u-1}} x_m x_{i_{u-2}} = x_{i_{u-1}} x_{i_{u-2}} x_m + x_{i_{u-1}}[x_m, x_{i_{u-2}}]$. Thus,

$$x_{i_1} \ldots x_{i_{u-1}} x_m x_{i_u} \ldots x_{i_{t-1}} =$$

$$x_{i_1} \ldots x_{i_{m-1}} x_m + \sum_{j=1}^{t-1} x_{i_1} \ldots x_{i_{j-1}} [x_m, x_{i_j}] x_{i_{j+1}} \ldots x_{i_{t-1}},$$

for each monomial in which x_m appears, and any polynomial $f(x_1, \ldots, x_m)$ can be written

$$f = p + g(x_1, \ldots, x_{m-1}) x_m + h,$$

where x_m does not appear in p, and h can be written in such a way as that x_m appears (at least) once in a Lie commutator. We continue this argument in turn with each appearance of x_m, so we may assume x_m appears in h only in Lie commutators. Now we repeat this procedure with x_{m-1}, and so forth, to yield the desired result. □

Corollary 6.2.2 (Specht's Lemma). *Every multilinear identity f of an algebra A (with 1) is a consequence of Spechtian identities.*

Proof. The proposition shows that $f = g(x_1, \ldots, x_{m-1}) x_m + h$, where h can be written in such a way as that x_m only appears in a Lie commutator. Thus $h(x_1, \ldots, x_{m-1}, 1) = 0$, implying $g(x_1, \ldots, x_{m-1}) = f(x_1, \ldots, x_{m-1}, 1)$ is an identity of A, so by induction on m is a consequence of Spechtian identities. Applying the same argument to x_{m-1} in h, and iterating, yields the desired result. □

Corollary 6.2.3. *Every identity of a $\mathbb{Q}$-algebra A is a consequence of Spechtian identities.*

6.2.1 Solution of Specht's problem for the Grassmann algebra

Unfortunately, so far there is no general effective way to attack Specht's Problem using Spechtian identities, but they do help us to solve an interesting special case: the Grassmann algebra G of a vector space V. Recall that G is generated by $\{e_i : i \in I\}$, satisfying the relations $e_i e_j = -e_j e_i$ and $e_i^2 = 0$. In Proposition 1.3.31 we noted that G satisfies the **Grassmann identity** $[[x_1, x_2], x_3]$, which happens to be a higher commutator, and we shall show that this generates $\mathrm{id}(G)$ as a T-ideal. Although this was first proved by Regev using the theory espoused in Chapter 3, there is a direct identity-theoretic proof, which we present here.

Remark 6.2.4. *The following identities hold in any associative algebra, seen by inspection:*

(i) $[x, yz] = [x, y]z + y[x, z]$.

(ii) $[xy, z] = x[y, z] + [x, z]y$.

(iii) $[x, yz] = [xy, z] + [zx, y]$.

Lemma 6.2.5 (Latyshev's Lemma). *Every algebra A satisfying the Grassmann identity satisfies the identity* $[x_1, x_2][x_3, x_4] + [x_1, x_3][x_2, x_4]$.

Proof. For any $a, b, c, d \in A$ we have (using (ii) repeatedly)

$$
\begin{aligned}
0 = [[ad, b], c] &= [a[d, b], c] + [[a, b]d, c] \\
&= (a[[d, b], c] + [a, c][d, b]) + ([a, b][d, c] + [[a, b], c]d) \\
&= 0 - [a, c][b, d] - [a, b][c, d] + 0.
\end{aligned}
$$

$\square$

Corollary 6.2.6.

(i) *Every algebra satisfying the Grassmann identity satisfies the identity* $[x_1, x_2][x_1, x_3]$.

(ii) *Every algebra satisfying the Grassmann identity satisfies the identity* $[x_1, x_2]x_4[x_1, x_3]$.

Proof. (i) Specialize $x_3 \mapsto x_1$ in Lemma 6.2.5 to get the identity $[x_1, x_2][x_1, x_4]$.

(ii) Immediate from (i), since any commutator is central. $\square$

Theorem 6.2.7. *Every Spechtian polynomial f can be reduced modulo the Grassmann identity to a sum of polynomials of the form*

$$
[x_{i_1}, x_{i_2}] \cdots [x_{i_{2k-1}}, x_{i_{2k}}],
$$

where $i_1 < i_2 < \cdots < i_{2k}$.

Proof. By definition, f can be written as a sum of products of higher commutators. But every higher commutator of length at least 3 vanishes as a consequence of the Grassmann identity, so f is a sum of products of commutators. The corollary to Latyshev's lemma enables us to assume that the letters in these commutators must be distinct, so f may be assumed to be the sum of multilinear polynomials. In view of Lemma 1.2.13, we may assume that f is multilinear.

Suppose $i_1 < i_2$ are the two lowest indices appearing in the indeterminates of f. By Lemma 6.2.5 we can always exchange the commutator $[x_{i_1}, x_j][x_{i_2}, x_k]$ with $-[x_{i_1}, x_{i_2}][x_j, x_k]$, so $f = [x_{i_1}, x_{i_2}]h$ and the result follows by induction. $\square$

Corollary 6.2.8. *Each multilinear identity of the Grassmann algebra G (over an infinite dimensional vector space over a field) is a consequence of the Grassmann identity.*

Proof. By Specht's lemma, any multilinear identity f can be assumed to be Spechtian and thus, by Theorem 6.2.7, reduced to the form

$$f = \alpha[x_1, x_2][x_3, x_4] \ldots [x_{2n-1}, x_{2n}],$$

which certainly is not an identity of G (seen by specializing $x_i \mapsto e_i$) unless $\alpha = 0$, i.e., f reduces to 0. $\square$

When V is finite dimensional, then so is G, implying it satisfies extra identities. But Theorem 6.2.7 also helps us out here, since the extra multilinear identities can be reduced to the form $[x_1, x_2][x_3, x_4] \ldots [x_{2n-1}, x_{2n}]$.

Corollary 6.2.9 (Regev). *In characteristic 0, the T-ideal of G is a consequence of the Grassmann identity.*

The same argument shows in characteristic 0 that any T-ideal properly containing $[[x_1, x_2], x_3]$ is generated by at most one other polynomial, $[x_1, x_2][x_3, x_4] \ldots [x_{2n-1}, x_{2n}]$ for some n.

This result also holds in all characteristics $\neq 2$, but is a bit trickier to prove, cf. Exercise 1, since Corollary 6.2.3 is no longer available. We shall study the Grassmann algebra in characteristic p in Chapter 9; one can prove readily that it satisfies the Frobenius identity $(x + y)^p = x^p + y^p$, which must then be deducible formally from $[[x_1, x_2], x_3]$, cf. Exercise 9.4. For $p = 2$, G is commutative, and so satisfies the identity $[x_1, x_2]$, which clearly is not a consequence of $[[x_1, x_2], x_3]$. But this case is rather trivial; in Chapter 9. We introduce a modification to the Grassmann algebra in characteristic 2, and now the Grassmann identity implies the identity $(x + y)^4 = x^4 + y^4$, cf. Exercise 9.9.

6.3 Kemer's PI-representability Theorem

For the remainder of this chapter, we assume that W is an affine PI-algebra over a field F, and in particular, its Jacobson radical $\mathrm{Jac}(W)$ is nilpotent by the Razmyslov-Kemer-Braun theorem (Theorem 4.0.1). We continue to use the abbreviation *f.d.* for finite dimensional algebras. The main step of Kemer's approach is proving:

Theorem 6.3.1 (Kemer). *Any affine PI-algebra W over a field F of characteristic 0 is **PI-representable**, in fact, PI-equivalent to some f.d. algebra over a purely transcendental extension of F.*

Remark 6.3.2. *Suppose L is a field extension of F. Then*

(i) $\mathrm{id}(W \otimes_F L) = \mathrm{id}(W)$, *for any PI-algebra W, by Proposition 1.2.19.*

(ii) *Any f.d. L-algebra A can be written as $A_0 \otimes_K \bar{F}$, where K is a finitely generated field extension of F, and A_0 is a f.d. K-algebra. (Indeed, take a base $b_1, \ldots, b_n$ of A over $\bar{F}$, and write*

$$b_i b_j = \sum_{k=1}^{n} \alpha_{ijk} b_k,$$

and let K be the field generated by the α_{ijk}.)

(iii) *In case L is the algebraic closure of F, A_0 is a K-algebra that is f.d. over F. (Indeed, $\mathrm{id}(A) = \mathrm{id}(A_0)$ and $[K : F] < \infty$, implying*

$$[A_0 : F] = [A_0 : K][K : F] < \infty.)$$

In view of Remark 6.3.2(ii), it suffices to prove Theorem 6.3.1 over a field containing a transcendence base of K over F, so we may assume that K is purely transcendental over F. Likewise, in the sequel, *we may assume throughout that the base field F is algebraically closed.*

Once one has this theorem, together with some classical results about f.d. algebras which we shall review below, it is not difficult to affirm Specht's conjecture, cf. Exercise 7, although the proof we shall give in the text is based on the techniques to be developed here. In this manner, the venerable theory of f.d. algebras is elevated to the level of current research, and we review it now.

6.3.1 Finite dimensional algebras

We fix our notation for the remainder of this chapter. A always denotes a f.d. algebra over an algebraically closed field F of characteristic 0, with $J = \mathrm{Jac}(A)$ and $\bar{A} = A/\mathrm{Jac}(A)$.

We write
$$\bar{A} = R_1 \oplus \cdots \oplus R_q,$$
each R_k being of the form $M_{n_k}(F)$, since F is algebraically closed.

A word about the use of $\times$ and $\oplus$: Strictly speaking, we could write

$$R_1 \times \cdots \times R_q$$

to emphasize this as a direct product of rings, the components respecting multiplication as well as addition. Having said this, we shall revert to the notation of $\oplus$ to avoid confusion with the notation in Remark 1.2.20 below.

Our main structure theorem about f.d. algebras is *Wedderburn's Principal Theorem*, Theorem 2.5.8 . Thus,

$$A = R_1 \oplus \cdots \oplus R_q \oplus J, \qquad (6.1)$$

where $\bar{A} = R_1 \oplus \cdots \oplus R_q$. We write e_k for the multiplicative unit of $R_k = M_{n_k}(F)$, for $1 \leq k \leq q$; then e_k is an idempotent and $e_k \bar{A} e_k = R_k$.

Remark 6.3.3. *As noted already in (2.3), A has a trace function given by the traces on the components of the semisimple part. In other words, if $a = (r_1, \ldots, r_q, u)$ for $r_k \in R_k = M_{n_k}(F)$ and $u \in J$, we define $\mathrm{tr}(a) = \sum_{k=1}^{q} n_k \, \mathrm{tr}(r_k)$. Since the semisimple part is invariant up to conjugation, cf. Theorem 2.5.8, this trace function does not depend on the particular isomorphism (6.1).*

Remark 6.3.4. *If A is representable, i.e., $A \subseteq M_n(K)$, then AK is f.d. over K, so much of the theory in this section presented for f.d. algebras also works for representable algebras. This will turn out to be a key observation, when we adjoin traces by means of Remark 6.3.3.*

Thus, the following "baby" version of Kemer's Representability Theorem is relevant:

Proposition 6.3.5. *The relatively free algebra $\mathcal{A}$ of a f.d. algebra A is representable over a rational function field K over F.*

Proof. In the notation of Example 1.8.15, let K be the field of fractions of $F[\Lambda]$, and take $\mathcal{A}$ inside $A \otimes_F K$. $\qquad\square$

There are two important numerical invariants of A:

(i) $t_A = \dim_F(\bar{A})$, the dimension of the semisimple part.

(ii) s_A, the nilpotence index of J.

Now $s_A = 1$ iff A is semisimple. On the other hand, one can have $s_A > q$ (of (6.1)), for example when $A = M_n(H)$ where H is a local commutative algebra with nonzero radical.

Example 6.3.6. $t_{\mathrm{UT}(n)} = n$ *and* $s_{\mathrm{UT}(n)} = n$.

We shall see that these two invariants can be described in terms of polynomials, and this procedure is perhaps the single most important aspect of the proof, providing a basis for induction.

In checking whether an n-linear polynomial $f(x_1,\ldots,x_n;y_1,\ldots,y_m)$ is an identity A, we may assume that each substitution for x_i is either in $\bar{A}$ or in J. We call such a substitution a **semisimple substitution** or a **radical substitution** , respectively.

We write $\{e_{i,j}^{(k)} : 1 \le i,j \le n_k\}$ for the matrix units of R_k. Thus, the $e_{i,i}^{(k)}$ are idempotents. We can (and shall) assume that any semisimple specialization is some matrix unit $e_{i,j}^{(k)}$. Since $J = \oplus_{j,k=1}^{n} e_k J e_\ell$, we shall assume that each radical substitution is in $e_k J e_\ell$ for suitable k,ℓ, and say such a radical substitution has **type** (k,ℓ) . (Note by passing to $\bar{A}$ that $e_k A e_\ell \subseteq J$ for $k \ne \ell$, implying $e_k A e_\ell = e_k J e_\ell$.)

Remark 6.3.7. *There is a well-known directed graph-theoretic interpretation of matrix units, in which one draws an edge from e_{ij} to $e_{i'j'}$ iff $j = i'$; thus, the paths correspond precisely to the nonzero products of matrix units. Viewed in this way, $\bar{A}$ is a disjoint union of q disjoint components, which we shall also call **stages**, corresponding to the matrix algebras $R_1,\ldots,R_q$, and this takes care of all semisimple substitutions. The situation becomes more interesting when we bring in radical substitutions. A type (k,ℓ) radical substitution connects the k and ℓ stages, and these are the only substitutions that can connect separate stages.*

Example 6.3.8. *When $q = 1$, the situation is well known. We take a set of $n_1 \times n_1$ matrix units of R_1 and write $A = M_{n_1}(A_1)$ for a suitable f.d. algebra A_1, cf. [Ro88b, Proposition 13.9]. But then letting $J_1 = \mathrm{Jac}(A_1)$, we have $A/M_{n_1}(J_1) \cong M_{n_1}(A_1/J_1) \cong M_{n_1}(F)$, implying $J = M_{n_1}(J_1)$. Thus, A is a matrix algebra over a local ring.*

Note: Although every f.d. algebra obviously is a finite subdirect product of irreducible f.d. algebras, this is not necessarily the case for PI-irreducibility of Definition 1.2.21, so we need to modify a definition from the original edition.

Definition 6.3.9. *A f.d. algebra A is **basic** if A is not PI-equivalent to a finite subdirect product A' of f.d. algebras A_k, $1 \le k \le m$, each satisfying $(t_{A_k}, s_{A_k}) < (t_{A'}, s_{A'})$.*

Exercise 1.21 also shows that $M_n(F)$ is basic. In general, one obtains basic algebras inductively, since if A is not basic, one can reduce (t_A, s_A) in the components of a subdirect product and this can only be done a finite number of times.

6.3.2   Sketch of Kemer's program

Theorem 6.3.1 will be proved in several steps, which we outline here. Let W be any affine PI-algebra, and let $\Gamma = \mathrm{id}(W)$ be its T-ideal (in an infinite number of indeterminates, so that we may make use of multilinear identities of arbitrarily large degree).

Step 1. $\Gamma \supseteq \text{id}(A)$ for some f.d. algebra A. This requires Lewin's Theorem 1.6.13.

Proposition 6.3.10. *For any affine PI-algebra W over a field F, there is a f.d. F-algebra A such that $\text{id}(A) \subseteq \text{id}(W)$.*

Proof. We may replace W by its relatively free algebra. Write $W = F\{X\}/I$. Also let $\text{Jac}(W) = J/I$, for $J \lhd F\{X\}$. Then, by Theorem 4.0.1, $\text{Jac}(W)^k = 0$ for some k, implying $J^k \subseteq I$. Furthermore, $F\{X\}/J$ is semiprime, and thus of some PI-class, but also relatively free PI so is an algebra of generic matrices and thus representable. Therefore $F\{X\}/J^k$ is representable by the highly nontrivial Corollary 1.6.14, and so is PI-equivalent to some f.d. algebra A; clearly $\text{id}(A) \subseteq \text{id}(W)$ since there is a surjection $F\{X\}/J^k \to W$. $\quad\square$

Step 2. We want to describe the two invariants t_A and s_A in terms of alternating polynomials. We define the **Kemer index** of Γ, denoted $\text{index}(\Gamma)$, to be a pair $(\beta_\Gamma, \gamma_\Gamma)$, ordered lexicographically, defined combinatorially in terms of the number of sets X_j of t_A- and $(t_A + 1)$-alternating indeterminates (in given nonidentities), which we call "folds."

When $\Gamma = \text{id}(A)$ for A irreducible f.d. basic, we will have

$$(\beta_\Gamma, \gamma_\Gamma) = (t_A, s_A). \tag{6.2}$$

Also it turns out that $\Gamma_1 \subseteq \Gamma_2$ implies $\text{index}(\Gamma_1) \geq \text{index}(\Gamma_2)$. Unfortunately, this correspondence is weaker than we might like (in the sense that one can have $\Gamma_1 \subset \Gamma_2$ with $\text{index}(\Gamma_1) = \text{index}(\Gamma_2)$).

Step 3. Folds enable us to restrict substitutions of the indeterminates X_j which yield nonzero values. When the substitutions are long enough, we can move certain indeterminates from one fold to another, thereby shortening the substitutions and providing an inductive procedure. This technique enables us to prove **Kemer's Lemmas**, the key to describing t_A (and later s_A) in Equation (6.2), by means of certain reasonable technical hypotheses. Surprisingly, q, the number of simple components of $\bar{A}$, plays only a secondary role. The proof of Kemer's Lemmas involve three kinds of polynomials:

- **Full polynomials**, which guarantee enough room to compute nonzero polynomial evaluations;

- **Kemer polynomials**, which are used to describe t_A and s_A;

- **Property K**, a technical and somewhat anti-intuitive condition used to build Kemer polynomials.

Unfortunately, each kind arises in a different context, but we need to use them in synchronization, which is one of the complications of the proof of Kemer's

Second Lemma. On the other hand, the "full" property comes unexpectedly easily for Kemer polynomials with enough "folds."

We need to work with the T-ideal generated by the Kemer polynomial, whose elements may lose these nice properties.

Definition 6.3.11. *A class $\mathcal{C}$ of polynomials is **T-hereditary** if for any polynomial f in $\mathcal{C}$, every polynomial in the T-ideal generated by f also is in $\mathcal{C}$.*

Property K is T-hereditary, but being a Kemer polynomial is not. To circumnavigate this difficulty, Kemer's Second Lemma enables us to prove the "Phoenix property" for basic algebras, that generates another Kemer polynomial, in fact infinitely many of them.

Step 4. Let Γ' be the T-ideal generated by Γ and the Kemer polynomials described in Step 4. Then $\text{index}(\Gamma') < \text{index}(\Gamma)$, so by induction on the Kemer index, there is a f.d. algebra A' with $\text{id}(A') = \Gamma'$. On the other hand, using the trace ring construction for representable algebras described in Definition 2.5.1, one can define a representable algebra $\hat{A}'$ with $\Gamma \subseteq \text{id}(\hat{A}')$, but whose Kemer polynomials are not identities of $\hat{A}'$. Being representable, $\hat{A}'$ is PI-equivalent to a f.d. algebra. Using the Phoenix property we will be able to conclude that

$$\Gamma = \Gamma' \cap \text{id}(\hat{A}) = \text{id}(A') \cap \text{id}(\hat{A}) = \text{id}(A \times A'),$$

so our desired algebra is $A \times A'$.

Concerning Step 1, by Theorem 1.6.13(ii) of Lewin, we can take $A = \text{UT}(n, \dots, n)$ of Definition 1.6.11, cf. Proposition 1.6.15.

Remark 6.3.12. *Step 1 is the only one which is not generalized to the nonassociative cases of alternative and Jordan algebras, as to be discussed in Volume II, because Lewin's Theorem is not known more generally, but nevertheless Specht's problem often has a positive solution in the affine case.*

Since Step 1 has been taken care of, we work out Steps 2, 3, 4, and 5.

Remark 6.3.13. *Suppose we could have $q = 1$ with $\text{Jac}(A)$ of nilpotence index 2. Then $A \approx M_n(H)$, with $H/\text{Jac}(H) = L$ central. But since $H = \text{Jac}(H) + L$, the fact that $\text{Jac}(H)^2 = 0$ implies H is commutative! Hence A is PI-equivalent to $M_n(F)$, so cannot be basic; the easiest nontrivial case is disposed of via a technicality!*

6.3.3 Theorems used for the proof

One of the indications of the depth of Kemer's PI-representability Theorem is the range of results that seem to be needed for its proof:

- The Wehrfritz-Beidar Theorem 1.6.22,

- Shirshov's Height Theorem 2.2.2,

- The Braun-Kemer-Razmyslov Theorem 4.0.1,

- Kemer's Capelli Theorem 5.1.8,

- Kemer's First Lemma 6.5.1,

- Kemer's Second Lemma 6.6.31.

6.3.4   Full algebras

Definition 6.3.14. *A multilinear polynomial* $f(x_1, \ldots, x_m)$ *is* **full** *on the f.d. algebra* A *if* $f(\bar{x}_1, \ldots, \bar{x}_m) \neq 0$ *for suitable substitutions* $\bar{x}_i$ *such that*

$$R_k \cap \{\bar{x}_1, \ldots, \bar{x}_m\} \neq \emptyset, \quad \forall 1 \leq k \leq q.$$

(In other words, some nonzero substitution "passes through" each stage R_k*.) This nonzero substitution is called the* **full substitution***.*

   The f.d. algebra A *is* **full** *if some multilinear monomial is full on it.*

Lemma 6.3.15. *The f.d. algebra* A *is full iff we have a permutation* π *of* $\{1, \ldots, q\}$ *such that*

$$R_{\pi(1)} J R_{\pi(2)} J \cdots J R_{\pi(q)} \neq 0. \tag{6.3}$$

Proof. ($\Leftarrow$) The monomial $x_1 \cdots x_{2q-1}$ is full on A.

   ($\Rightarrow$) Suppose the monomial f is full on A. Then the substitutions pass through each component, and are connected by the radical substitutions.   □

Remark 6.3.16. *For computational purposes, let us make Definition 6.3.14 more specific. Assume that* f *is full on* A*. Rearranging the* x_k *if necessary we assume that* $\bar{x}_k \in R_k$*. Put*

$$a = f(\bar{x}_1, \ldots, \bar{x}_q, \bar{x}_{q+1}, \ldots, \bar{x}_m) \neq 0.$$

As noted earlier, we may assume that each $\bar{x}_k$ *has the form* $e_{ij}^{(k)} \in R_k$*. But*

$$e_{ij}^{(k)} = e_{i1}^{(k)} e_{11}^{(k)} e_{1j}^{(k)},$$

so defining

$$g(x_1, \ldots, x_m; y_1, \ldots, y_q; z_1, \ldots, z_q) = g(y_1 x_1 z_1, \ldots, y_q x_q z_q, x_{q+1}, \ldots x_m)$$

and taking $\bar{y}_k = e_{i1}^{(k)}$, $\bar{z}_k = e_{1j}^{(k)}$, $1 \leq k \leq q$, *we have*

$$g(e_{11}^{(1)}, \ldots, e_{11}^{(q)}, \bar{x}_{q+1}, \ldots, \bar{x}_m; \bar{y}_1, \ldots, \bar{y}_q; \bar{z}_1, \ldots \bar{z}_q)$$
$$= f(\bar{x}_1, \ldots, \bar{x}_q, \bar{x}_{q+1}, \ldots, \bar{x}_m) = a.$$

Thus, if we replace f *by* g *(which is in the* T*-ideal generated by* f*), we may assume in Definition 6.3.14 that each* $\bar{x}_k = e_{11}^{(k)}$*.*

Definition 6.3.17. *Let e_i denote the unit element of R_i, thereby yielding the Peirce decomposition $A = \oplus_{i,j=1}^{q} e_i A e_j$. Define $\tilde{A}_k = \oplus_{i,j \neq k} e_i A e_j$, and $\tilde{A} = \prod_{k=1}^{q} \tilde{A}_k$.*

This definition only makes sense when $q > 1$. Clearly $e_i A e_j \subset J$ for $i \neq j$, since its image is 0. In the case of algebras without 1, we can proceed as in Exercise 1.1, or we can take e_0 as defined in Exercise 8, and take the direct sum from 0 to q.

Remark 6.3.18. *(i) Suppose $q > 1$. Then $t_{\tilde{A}_k} < t_A$ for each k.*

(ii) If $e_{i_1} J e_{i_2} \ldots e_{i_{q-1}} J e_{i_q} \neq 0$ for $i_1, \ldots, i_q$ distinct, then A is full; otherwise,

$$e_{i_1} A e_{i_2} \ldots e_{i_{q-1}} A e_{i_q} = e_{i_1} J e_{i_2} \ldots e_{i_{q-1}} J e_{i_q} = 0$$

whenever $i_1, \ldots, i_q$ are distinct.

Lemma 6.3.19. *(char$(F) = 0$.) If A is not full, then $id(\tilde{A}) \subseteq id(A)$. Furthermore, if every consequence of a nonidentity $f(x_1, \ldots, x_n)$ of A is not full, then f is also a nonidentity of $\tilde{A}$.*

Proof. It is enough to prove the second part and assume that f is multilinear. Consider a nonzero substitution $x_i \to \bar{x}_i \in R_1 \cup \cdots \cup R_q \cup \text{Jac}(A) \subseteq A$. Since f is not full, at least one of the R_i (say for $i = 1$) does not contain any of the $\bar{x}_j$.

If $(e_1 \text{Jac}(A) \cup \text{Jac}(A) e_1) \cap \{x_1, \ldots, x_n\} = 0$ then the assertion is clear since we take the component $\oplus_{i,j \neq 1} e_i A e_j$ of $\tilde{A}$. Thus, we can take some

$$\bar{x}_j \in e_1 \text{Jac}(A) \cup \text{Jac}(A) e_1,$$

say $\bar{x}_1 \in \text{Jac}(A) e_1$. We consider

$$f_1(x_1, x_2, \ldots, x_n, y_1) = f(x_1 y_1, x_2, \ldots, x_n),$$

a consequence of f which has the substitution $x_i \to \bar{x}_i, y_1 \to \bar{y}_1 = e_1$, thereby also entering R_1 with $\bar{f}_1 = \bar{f} \neq 0$.

Continuing in this fashion we eventuality get some $f_m(x_1, \ldots, x_n, y_1, \ldots, y_m)$ with a nonzero substitution passing through each component $\tilde{A}_k$. Therefore $f \notin id(\tilde{A})$. $\qquad \square$

Our first application of this construction relates directly to fullness.

Proposition 6.3.20. *Any f.d. algebra A that is not full is PI-reducible to a product of full algebras each having fewer simple components than A.*

Proof. Induction on the number q of simple components. An algebra with one simple component clearly is full, so we may assume that $q > 1$. Define $\tilde{A}$ as in Definition 6.3.17.

We claim that $A \sim_{\text{PI}} \tilde{A}$. Clearly

$$id(A) \subseteq id(\tilde{A}),$$

so it suffices to prove that any nonidentity f of A also is not an identity of $\tilde{A}$; this follows from Lemma 6.3.19, since A lacks full polynomials by hypothesis.

$\square$

Remark 6.3.21. *Although our exposition deals with algebras with 1, let us digress for a moment to discuss how to generalize this for an algebra A without 1. In this case, Wedderburn's Principal Theorem still holds, enabling us to write $A = \bar{A} \oplus J$ where $\bar{A}$ is a semisimple subalgebra of A. Of course, the unit element e of $\bar{A}$ is not the unit of A. Nevertheless, the Peirce decomposition can be carried out for algebras without 1, cf. Exercise 1.1.*

6.4 Multiplying Alternating Polynomials, and the First Kemer Invariant

We start with some important invariants described in terms of polynomials. Recall that we are looking for a way of describing $t_A = \dim_F(\bar{A})$ and s_A, the nilpotence index of $\operatorname{Jac}(A)$, in terms of polynomials. In view of Theorem J of §1.5, our description should involve alternating nonidentities. But in general one set of alternating indeterminates might involve the "wrong" substitutions, for example radical substitutions. In order to force a usable set of semisimple substitutions we must consider nonidentities that alternate simultaneously in many sets of indeterminates.

Since we consider disjoint subsets $X_1, \ldots, X_\mu$ of indeterminates inside X, it is convenient to rewrite some of the x_i as $x_{k,1}, x_{k,2}$, to indicate that they belong to X_k.

One might think that all of the semisimple substitutions in a polynomial $f(X_1, \ldots, X_\mu; \bar{y})$ must be in the same component R_j, since $R_j R_{j'} = 0$ for all $j \neq j'$. However, this is not so, since the specializations of the y_j could transfer us from stage to stage.

The following definition generalizes Definition 1.2.22.

Definition 6.4.1. *Given disjoint sets $X_1, \ldots, X_\mu$ of indeterminates, say*

$$X_k = \{x_{k,1}, \ldots, x_{k,t_k}\},$$

*suppose the polynomial $f(X_1, \ldots, X_\mu; \bar{y})$ is linear in all of these indeterminates as well as perhaps in extra indeterminates denoted Y. We say that f is $(t_1, \ldots, t_\mu)$-**alternating** if f is t_k-alternating in the indeterminates of X_k, for each $1 \leq k \leq \mu$.*

*If $t_1 = \cdots = t_\mu = t$, we say that f is μ-**fold** t-**alternating**, and call the X_k the **folds** of f. The indeterminates $\cup X_k$ are called the **designated indeterminates**. More generally, if f is $(t_1, \ldots, t_{\mu+\mu'})$-alternating where*

$t_1 = \cdots = t_\mu = l$ *and* $t_{\mu+1} = \cdots = t_{\mu+\mu'} = t'$ *we say that* f *has* μ *folds of* t *indeterminates and* μ' *folds of* t' *indeterminates, and has* $\mu t + \mu' t'$ *designated indeterminates.*

A polynomial with specified alternating folds, cf. Definition 6.4.1 is **monotonic** *if the undesignated indeterminates occur in the same order in each monomial of* f.

For example, the Capelli polynomial c_n, viewed as a 1-fold t-alternating polynomial, is monotonic since the y_j always occur in the order $y_1, \ldots, y_n$.

Our goal is to translate Wedderburn's Principal Theorem into the combinatorics of polynomials, thereby producing non-identities that are alternating in several folds, our main tool being the t-alternator. As in the alternator of Definition 1.2.31, there is an effective way of producing μ-fold t-alternating polynomials. First we focus on a single fold, taking $X = \{x_{i_1}, \ldots, x_{i_t}\}$.

Definition 6.4.2. *We say that a permutation* $\pi \in S_t$ *acts on the indeterminates of* X *when* $x_{i_{\pi(k)}}$ *is substituted for* $x_{i_k} \in X_i$, *for all* $1 \le k \le t$. *Then we write* $f_\pi(X,Y) := f(\pi(X),Y)$. *For example, if* $f = h_1 x_1 h_2 x_2 \cdots h_m x_m h_{m+1}$ *is a monomial, then* $\pi(f) = h_1 x_{\pi(1)} h_2 x_{\pi(2)} \cdots h_m x_{\pi(m)} h_{m+1}$.

Likewise, if $X = X_1 \cup \cdots \cup X_k$ *with* $|X_k| = t_k$, $1 \le k \le \mu$, *and* $\pi_k \in S_{t_k}$, *we can define*

$$f_{\pi_k}(X_1, \ldots, X_\mu, Y) = f(X_1, \ldots, X_{k-1}, \pi_k(X_k), X_{k+1}, \ldots, X_\mu, Y),$$

and likewise

$$f_{\pi_1, \ldots \pi_\mu}(X_1, \ldots, X_\mu, Y) = f(\pi_1(X_1), \ldots, \pi_\mu(X_\mu), Y).$$

Note that the X_k do not need to appear in succession in f. For example, we could have $X_1 = \{x_{11}, x_{12}, x_{13}\}$ and $X_2 = \{x_{21}, x_{22}\}$, and

$$f = x_{11}y_1 x_{21} x_{12} y_2 x_{22} x_{13} y_3 y_4.$$

If $\pi_1 = (132)$ and $\pi_2 = (12)$, then

$$f_{\pi_1 \pi_2} = x_{13}y_1 x_{22} x_{11} y_2 x_{21} x_{12} y_3 y_4.$$

Remark 6.4.3. *Embed* $S_{t_1} \times \cdots \times S_{t_\mu}$ *naturally into* S_t *where* $t = t_1 + \cdots + t_\mu$. *Now we can identify* $f_{\pi_1, \ldots, \pi_\mu}(X_1, \ldots, X_\mu, Y)$ *with* $f_\pi(X_1, \ldots, X_\mu, Y)$, *where* $\pi \in S_t$ *is the permutation corresponding to* $(\pi_1, \ldots, \pi_\mu)$.

Definition 6.4.4. *Recall (Definition 1.27) that the alternator* $f_{A(X)}$ *of a polynomial* f *in a set of indeterminates* $X = \{x_{i_1}, \ldots, x_{i_t}\}$ *is* $\sum_{\pi \in S_t} \text{sgn}(\pi) f_\pi$. *(Up to sign, this is independent of the order of* $i_1, \ldots, i_t$.) *Likewise, the* $(t_1, \ldots, t_\mu)$-**alternator** *in disjoint subsets* X_k *of* X *is*

$$f_{A(X_1)A(X_2)\ldots A(X_\mu)}.$$

Note that the order in which we apply the $\mathcal{A}(X_k)$ is irrelevant. In most instances, f will be a monomial.

Remark 6.4.5. *We need some properties of multiply alternating polynomials.*
  (i) If f is $(t_1,\dots,t_\mu)$-alternating, and we take f_0 to be the sum of monomials of f in which the designated x_i appear in a given order, then f is the $(t_1,\dots,t_\mu)$- alternator of f_0.
  (ii) If f is monotonic in the non-designated indeterminates, then f is the $(t_1,\dots,t_\mu)$-alternator of some monomial. (Indeed, this follows from (i), since $y_1, y_2, \dots$ also occur in a fixed order.)

Remark 6.4.6. *Embedding $S_{t_1} \times \cdots \times S_{t_\mu}$ naturally into S_t as in Remark 6.4.3, where $t = t_1 + \cdots + t_\mu$, we take a transversal $T = \{\sigma_1,\dots,\sigma_m\}$, where $m = \frac{t!}{t_1!\cdots t_\mu!}$. Then*

$$f_{\mathcal{A}(X)} = \sum_{\sigma \in T} \operatorname{sgn}(\sigma)\sigma(f)_{\mathcal{A}(X_1)\mathcal{A}(X_2)\dots\mathcal{A}(X_\mu)},$$

This gives us a way of "combining" $(t_1,\dots,t_\mu)$-alternating polynomials into t-alternators.

Definition 6.4.7. *For any affine PI-algebra W, $\beta(W)$ is the largest t such that for any μ, there is a μ-fold t-alternating polynomial*

$$f(X_1,\dots,X_\mu;Y) \notin \operatorname{id}(W).$$

Our motivating examples are:

Example 6.4.8. *If $A = M_n(F)$, then every (n^2+1)-alternating polynomial is in $\operatorname{id}(A)$, whereas $c_{n^2}(x_1,\dots,x_{n^2};y_1,\dots,y_{n^2})^\mu$ takes on the value $e_{11}^\mu = e_{11}$, for any μ. Thus, $\beta(M_n(F)) = n^2$ in this case.*

Example 6.4.9. *Take $A = \operatorname{UT}(2)$, and*

$$f(x_{1,1},x_{1,2},y,x_{2,1},x_{2,2}) = x_{1,1}x_{2,1}yx_{1,2}x_{2,2}.$$

Then $f(e_{1,1},e_{1,1},e_{1,2},e_{2,2},e_{2,2}) = e_{1,2} \neq 0$, whereas taking $X_k = \{x_{k,1},x_{k,2}\}$ for $k=1,2$ we get $f_\pi(e_{1,1},e_{1,1},e_{1,2},e_{2,2},e_{2,2}) = 0$ for every $\pi \neq (1)$ in $S_2 \times S_2$ (acting naturally on $X_1 \cup X_2$), so $f_{\mathcal{A}(X_1)\mathcal{A}(X_2)} \notin \operatorname{id}(A)$. This same argument works for arbitrary μ instead of 2, i.e., taking $X_k = \{x_{k,1},x_{k,2}\}$, $1 \leq k \leq \mu$, and

$$f = x_{1,1} \cdots x_{k,1}yx_{1,2} \cdots x_{k,2}.$$

Thus we conclude that $\beta(A) \geq 2 = \dim(F \oplus F) = \dim \bar{A}$. But clearly $\beta(A) < 3$, which also follows as an application of Lemma 6.5.1, so $\beta(A) = 2$.

Example 6.4.10. *Let us try to get some intuition. Given a f.d. full algebra A, we assume for ease of notation that $R_1 J \cdots R_{q-1} J R_q \neq 0$. Let*

$$u_{j,1},\dots,u_{j,t_j};v_{j,1},\dots,v_{j,t_j} \in R_j$$

be as in the easy Remark 1.4.3, that is, compatible with $x_1 y_1 \ldots x_{t_j} y_{t_j}$, where $t_j = n_j^2$, and whose product is e_{11} in R_j. Taking type $(j, j+1)$ radical substitutions r_j, we claim that

$$u_{1,1}, \ldots, u_{1,t_1}, u_{2,1}, \ldots, u_{2,t_2}, \ldots, u_{q,1}, \ldots, u_{q,t_q};$$
$$v_{1,1}, \ldots, v_{1,t_1} r_1, v_{2,1}, \ldots, v_{2,t_2} r_2, \ldots, v_{q,1}, \ldots, v_{q,t_q} \tag{6.4}$$

is a 1-fold $(\dim_F \bar{A} + q - 1)$-compatible substitution (when the r_i are chosen to make it nonzero). Thus, A has a 1-fold $(\dim_F \bar{A} + q - 1)$-alternating polynomial that is not an identity. (Here the r_k are counted among the alternating positions.) But this only happens once, not arbitrarily many times, so it does not give us $\beta(A)$.

Although, as we just saw, $\beta(W)$ exists for affine algebras W, Equation (1.17) shows that any product of Capelli identities is not an identity of the Grassmann algebra G. So $\beta(G)$ does not exist. This is the main reason whereby in this chapter we can only prove Kemer's PI-representability Theorem for affine algebras, and need more theory to get the full solution to Specht's question in characteristic 0, in Chapter 7.

6.4.1   Compatible substitutions

Fortunately, there is one case when the t-alternator clearly is not an identity, namely when we have a t-compatible set of matrix units as defined in Remark 1.4.3. Since we shall rely heavily on this idea, let us generalize it.

Definition 6.4.11. *A set of elements* $u_1, \ldots, u_t; v_1, \ldots, v_m$ *is* **compatible with** *a polynomial*

$$h(x_1, \ldots, x_t; y_1, \ldots, y_m)$$

if $h(u_1, \ldots, u_t; v_1, \ldots, v_m) \neq 0$ *but*

$$h(u_{\pi(1)}, \ldots, u_{\pi(t)}; v_1, \ldots, v_m) = 0, \quad \forall \pi \neq (1).$$

More generally, a set of semisimple and radical elements $\{u_{i,j} : 1 \leq i \leq t, 1 \leq j \leq \mu\}$, $\{v_1, \ldots, v_m\}$ *in* A *is* μ-**fold** t-**compatible** *in a polynomial* $h(X_1, \ldots, X_\mu; Y)$ *in* μ *sets of variables* $X_j = \{x_{1,j}, \ldots, x_{t,j}\}$, $1 \leq j \leq \mu$, *and in* $y_1, \ldots, y_m$ *if*

$$h(u_{1,1}, \ldots u_{t,1}, \ldots, u_{1,\mu}, \ldots, u_{t,\mu}; v_1, \ldots, v_m) \neq 0,$$

but the specialization

$$h(u_{\pi_1(1),1}, \ldots, u_{\pi_1(t),1}, \ldots, u_{\pi_\mu(1),\mu}, \ldots, u_{\pi_\mu(t),\mu}; v_1, \ldots, v_m)$$

is 0 *whenever some* $\pi_j \neq (1)$. *We call* $\{u_{ij}\}$ *and* $\{v_1, \ldots, v_m\}$ *an* μ-**fold** t-**compatible substitution**; *the* $\{u_{ij}\}$ *occupy* **alternating positions**.

Example 6.4.12. *The special case we encountered in Remark 1.4.3 was for the Capelli polynomial $h = c_t(x_1 y_1 \cdots x_t y_t)$.*

The interplay between full and compatible lies at the heart of our discussion.

Lemma 6.4.13. *If a polynomial $h(X_1, \ldots, X_\mu; Y)$ has a full μ-fold t-compatible set of substitutions in $\bar{A}$ which runs through all the R_i, then the μ-fold t-alternating polynomial*

$$\tilde{h} = \sum_{\pi_1, \ldots, \pi_\mu \in S_t} \mathrm{sgn}\pi_1 \cdots \mathrm{sgn}\pi_\mu h(\pi_1(X_1), \ldots, \pi_\mu(X_\mu); Y)$$

is full, and this set of substitutions is full in $\tilde{h}$.

Proof. Only one substitution is nonzero, and this is the original one. $\square$

(Note that this argument is characteristic free.)

6.5   Kemer's First Lemma

Recall $\beta(A)$ from Definition 6.4.7.

Lemma 6.5.1. $\beta(A) \leq t_A$.

Proof. Suppose the polynomial $f(X_1, \ldots, X_\mu; y_1, \ldots, y_m)$ is μ-fold and t-alternating, for $X_j = \{x_{j,1}, \ldots, x_{j,t}\}$ and is nonzero under the specializations $x_{j,i} \mapsto \bar{x}_{j,i}$ and $y_i \mapsto \bar{y}_i$. At most $\dim \bar{A}$ substitutions of every X_j are semisimple. So if $t > \dim \bar{A}$, then every nonzero substitution in f must have a radical substitution. But this can happen at most $s - 1$ times in the various folds when $J^s = 0$, so taking $\mu > s$ we conclude that $\beta(A) < t$. $\square$

Proposition 6.5.2 (Kemer's First Lemma). *If F is algebraically closed and A is f.d. full, then $\beta(A) = t_A$.*

Proof. Let $t = t_A$. $\beta(A) \leq t$, by Lemma 6.5.1. To show that $\beta(A) = t$, for any μ we must find a μ-fold t-alternating polynomial that is not an identity of A. We do this by finding a μ-fold t-compatible substitution for a suitable monomial $h(X_1, \ldots, X_\mu; Y; Z)$ of f, where each $|X_j| = t$, $|Y| = \mu t$, and $|Z| = q - 1$. (We have introduced a new set of variables Z that are not involved in the alternation, but play a special role that we want to specify for further use.)

The idea of the proof is to build folds inside idempotent substitutions. Using Remark 6.3.16, we may modify the full polynomial $f(x_1, x_2, \ldots)$ that we are constructing, so that we can replace the various $\bar{x}_k$ by the idempotent $e_{11}^{(k)}$ and still get the same (nonzero) value. Writing this $e_{11}^{(k)}$ itself as an evaluation of a Capelli polynomial, permits us to attach μ-folds to f and still have a nonidentity.

For example, recall from the easy Remark 1.4.3 that

$$c_4(e_{11}, e_{12}, e_{21}, e_{22}; e_{11}, e_{22}, e_{12}, e_{21}) = e_{11},$$

implying $c_4 \notin \mathrm{id}(M_2(\mathbb{Q}))$. But then we see that

$$c_4(x_1 c_4(x_5, x_6, x_7, x_8; y_5, y_6, y_7, y_8), x_2, x_3, x_4; y_1, y_2, y_3, y_4) \notin \mathrm{id}(M_2(\mathbb{Q}),$$

and we could continue this ad infinitum. The case $q > 1$ is more complicated since we may need to interweave substitutions in several different simple components $R_1, \ldots, R_q$ of $\bar{A}$, but the idea always remains the same.

Take the substitution of Example 6.4.12. By definition,

$$u_{k,1} v_{k,1} \cdots u_{k,t_k} v_{k,t_k} = e_{11}^{(k)} \in R_k;$$

since $e_{11}^{(1)}$ is idempotent, we could take μ copies of these and label them with j, for $1 \leq j \leq \mu$, i.e.,

$$u_{k,1,j} v_{k,1,j} \cdots u_{k,t_k,j} v_{k,t_k,j} = e_{11}^{(k)}$$

for each j, so we still have

$$u_{k,1,1} v_{k,1,1} \cdots u_{k,t_k,1} v_{k,t_k,1} \cdots u_{k,1,\mu} v_{k,1,\mu} \cdots u_{k,t_k,\mu} v_{k,t_k,\mu} = e_{11}^{(k)}.$$

From the graph-theoretic point view of Remark 6.3.7, all substitutions into the stage R_k still must occur together to achieve a nonzero value. By the same argument as Lemma 1.4.3, this is t-compatible, since any non-identity permutation of the $u_{k,i,j}$ (for any given j, k) would yield the value 0.

To write the monomial h explicitly, recall that $\bar{A} = \oplus_{k=1}^{q} M_{n_k}(F)$ and $t_k = n_k^2$; write

$$X_j = \{x_{k,i,j} : 1 \leq k \leq q, \quad 1 \leq i \leq t_k\}.$$

Thus,

$$|X_j| = \sum_{k=1}^{q} t_k = \sum_{k=1}^{q} t_k = \dim_F \bar{A} = t.$$

Write

$$Y = \{y_{k,i,j} : 1 \leq k \leq q, \ 1 \leq i \leq t_k, \ 1 \leq j \leq \mu\}; \quad Z = \{z_k : 1 \leq k \leq q - 1\},$$

and

$$h_k = x_{k,1,1} y_{k,1,1} \cdots x_{k,t_k,1} y_{k,t_k,1} \cdots x_{k,1,\mu} y_{k,1,\mu} \cdots x_{k,t_k,\mu} y_{k,t_k,\mu}, \quad 1 \leq k \leq q;$$

consider

$$h_1 h_2 \cdots h_q \, _{\mathcal{A}(X_1) \ldots \mathcal{A}(X_\mu)}. \tag{6.5}$$

Specializing $x_{k,i,j} \mapsto u_{k,i,j}$, $y_{k,i,j} \mapsto v_{k,i,j}$, $z_k \mapsto r_k$ yields the desired μ-fold t-compatible substitution for the sets of variables $X_1, \ldots, X_\mu; Y \cup Z$. (Switching $u_{k,i,j}$ with $u_{k',i,j}$ for $k \neq k'$ yields 0 since we are jumping from one stage to another.) $\qquad\square$

Remark 6.5.3. *The above proof shows that we can tack an extra fold of t-alternating variables onto any non-identity of A, whenever $\beta(A) \geq t$.*

6.6   Kemer's Second Lemma

Kemer's Second Lemma is subtler than the first, and requires considerably more preparation. The radical substitutions come into play more as we describe intrinsically the second invariant s_A.

Remark 6.6.1. *For any f.d. algebra A, any s_A-fold (t_A+1)-alternating polynomial $f(X_1, \ldots, X_{s_A}; Y)$ is an identity of A. Indeed, in view of Example 6.4.12, any nonzero specialization of f must involve at least one radical substitution in each X_j, so has at least s_A radical substitutions in all. But then every monomial must have elements of J appearing in at least s_A positions, so belongs to $J^{s_A} = 0$, contradiction. (Compare with Example 6.4.12.)*

To continue, we must refine our previous analysis — fullness requires $q \leq s_A$ in the above notation, but does not give us enough information when $s_A > q$. Toward this end, we need a few more definitions focusing on the invariants t_A and s_A.

Having defined $\beta(W)$ in Definition 6.4.7, we are led naturally to a refinement:

Definition 6.6.2. *$\gamma(W)$ is the largest number s such that for each μ, there is a nonidentity of W which is $(s-1)$-fold $(t+1)$-alternating with another μ folds that are t-alternating.*

In the literature, $\gamma(W)$ is defined to be 1 less, so as to be 0 for semisimple PI-algebras. We chose this definition so that Kemer's Second Lemma below would often give equality with the nilpotence index of the Jacobson radical.

Definition 6.6.3. *The **Kemer index** of an affine PI-algebra W, denoted index(W), is the ordered pair $(\beta(W), \gamma(W))$. For a T-ideal $\Gamma = \mathrm{id}(W)$, we also write index(Γ) for the Kemer index of W, and call it the **Kemer index** of Γ.*

Remark 6.6.4. *Let us note that, as with Zubrilin's theory of Chapter 2, the definition of Kemer index is purely combinatoric, and does not rely on the base ring being a field. Thus, it can be used as a tool for proving theorems about PI-algebras over arbitrary commutative rings satisfying a Capelli identity, such as Theorem 10.1.2 below.*

Example 6.6.5. index($M_n(F)$) = $(n^2, 1)$. *Thus, the same holds for any algebra of PI-class n.*

Our first objective with the Kemer index is to find a f.d. algebra over an infinite field F of characteristic 0, having the same Kemer index as an arbitrary affine PI-algebra W. Toward this end, we want to describe the Kemer index of a f.d. algebra A in terms of t_A and s_A.

Latent in the definition of Kemer index is another PI-invariant.

Definition 6.6.6. $\omega(W)$ *is the smallest μ such that every polynomial which is $\gamma(W)$-fold, $(\beta(W) + 1)$-alternating and μ-fold $\beta(W)$-alternating is in* id(W).

In the relevant applications, we require μ to be sufficiently larger than $\omega(W)$. We write $\omega(\Gamma)$ for $\omega(W)$ where id$(W) = \Gamma$. The Kemer index is measured in terms of a certain class of polynomials.

Definition 6.6.7. *Suppose $\beta(W) = t$ and $\gamma(W) = s$. A μ-**Kemer polynomial** for W is a polynomial $f \notin$ id(W), in $s + \mu - 1$ sets of designated indeterminates X_j (and may could also involve other indeterminates), which is $(s-1)$-fold $(t+1)$-alternating with another μ folds that are t-alternating. In other words,*

$$|X_1| = \cdots = |X_{s-1}| = t + 1; \qquad |X_s| = |X_{s+1}| = \cdots = |X_{s+\mu-1}| = t,$$

and f is alternating in each set of X_j.

If $\Gamma = $ id(W), a μ-Kemer polynomial for W is also called a μ-Kemer polynomial for Γ.

We take μ "large enough" for the proof to work, in line with the following observation:

Lemma 6.6.8. *If $\mu \geq s_A$ and A is full, then any μ-Kemer polynomial f of A is full.*

Proof. Otherwise for any nonzero evaluation of f, any t_A-alternating fold of f has at most $t_A - 1$ semisimple substitutions, and thus at least one radical substitution. Altogether there are at least $\mu > s_A$ radical substitutions, thereby yielding the value 0, a contradiction. □

Lemma 6.6.9. *If* index$(W) = \kappa = (\beta_W, \gamma_W)$ *and I is the ideal generated by all evaluations in W of μ-Kemer polynomials of W, then* index$(W/I) < \kappa$.

Proof. Clearly, index$(W/I) \leq \kappa$. But if index$(W/I) = \kappa$, then there would be a μ-Kemer polynomial, which by hypothesis is in id(W/I), contradiction. □

Lemma 6.6.10. *If $\Gamma_1 \subseteq \Gamma$, then* index$(\Gamma) \leq$ index(Γ_1).

Proof. Each Kemer polynomial for Γ is not in Γ_1, so we first note that $\beta(\Gamma) \leq \beta(\Gamma_1)$ and then $\gamma(\Gamma) \leq \gamma(\Gamma)$.

An immediate consequence from Step 1 is that index(W) exists for every affine PI-algebra W, since it exists for f.d. algebras.

□

Kemer indices are ordered via the lexicographic order on pairs.

Proposition 6.6.11. *Suppose for each i that* index(Γ_i) < index(Γ) *for T-ideals Γ and $\Gamma_1, \ldots, \Gamma_m$. Then there is $\mu \in \mathbb{N}$ such that **every** μ-Kemer polynomial for Γ is in $\bigcap_{i=1}^m \Gamma_i$.*

Proof. This observation follows at once from the definitions. Write index $\Gamma = (t, s)$ and index(Γ_i) $= (t_i, s_i)$. If $t_i < t$, then by definition there is some number $\mu_i = \omega(\Gamma_i)$ such that any μ_i-fold t-alternating polynomial is in Γ_i. If $t_i = t$, then by definition any μ_i-fold t-alternating and $\gamma(\Gamma) - 1$-fold $(t+1)$-alternating polynomial is in Γ_i. Hence, taking

$$\mu = \max\{\mu_1, \ldots, \mu_m\},$$

we see by definition that each μ-Kemer polynomial is in each Γ_i, and thus the T-ideal they generate also is in Γ_i. $\qquad\square$

Corollary 6.6.12. index($W_1 \times \cdots \times W_m$) $= \max_{i=1}^m$ index(W_i), *for any PI-algebras W_i on which the Kemer index is defined.*

Nevertheless, we must be careful.

Example 6.6.13. UT(n) *has Kemer index (n, n), but does not satisfy the standard identity s_{2n-2}, which holds in $M_{n-1}(F)$ (which has Kemer index $(n^2 - 1, 1)$). Thus the Kemer index does not correlate certain celebrated identities.*

6.6.1 Computing Kemer polynomials

Since Kemer polynomials will comprise the backbone of our theory, let us digress to see how they can be constructed.

Remark 6.6.14. *Let f be a Kemer polynomial. If we call the undesignated indeterminates $y_1, \ldots, y_m$, then for $\sigma \in S_m$, we could define f_σ to be the sum of those monomials in which the y_j appear in the order*

$$y_{\sigma(1)}, \ldots, y_{\sigma(m)}.$$

*Then clearly $f = \sum_{\sigma \in S_m} f_\sigma$, so some $f_\sigma \notin \mathrm{id}(A)$, and any such f_σ is also a Kemer polynomial. (One such f_σ is Kemer iff all are Kemer.) For future use (after we prove Kemer's Theorem) we call the f_σ the **monotonic components** of the Kemer polynomial f.*

We want to apply the same idea to Kemer polynomials (or more precisely, their monotonic components), and so we must consider polynomials which alternate in several folds, cf. Definition 6.4.4.

Definition 6.6.15. *Suppose* $\kappa = (t, s)$. *We define the* μ-**alternator** *(for* **Kemer index** κ) *to be the application of* $s - 1$ *folds of* $(t+1)$-*alternators and* μ *folds of* t-*alternators, or in the notation of the previous paragraph,*

$$\hat{f}_{\mathcal{A}(X_1)\ldots\mathcal{A}(X_s)\mathcal{A}(X_{s+1})\ldots\mathcal{A}(X_{s+\mu})},$$

where $|X_k| = t + 1$ *for* $1 \le k \le s$ *and* $|X_k| = t$ *for* $s + 1 \le k \le s + \mu$.

Remark 6.6.16. *Any* μ-*Kemer polynomial for index* $\kappa = (t, s)$ *is the sum of* μ-*alternators (appropriate to Kemer index* κ) *of suitable monomials. (Indeed, the polynomial is a sum of its monotonic Kemer components, so apply Remark 6.4.5(ii).)*

Definition 6.6.17. *A f.d. algebra A is* **basic** *if A is not PI equivalent to an algebra $A_1 \times \cdots \times A_m$, for f.d. algebras A_i (over F) with* index(A_i) < index(A), $1 \le i \le m$.

Since we are at the level of identities, we may assume that the A_i all are basic. Conceivably A could be reducible but still basic since the decomposition could involve an algebra of smaller dimension but still of the same Kemer index.

Remark 6.6.18. *Any basic algebra is full, since otherwise it is reducible into algebras of smaller Kemer index, by Proposition 6.3.20.*

The basic algebras are the building blocks of our theory, as indicated by our next observation.

Proposition 6.6.19. *Any f.d. algebra A is PI-reducible to finitely many basic f.d. algebras.*

Proof. Induction on the ordered pair (t_A, s_A), where we write A as a subdirect product of irreducible f.d. algebras A_i. $\qquad\square$

6.6.2 Property K

We turn to our next important kind of polynomial.

Definition 6.6.20. *A multilinear polynomial $f(x_1, \ldots, x_m)$ has* **Property K** *on A if f vanishes under any specialization $\bar{x}_1, \ldots, \bar{x}_m$ with fewer than $s_A - 1$ radical substitutions.*

This is anti-intuitional when $f \notin \mathrm{id}(A)$, since we know f must vanish under any specialization $\bar{x}_1, \ldots, \bar{x}_m$ with $\ge s_A$ radical substitutions (because $J^{s_A} = 0$). Since $f \notin \mathrm{id}(A)$, one concludes that f must not vanish with respect to some specialization with precisely $s_A - 1$ radical substitutions. But this is precisely the case that will be of interest to us.

Here is a very useful construction. Since we need a related formulation later, we start with the underlying idea.

Definition 6.6.21. *Suppose we are given a T-ideal $\Gamma \supset \mathrm{id}(A)$. Take $A' = \bar{A}\langle x_1, \ldots, x_\nu \rangle$ of Definition 1.9.3, where ν is some fixed large number, at least $\dim \mathrm{Jac}(A) = \dim A - \dim \bar{A}$. Let $\mathcal{I}_1$ be the ideal generated by all evaluations in A' of polynomials in Γ, and L be the ideal generated by $x_1, \ldots, x_\nu$, and define*

$$\hat{A}_{u,\nu;\Gamma} = A'/(\mathcal{I}_1 + \mathcal{I}_2), \tag{6.6}$$

where $\mathcal{I}_2 \supseteq L^u$. Since Γ is understood, we omit it from the notation and write merely $\hat{A}_{u,\nu}$.

Remark 6.6.22. *In other words, $\hat{A}_{u,\nu}$ contains $\bar{A}$ and satisfies the identities from Γ, and the radical has nilpotence index at most u.*

Remark 6.6.23. *When the ideal I_2 is homogeneous under the natural $\mathbb{N}$-grading, Equation (6.6) enables us to view this construction as an instance of the relatively free product, with a natural grade on the $\hat{x}_j$.*

This provides an important tool.

Definition 6.6.24. *The u-generic algebra $\hat{A}_u$ of a f.d. algebra A is defined as in Definition 6.6.21, where $\Gamma = \mathrm{id}(A)$ and $I_2 = L^u$.*

Remark 6.6.25. *$\hat{A}_u$ is the "freest" algebra containing $\bar{A}$ and satisfying the identities from $\mathrm{id}(A)$, such that the radical J is generated by ν elements and has nilpotence index u.*
   Note that $\mathrm{Jac}(\hat{A}_u)$ is the image $\hat{I}_2$ of I_2, since $\hat{A}_u/\hat{I}_2 \cong \bar{A}$, which is semisimple, and $\hat{I}_2$ is nilpotent. Furthermore, letting $\hat{x}_i$ denote the image of x_i, and taking a base $\{b_1, \ldots, b_k\}$ of $\bar{A}$, we see that $\hat{A}_u$ is spanned by all terms of the form

$$b_{i_1}\hat{x}_{j_1}b_{i_2}\hat{x}_{j_2}\cdots\hat{x}_{j_{u'-1}}b_{i_{u'}}, \quad u' \le u, \tag{6.7}$$

and thus is f.d. over F. By definition, $\mathrm{id}(A) \subseteq \mathrm{id}(\hat{A}_u)$ and $(t_{\hat{A}_u}, s_{\hat{A}_u}) \le (t_A, s_A)$.

Remark 6.6.26.

(i) *Equation (6.6) enables us to view $\hat{A}_u$ as an instance of the relatively free product, thereby yielding a natural map $\psi_{u',u} : \hat{A}_{u'} \to \hat{A}_u$ for any $u < u'$, since $J^u = 0$ implies $J^{u'} = 0$ in $\hat{A}_{u'}$.*

(ii) *One can find a surjection $\hat{A}_{s_A} \to A$ by means of the Wedderburn decomposition of A (sending the "generic nilpotent" elements to a base of $\mathrm{Jac}(A)$). Thus, $\hat{A}_u \sim_{\mathrm{PI}} A$ for all $u \ge s_A$.*

(iii) *When $\mathrm{id}(A)$ is generated by completely homogeneous identities, $\hat{A}_u$ is $\mathbb{N}$-graded, according to the total degree in the "radical" indeterminates. But in characteristic 0 every T-ideal is generated by multilinear identities, because of multilinearization. (Thus the graded theory is available for these algebras.)*

(iv) Every polynomial in $\mathrm{id}(\hat{A}_{s_A-1}) \setminus \mathrm{id}(\hat{A}_{s_A})$ satisfies Property K, in view of (iii), since it vanishes on fewer than $s_A - 1$ radical substitutions.

Lemma 6.6.27. *The natural map $\psi_{u',u} : A_{u'} \to \hat{A}_u$ preserves the $\mathbb{N}$-graded vector space $V = \sum_{j=1}^{u-1} (\hat{A}_u)_j$.*

Proof. The homogeneous components of V are preserved, in view of Remark 6.6.26(iii). $\qquad\square$

6.6.3 The second Kemer invariant

We return to the second Kemer invariant, $\gamma(A)$.

Remark 6.6.28. *In Remark 6.6.1, we saw that $\gamma(A) \leq s_A$, for any f.d. algebra A.*

To determine when $\gamma(A) = s_A$, we need to refine our argument of Kemer's First Lemma. One of the key steps is a technical-sounding result, which enables us both to calculate $\gamma(A)$ and also get a good hold on Kemer polynomials.

Definition 6.6.29. *For any T-ideal Γ of Kemer index (t, s),*

$$\nu_\Gamma(\mu) = (t+1)(s-1) + t.$$

When $\Gamma = \mathrm{id}(A)$ we write $\nu_A(\mu)$ for $\nu_\Gamma(\mu)$.

Lemma 6.6.30. *For any full nonidentity f of A satisfying Property K, and arbitrary μ, the T-ideal generated by f contains a full μ-Kemer polynomial of degree $\deg f + \nu_A(\mu)$.*

Proof. Let $s = s_A$. We know that $\gamma(A) \leq s$, so to prove the assertion it suffices to find a μ-fold t-alternating and $(s-1)$-fold $(t+1)$-alternating polynomial

$$g(X_1, \ldots, X_{\mu+s-1}; Y) \notin \mathrm{id}(A).$$

As in the proof of Kemer's First Lemma, we want to achieve this by finding an $(s-1)$-fold $(t+1)$-compatible substitution for a suitable polynomial. (In fact, we want to add another μ folds that are t-compatible, in order to obtain a μ-Kemer polynomial.)

In view of f being full, we can choose $r_k \in e_{\ell_k} J e_{\ell'_k}$ for $1 \leq k \leq s-1$ and $b_1, \ldots, b_m \in A$ yielding a nonzero full substitution for f, and we rewrite f as $f(z_1, \ldots, z_{s-1}; \vec{x}; \vec{y})$, in order to pinpoint the variables $z_1, \ldots, z_{s-1}$ that are to have the radical substitutions. Thus,

$$f(r_1, \ldots, r_{s-1}; \vec{a}; b_1, \ldots, b_m) \neq 0. \tag{6.8}$$

As in the proof of Kemer's First Lemma, we shall specialize the z_k to more complicated expressions involving Capelli polynomials, in order to create new folds.

Recall that $\bar{A} = R_1 \oplus \cdots \oplus R_q$, where $R_k = M_{n_k}(F)$. Write $t_k = n_k^2$.

Case 1. $q > 1$. Similar to the proof of Kemer's First Lemma, since A is full, the graph touches each stage (cf. Remark 6.3.7). Thus, $q \leq s$. We write

$$r_k \in e_{\iota_k} J e_{\iota'_k}, \quad 1 \leq k < q \tag{6.9}$$

where $\iota_k \neq \iota'_k$ and together they cover all the indices from 1 to q. For notational simplicity we assume from now on that $\iota_k = k$, $\iota'_k = k+1$ for $k < q$, and $\iota'_q = q$. Note that we may have $q < s$, in which case there are "extra" r_k.

For $k < q - 1$ we then have $r_k = e^{(k)}_{\ell\ell} r_k$, for a suitable matrix unit $e^{(k)}_{\ell\ell}$ in R_k; we assume that $\ell = 1$, so $r_k = e^{(k)}_{11} r_k$.

Likewise, for $k = q - 1$, we may rewrite $r_1, \ldots, r_q$ as $r_{q-1} = e^{(q)}_{11} r_{q-1} e^{(q)}_{11}$.

Thus, we have pasted appearances of $e^{(k)}_{11}$ in each monomial of f, for each $1 \leq k \leq q$. Let us write $e^{(k)}_{11}$ as a value of a suitable Capelli polynomial $c_{n_{t_k}}$, say

$$e^{(k)}_{11} = c_{t_k}(u_{k,1}, \ldots, u_{k,t_k}; v_{k,1}, \ldots, v_{k,t_k}),$$

under a suitable compatible substitution $u_{k,i}, v_{k,i}, 1 \leq i \leq t_k$ of matrix units in R_k, cf. Remark 1.4.3. These are all semisimple substitutions, by definition.

Replacing z_k $(1 \leq k \leq q - 1)$ throughout by the monomials

$$y_{k,0}(x_{k,1,1} y_{k,1} \cdots x_{k,t_k,1} y_{k,t_k}) \cdots (x_{k,1,s+\mu-1} y_{k,1} \cdots x_{k,t_k,s+\mu-1} y_{k,t_k}) z_k,$$

and then once again replacing z_{q-1} by

$$z_{q-1}(x_{q,1,1} y_{q,1} \cdots x_{q,t_q,1} y_{q,t_q}) \cdots (x_{q,1,s+\mu-1} y_{q,1} \cdots x_{q,t_q,s+\mu-1} y_{q,t_q}) y_{q,0},$$

yields a new polynomial f', clearly in the T-ideal generated by f, for which the substitutions

$$x_{k,i,j} \mapsto u_{k,i}, \quad y_{k,i} \mapsto v_{k,i}, \quad 1 \leq i \leq t_k, \quad y_{k,0} \mapsto e^{(k)}_{11}$$

yield the original (nonzero) value of f, but any other substitution that sends $x_{k,i,j} \mapsto u_{k',i'}$ for different k' or i' (and still sending $y_{k,i} \mapsto v_{k,i}$) is 0. In other words, f' is $(s + \mu - 1)$-fold t-compatible. (Note that we tacked expressions on both sides of z_{q-1}.)

Letting $X_j = \{x_{k,1,j}, \ldots, x_{k,t_k,j} : 1 \leq k \leq q\}$ for each $1 \leq j \leq \mu$, we obtain an $(s+\mu-1)$-fold t_A-alternating polynomial $f'' = f'_{A(X_{k,j}), \; 1 \leq k \leq q, 1 \leq j \leq \mu}$ whose degree is $\deg f + \nu_A(\mu)$.

To complete the proof of Case I, we need to increase $|X_j|$ by 1, for each $j \leq s - 1$. Toward this end, we reassign z_j to X_j, now putting

$$X_j = \{z_j, x_{k,i,j} : 1 \leq k \leq q, \quad 1 \leq i \leq t_k\}$$

for $1 \leq j \leq s - 1$.

To prove compatibility, we need to show that in our evaluation of f', any permutation of the r_j and the $u_{k,i,j}$ produces 0 (where we are working with the specializations $u_{k,i,j} := u_{k,j}$ of $x_{k,i,j}$). Since we have already seen this with permutations fixing r_j, it remains to consider the situation where we switch r_j with some $u_{k,i,j}$. But in this case, we get 0 unless $r_j \in e_k A e_k$ (because it is now bordered on either side by semisimple substitutions from R_k). In particular, $j \geq q$ in view of (6.9). The effect in our original evaluation of f in (6.8) has been to incorporate r_j into the substitution for z_k (which now is thus in J^2), and turn the substitution of z_j into $u_{k,i,j}$, a semisimple substitution. This switch has lowered the number of radical substitutions to $< s - 1$, so by the hypothesis of property K on f, the corresponding specialization of f must be 0. But the new specialization of f' (under switching r_j with $u_{k,i,j}$) has the same value as this new specialization of f, i.e., it also is 0, as desired.

Case 2. $q = 1$. This is conceptually easier, but less convenient, than the previous case. Now we do not have any place to tag on the extra substitutions; we have no r_k for $k \leq q - 1$ because $q - 1 = 0$. On the other hand, the structure of A is much nicer: $\bar{A} = M_n(F)$ where $n^2 = t = \dim_F \bar{A}$, so we can lift the matrix units and get $A = M_n(H)$ for some algebra H for which $H/\operatorname{Jac}(H) \cong F$. (This is not strictly relevant to the proof, but is included to provide intuition and keep the notation as easy as possible.) So suppose $f(r_1, \dots, r_{s-1}, b_1, \dots, b_m) = a e_{ij}$ for some $a \in H$. Reordering the indices, we may assume $e_{ij} = e_{1j}$, and again we take $s + \mu - 1$ t-compatible substitutions of matrix units for c_t, where

$$c_t(u_{1,j}, \dots, u_{t,j}; v_1, \dots, v_t) = e_{11}.$$

Take

$$f' = c_t(x_{1,1}, \dots x_{t,1}, y_1, \dots, y_t) \cdots c_t(x_{1,s+\mu-1}, \dots, x_{t,s+\mu-1}, y_1, \dots y_t) f.$$

Clearly, this is $(s + \mu - 1)$-fold t-compatible, and as before, we claim that we may enlarge the first $s - 1$ sets of variables, i.e., put

$$X_j = \{x_{1,j}, \dots, x_{t,j}, z_j\}$$

for $1 \leq j \leq s - 1$. Again, we need to show that interchanging r_j and some $u_{i,j}$ gives 0. But this is for the same reason: The radical substitution for z_j has been changed to a semisimple substitution, so the specialization of f now has fewer than $s - 1$ radical substitutions, so is 0, and this is also the specialization of f', as desired.

Our polynomial is full, by Lemma 6.6.8. $\qquad\square$

(The possibility that $q < s$ made this proof more complicated than that of Kemer's First Lemma.) We are ready for a key result.

Proposition 6.6.31 (Kemer's Second Lemma). *(Requires $\operatorname{char}(F) = 0$.) If A is a basic f.d. algebra, then $\gamma(A) = s_A$. In this case, A has a μ-Kemer* polynomial for every μ. In fact, for any full polynomial f satisfying property K, the T-ideal generated by f contains a μ-Kemer polynomial for A whose degree is $\deg f + \nu_A(\mu)$.

Proof. Consider the algebra $B = \mathcal{A}_{s-1,\nu} \times \tilde{A}$, where $s = s_A, \nu = \dim Jac(A)$, and $\tilde{A}$ is the algebra constructed in Definition 6.3.17. Note that $id(A) \subset \times \tilde{A}$ and $id(A) \subset \mathcal{A}_{s-1,\nu}$, implying $id(A) \subseteq id(B)$.

We claim that if there are no multilinear nonidentities of A which are both full and satisfy property K, then $id(B) = id(A)$, contrary to the hypothesis that A is basic. Indeed, since $char(F) = 0$, it suffices to prove that every multilinear nonidentity f of A is a nonidentity of B.

If every consequence of f is not full, then by Lemma 6.3.19, f also is a nonidentity of $\tilde{A}$ and thus of B.

Thus, we may assume that f has a consequence g which is multilinear and full. Hence, g does not satisfy property K, implying that f also does not satisfy property K. By Remark 6.6.26(iv), f is a nonidentity of $\mathcal{A}_{s-1,\nu}$. We conclude that f is a nonidentity of B. $\qquad\square$

6.7   Significance of Kemer's First and Second Lemmas

As always in this chapter, F is an algebraically closed field of characteristic 0, and A is a f.d. PI-algebra (with $\bar{A} = A/Jac(A)$). A is a finite subdirect product of basic f.d. algebras, which we can replace by their direct product. So we will assume that

$$A = A_1 \times \cdots \times A_m \qquad\qquad (6.10)$$

for each A_k basic f.d. Note that $\bar{A} = \overline{A_1} \times \cdots \times \overline{A_m}$, so the components of $\bar{A}$ are the direct products of the matrix components of the $\bar{A}_k$.

Remark 6.7.1.

(i) For each $1 \le k \le m$, $\beta(A_k) = t_k = \dim_F \bar{A}_k$, and $\gamma(A_k) = s_k$, the nilpotence index of $Jac(A_k)$.

(ii) For any μ, A_k has a μ-Kemer polynomial whose first $s_k - 1$ folds are $(t_k + 1)$-alternating and has μ extra t_k-alternating folds (but is not in $id(A)$ by definition). We also recall that by definition of $\gamma(A)$, any s_k-fold $(t_k + 1)$-alternating polynomial is in $id(A_k)$.

(iii) Any nonzero specialization of a μ-Kemer polynomial f of A has precisely t semisimple substitutions in each of the first $s_k - 1$ folds. (Indeed, if there were $t_k + 1$ semisimple substitutions in some fold, then that fold would yield 0. Hence each fold has at least one radical substitution, and if any fold had fewer than t_k semisimple substitutions it would have two radical substitutions, the evaluations would be in $J^{s_k} = 0$.

(iv) *Any non-zero specialization of a non-full μ-Kemer polynomial f has a radical substitution in each of the subsequent t_k-folds, so there are at most $s_k - 1$ of these; i.e., $\mu \leq s_k - 1$.*

(v) *Applying (iii) to Theorem J of §1.5, we see that if f is any Kemer polynomial for any A_k, and α is a characteristic coefficient of an element of a matrix component of $\bar{A}$ of maximal size, then $\alpha f(\bar{A}_k) \subseteq f(\bar{A}_k)$, and in fact (1.22) shows that $\alpha f(A) \subseteq f(\bar{A}_k)$, since any extra radical substitution would yield s_A radical substitutions in the right side and thus evaluate as 0.*

(vi) *We view an element of $\bar{A}$ as an m-tuple of matrices), and view its "characteristic coefficients" as m-tuples in $F^{(m)}$) of the characteristic coefficients of its components (in $\overline{A_1} \times \cdots \times \overline{A_m}$). If $\mathcal{S}$ is a T-ideal generated by a set of Kemer polynomials for A, then letting $\mathcal{I} = \{ f(A) : f \in \mathcal{S} \}$, we have $\alpha \mathcal{I} \subseteq \mathcal{I}$ for any characteristic coefficient α of an element of $\bar{A}_k$. (This is seen by applying (v) to each substitution in a Kemer polynomial.) In particular, we may take $\mathcal{I}$ to be the set of evaluations of some monotonic component of a Kemer polynomial.*

In our treatment here, since we are working in characteristic 0, in view of Newton's formulas, the only characteristic coefficients we need consider are traces of powers of matrices. In this case, we define traces on A_k as in Remark 2.5.9.

The next example shows that the property of being a Kemer polynomial is not T-hereditary, cf. Definition 6.3.11.

Example 6.7.2. *Technical problems arise when the names of the indeterminates do not match up; for example,*

$$f = [x_1, x_2][x_3, x_4] + [x_1, x_4][x_2, x_3]$$

does not alternate in any pair of indeterminates, even though each of the summands is a Kemer polynomial for the algebra UT(3).

This is a serious obstacle, since we will need Kemer polynomials to build representable algebras. The following observation bypasses this obstacle.

Lemma 6.7.3. *Suppose f is any nonidentity of a basic algebra A which is a consequence of a μ-Kemer polynomial $g(X_1, ..., X_\mu, X_{\mu+1}, ..., X_{\mu+s_A-1}; y))$ of A. Then f has a consequence which is both full and satisfies property K, so, by Kemer's Second Lemma, f has a consequence which is a μ'-Kemer polynomial of A for any μ'.*

Proof. Since the property K is T-hereditary, clearly f satisfies property K. Again, we may replace f by its multilinearization. We write $f(x_1, ..., x_n) = \sum_i h_i g(q_{i,1}, ..., q_{i,m}) h'_i$, where $h_i, h'_i, q_{i,j}$ are monomials and $m = \deg g$. Consider a nonzero substitution $x_i \to \tilde{x}_i$ by elements of $R_1 \cup \cdots \cup R_q \cup \text{Jac}(A)$. Denote by I all the indexes i for which the expression $\bar{h}_i g(\bar{q}_{i,1}, ..., \bar{q}_{i,m}) \bar{h}'_i$ is nonzero. Since g is Kemer for the basic algebra A and $\bar{q}_{i,j} \in R_1 \cup \cdots \cup R_q \cup \text{Jac}(A)$, the elements $\bar{q}_{i,1}, ..., \bar{q}_{i,t_A} (i \in I)$ all must be semisimple. Due to the alternating property these elements must be a basis of $\bar{A}$, which is possible only when f is full. The Lemma now follows.   □

In other words, starting with a full Kemer polynomial f of A, we can recover new Kemer polynomials with respect to arbitrarily large μ' as consequences of f. This important observation, which one could call the **Phoenix property** , is then transferred to our T-ideal Γ and is essential to all proofs that we know of Kemer's theorem.

Example 6.7.4. *Conversely to Example 6.6.5, suppose A is a basic f.d. algebra of Kemer index $(t, 1)$. By Kemer's First and Second Lemmas, $J = 0$ and $t = \dim \bar{A}$. But the only way for a t-alternating polynomial to be a non-identity of a semisimple algebra of dimension t is for the algebra to be simple. In particular, A has PI-class n, where $n^2 = t$, so any algebra of the same Kemer index as A also must have PI-class n.*

Let us formulate the situation for f.d. algebras which need not be basic, refining (6.10).

Proposition 6.7.5. *Any f.d. algebra A is PI-equivalent to a direct product of f.d. algebras*

$$\tilde{A} \times A_1 \times \cdots \times A_m,$$

where $A_1, \ldots, A_m$ are basic, and where $\tilde{A}$ is a direct product of basic algebras $\tilde{A}_1, \ldots, \tilde{A}_u$, each of the same Kemer index as A, such that

$$\text{index}(\tilde{A}_1) > \text{index}(A_1) \geq \text{index}(A_2) \geq \cdots \geq \text{index}(A_m).$$

There exists a polynomial f which is Kemer for each of $\tilde{A}_1, \ldots, \tilde{A}_u$ but is an identity for $A_1 \times \cdots \times A_m$.

Proof. For each $j \leq m$, let f_j denote a Kemer polynomial of $\tilde{A}_j$ that is an identity of $A_1 \times \cdots \times A_m$, by Proposition 6.6.11. But, matching the indeterminates in the folds, it is easy to find $\alpha_i \in F$ such that

$$f := \alpha_1 f_1(x_1, \ldots, x_n) + \alpha_2 f_2(x_1, \ldots, x_n) + \cdots + \alpha_u f_u(x_1, \ldots, x_n) y,$$

is Kemer for $\tilde{A}_1, \ldots, \tilde{A}_u$, but an identity of $A_1 \times \cdots \times A_m$. (For example, one could use a Vandermonde argument (Remark 1.1.7) for the powers of the α_i.)   □

Remark 6.7.6. *Suppose $A = A_1 \times \cdots \times A_m$, where each A_i have the same Kemer index (t, s). A μ-fold t-alternating and $s - 1$-fold $(t + 1)$-alternating polynomial f is μ-Kemer for A iff f is not an identity (and thus is μ-Kemer) for some A_i.*

6.8 Manufacturing Representable Algebras

As noted in Remark 6.3.4, one can easily get a f.d. algebra PI-equivalent to a given representable algebra. In order to utilize Proposition 6.7.5, we bring in Shirshov's program to generate new representable algebras. In order for Shirshov's Theorem to be operative, we need to work with affine relatively free algebras, i.e., with a finite set of $\bar{y}_1, \ldots, \bar{y}_\ell$ for some ℓ.

Proposition 6.8.1. *Suppose $\mathcal{A} = F\{\bar{y}_1, \ldots, \bar{y}_\ell\}$ is the relatively free algebra of a f.d. algebra A over an algebraically closed field of characteristic 0, and let $\mathcal{S}$ be a set of Kemer polynomials for A. Let I denote the ideal generated by all $\{f(\mathcal{A}) : f \in \mathcal{S}\}$. Then $\mathcal{A}/I$ is representable.*

Proof. Take the relatively free algebra $\mathcal{A}$ as described in Example 1.8.15. Since

$$\mathcal{A} \subseteq A_1[\Lambda] \times \cdots \times A_m[\Lambda]$$

for $\Lambda = \{\lambda_1, \ldots, \lambda_{\ell \dim_F A}\}$, we can extend the trace linearly to $\mathcal{A}$, as in Remark 2.5.9, to take on values in $F[\Lambda]^{(m)} \subset F(\Lambda)^{(m)}$. As in Definition 2.5.13, we form the trace ring: Letting C denote the commutative affine subalgebra of $F[\Lambda]^{(m)}$ consisting of all the traces of powers (up to $\beta(A)$) of a Shirshov base of $\mathcal{A}$, we define $\hat{\mathcal{A}} = \mathcal{A}C \subseteq A_1[\Lambda] \times \cdots \times A_m[\Lambda]$.

By Proposition 2.5.14, $\hat{\mathcal{A}}$ is finite over C, and in particular is a Noetherian algebra. In the notation of Remark 6.7.1(vi), I is also closed under multiplication by elements of C, so in fact, I is a common ideal of $\mathcal{A}$ and $\hat{\mathcal{A}}$. Thus, $\mathcal{A}/I \subseteq \hat{\mathcal{A}}/I$, which is finite over the image of the commutative affine algebra C, so $\mathcal{A}/I$ is representable, by Proposition 1.6.22. $\square$

6.8.1 Matching Kemer indices

Half of the proof of Kemer's PI-representability Theorem is to find a f.d. algebra A having the same Kemer index as a T-ideal Γ, such that $\Gamma \supseteq \mathrm{id}(A)$.

There is one technical point that we have not yet addressed: We work with the relatively free images of a free affine algebra $F\{x_1, \ldots, x_\ell\}$ but work with multilinear polynomials of arbitrary degree. Toward this end, we need a result enabling us to reduce from the free algebra $F\{X\}$ to the free affine algebra $F\{x_1, \ldots, x_m\}$ for some m. Recalling Kemer's Capelli Theorem 5.1.8, we achieve this objective with:

Theorem 6.8.2. *Let W be an algebra satisfying the ℓ-th Capelli identity c_ℓ. Then $\mathrm{Id}(W) = \mathrm{Id}(\mathcal{W})$ where $\mathcal{W}$ is the relatively free algebra of W generated by $\ell - 1$ indeterminates.*

Proof. Clearly $\mathrm{Id}(W) \subset \mathrm{Id}(\mathcal{W})$. The converse utilizes the representation theory of Chapter 3. Suppose f is a multilinear nonidentity of W of degree n. Then there is a partition $\lambda = (\lambda_1, \ldots \lambda_k)$ of n and a tableau T_λ such that $g = e_{T_\lambda} f \notin \mathrm{id}(W)$. Thus, we can replace f by g.

In the terminology of §3.2.2.3, let $g_0 = C_{T_\lambda}^- = \sum_{\tau \in C_{T_\lambda}} \mathrm{sgn}(\tau)\tau$. In case $k \geq \ell$, g_0 is alternating on the indeterminates of the first column and hence by assumption is an identity of W. But in that case also the polynomial $g = \sum_{\sigma \in R_{T_\lambda}} \sigma \cdot g_0 \in \mathrm{Id}(W)$, contradicting our assumption. Hence $k < \ell$.

Since $g = \sum_{\sigma \in R_{T_\lambda}} \sigma \cdot g_0$, it is symmetric in the indeterminates corresponding to any row of T_λ. Hence, replacing each of the indeterminate in g corresponding to the i-th row by a single indeterminate x_i for each $i = 1, \ldots, k$, we obtain a polynomial $\hat{g}$ whose multilinearization is g. But $\hat{g} \notin \mathrm{Id}(W)$, since $\mathrm{char}(F) = 0$, and $\hat{g}$ can be viewed in $F\{x_1, \ldots, x_\ell\}$, so $\hat{g} \notin \mathrm{Id}(\mathcal{W})$, implying $g \notin \mathrm{Id}(\mathcal{W})$. $\square$

Corollary 6.8.3. *Let W be a PI-algebra satisfying c_ℓ, and let $\mathcal{W} = F\langle x_1, \ldots, x_n \rangle / \widehat{\mathrm{Id}(W)}$ be an affine relatively free algebra, where $n \geq \ell - 1$. Let I be any T-ideal and denote by $\hat{I}$ the ideal of $\mathcal{W}$ generated (or consisting rather) by all evaluation on $\mathcal{W}$ of elements of I. Then $\mathrm{Id}(\mathcal{W}/\hat{I}) = \mathrm{Id}(W) + I$.*

Proposition 6.8.4. *Suppose Γ is a T-ideal properly containing $\mathrm{id}(A)$ for some f.d. algebra A. Then there is a f.d. algebra A' such that $\Gamma \supseteq \mathrm{id}(A')$, but $\mathrm{index}(A') = \mathrm{index}(\Gamma)$ and A' and Γ share the same μ-Kemer polynomials for large enough μ.*

Proof. First we construct a representable algebra $\tilde{A}$ with $\mathrm{Id}(\tilde{A}) \subseteq \Gamma$ and $\mathrm{index}(\tilde{A}) = \mathrm{index}(\Gamma)$. We are done unless $\mathrm{index}(A) > \mathrm{index}(\Gamma)$. It follows that there exists μ, such that any μ-Kemer polynomial of A is in Γ. Let I be the T-ideal generated by all μ-Kemer polynomials of A and let $\hat{I}$ be the ideal of $\mathcal{A}$ consisting of all evaluations of polynomials in I on $\mathcal{A}$. By Proposition 6.8.4 and Corollary 6.8.3, $\mathcal{A}/\hat{I}$ is representable and

$$\mathrm{Id}(\mathcal{A}/\hat{I}) = \mathrm{Id}(A) + I \subseteq \Gamma.$$

We claim that $\mathrm{index}(\mathcal{A}/\hat{I}) < \mathrm{index}(A)$. Clearly,

$$\mathrm{index}(\mathcal{A}/\hat{I}) \leq \mathrm{index}(\mathcal{A}) = p_A.$$

Suppose on the contrary that $\mathrm{index}(\mathcal{A}/\hat{I}) = \mathrm{index}(A)$. Let f be a μ-Kemer polynomial of $\mathcal{A}/\hat{I}$. Then $f \notin \mathrm{id}(A)$ since $\mathrm{Id}(\mathcal{A}/\hat{I}) \supseteq \mathrm{Id}(A)$. Hence f is a

Kemer polynomial of A and thus in I; hence $f \in \mathrm{Id}(\mathcal{A}/\hat{I})$, a contradiction. Repeating this process finitely many times we obtain a representable algebra $\tilde{A}$ with $\mathrm{Id}(\tilde{A}) \subseteq \Gamma$ and $\mathrm{index}(\tilde{A}) = \mathrm{index}(\Gamma)$, as claimed.

Next, let I be the T-ideal generated by all Kemer polynomials of $\tilde{A}$ which are contained in Γ and let $\hat{I}$ the corresponding ideal of $\mathcal{A}$. The algebra $A' := \mathcal{A}/\hat{I}$ is representable and has the same Kemer polynomials as Γ. $\qquad\square$

Let $\langle f \rangle_T$ denote the T-ideal generated by a polynomial f.

Theorem 6.8.5. *Any Kemer polynomial f of a T-ideal Γ satisfies the Phoenix property.*

Proof. Let $h \in \langle f \rangle \setminus \Gamma$. We need to show, for any μ' that there is $f' \in \langle h \rangle$ which is μ'-Kemer for Γ. By Proposition 6.8.4 there is a f.d. algebra A with $\mathrm{Id}(A) \subseteq \Gamma$, whose Kemer polynomials are precisely those of Γ. Hence f is a Kemer polynomial of A, and taking $A = A_1 \times \cdots \times A_m$ for A_k basic, the polynomial f is Kemer with respect to each basic algebra A_k for which $f \notin \mathrm{Id}(A_j)$. Applying the Phoenix property (Lemma 6.7.3) for Kemer polynomials of the basic algebras A_k, we have a μ'-Kemer polynomial f' for A. Applying Proposition 6.8.4 shows that f' is μ'-Kemer for Γ. $\qquad\square$

6.9 Kemer's PI-Representability Theorem Concluded

We are finally ready to prove Theorem 6.3.1, which we rephrase for convenience.

Theorem 6.9.1 (Kemer). *Suppose* $\mathrm{char}(F) = 0$. *Any T-ideal Γ of an affine algebra is the set of identities of a suitable f.d. algebra over a field that is purely transcendental of finite transcendence degree over F.*

Proof. Let $\kappa = \mathrm{index}(\Gamma)$. $\Gamma \supseteq \mathrm{id}(A)$ for suitable f.d. A, cf. Proposition 6.3.10. By Proposition 6.8.4, we may assume that $\mathrm{index}(A) = \kappa$, and furthermore, for μ large enough, Γ and A satisfy precisely the same μ-Kemer polynomials.

Let $\mathcal{S}_\mu$ be the set of μ-Kemer polynomials of Γ, and $\langle \mathcal{S}_\mu \rangle_T$ the T-ideal generated by $\mathcal{S}_\mu$. and let $\Gamma' = \Gamma + \langle \mathcal{S}_\mu \rangle_T$. Writing $\kappa = (t, s)$, we see by definition that

$$\mathrm{index}(\Gamma') < \kappa,$$

so by induction there is some f.d. algebra A' with $\mathrm{id}(A') = \Gamma'$.

Our objective is to find some f.d. algebra A'' for which all elements of Γ are identities, but none of the $\mathcal{S}_\mu$ are identities. Once we have such an algebra, we claim that

$$\Gamma = (\Gamma + \langle \mathcal{S}_\mu \rangle_T) \cap \mathrm{id}(A'')(= \mathrm{id}(A' \times A'')),$$

as desired.

Indeed, ($\subseteq$) is clear. For ($\supseteq$) suppose $f \in \text{id}(A'')$ and $f = g + h$ for $g \in \Gamma$ and $h \in \langle \mathcal{S}_\mu \rangle_T$. Then $h = f - g \in \text{id}(A'')$. By the Phoenix property, if $h \notin \Gamma$, then it has some consequence $h' \in \langle \mathcal{S}_\mu \rangle_T$, so $h' \in \text{id}(A'')$, contrary to hypothesis on A''. We then conclude that $h \in \Gamma$, yielding $f \in \Gamma$.) $\square$

In fact, it is enough to find such A' representable, since if $A' \subseteq M_n(K)$, we can replace A' by $A'K$. There are two ways to find A'', the first ad hoc but quicker, and the second using Zubrilin's method of Chapter 4. We present the first approach in the main text, and the second in the exercises.

6.9.1   Conclusion of proof — Expediting algebras

We have not finished with the idea of Proposition 6.8.1.

Definition 6.9.2. *Given a f.d.-algebra A, consider the algebra $\hat{A}_{s_A,\nu}$ of Definition 6.6.24, where $\nu = \max\{\dim \text{Jac}(A), \beta(\Gamma) = \beta(A), \nu_A(\mu)\}$ and ν is taken large enough for A and Γ to have the same μ-Kemer polynomials. $\hat{A}_{s_A,\nu}$ being f.d. and graded by Remark 6.6.26, we let $\mathcal{A}$ be the algebra generated by $l = \beta(\Gamma)$ elements, which we consider as a subalgebra of $\hat{A}_{s_A,\nu}[\Lambda = \{\lambda_{i,j} : i = 1, \ldots, \ell, \ j = 1, \ldots, \dim_F \bar{A}\}]$, where the i'th element is identified with $\sum_j \lambda_{i,j} a_j + \bar{x}_i$. (Here $a_1, \ldots, a_{\dim_F \bar{A}}$ is some base of $\bar{A}$.) Note that $\text{id}(\mathcal{A}) = \text{id}(\hat{A})$.*

Let $\mathcal{A}_1$ be the subalgebra of $\hat{A}_{s_A,\nu}[\Lambda]$ generated by $\mathcal{A}$ and $\{\bar{x}_i : 1 \leq i \leq \nu\}$.

We define a trace on $A[\Lambda]$ as in Remark 6.7.1(vi), and extend it to $\mathcal{A}_1$ by declaring that the trace of every word including some $\bar{x}_i$ is 0. $\mathcal{A}_2$ is the subalgebra of $\hat{A}_{s_A,\nu}[\Lambda]$ obtained by adjoining the traces of $\mathcal{A}$ to $\mathcal{A}_1$.

*We call $\mathcal{A}_1$ the **expediting algebra** of A, and $\mathcal{A}_2$ the **trace expediting algebra**, since they expedite our proof.*

Remark 6.9.3. *$\mathcal{A}_1$ is $\mathbb{N}$-graded as in Remark 6.6.26, and $\text{Jac}(\mathcal{A}_1)$ contains the ideal J_0 generated by the $\bar{x}_i$, and $\mathcal{A}_1/J_0$ is isomorphic to $\mathcal{A}$. Then*

$$\text{id}(A) = \text{id}(\hat{A}_{s_A,\nu}) = \text{id}(\hat{A}_{s_A,\nu}[\Lambda]) = \text{id}(\mathcal{A}_1) = \text{id}(\mathcal{A}_2).$$

Proposition 6.9.4. *Let Γ_2 be the ideal of $\mathcal{A}_2$ obtained by all evaluations of Γ on $\mathcal{A}_2$. Then $\mathcal{A}' := \mathcal{A}_2/\Gamma_2$ is finite over its center (and thus is representable), and $\text{id}(\mathcal{A}'') \cap S_\nu = \emptyset$.*

Proof. Let $\mathcal{S}_\mu$ be the set of μ-Kemer polynomials of A. In constructing $\mathcal{A}_1$, recall that we have a set of nilpotent indeterminates $\bar{x}_j : 1 \leq j \leq \nu$; we took this set large enough to include all of the designated alternating indeterminates in a μ-Kemer polynomial. (Thus, $\nu \geq \nu_\Gamma(\mu)$.) By Corollary 6.8.3 we can find a specialization of every $f \in \mathcal{S}_\mu$ to $\mathcal{A}_1$, for which each designated indeterminate x_i is sent to $\bar{x}_i$ and $f \notin \Gamma_1$, where Γ_1 is the ideal of $\mathcal{A}_1$ consisting all evaluations of polynomials of Γ on $\mathcal{A}_1$. Denote the set of all such specializations by $\overline{\mathcal{S}_\mu}$.

Claim. *We claim that* $\Gamma_2 \cap \overline{S_\mu}$ *is empty. Indeed suppose*

$$f = \sum_j v_j w_j \in \Gamma_2 \cap S_\nu,$$

where v_j *is a product of traces of elements of* $\overline{A}[\Lambda]$ *and* $w_j \in \Gamma_1$. *Since* $\mathcal{A}_1$ *is* $\mathbb{N}$*-graded by the* $\bar{x}_i$*'s, we may assume that every* $\bar{x}_i$ *appears exactly once in any monomial of every* w_j. *Applying the alternator (to one of the t-alternating folds) to* f *yields a nonzero multiple of* f, *but on the right side we get* $\sum_j v_j w'_j$. *Since* v_j *is a product of traces of elements from* $\mathcal{A}$, *iterating Theorem I of Section 1.5 shows that* $v_j w'_j$ *is an element of* Γ_1. *Hence* $f \in \Gamma_1$, *so* $f \notin \overline{S_\mu}$, *a contradiction.*

Thus, $\mathrm{id}(\mathcal{A}'') \supseteq \Gamma$. The claim shows that the Kemer polynomials S_μ remain nonidentities of $\mathcal{A}''$.

On the other hand, by construction and Shirshov's Theorem, $\mathcal{A}''$ is finite over its center and thus, by Proposition 1.6.22 is representable. $\qquad\square$

Thus, we have concluded the proof of Theorem 6.3.1. Here is a related result.

Proposition 6.9.5. *Any affine algebra* W *satisfies one of the following two cases:*

(i) *There exists a Kemer polynomial* f *such that* $f(W)Wf(W) = 0$ *(and thus* $\langle f(W) \rangle$ *is nilpotent); or*

(ii) $\mathrm{index}(W) = (t, 1)$ *for some* t. *In this case, if* $\mathrm{char}(F) = 0$, W *has some PI-class* n, *and in particular* W *has an* n^2*-alternating central polynomial* $\hat{f}$ *in the T-ideal generated by* f, *taking on non-nilpotent values.*

Proof. Let $\kappa = (t, s) = \mathrm{index}(W)$. First assume that $s > 1$. Then any Kemer polynomial $f(x_1, \ldots, x_m)$ is $(s-1)$-fold $(t+1)$-alternating, so any polynomial in

$$f(x_1, \ldots, x_m)F\{X\}f(x_{m+1}, \ldots, x_{2m})$$

has at least $2s - 2$ folds that are $(t+1)$-alternating, and by definition is in $\mathrm{id}(W)$ since $2s - 2 \geq s$. Thus, $f(W)Wf(W) = 0$, and we have (i).

Since one cannot have $s = 0$, we are left with $s = 1$. In this case, some t-alternating polynomial is a Kemer polynomial; since this is a consequence of the Capelli polynomial c_t, we see that c_t itself must be a Kemer polynomial.

Let A be a basic algebra of the same Kemer index $(t, 1)$. By Example 6.7.4, A is simple of some PI-class n. Thus, W has the same central polynomials as A, and the values are non-nilpotent (or else we pass back to (i)). $\qquad\square$

Remark 6.9.6. *This result underlines the difference between the combinatoric approach to the PI-theory of this monograph and the structural approach of, say, [Ro80], whose foundation was central polynomials. This is our first use* *of central polynomials. Note that* $\text{index}(A) = (t, 1)$ *implies* c_{t+1} *is an identity of* A, *and localizing* A *by a regular value of a* t-*alternating central polynomial yields an Azumaya algebra of rank* t, *cf.* [Ro80, Theorem 1.8.48]. *So one might well say that Proposition 6.9.5 is where the East (Kemer's theory) meets the West (Artin-Procesi theorem).*

Here is a nice application.

Corollary 6.9.7. *Any relatively free affine PI-algebra* U *(in characteristic* 0*) is representable.*

Proof. By Theorem 6.3.1, U is PI-equivalent to a f.d. algebra A, so Example 1.8.15 enables us to view U as a subalgebra of $A[\Lambda] \subset A \otimes_F K$, where $K = F(\Lambda)$. □

Remark 6.9.8. *It is natural to ask for analogs to Kemer's PI-representability Theorem for other varieties of algebras. We formulate this in general in the context of universal algebra in Volume II, providing a solution for algebras over arbitrary Noetherian rings in various related situations.*

6.10 Specht's Problem Solved for Affine Algebras

The solution to Specht's Problem is an easy corollary of Kemer's PI-representability Theorem. Kemer [Kem91, pp. 66–67] proves the ACC on T-ideals by modifying the original chain and appealing to semisimple substitutions; his argument is given in Exercise 7. However, we choose a more structural method based on the techniques already developed.

Remark 6.10.1. *Let us refine Proposition 6.7.5 even further, using the proof of Lemma 6.6.30. Given any f.d. algebra* A *of Kemer index* $\kappa = (t, s)$, *we replace* A *by a direct product of basic algebras, notation as in Proposition 6.7.5, so* $\text{index}(\tilde{A}_1) = \kappa$. *We call* $\tilde{A} = \tilde{A}_1 \times \cdots \times \tilde{A}_u$ *the **leading component** of* A.

Write $B = \tilde{A}_1 / \text{Jac}(\tilde{A}_1)$, *which we call a **foundation** of* A. *Then* $\dim B = t$. *Since we have assumed all along that* F *is algebraically closed,* B *is a direct product of matrix algebras the sum of whose dimensions is* t. *Thus, the algebra* B *is determined completely by the way* t *is partitioned into a sum of squares. Noting there are only a finite number of such partitions, and thus a finite number of possible foundations.*

Now, analogously to the proof of Lemma 6.6.30, we take the free product $B' = B * F\{x_1, \ldots, x_\nu\}$, *and*

$$\hat{B}_s = B'/(I_1 + I_2^s),$$

where I_1 *is the ideal generated by all evaluations in* B' *of polynomials in* $\text{id}(A)$, and I_2 is the ideal generated by $x_1, \ldots, x_{s-1}$. Letting $\bar{x}_i$ denote the image of x_i, and taking a base $\{b_1, \ldots, b_t\}$ of $\bar{A}$, we see that $\hat{B}_s$ is spanned by all terms of the form

$$b_{i_1} \bar{x}_{j_1} b_{i_2} \bar{x}_{j_2} \ldots \bar{x}_{j_{k-1}} b_{j_k}, \quad , 1 \leq k \leq s, \tag{6.11}$$

and thus is f.d.

By definition, $\mathrm{id}(A) \subseteq \mathrm{id}(\hat{B}_s)$. Conversely, note that $\gamma(\tilde{A}_i) = s$ for each $1 \leq i \leq u$. If $f \in \mathrm{id}(\hat{B}_s)$ is multilinear, then f vanishes whenever there are s radical substitutions, which implies $f \in \mathrm{id}(\tilde{A}_i)$ for each i such that $\tilde{A}_i$ has the same foundation as $\tilde{A}_1$, by the same argument as in Lemma 6.6.30. Hence, we can replace each such $\tilde{A}_i$ by $\hat{A}_1$. Thus, the leading component may be assumed to be a direct product of the $\hat{B}_s$, formed in this way, each arising from a different foundation.

Under this construction, if $\mathrm{id}(A) \subset \mathrm{id}(A')$ and A' has some $\tilde{A}'_i$ with the same foundation as $\tilde{A}_i$, then $\hat{A}'_i$ is a homomorphic image of $\hat{A}_i$ and thus, either $\hat{A}'_i = \hat{A}_i$ or $\dim \hat{A}'_i < \dim \hat{A}_i$. The number of such inequalities is finite (bounded by $t^s(s-1)^s$); we conclude that for any infinite ascending chain of T-ideals of f.d. algebras, infinitely many of them must have the same leading component.

We finally have reached the end of our journey.

Theorem 6.10.2. *Any T-ideal of $F\{x_1, \ldots, x_\ell\}$ is finitely based (assuming $\mathrm{char}(F) = 0$).*

Proof. In view of Remark 1.1.28, it suffices to verify the ACC for T-ideals, i.e., any chain

$$\mathrm{id}(A_1) \subseteq \mathrm{id}(A_2) \subseteq \mathrm{id}(A_3) \subseteq \ldots \tag{6.12}$$

of T-ideals terminates. In view of Theorem 6.9.1, we may assume that each A_i is f.d. Since ordered pairs of natural numbers satisfy the descending chain condition, we may start far enough along our chain of T-ideals and assume that all of the A_j in (6.12) have the same Kemer index κ. We do this for κ minimal possible, and induct on κ.

By Proposition 6.7.5, we can write each

$$A_j = \tilde{A}_j \times A_{j,1} \times \cdots \times A_{j,t_j},$$

where $\tilde{A}_j$ is the leading component and each $\mathrm{index}(A_{j,i}) < \kappa$. Furthermore, by Remark 6.10.1, some particular leading component must repeat infinitely often throughout the chain. Selecting only these entries, we may assume that $\tilde{A}_j = \tilde{A}$, where $\tilde{A}$ is a direct product of basic algebras of Kemer index κ.

Let $\mathcal{I}$ be the T-ideal generated by all Kemer polynomials of $\tilde{A}$ that are identities of $A_{1,1} \times \cdots \times A_{1,t}$. As we argued before, $\tilde{A}/\mathcal{I}(\tilde{A})$ cannot contain any Kemer polynomials (for κ) that are identities of $A_{1,1} \times \cdots \times A_{1,t}$, so its Kemer index must decrease and the chain

$$\mathrm{id}(A_1) + \mathcal{I} \subseteq \mathrm{id}(A_2) + \mathcal{I} \subseteq \mathrm{id}(A_3) + \mathcal{I} \subseteq \ldots \tag{6.13}$$

must stabilize. On the other hand, for each j,

$$\mathcal{I} \cap \mathrm{id}(\tilde{A}) = \mathcal{I} \cap \mathrm{id}(A_1) = \mathcal{I} \cap \mathrm{id}(A_j) \subseteq \mathcal{I} \cap \mathrm{id}(\tilde{A}),$$

so equality must hold at each stage, i.e., $\mathcal{I} \cap \mathrm{id}(A_1) = \mathcal{I} \cap \mathrm{id}(A_j)$ for each j. But coupled with (6.13), this implies that (6.12) stabilizes, contradiction. $\quad\square$

We also have a useful finiteness test for the Kemer index.

Remark 6.10.3. *Suppose S is the set of monotonic components of prospective μ-Kemer polynomials of a PI-algebra A, i.e., having $s - 1$ $(t + 1)$-alternating folds and μ t-alternating folds. Let $\tilde{A}'$ be the algebra described in Remark 4.6.7, and let I be the ideal of A generated by all $\{f(A) : f \in S\}$. Then I can also be viewed as an ideal in $\tilde{A}'$. Since $\tilde{A}'$ is Noetherian, $\bar{I}$ is a finite $\tilde{A}'$-module. Write $\bar{I} = \sum_{i=1}^{t} \tilde{A}' f_i$. If we can show that $f_1, \ldots, f_t \in \mathrm{id}(A)$, then $S \subseteq \mathrm{id}(A)$, contradiction.*

To summarize, for any potential Kemer index κ and any PI-algebra A, a finite set of monotonic polynomials tests whether or not index$(A) \geq \kappa$.

Although we assumed that the base field is algebraically closed, we would like to have the tools of Kemer's theory available also for non-algebraically closed fields of characteristic 0. This is done in Exercises 14ff. (Clearly one can define the Kemer index for any affine PI-algebra or T-ideal.)

6.11 Pumping Kemer Polynomials

The Kemer index provides a powerful tool in conjunction with pumping, to be applied in Chapter 10 to affine Noetherian PI-algebras. In order to work effectively with Kemer polynomials, we pump them to make them look more like Capelli polynomials. Recall by Kemer's Capelli Theorem 5.0.1 that any affine algebra A satisfies a suitable Capelli identity c_d. Let us see how the pumping procedure fits into this. Note that this discussion is characteristic free, and does not depend on the base ring. We modify the notion of d-identities (Definition 5.2.2), by considering substitutions inside folds.

We call a specialization **short** if it has length $\leq d$; the specialization is **long** if it has length $> d$. We call one of the folds X_j **acceptable** if all of the $\bar{x}_{j,i}$ of this fold are short.

Definition 6.11.1. *A μ-fold t-alternating polynomial is k-**acceptable** if it has a nonzero specialization with k acceptable folds.*

Proposition 6.11.2. *Suppose an algebra A has Kemer index $\kappa = (t, s)$ and satisfies c_d. Then for any $\mu \geq \max\{\omega(A), d\}$, every μ-Kemer polynomial of A is $\mu - d + 1$-acceptable.*

Proof. Take any μ-Kemer polynomial f for A. By definition f is μ-fold t-alternating and s-fold $(t+1)$-alternating. Let us write $X_j = \{x_{j,1}, x_{j,2}, \dots\}$ for the designated indeterminates in the j-th fold. Thus, $|X_j| = t$ for $j \leq \mu$, and $|X_j| = t+1$ for $\mu < j \leq \mu + s - 1$. So we write $f = f(X_1, \dots, X_{\mu+s-1}; Y)$.

Our objective is to make as many folds acceptable as possible. Namely, we need to prove that all X_k can be made to have specializations of length $\leq d+1$, and then reapply our method, to make all the specializations of X_j short, for $1 \leq j \leq \mu - d + 1$.

Toward this end we assume that, given $k > \mu$, we have a Kemer polynomial that is nonzero with respect to as few as possible long specializations for the X_k, and induct on the length of such long specializations. Let us call a specialization **moderate** if it has length precisely $d+1$. We claim that we can reduce the length of all long specializations in the X_k to moderate or short specializations. So we assume this is impossible — i.e., any nonzero specialization for f must involve a specialization of length $> d+1$ in some X_k.

We want to switch the sizes of the folds. We focus on some X_j and X_k, where $j \leq \mu - d + 1$ and $\mu < k \leq \mu + s - 1$. In other words, suppose X_k has a long specialization of length $> d + 1$, which we may as well assume is $\bar{x}_{k,t+1}$. Write

$$\bar{x}_{k,t+1} = \bar{x}'_{k,t+1}\bar{z},$$

where $\bar{x}'_{k,t+1}$ is moderate. Let f' be the polynomial obtained from f by replacing $x_{k,t+1}$ by $x'_{k,t+1}z$, and

$$h_i = f'(x_{j,1}, \dots, x_{j,i-1}, x'_{k,t+1}, x_{j,i+1}, \dots, x_{j,t}, x_{k,1}, \dots, x_{k,t}, x_{j,i}z; y).$$

Then $g = f' + \sum_{i=1}^{t} h_i$ is $(\mu - 1)$-fold, t alternating and s-fold, $(t+1)$ alternating, whereas each h_i is $t, t+1$ alternating in $X_j \cup \{x'_{k,t+1}\} \setminus \{x_{j,i}\}$ and $X_k \cup \{x_{j,i}z\} \setminus \{x_{k,t+1}\}$.

In each h_i we have interchanged the specialization $\bar{x}_{k,t+1}$ with the moderate specialization $\bar{x}_{j,i}$, contrary to our assumption unless the specialization of h_i is 0. But since

$$f = g - \sum h_i \quad (\text{mod id}(A)),$$

we must have a nonzero specialization for g. In g we have created a new fold using one specialization of length $d+1$ and t short specializations, contrary to hypothesis.

Take a nonzero specialization $x_{j,i} \mapsto \bar{x}_{j,i}$ and $y_i \mapsto \bar{y}_i$ (where the x_i are the designated indeterminates). We select one $x_{j,i}$ from each set X_j, and to emphasize their special role, we write $v_{j,i}$ for $\bar{x}_{j,i}$, $1 \leq i \leq \mu$; $\bar{x}$ denotes the specializations of the other $x_{j,i}$. By Corollary 5.2.6, we may write

$$f(v_1, \dots, v_\mu; \bar{x}; \bar{y})$$

as a linear combination of elements of the form $f(v'_1, \dots, v'_\mu; \bar{x}; \bar{y})$ where at most $d-1$ of the v'_i have length $\geq d$, and the $\bar{y}$ remain unchanged. Thus, replacing v_i by v'_i, we have made all but $d-1$ of these specializations short, without affecting the other specializations. Using this procedure wherever possible, we may assume that *every* specialization in the components $X_1, \ldots, X_\mu$ is short, except for at most $\mu - d + 1$ components; indeed, if d components contain long specializations, we choose the indeterminates of those long specializations and apply the pumping procedure while leaving the other specializations alone.

Thus, renumbering the X_j, we may assume that every specialization in $X_1, \ldots, X_{\mu-d+1}$ is short. $\qquad\square$

Corollary 6.11.3. *Suppose A is an affine algebra, of Kemer index κ. Taking d such that A satisfies the Capelli identity c_d, let $\mu \geq \max\{\omega(A), d\}$ be arbitrary and let I be the ideal generated by all evaluations of μ-Kemer polynomials, where each designated indeterminate is specialized to a word in A of length $\leq d + 1$. Then $\mathrm{index}(A/I) < \kappa$.*

Proof. For any ν we take $\mu' = \mu + d - 1$, obtain a μ'-Kemer polynomial via the Phoenix property, and then go back to μ with the theorem. $\qquad\square$

This result provides an inductive procedure, based on the Kemer index, that will be used to prove Theorem 10.1.2, and to obtain a converse to Shirshov's Theorem in Exercise 11.27.

6.12   Appendix: Strong Identities and Specht's Conjecture

We already investigated strong identities (Definition 5.2.4.1) in the context of Capelli identities. Their advantage over the Capelli identities is that they exist for *all* PI-algebras (Theorem 5.2.14), whereas we recall that arbitrary PI-algebras (such as the Grassmann algebra) need not satisfy a Capelli identity.

Given the important combinatoric properties of strong identities already exploited, we welcome the following cheap version of Specht's conjecture. Formanek-Lawrence [ForL76] proved for any field F of characteristic 0 that $F[S_\infty]$, the group algebra of the infinite symmetric group, is weakly Noetherian, i.e., satisfies the ACC on two-sided ideals. Since, in characteristic 0, strong identities generate two-sided ideals in the group algebra, the Formanek-Lawrence result gives a quick proof of:

Proposition 6.12.1. *Any T-ideal in characteristic 0 generated by strong identities is finitely based (by strong identities).*

We expect that this weaker result holds in arbitrary characteristic despite the counterexamples given in Chapter 9 to Specht's problem in nonzero characteristic.

Exercises for Chapter 6

1. In any characteristic $p \neq 2$, the T-ideal of the Grassmann algebra G over an infinite dimensional vector space is generated by the Grassmann identity $[[x_1, x_2], x_3]$. (Hint: Applying Proposition 6.2.1 to Theorem 6.2.7, reduce any element of $\mathrm{id}(G)$ to the form

$$f = \sum \alpha_u [x_{i_1}, x_{i_2}] \dots [x_{i_{2k-1}}, x_{i_{2k}}] x_1^{u_1} \dots x_m^{u_m}$$

where $i_1 < i_2 < \dots i_{2k}$. Let f_1 be the sum of terms in which $i_1 = 1$. Then $f = f_1 + f_2$ where x_1 does not appear in any of the commutators used in f_2. Taking u to be the largest value for u_1, and specializing $x_1 \mapsto e_1 e_2 + \dots + e_{2u-1} e_{2u}$, a central element of G, show that any specialization of $x_2, \dots, x_m$ sends $f_1 \mapsto 0$, implying $f_2 \in \mathrm{id}(G)$, so $f_1 = f - f_2 \in \mathrm{id}(G)$. By induction on the number of monomials in f, assume that $f = f_1$ or $f = f_2$, and continuing in this way, assume that

$$f = [x_{i_1}, x_{i_2}] \dots [x_{i_{2k-1}}, x_{i_{2k}}] \sum \alpha_u x_1^{u_1} \dots x_m^{u_m},$$

i.e., the same $i_1, i_2, \dots, i_{2k}$ appears in each term.

If $f \neq 0$, one can obtain the contradiction $f \notin \mathrm{id}(G)$. Indeed, specializing x_j to 1 for all $j \neq i_1, \dots, i_{2k}$ and renumbering, assume that

$$f = [x_1, x_2] \dots [x_{2k-1}, x_{2k}] \sum \alpha_u x_1^{u_1} \dots x_m^{u_m}.$$

Now specialize x_i to a sum of one odd and $u_i - 1$ even base elements. Then specialize to the product of the commutators of the odd parts; f has reverted to the even part, and the usual Vandermonde argument shows that this product is a specialization, contradiction. Thus, $f = 0$, as desired.) The result also holds in characteristic 2, as shown in Chapter 9.

2. The standard identity s_{2n} does not generate $\mathrm{id}(M_n(F))$, for $n \geq 2$. (Hint: Confront Exercise 1.20 with the smallest degree obtained by substituting monomials in x, y into s_4.)

Kemer index

3. $\beta(A)$ is the PI-exponent of A, in the sense of Giambruno and Zaicev [GiZa03c].

4. An example of an irreducible algebra that is neither full nor basic: Let A be the subalgebra of $M_4(F)$ spanned by the matrix units

$$e_{11}, e_{22}, e_{33}, e_{44}, e_{12}, e_{24}, e_{13}, e_{34}, e_{14}.$$

Any ideal contains e_{14}, so A is irreducible.

The last five matrix units span $J = \mathrm{Jac}(A)$, so $J^3 = 0$. But $\bar{A} = F^{(4)}$, i.e., $q = 4$, so A is not full.

Show that $\mathrm{id}(A) = \mathrm{id}(A_1) \cap \mathrm{id}(A_2)$ where the subalgebra A_1 is spanned by the matrix units $e_{11}, e_{22}, e_{33}, e_{44}, e_{12}, e_{24}, e_{14}$ and A_2 by $e_{11}, e_{22}, e_{33}, e_{44}, e_{13}, e_{34}, e_{14}$.

5. For any κ, there exists a f.d. algebra A with Kemer index κ and $\delta(A)$ arbitrarily large. (Hint: Take the direct product of an algebra with Kemer index κ and $\mathrm{UT}(n)$ for large n.)

6. If f is a μ-Kemer polynomial of $\Gamma + \mathrm{id}(A)$, then Γ contains a μ-Kemer polynomial of A. (Hint: Write $f = g + h$ where $g \in \Gamma$ and $h \in \mathrm{id}(A)$. Apply the alternator (cf. Definition 1.2.31) corresponding to each of the folds of f in turn.)

7. Fill in the details of Kemer's original argument (given below) that any T-ideal in characteristic 0 is finitely based, given Theorem 6.3.1: Suppose on the contrary that there is an infinitely based T-ideal generated by multilinear polynomials $f_1, f_2, f_3, \ldots$, such that $\deg f_{j-1} \leq \deg f_j$, and, for each j, f_j is not a consequence of $\{f_1, \ldots, f_{j-1}\}$. Assume that each f_j is Spechtian, cf. Proposition 6.2.1. Let $n_j = \deg f_j$, and let $\mathcal{T}_j$ denote the T-ideal generated by all

$$f_j(x_1, \ldots, x_{i-1}, [x_i, x_{n_j+1}], x_{i+1}, \ldots, x_{n_j}), \quad 1 \leq i \leq n_j.$$

Then $f_j \notin \mathcal{T}_j$, and by hypothesis $\mathcal{T}_j = \mathrm{id}(A)$ for some f.d. algebra A. Thus, there are semisimple or radical substitutions $a_1, \ldots, a_{n_j}$ such that $f_j(a_1, \ldots, a_{n_j}) \neq 0$. Taking j larger than the index of nilpotence of $\mathrm{Jac}(A)$, some substitution a_i must be semisimple, say in some simple component R_k of $\bar{A} = A/\mathrm{Jac}(A)$. But $R_k = F \oplus [R_k, R_k]$ and

$$f_j(a_1, \ldots, a_{i-1}, [R_k, R_k], a_{i+1}, \ldots, a_{n_j})$$

is a specialization of $\mathcal{T}_j(A)$, and thus is 0. This implies

$$f_j(a_1, \ldots, a_{n_j}) = 0,$$

a contradiction.

8. Alternatively to Exercise 1.1, suppose that A is a f.d. algebra without 1. Write $A = (\bar{A} \oplus J)$, where $J = \mathrm{Jac}(A)$ and $\bar{A}$ is semisimple and thus has a unit element e which is an idempotent in A. Adjoin a unit element 1 to A in the notation of Remark 1.1.1, and note that $\hat{1} = (0, 1)$ is the multiplicative unit of A_1. Furthermore we can write $A_1 = (\bar{A} \oplus J) \oplus F$, and put $e_0 = \hat{1} - e$. Now we embed our Peirce decomposition of A into

$$A_1 = (e + e_0)A_1(e + e_0)w = eA_1e \oplus eA_1e_0 \oplus e_0A_1e \oplus e_0A_1e_0,$$

where the last three components belong to J. Thus, $J = e_0A$.

9. Suppose A is a basic algebra without 1, and in the terminology of Exercise 8, $e_0 A = A$. Then $A = \text{Jac}(A)$; in this case, $\beta(A) = 0$, and $\gamma(A)$ equals the index of nilpotence of $\text{Jac}(A)$, so Kemer's Second Lemma holds. Use this observation to prove Kemer's PI-representability Theorem for algebras without 1.

10. Suppose $h = f_{A(1,\ldots,t)}$ is a t-alternating polynomial, and $I_1, \ldots, I_\mu$ a partition of $\{1, \ldots, t\}$. Then there are permutations $\sigma_1 = 1, \sigma_2, \ldots, \sigma_m$ for some m, such that

$$h = \sum_{j=1}^{m} \text{sgn}(\sigma_j) \sigma_j f_{A(I_1)\ldots A(I_\mu)}. \tag{6.14}$$

(Hint: Writing each I_k as $\{i_1, \ldots, i_{t_k}\}$, let S_{I_k} denote the group of permutations in $i_1, \ldots, i_{t_k}$, viewed naturally as a subgroup of S_t. Since the $I_1, \ldots, I_\mu$ are disjoint, S_{I_k} commutes with $S_{I_{k'}}$ for each k, k', so $H = S_{I_1} \cdots S_{I_\mu}$ is a subgroup of S_t. Take a transversal $\sigma_1 = 1, \sigma_2, \ldots, \sigma_m$ of H in S_t, i.e., $S_t = \cup \sigma_i H$. Then match the terms of (6.14) permutation by permutation.)

11. Define algebras A, A' to be Γ-**equivalent**, written $A \sim_\Gamma A'$, if

$$\text{id}(A') \cap \Gamma = \text{id}(A) \cap \Gamma.$$

Also, an algebra A is Γ-**full** if A is full with respect to a polynomial in Γ.

Define $\beta_\Gamma(A) = 0$ whenever $\Gamma \subseteq \text{id}(A)$; if $\Gamma \not\subseteq \text{id}(A)$, then $\beta_\Gamma(A)$ is the largest t such that for any μ there is a μ-fold t-alternating polynomial $f(X_1, \ldots, X_s; Y) \in \Gamma \setminus \text{id}(A)$. Likewise, $\gamma_\Gamma(A) = 0$ whenever $\Gamma \subseteq \text{id}(A)$; if $\Gamma \not\subseteq \text{id}(A)$ then $\gamma_\Gamma(A)$ is the largest μ such that some $(s-1)$-fold $(\beta_\Gamma(A)+1)$-alternating polynomial $f(X_1, \ldots, X_{\mu-1}; Y) \in \Gamma \setminus \text{id}(A)$. The pair $(\beta_\Gamma(A), \gamma_\Gamma(A))$ is called the **relative Kemer index**.

Show that $\beta_\Gamma(A) \leq \beta(A)$; $\gamma_\Gamma(A) \leq \gamma(A)$. Prove the relative versions of Kemer's first and second lemmas in this setting.

12. In Theorem 6.8.2, the proof for $\ell = 2$ works for arbitrary characteristic.

Alternate conclusion of the proof of Theorem 6.3.1, via Zubrilin

13. Write an alternate conclusion to the proof of Theorem 6.3.1 using representable spaces. (Hint: Taking $\mu > 2$, let $\mathcal{A}$ be the relatively free algebra of the T-ideal Γ. Then let $\nu = \nu(s, \mu)$ and, in the notation of Definition 6.6.21, take

$$\hat{A} = A'/(I_1 + I_2),$$

where I_2 is the ideal generated by all $x_i A' x_i$. This is to make the image $\bar{x}_i$ have degree at most 1 in any nonzero evaluation of a polynomial; also note that the ideal generated by the $\bar{x}_i$ is nilpotent of index at most ν, since taking the product of $\nu + 1$ words means some $\bar{x}_i$ must have degree at least 2, so is zero.

Make the elements in the image $\bar{A}$ of $\mathcal{A}$ integral by the formal procedure of Remark 4.4.1. Thus, there is a map $\phi : \hat{A} \to R$ where R is representable, and where $\ker \phi$ is the obstruction to integrality of the elements of $\bar{A}$. On the other hand, define Zubrilin traces by means of μ-Kemer polynomials, which provide integrality. Thus, $\ker \phi$ has trivial intersection with the space $\mathcal{S}$ generated by evaluations of the Kemer polynomials in which all the indeterminates of the folds are specialized to the $\bar{x}_i$. Now return from Γ' to Γ.)

Kemer's theory over non-algebraically closed fields of characteristic 0

14. A *K-admissible* affine algebra $W = F\{a_1, \ldots, a_\ell\}$ is a subalgebra of $M_n(K)$. Verify the following examples of K-admissible algebras for K algebraically closed:

(i) Any prime affine PI-algebra W is K-admissible as a K-algebra, for K the algebraic closure of the field of fractions of W.

(ii) Any relatively free PI-algebra $\tilde{W}$ that is PI-irreducible is K-admissible.

(Hint for (ii): Pass to the algebraic closure of F and assume that F is algebraically closed. By Kemer's Theorem 6.3.1, $\tilde{W}$ is PI-equivalent to a f.d. algebra A. Replace A by the algebra constructed in the proof of Lemma 6.6.30. $\tilde{A}$ is the relatively free algebra of A, so by Example 1.8.15, $\tilde{A}$ can be constructed as a subalgebra of $A[\Lambda]$, which can be viewed as a subalgebra of $A \otimes_F K$, where K is the algebraic closure of the field of fractions of $F(\Lambda)$.)

15. Suppose $\mathrm{char}(F) = 0$. Define the *K-trace ring* of a K-admissible affine algebra $W = F\{a_1, \ldots, a_\ell\}$ as follows:

Let $\bar{W} = WK/\mathrm{Jac}(WK)$, written as a direct sum of matrix subalgebras $R_1 \oplus \cdots \oplus R_q$, where $R_k = M_{n_k}(K)$. Define tr_k to be the trace in the k component and then for any $a = (a_1, \ldots, a_q) \in \bar{W}$, define

$$\mathrm{tr}(a) = \sum n_k \, \mathrm{tr}_k(a_k) \in K.$$

Consider the relatively free algebra $\mathcal{A}$ of W as constructed in Example 1.8.15. Since $\mathcal{A} \subseteq A[\Lambda]$ for $\Lambda = \{\lambda_1, \ldots, \lambda_t\}$, the trace extends linearly to $\mathcal{A}$, taking values in $K[\Lambda]$. Let $\hat{F}$ denote the commutative affine subalgebra of $K[\Lambda]$ consisting of all the traces of powers ($\leq n_k$) of products (of length up to the Shirshov height) of the generators of $\mathcal{A}$, let $\hat{A} = \mathcal{A}\hat{F} \subseteq A[\Lambda]$. As in Proposition 6.8.1, $\hat{A}$ is finite over the commutative affine algebra $\hat{F}$.

16. In the notation of Exercise 15 and Remark 6.7.1(vi), the algebra $A/\mathcal{I}$ is representable. $\mathcal{I}$ is also closed under multiplication by elements of $\hat{F}$, so in fact, $\mathcal{I}$ is a common ideal between A and $\hat{A}$. Thus, $A/\mathcal{I} \subseteq \hat{A}/\mathcal{I}$ is representable. (Hint: Proposition 1.6.22.)

Applying Razmyslov-Zubrilin traces to Kemer polynomials

17. If f is a μ-Kemer polynomial for a PI-algebra A, where $\mu > \omega(A)$, cf. Definition 6.6.6, then the polynomial $\tilde{f}$ is in id(A). (Hint: If index$(A) = (t, s)$, then $\tilde{f}$ is $t+1$-fold $s+1$-alternating and t-fold $\mu-1$-alternating.)

18. Suppose $a_1, \ldots, a_m$ is a finite set of evaluations of Kemer polynomials of a PI-algebra A. Then $L = \sum_{i=1}^{t} A a_i$ is a Noetherian A-module, and $I = \sum_{i=1}^{t} A a_i A$ is Noetherian as an A, A bimodule. (Hint: Suppose we have an infinite ascending chain $L_0 \subseteq L_1 \subseteq \ldots$ of left ideals contained in L. Let $\tilde{A}'$ be the algebra described in Remark 4.6.7, and let $L'_i = L_i \tilde{A}'$. There is a natural map $L'_i/L'_{i+1} \to L_i/L_{i+1}$.)

Chapter 7

Superidentities and Kemer's Solution for Non-Affine Algebras

As indicated in §1.3.3, "super" means "$\mathbb{Z}_2$-graded." In this chapter we develop a theory of superidentities of the superalgebra $A = A_0 \oplus A_1$, and present Kemer's "Grassmann trick," (Theorem 7.2.1) that any PI-algebra in characteristic 0 corresponds naturally to a PI-superalgebra that is PI_2-equivalent to a suitable affine superalgebra. This assertion is false in nonzero characteristic. (Indeed, if true, it would verify Specht's conjecture, but we shall present counterexamples in Chapter 9.)

Following Kemer, we then backtrack and modify the results of Chapter 6 for superalgebras. As in Chapter 6, the main step is what we call Kemer's Super-PI Representability Theorem:

Theorem 7.0.1. *Every affine ungraded-PI superalgebra is PI_2-equivalent to a finite dimensional superalgebra.*

Together, these results yield Specht's conjecture for all algebras (not necessarily affine) of characteristic 0.

We need the following key observation about superalgebras.

Remark 7.0.2. *If A, R are F-superalgebras, then $A \otimes_F R$ is also a superal-*

gebra, via

$$(A \otimes R)_0 = A_0 \otimes R_0 \oplus A_1 \otimes R_1; \qquad (A \otimes R)_1 = A_0 \otimes R_1 \oplus A_1 \otimes R_0.$$

In particular, viewing any algebra A as a superalgebra $A = A_0$ under the trivial grade, $A_0 \otimes R$ becomes a superalgebra under the grade

$$(A_0 \otimes R)_i = A_0 \otimes R_i.$$

In particular, if K is any field extension of F, then we may view $A \otimes_F K$ as a superalgebra by grading K trivially.

7.1 Superidentities

We shall develop a general theory of identities in Volume II that also encompasses the graded situation. Meanwhile, let us do this in an ad hoc fashion, motivated by the need of having the identity keep track both of the even and odd parts. Consider polynomials in two sets of indeterminates, $Y = \{y_1, y_2, \ldots\}$ and $Z = \{z_1, z_2, \ldots\}$. We view F in A_0. We say that the y's are even and the z's are odd. Then $F\{Y, Z\} = F\{y_1, y_2 \ldots; z_1, z_2 \ldots\}$ is the free superalgebra, in the sense that for any superalgebra $A = A_0 \oplus A_1$ and $a_i \in A_0$, $b_i \in A_1$, there is a homomorphism to superalgebras sending $y_i \mapsto a_i$ and $z_i \mapsto b_i$. The elements of $F\{Y, Z\}$ are called **superpolynomials**.

Definition 7.1.1. *A **superideal** of a superalgebra is an ideal $\mathcal{I}$ each of whose elements is a sum of homogeneous elements of $\mathcal{I}$, i.e., $\mathcal{I} = \mathcal{I}_0 \oplus \mathcal{I}_1$ where $\mathcal{I}_j = I \cap A_j$. In this case we write $\mathcal{I} \triangleleft_2 A$.*

Definition 7.1.2.

1. *The superpolynomial $p(y_1, \ldots, y_k; z_1, \ldots, z_\ell)$ is a **superidentity** of the superalgebra A if for any $\bar{y}_i \in A_0$ and $\bar{z}_j \in A_1$,*

$$p(\bar{y}_1, \ldots, \bar{y}_k; \bar{z}_1, \ldots, \bar{z}_\ell) = 0.$$

   *We denote by $\mathrm{id}_2(A) \subseteq F\{Y, Z\}$ the set of superidentities of A; we say superalgebras A_1 and A_2 are PI_2-**equivalent**, written $A_1 \sim_{PI_2} A_2$, if $\mathrm{id}_2(A_1) = \mathrm{id}_2(A_2)$.*

2. *Similarly, the multilinear superidentities are the multilinear superpolynomials that are superidentities.*

3. *Define a T_2-**ideal** of $F\{Y, Z\}$ to be a superideal that is invariant under all graded homomorphisms $F\{Y, Z\} \to F\{Y, Z\}$ (sending even elements to even elements and odd elements to odd elements). Then $\mathrm{id}_2(A)$ is a T_2-ideal of $F\{Y, Z\}$, for any superalgebra A.*

Example 7.1.3. *Grade the free algebra $F\{X\}$ with the trivial grading, i.e., every element is even. Then $F\{X\}$ satisfies the odd identity Z. If this example seems too trivial, one could adjoin one odd indeterminate, and then $[z_1, z_2]$ would be a superidentity.*

We are only interested in PI-superalgebras, i.e., superalgebras that also satisfy an ungraded PI.

Remark 7.1.4.

1. *Paralleling the ungraded case, given any T_2-ideal $\mathcal{I}$, we have the* ***relatively free*** *superalgebra*

$$\mathcal{U} = F\{\bar{Y}, \bar{Z}\} = F\{Y, Z\}/\mathcal{I},$$

   which is generated by even-generic elements $\bar{y}_i$ and odd-generic elements $\bar{z}_j$. Clearly, $\mathrm{id}_2(\mathcal{U}) = \mathcal{I}$.

2. *Often we want to view some* ***ungraded*** *polynomial $f = f(x_1, \ldots, x_n)$ as a superpolynomial; to do this, we specify $I \subseteq \{1, \ldots, n\}$. Then for $i \notin I$, we rewrite y_i for x_i as an even indeterminate, and for $i \in I$, we rewrite z_i for x_i as an odd indeterminate. Denote this by $\tilde{f}_I = f_I(y; z)$. Then f is an ungraded identity of a superalgebra A iff $\tilde{f}_I$ is a superidentity of A for every $I \subseteq \{1, \ldots, n\}$. Or put differently, an ungraded polynomial $f \in \mathrm{id}(A)$ iff $f(y_1 + z_1, \ldots, y_n + z_n) \in \mathrm{id}_2(A)$. Thus, we obtain an embedding of the ungraded free algebra $F\{X\}$ into the superfree algebra $F\{Y, Z\}$. As a result we can view T-ideals as T_2-ideals.*

3. *At times we shall want to permit multilinear superidentities to include ungraded indeterminates, i.e.,*

$$f(x_1, \ldots, x_j; y_1, \ldots, y_k; z_1, \ldots, z_\ell)$$

   vanishes for all substitutions of x_i, all even substitutions of y_i, and all odd substitutions of z_i. We call this a ***mixed superidentity***. *Since any substitution for x_i is the sum of its even and odd parts, we see that f is an identity of A iff every superpolynomial obtained by substituting each x_i to an even or odd indeterminate is in $\mathrm{id}_2(A)$. (To express all of these superidentities, we need $j + k$ odd indeterminates and $j + \ell$ even indeterminates.) Abusing language slightly, we get the corresponding relatively free mixed superalgebra $F\{X, Y, Z\}/\mathrm{id}_2(A)$.*

4. *One gets a theory of mixed superidentities for affine superalgebras by taking finite sets of indeterminates X, Y, Z.*

Most of the elementary facts about ordinary identities extend easily to (mixed) superidentities. This in particular includes the process of multilinearization described in (1.3) of Chapter 1 (and in more detail in Chapter 3). The operator Δ is defined exactly as before, with the proviso that odd elements linearize to odd elements, and even elements linearize to even elements.

Remark 7.1.5. *Let $w = w(y; z) = w(y_1, \ldots, y_k; z_1, \ldots, z_\ell)$ be any word in the y's and z's. For any superalgebra $A = A_0 \oplus A_1$, specialize $y_i \mapsto \bar{y}_i \in A_0$ and $z_j \mapsto \bar{z}_j \in A_1$, and denote $\bar{w} = w(\bar{y}; \bar{z})$. Let d be the total z-degree of w. (For example, if $w = z_1 y_2 z_2 y_1 z_1$, then $d = 3$). If d is even (resp. odd), then $\bar{w} \in A_0$ (resp. $\bar{w} \in A_1$).*

7.1.1   The role of odd elements

One of the main roles is played, perhaps surprisingly, by the Grassmann algebra G on a countably infinite dimensional F-vector space with base $e_1, e_2, \ldots$, cf. Example 1.3.27.

Example 7.1.6.

(i) *The Grassmann identity $[[x, y], z]$ can be translated to the superidentities*

$$[y_1, y_2], \qquad [y_1, z_1], \qquad z_1 z_2 + z_2 z_1, \qquad (7.1)$$

for even y_i and odd z_i.

(ii) *Remark 1.3.29 translates into the superidentities*

$$z_1 x z_2 + z_2 x z_1 \qquad (7.2)$$

for odd indeterminates z_i.

Taking $z_1 = z_2$, we get the superidentity $z_1 x z_1$ whenever $\mathrm{char}(F) \neq 2$.

(iii) *Remark 1.4.12 yields the superidentity z_1^{2n} for $M_n(G)$, for odd z_1.*

Kemer's contribution here was to incorporate this structure of G into the PI-theory, by passing from A_0 to the superalgebra $A_0 \otimes G$, cf. Remark 7.0.2. We study the superidentities in a procedure formalized by Berele [Ber85b]. Recall from Remark 1.3.27 the base

$$B = \{ e_{i_1} \cdots e_{i_m} \mid i_1 < \cdots < i_m, \quad m = 1, 2, \ldots \} \qquad (7.3)$$

of G. Generalizing Example 7.1.6, we introduce the sign function $\varepsilon(\sigma, I)$.

Remark 7.1.7. *Given an n-tuple $(b) = (b_1, \ldots, b_n)$ of such base elements such that $b_1 \cdots b_n \neq 0$, denote $I = \mathrm{Odd}(b) = \{ i \mid b_i \text{ is odd} \}$. Then for any $\sigma \in S_n$, the equation*

$$b_{\sigma(1)} \cdots b_{\sigma(n)} = \varepsilon(\sigma, I) b_1 \cdots b_n \qquad (7.4)$$

uniquely defines a function $\varepsilon(\sigma, I)$, independent of the particular choice of the b_i; note that $\varepsilon(\sigma, I) = \pm 1$.

Remark 7.1.8. *If* $\mathrm{Odd}(b) = \mathrm{Odd}(b_1, \ldots, b_n) = I$, *then*

$$\mathrm{Odd}(b_{\sigma(1)}, \ldots, b_{\sigma(n)}) = \sigma^{-1}(I),$$

where $\sigma^{-1}(I) = \{\sigma^{-1}(i) \mid i \in I\}$. *Indeed,* $j \in \mathrm{Odd}(b_{\sigma(1)}, \ldots, b_{\sigma(n)})$ *if and only if* $b_{\sigma(j)}$ *is odd. Write* $j = \sigma^{-1}(i)$. *Then* $\sigma^{-1}(i) \in \mathrm{Odd}(b_{\sigma(1)}, \ldots, b_{\sigma(n)})$ *if and only if* b_i *is odd, namely, if and only if* $i \in I$; *that is,* $\sigma^{-1}(i) \in \sigma^{-1}(I)$.

We can show now that $\varepsilon(-, I) : S_n \to \{\pm 1\}$ is almost a homomorphism: It is, precisely when either I is empty or $I = \{1, \ldots, n\}$.

Lemma 7.1.9. *Let* $\sigma, \eta \in S_n$ *and* $I \subseteq \{1, \ldots, n\}$, *then*

$$\varepsilon(\sigma\eta, I) = \varepsilon(\eta, \sigma^{-1}(I))\varepsilon(\sigma, I).$$

Proof. Choose $(b) = (b_1, \ldots, b_n)$, where $b_1, \ldots, b_n \in B$ are base elements such that $b_1 \cdots b_n \neq 0$ and $\mathrm{Odd}(b) = I$. Denote $a_j = b_{\sigma(j)}$; then $\mathrm{Odd}(a_1, \ldots, a_n) = \sigma^{-1}(I)$ and also $b_{\sigma\eta(i)} = a_{\eta(i)}$. Thus,

$$\varepsilon(\sigma\eta, I)b_1 \cdots b_n = b_{\sigma\eta(1)} \cdots b_{\sigma\eta(n)} = a_{\eta(1)} \cdots a_{\eta(n)}$$
$$= \varepsilon(\eta, \sigma^{-1}(I))a_1 \cdots a_n = \varepsilon(\eta, \sigma^{-1}(I))b_{\sigma(1)} \cdots b_{\sigma(n)} \quad (7.5)$$
$$= \varepsilon(\eta, \sigma^{-1}(I))\varepsilon(\sigma, I)b_1 \cdots b_n.$$

Matching coefficients yields the desired result. $\qquad\qquad\square$

7.1.2 The Grassmann involution on polynomials

Assume throughout that $\mathrm{char}(F) = 0$.

Definition 7.1.10. *Given a multilinear polynomial*

$$p = p(x_1, \ldots, x_n) = \sum_{\sigma \in S_n} \alpha_\sigma x_{\sigma(1)} \cdots x_{\sigma(n)}$$

and $I \subseteq \{1, \ldots, n\}$, *define*

$$p_I^* = \sum_{\sigma \in S_n} \alpha_\sigma \varepsilon(\sigma, I) x_{\sigma(1)} \cdots x_{\sigma(n)}.$$

We call (∗) the **Grassmann involution** *(although it also depends on our choice of* I*).*

Theorem 7.1.11. *Let* A_0 *be a PI-algebra, and let*

$$p = p(x_1, \ldots, x_n) = \sum_{\sigma \in S_n} \alpha_\sigma x_{\sigma(1)} \cdots x_{\sigma(n)}$$

be a multilinear polynomial. Then the following assertions are equivalent:

(i) $p \in \mathrm{id}(A_0)$.

(ii) $\tilde{p}_I^*$ is a superidentity of $A = A_0 \otimes G$ for some subset $I \subseteq \{1, \ldots, n\}$.

(iii) $\tilde{p}_I^*$ is a superidentity of $A = A_0 \otimes G$ for every subset $I \subseteq \{1, \ldots, n\}$.

Proof. It clearly is enough to show that *(i)* $\Leftrightarrow$ *(ii)*, since then we apply this for each I to get (iii). Let B denote the canonical base of G given in (7.3). Both directions of the proof follow easily from the equality

$$\tilde{p}_I^*(a_1 \otimes b_1, \ldots, a_n \otimes b_n) = p(a_1, \ldots, a_n) \otimes b_1 \cdots b_n \qquad (7.6)$$

whenever $a_1, \ldots, a_n \in A_0$ and $b_1, \ldots, b_n \in B$ with $\mathrm{Odd}(b_1, \ldots, b_n) = I$. This holds since

$$\tilde{p}_I^*(a_1 \otimes b_1, \ldots, a_n \otimes b_n) = \sum_\sigma \varepsilon(\sigma, I)\alpha_\sigma(a_{\sigma(1)} \otimes b_{\sigma(1)}) \cdots (a_{\sigma(n)} \otimes b_{\sigma(n)})$$

$$= \sum_\sigma \varepsilon(\sigma, I)\alpha_\sigma(a_{\sigma(1)} \cdots a_{\sigma(n)}) \otimes (b_{\sigma(1)} \cdots b_{\sigma(n)})$$

$$= \left(\sum_\sigma \alpha_\sigma a_{\sigma(1)} \cdots a_{\sigma(n)} \right) \otimes b_1 \cdots b_n.$$

$$(7.7)$$

Hence, if $p \in \mathrm{id}(A_0)$, then each $\tilde{p}_I^* \in \mathrm{id}_2(A_0 \otimes G)$.

Conversely, assume that $\tilde{p}_I^* \in \mathrm{id}_2(A_0 \otimes G)$. Pick $b_1, \ldots, b_n \in B$ such that $\mathrm{Odd}(b_1, \ldots, b_n) = I$ and $b_1 \cdots b_n \neq 0$; then (7.6) shows that $p(a_1, \ldots, a_n) = 0$ for any elements $a_1, \ldots, a_n \in A_0$. $\square$

Corollary 7.1.12. *Suppose* $\mathrm{char}(F) = 0$. *There is an inclusion-preserving map* $\Phi : \{T\text{-ideals}\} \to \{T_2\text{-ideals}\}$ *given by* $\mathrm{id}(A_0) \to \mathrm{id}_2(A_0 \otimes G)$.

Proof. Every T-ideal is determined by its multilinear polynomials. $\square$

Better yet, Φ could be viewed as a functor from varieties to supervarieties; this can be done in any characteristic, cf. Exercise 2.

7.1.3   The Grassmann envelope

Kemer's next observation was to find an inverse functor, thereby elevating the Grassmann algebra to a position of prominence in the PI-theory.

Definition 7.1.13. *The **Grassmann envelope** of a superalgebra A is the subalgebra*

$$G(A) = (A \otimes G)_0 = A_0 \otimes G_0 \oplus A_1 \otimes G_1.$$

Theorem 7.1.14. *Let A be a PI superalgebra, and let $p = p(x_1, \ldots, x_n) = \sum_{\sigma \in S_n} \alpha_\sigma x_{\sigma(1)} \cdots x_{\sigma(n)}$ be a multilinear polynomial. Then $p \in \mathrm{id}(G(A))$ if and only if $\tilde{p}_I^* \in \mathrm{id}_2(A)$ for every subset $I \subseteq \{1, \ldots, n\}$.*

Proof. As in the proof of Theorem 7.1.11. To check whether $p \in \mathrm{id}(G(A))$ we check substitutions $a \otimes b$, where $a \in A$ and $b \in G$ are both homogeneous of the same parity (i.e., $a_i \otimes b_i$ is in $A_u \otimes G_u$ for $u \in \{0, 1\}$). But then the following analog of Equation (7.6) is applicable, where $I = \{i : a_i \text{ is odd}\}$:

$$p(a_1 \otimes b_1, \dots, a_n \otimes b_n) = \sum_\sigma \alpha_\sigma (a_{\sigma(1)} \otimes b_{\sigma(1)}) \cdots (a_{\sigma(n)} \otimes b_{\sigma(n)}) \qquad (7.8)$$

$$= \sum_\sigma \alpha_\sigma (a_{\sigma(1)} \cdots a_{\sigma(n)}) \otimes (b_{\sigma(1)} \cdots b_{\sigma(n)})$$

$$= \left(\sum_\sigma \varepsilon(\sigma, I) \alpha_\sigma a_{\sigma(1)} \cdots a_{\sigma(n)} \right) \otimes b_1 \cdots b_n, \qquad (7.9)$$

i.e.,

$$p(a_1 \otimes b_1, \dots, a_n \otimes b_n) = \tilde{p}_I^*(a_1, \dots, a_n) \otimes b_1 \dots b_n, \qquad (7.10)$$

which now shows that if each $\tilde{p}_I^* \in \mathrm{id}_2(A)$, then $p \in \mathrm{id}(G(A))$. Conversely, assume that $p \in \mathrm{id}(G(A))$ and let $I \subseteq \{1, \dots, n\}$. Pick $b_1, \dots, b_n \in B$ such that $\mathrm{Odd}(b_1, \dots, b_n) = I$ and $b_1 \cdots b_n \neq 0$; then Equation (7.10) shows that $p_I^*(a_1, \dots, a_n) = 0$ for any homogeneous $a_1, \dots, a_n \in A$ where a_i is odd iff $i \in I$. $\qquad \square$

Theorem 7.1.15. *There is a correspondence Ψ from $\{$varieties of superalgebras$\}$ to $\{$varieties of algebras$\}$ given by $A \mapsto G(A)$, which is the inverse of Φ of Corollary 7.1.12, in the sense that*

$$A_0 \sim_{\mathrm{PI}} G(A_0 \otimes G), \qquad A_0 \text{ any algebra} \qquad (7.11)$$

and

$$A \sim_{\mathrm{PI}_2} G(A) \otimes G, \qquad A \text{ any superalgebra.} \qquad (7.12)$$

Proof. We have proved everything except that these two correspondences are inverses. But

$$G(A_0 \otimes G) \cong ((A_0 \otimes G_0) \otimes G_0) \oplus ((A_0 \otimes G_1) \otimes G_1) \cong A_0 \otimes (G_0 \otimes G_0 \oplus G_1 \otimes G_1)).$$

Now $G_0 \otimes G_0 \oplus G_1 \otimes G_1$ is a commutative algebra, yielding (7.11) in view of Proposition 1.2.16.

To see (7.12), it suffices by Theorem 7.1.14 to show that

$$G(A) \sim_{\mathrm{PI}} G(G(A) \otimes G) \cong G(A) \otimes (G_0 \otimes G_0 + G_1 \otimes G_1),$$

which again is true by Proposition 1.2.16. $\qquad \square$

     Theorem 7.1.15 is so important that, in the literature, $M_{k,\ell}$ often refers to the Grassmann envelope of the superalgebra that we have called $M_{k,\ell}(C)$. This correspondence also has led Zelmanov to the sublime meta-definition that a superalgebra A (not necessarily associative) has property **super-P** if $G(A)$ has property P. This leads one to super-Lie algebras and super-Jordan algebras.

7.1.4   The •-action of S_n on polynomials

In order to understand the map Φ of Corollary 7.1.12 and its inverse Ψ, we introduce a new action of the symmetric group, on the multilinear polynomials $T^n(V)$, cf. §3.3, but taking into account the indices I of odd (i.e., strictly anticommuting) Grassmann substitutions, by means of the sign function $\varepsilon(\sigma, I)$.

Definition 7.1.16.

(i) *Let $V = V_0 \oplus V_1$ be vector spaces with $\dim V_0 = k$ and $\dim V_1 = \ell$. For $i \in \{0,1\}$, we say that $x \in V$ is **homogeneous of degree** i if $x \in V_i$, and then denote $\delta(x) = i$. An element $w \in T^n(V)$ is **homogeneous** if $w = x_1 \otimes \cdots \otimes x_n = x_1 \cdots x_n$ where each x_j is homogeneous; in that case, denote $\delta(w) = \delta(x_1) + \cdots + \delta(x_n) \pmod 2$.*

(ii) *Given $(x) = (x_1, \ldots, x_n)$ where $x_1, \ldots, x_n \in V_0 \cup V_1$, denote*

$$\mathrm{Odd}(x) = \{i \mid \delta(x_i) = 1\}.$$

(iii) *Define the •-action of S_n on $T^n(V)$ as follows. Take homogeneous elements $x_1 \cdots x_n \in T^n(V)$ with $I = \mathrm{Odd}(x)$, and let $\sigma \in S_n$. Then*

$$\sigma \bullet (x_1 \cdots x_n) = \varepsilon(\sigma, I)(x_{\sigma(1)} \cdots x_{\sigma(n)}).$$

Extend that action to all of $T^n(V)$ by linearity. In fact, we can choose bases $y_1, \ldots, y_k \in V_0$, $z_1, \ldots z_\ell \in V_1$, and replace $x_1, \ldots, x_n$ by elements in $\{y_1, \ldots, y_k, z_1, \ldots, z_\ell\}$; then extend by linearity.

Remark 7.1.17. *Notice that here $\varepsilon(\sigma, I) = \pm 1$ is calculated as follows: σ induces a permutation τ on the elements of degree 1, and $\varepsilon(\sigma, I) = \mathrm{sgn}(\tau)$.*

For example, let $w = x_1 x_2 x_3$ and let σ be the 3-cycle $\sigma = (1,2,3)$, so that $\sigma w = x_2 x_3 x_1$. If $\delta(x_1) = \delta(x_2) = 1$ and $\delta(x_3) = 0$, then $\tau = (1,2)$; hence, $\varepsilon = -1$ and $\sigma \bullet w = -\sigma w = -x_2 x_3 x_1$. However, if $\delta(x_2) = \delta(x_3) = 1$ and $\delta(x_1) = 0$, then $\tau = 1$ and $\varepsilon = +1$.

Proposition 7.1.18. *Let $\sigma, \eta \in S_n$ and let $w \in V = V_0 \oplus V_1$. Then*

$$(\sigma\eta) \bullet w = \eta \bullet (\sigma \bullet w).$$

Hence the •-action makes $T^n(V)$ into a left module over the opposite algebra $(F[S_n])^{op}$.

Proof. By linearity we may assume that $w = x_1 \cdots x_n$ with $x_1, \ldots, x_n$ homogeneous. Let $I = \mathrm{Odd}(x_1, \ldots, x_n)$. As in Remark 7.1.8,

$$\mathrm{Odd}(x_{\sigma(1)}, \ldots, x_{\sigma(n)}) = \sigma^{-1}(I).$$

Now

$$(\sigma\eta) \bullet (x_1 \cdots x_n) = \varepsilon(\sigma\eta, I) x_{\sigma\eta(1)} \cdots x_{\sigma\eta(n)},$$

while
$$\eta \bullet (\sigma \bullet (x_1 \cdots x_n)) = \varepsilon(\sigma, I)(\eta \bullet (x_{\sigma(1)} \cdots x_{\sigma(n)}))$$
$$\varepsilon(\sigma, I)\varepsilon(\eta, \sigma^{-1}(I))x_{\sigma\eta(1)} \cdots x_{\sigma\eta(n)}. \tag{7.13}$$

The proof now follows from Lemma 7.1.9. □

Corollary 7.1.19. *The map Φ of Corollary 7.1.12 is a homomorphism; i.e., if $\mathcal{J}_1, \mathcal{J}_2$ are T-ideals, then $\Phi(\mathcal{J}_1\mathcal{J}_2) = \Phi(\mathcal{J}_1)\Phi(\mathcal{J}_2)$.*

This •-action has a very important property:

Remark 7.1.20. *For σ a transposition (i, j), if $i, j \in I$, then $\sigma \bullet f = -\sigma f$; otherwise, $\sigma \bullet f = +\sigma f$.*

7.2 Kemer's Super-PI Representability Theorem

The next step is to pass from the Grassmann envelope to superalgebras. Our approach takes aspects both from Kemer's original proof and from Berele [Ber85b]. $H(k, \ell; n)$ from Definition 3.4.19 is called the (k, ℓ)-**hook**.

Theorem 7.2.1. *Let A be a PI-algebra over a field of characteristic 0. Then $\mathrm{id}_2(A \otimes G) = \mathrm{id}_2(\tilde{A})$, for a suitable affine superalgebra $\tilde{A}$ to be described in the proof.*

Proof. By Proposition 3.4.20, there are integers k, ℓ such that the Young frame of every multilinear nonidentity of A lies in the (k, ℓ)-hook. We already saw that $\mathcal{I} = \mathrm{id}_2(A \otimes G)$ is the T_2-ideal determined by the multilinear superpolynomials $\{\tilde{f}_I^* : f \in \mathrm{id}(A)\}$. Let

$$\mathcal{A} = F\{Y, Z\}/\mathcal{I}$$

be the relatively free superalgebra determined by these superpolynomials, which by definition is T_2-equivalent to $A \otimes G$, and let $\tilde{A}$ be the affine superalgebra generated by $\bar{y}_1, \ldots, \bar{y}_{k\ell+k}, \bar{z}_1, \ldots, \bar{z}_{k\ell+\ell}$.

In other words,

$$\tilde{A} = F\{\bar{y}_1, \ldots, \bar{y}_{k\ell+k}, \bar{z}_1, \ldots, \bar{z}_{k\ell+\ell}\}/(\mathcal{I} \cap F\{\bar{y}_1, \ldots, \bar{y}_{k\ell+k}, \bar{z}_1, \ldots, \bar{z}_{k\ell+\ell}\}).$$

We claim that $\tilde{A} \sim_{\mathrm{PI}_2} \mathcal{A}$.

Since any superidentity of $\mathcal{A}$ obviously is a superidentity of the subsuperalgebra $\tilde{A}$, it suffices to prove that $\mathrm{id}_2(\tilde{A}) \subseteq \mathrm{id}_2(\mathcal{A})$, or equivalently, that $\mathrm{id}(G(\tilde{A})) \subseteq \mathrm{id}(A)$. So given multilinear $f(x_1, \ldots, x_n) \in \mathrm{id}(G(\tilde{A}))$, we need to prove that f is an identity of A. We bring in the Young theory. Suppose f corresponds to a Young tableau T_λ. As noted above, we are done unless T_λ lies in the (k, ℓ)-hook. We subdivide $T_\lambda = T_0 \cup T_1 \cup T_2$, where:

1. T_0 is the horizontal strip of width at most k lying past the ℓ column.

2. T_1 is the vertical strip of width at most ℓ lying past the k row.

3. T_2 is the part of T lying in ℓ^k, the $k \times \ell$ rectangle.

We may replace f by any element in the left ideal it generates in $F[S_n]$, and thus may take f to be the semi-idempotent

$$e_{T_\lambda} = \sum_{q \in C_{T_\lambda}} \sum_{p \in R_{T_\lambda}} \operatorname{sgn}(q)qp = C_{T_\lambda}^- R_{T_\lambda}^+ .$$

In particular, if i, j lie in the same row in T_0, then $(i,j)f = f$; likewise if i, j lie in the same column in T_1, then $(i,j)f = -f$. Let I be the collection of indices labeling boxes in T_1. Let $g = \tilde{f}_I^*$. In view of Theorem 7.1.11(ii), it suffices to prove $g \in \operatorname{id}_2(A \otimes G)$, i.e., $g(a_1 \otimes b_1, \ldots, a_n \otimes b_n) = 0$ where the b_i are odd iff i appears in T_1. But let us see what happens with $(i,j) \bullet g$.

If i, j lie in the same row in T_0, then $(i,j) \bullet g = g$, as before.

If i, j lie in the same column in T_1, then $(i,j) \bullet g = -(-g) = g$.

Let h denote the superpolynomial obtained from g by specializing all indeterminates in the i row of T_0 to y_i, for $1 \le i \le k$, and specializing all indeterminates in the j column of T_1 to z_j, for $1 \le j \le \ell$. Then g is the multilinearization of h (up to a scalar multiple), by Remark 1.2.34. But h involves at most $k\ell$ indeterminates in x, as well as at most k indeterminates in y and $\ell\ell$ in z. Thus, $h \in \operatorname{id}_2(\tilde{A})$ iff $h \in \operatorname{id}_2(\mathcal{A})$.

Now we put everything together. By hypothesis $g \in \operatorname{id}_2(\tilde{A})$, so $h \in \operatorname{id}_2(\tilde{A})$, implying $h \in \operatorname{id}_2(\mathcal{A}) = \operatorname{id}_2(A \otimes G)$, so its multilinearization $g \in \operatorname{id}_2(A \otimes G)$, as desired. $\qquad\qquad\square$

To illustrate the power of this theorem, let us present some corollaries.

Corollary 7.2.2. *(In characteristic 0), any PI-algebra A satisfies the identities of $M_n(G)$ for suitable n.*

Proof. The theorem shows that $A \sim_{\text{PI}} G(\tilde{A})$ for a suitable affine superalgebra $\tilde{A}$. But Theorem 4.0.1 says $\operatorname{Jac}(\tilde{A})$ is nilpotent, implying by Lewin's Theorem that $\tilde{A}$ satisfies all identities of $M_n(F)$ for some n, so is a homomorphic image of some generic matrix algebra R. Then $G(\tilde{A}) \subset \tilde{A} \otimes G$ satisfies all identities of $R \otimes G$, which satisfies all identities of $M_n(G)$. $\qquad\square$

We shall improve this result in Exercise 5. A stronger result holds in characteristic p, cf. Theorem 8.4.1.

Corollary 7.2.3. *Any relatively free algebra has a unique largest nilpotent T-ideal.*

This in fact follows at once from Kemer's solution to Specht's problem, so we leave it as an exercise for the time being.

7.2.1   The structure of finite dimensional superalgebras

Since superalgebras play such a crucial role in the theory, let us review their structure.

Remark 7.2.4. *Any superalgebra* $A = A_0 \oplus A_1$ *has an automorphism* σ *of order 2, given by* $a_0 + a_1 \mapsto a_0 - a_1$. *(Indeed,*

$$\begin{aligned} \sigma((a_0 + a_1)(b_0 + b_1)) &= \sigma(a_0 b_0 + a_1 b_1 + a_0 b_1 + a_1 b_0) \\ &= a_0 b_0 + a_1 b_1 - a_0 b_1 - a_1 b_0 = (a_0 - a_1)(b_0 - b_1), \end{aligned} \tag{7.14}$$

proving σ *is a homomorphism, and clearly* $\sigma^2 = 1_A$.)

Conversely, if $\frac{1}{2} \in A$ *and* A *has an automorphism* σ *of order 2, then we can define a 2-grade on* A *by putting*

$$r_0 = \frac{r + \sigma(r)}{2}, \qquad r_1 = \frac{r - \sigma(r)}{2}, \qquad \forall r \in A.$$

Thus, $r \in A_0$ *iff* $\sigma(r) = r$ *and* $r \in A_1$ *iff* $\sigma(r) = -r$.

The following is a special case of [Ro88b, Lemma 2.5.39].

Lemma 7.2.5. *If* A *is a superalgebra and* $\frac{1}{2} \in A$, *then* $J = \mathrm{Jac}(A)$ *is a superideal, and likewise the upper nilradical* $\mathrm{Nil}(A)$ *is a superideal.*

Proof. Suppose $r_0 + r_1 \in J$. Notation as in Remark 7.2.4,

$$r_0 - r_1 \in \sigma(J) = J,$$

so $2r_0 \in J$, implying $r_0 \in J$, and hence $r_1 = r - r_0 \in J$. The same argument holds for $\mathrm{Nil}(A)$ since $\sigma(\mathrm{Nil}(A)) \subseteq \mathrm{Nil}(A)$. $\qquad\square$

Corollary 7.2.6. $A/\mathrm{Jac}(A)$ *is also a superalgebra, under the induced grading.*

We would like to determine its structure in the finite dimensional case. We say A is **supersimple** if A has no proper nonzero superideals. By Lemma 1.3.18 and Remark 1.3.19, this implies $\mathrm{Cent}(A_0)$ is a field (although A_0 need not be simple, cf. Example 1.3.17).

Here is an example of a supersimple superalgebra that is not simple.

Example 7.2.7. $A = M_n(F[c])$, *where* $A_0 = M_n(F)$, $A_1 = M_n(F)c$, *and* $c^2 = 1$. *Any nonzero superideal* $\mathcal{I}$ *must contain some* a *or* ac, *for* $a \in A_0$.

If $a \in \mathcal{I}$, *then, since* A_0 *is simple,* $A_0 = A_0 a A_0 \subseteq \mathcal{I}$, *implying* $1 \in \mathcal{I}$; *if* $ac \in \mathcal{I}$, *then* $A_0 c = (A_0 a A_0)c = A_0 a c A_0 \in \mathcal{I}$, *implying* $c \in \mathcal{I}$, *so* $1 = c^2 \in \mathcal{I}$.

Hence $\mathcal{I} = A$ *in either case, implying* A *is supersimple. On the other hand,* A *is not simple, since its center* $F[c]$ *is not a field.*

In one important case, this is the only example.

Proposition 7.2.8. *Suppose* F *is an algebraically closed field.*

(i) *The only nontrivial $\mathbb{Z}/2$-gradings on $M_n(F)$ yield superalgebras isomorphic to those in Example 1.3.14.*

(ii) *Any f.d. supersimple superalgebra is either as in (i) or as in Example 7.2.7.*

(iii) *Any f.d. semisimple superalgebra A is isomorphic (as superalgebra) to a direct product $\prod_{i=1}^k S_i$ of supersimple superalgebras of form (i) or (ii).*

Proof. We write $Z := \operatorname{Cent} A$ as $Z = Z_0 \oplus Z_1$, cf. Lemma 1.3.18.

(i) Write $A = M_n(F) = A_0 \oplus A_1$, and assume $A_1 \neq 0$. $F^{(n)}$ cannot be a simple A_0-module, since otherwise by the density theorem $A_0 \cong \operatorname{End} F_D^{(n)}$ for some f.d. division algebra D over F; $D = F$ since F is algebraically closed, and we would conclude that $A_0 = M_n(F)$.

So let V_0 be a proper A_0-submodule of $F^{(n)}$, and $V_1 = A_1 V_0$. Then $V_0 + V_1$ and $V_0 \cap V_1$ are A-submodules of $F^{(n)}$, which is a simple A-module, so

$$V_0 + V_1 = F^{(n)} \quad \text{and} \quad V_0 \cap V_1 = 0.$$

Thus, since $1 \in A_0$, $A_0 V_0 = V_0$ and $A_0 V_1 = V_1$. Furthermore, $A_1^2 + A_1$ is a superideal of A, implying $A_1^2 = A_0$. Hence, $A_1 V_1 = A_0 V_0 = V_0$. Writing this all in terms of bases of V_0 and V_1 yields the desired result.

(ii) If a f.d. supersimple superalgebra A is simple, then it is as in (i), so we assume that A is not simple. Hence Z is not a field. Let $0 \neq I \lhd \operatorname{Cent} A$. Since $\operatorname{Cent} A$ is supersimple, we have $I \not\subseteq A_1$, so I contains some element $a = a_0 + a_1$ for $a_i \in Z_i$, $a_0 \neq 0$. Replacing a by $a_0^{-1} a$ since Z_0 is a field, we may assume that $a_0 = 1$. Let $c = a_1 \in Z_1$, so $1 + c \in I$. Then

$$1 - c^2 = (1 - c)(1 + c) \in I \cap Z_0 = 0$$

(since Z_0 is a field), implying $c^2 = 1$. In particular, c is invertible. Furthermore,

$$A_1 = A_1 c^2 = (A_1 c)c \subseteq A_0 c \subseteq A_1,$$

implying $A_1 = A_0 c$ and $A = A_0 + A_0 c$; we have recovered Example 7.2.7.

(iii) Z_0 is semisimple by Corollary 1.3.22. Writing $Z_0 = F e_1 \times \cdots \times F e_k$ for each $F e_i \cong F$, i.e., the e_i are central idempotents with $\sum e_i = 1$, we see that $R = R e_1 \times \cdots \times R e_k$ is a direct product of superalgebras whose even part of the center is a field, so we are done by (ii). $\square$

In order to pass up the radical, we need to check that the standard process of lifting idempotents ([Ro88b, Proposition 1.1.25ff]) respects the grade. Note that any homogeneous idempotent $e \neq 0$ must be even, since $e = e^2$.

Lemma 7.2.9. *If $\mathcal{I}$ is a nil superideal of a superalgebra A, then any even idempotent e of $A/\mathcal{I}$ (under the induced grade) can be lifted to an even idempotent of A.*

Proof. Clearly, $e = a + \mathcal{I}$ for some $a \in \mathcal{I}_0$. Then $(a^2 - a)^n = 0$ for some n, so a^n is a sum of higher powers of a. Conclude as in [Row88, Proposition 1.1.26], noting that the idempotent constructed there will be even.   □

Proposition 7.2.10. *If $\mathcal{I}$ is a nil superideal of a superalgebra A, then any set of homogeneous matrix units $\{e_{ij} : 1 \leq i, j \leq n\}$ of $A/\mathcal{I}$ can be lifted to homogeneous matrix units of A of the same parity.*

Proof. First note the orthogonal idempotents $e_{11}, \ldots, e_{nn}$ can be lifted to orthogonal (even) idempotents $x_{11}, \ldots, x_{nn}$ of A, since all the elements used in the proof of [Ro88b, Proposition 1.1.25(i)] are even. But for $i \neq j$ lift e_{ij} to b_{ij} of A of the same parity, mimicking the proof of [Ro88b, Proposition 1.1.25(iii)] (noting that the elements computed there are always homogeneous).   □

Theorem 7.2.11 (Wedderburn's Principal Supertheorem). *Any f.d. superalgebra A over an algebraically closed field has a semisimple super-subalgebra $\bar{A}$ such that $A = \bar{A} \oplus \mathrm{Jac}(A)$.*

Proof. We only sketch the proof, since it is completely analogous to the standard ungraded proofs, cf. [Ro88b, Theorem 2.5.37], for example. The reduction to the case $J^2 = 0$ is standard. In this case, $\bar{A}$ is a superalgebra by Corollary 7.2.6, so we can write $\bar{A} = \prod_{i=1}^k S_i$, where each S_i is supersimple, i.e., has the form of Proposition 7.2.8(ii). Write $\bar{e}_i$ for the multiplicative unit of S_i; then $e_1, \ldots, e_k$ are orthogonal central idempotents of $\bar{A}$ which we lift to orthogonal (but not necessarily central) even idempotents $e_1, \ldots, e_k$ of A. We need to show that S_i is isomorphic to a super-subalgebra of $A_i = e_i A e_i$, for $1 \leq i \leq k$. If S_i is simple, then it is $M_{n_i}(F)$ under some grade, and the matrix units can be lifted by Proposition 7.2.10, so we have a corresponding copy of $M_{n_i}(F)$ in A_i.

Thus, we may assume that S_i is not simple, i.e., is as in Example 7.2.7, i.e., $S_i = M_{n_i}(F[\bar{c}])$ where $\bar{c}^2 = 1$. Again we lift the matrix units to a set of even matrix units $\{e_{uv} : 1 \leq u, v \leq n\}$ of A; we also take odd $c_1 \in A_i$ such that $c_1 + J = \bar{c}$. Thus, $c_1^2 = a + 1$ for some $a \in J$, clearly even. Also $a^2 \in J^2 = 0$. Note that

$$[a, c_1] = [c_1^2 - 1, c_1] = 0.$$

We define

$$c_2 = c_1 - \frac{a}{2} c_1.$$

Since $J^2 = 0$, we have

$$c_2^2 = c_1^2 - a c_1^2 = (1+a)(1-a) = 1 - a^2 = 1.$$

Since c_2 need not be scalar, we put

$$c = \sum e_{u1} c_2 e_{1u},$$

which clearly commutes with each matrix unit. Furthermore $[c_2, e_{uv}] \in J$ since $\bar{c}_2 = \bar{c}_1$ is scalar, so we see

$$
\begin{aligned}
c^2 &= \sum_{u,v=1}^{n_i} e_{u1}c_2e_{1u}e_{v1}c_2e_{1v} = \sum_{u=1}^{n} e_{u1}c_2e_{11}c_2e_{1u} \\
&= \sum_{u=1}^{n} e_{u1}c_2c_2e_{1u} + \sum_{u=1}^{n} e_{u1}[c_2, e_{11}]c_2e_{1u} \\
&= 1 + \sum_{u=1}^{n} e_{u1}[c_2, e_{11}]([c_2, e_{1u}] + e_{1u}c_2) \\
&= 1 + \sum_{u=1}^{n} e_{u1}[c_2, e_{11}]e_{1u}c_2 \\
&= 1 + \sum (e_{u1}c_2e_{1u} - e_{u1}c_2e_{1u})c_2 = 1 + 0 = 1.
\end{aligned}
\tag{7.15}
$$

$\square$

When using the Wedderburn Principal Supertheorem, we recall from Proposition 7.2.8 that the idempotents used in constructing $\bar{A}$ may all be taken to be homogeneous, and clearly any radical substitution is the sum of an odd and even radical substitution, so in checking whether or not a superpolynomial f is an identity of a f.d. superalgebra A, we may confine our attention to homogeneous semisimple and radical substitutions.

We also shall need the super-analog of Theorem J of §1.5, a way of absorbing traces into superpolynomials. Given a f.d. superalgebra $A = A_0 \oplus A_1$, we define its **superdimension** $\dim_2 A := (t_0, t_1)$ where $t_u = \dim A_u$, $u = 1, 2$.

A superpolynomial f is (t_0, t_1)-**alternating** if

$$
f = f(X; y_1, \ldots, y_{t_0}, z_1, \ldots, z_{t_1})
$$

is alternating in the y and also is alternating in the z. (There is no alternation required between the even and the odd elements.) Such a superpolynomial will clearly be a superidentity of any superalgebra of superdimension (t_0', t_1') unless $t_0' \geq t_0$ and $t_1' \geq t_1$; this is seen by checking homogeneous bases.

Theorem 7.2.12. *Suppose that A is a f.d. superalgebra of superdimension (t_0, t_1). Consider $V = A_0$ as a vector space. For any C-linear homogeneous map $T : V \to V$, letting*

$$
\lambda^{t_0} + \sum_{k=0}^{t_0 - 1} (-1)^k \alpha_k \lambda^{t_0 - k}
$$

denote the characteristic polynomial of T (as a $t_0 \times t_0$ matrix), then for any (t_0, t_1)-alternating superpolynomial f, and any even homogeneous $a_1, \ldots, a_{t_0}$ in V_0, we have

$$
\alpha_k f(a_1, \ldots, a_{t_0}, r_1, \ldots, r_m) = \sum f(T^{k_1}a_1, \ldots, T^{k_t}a_{t_0}, r_1, \ldots, r_m),
$$

where $r_1, \ldots, r_m \in A$ *and summed over all vectors* $(k_1, \ldots, k_{t_0})$ *for which* $k_1 + \cdots + k_{t_0} = k.$

Proof. As for Theorem J of §1.5.   □

7.2.2   Proof of Kemer's Super-PI Representability Theorem

Having the general preliminaries in hand, let us go back to modify Kemer's proof of Specht's conjecture in the affine case, cf. Chapter 6, to prepare for the proof of Specht's conjecture for arbitrary T-ideals in characteristic 0. In view of Theorem 7.2.1, it suffices to prove the ACC for T_2-ideals of affine superalgebras. Fortunately, the same program we followed in §6.3.2 for proving Theorem 6.3.1 can be followed here, including the parallel super-PI representability theorem:

Theorem 7.2.13 (Kemer). *Any affine PI-superalgebra* W *over a field* F *of characteristic 0 is* **super-PI representable,** *i.e.,* $\sim_{\mathrm{PI}_2}$-*equivalent to some f.d. superalgebra over a purely transcendental extension of* F.

Unfortunately, everything is more complicated than before because we need to deal with more classes of identities, and the notation doubles (at least). So we go through the same steps, with some care. We start with some preliminaries. First we consider the super-version of the T-ideals of matrices. We view $\mathcal{M}_{n,F} = \mathrm{id}(M_n(F))$, as superidentities via Definition 7.1.2.

Definition 7.2.14. $\mathcal{M}_{k,\ell} = \mathrm{id}_2(M_{k,\ell}(F))$, *cf. Example 1.3.14.* $\mathcal{M}'_n = \mathrm{id}_2(M_n(F[c]))$ *of Example 7.2.7. (We suppressed* F *in order to unclutter the notation.) Define*

$$\tilde{\mathcal{M}}_n = \left(\bigcap_{k=1}^{n-1} \mathcal{M}_{k,n-k} \right) \cap \mathcal{M}_{n,F} \cap \mathcal{M}'_n = \mathrm{id}_2 \left(\prod M_{k,n-k} \times M_n(F) \times M_n(F[c]) \right).$$

Whereas in the ungraded case, any simple PI-algebra of PI-class n has T-ideal $\mathcal{M}_{n,F}$, the graded case also includes $\mathcal{M}_{k,\ell}$, for $\ell = n - k$, $1 \leq k \leq n - 1$, and $\mathcal{M}'_n$, a total of $n+1$ possibilities for each n. These are all distinct and have no inclusions among themselves.

Fortunately, we still have

Remark 7.2.15.

(i) $\mathcal{M}_{k,\ell} \subseteq \mathcal{M}_{k',\ell}$ *for* $k \geq k'$; *likewise* $\mathcal{M}_{k,\ell} \subseteq \mathcal{M}_{k,\ell'}$ *for* $\ell \geq \ell'$.

(ii) $\mathcal{M}'_n \subseteq \mathcal{M}'_m$ *for* $n \geq m$.

Remark 7.2.16. *Suppose that a superalgebra* A *is semiprime (as an ungraded algebra) of PI-class* n. *Then*

$$\mathrm{id}_2(A) \supseteq \tilde{\mathcal{M}}_n,$$

because we have natural embeddings of matrices that also respect the grade.

Steps in proving Theorem 7.0.1.

Step 1. Since W is ungraded-PI, we have a f.d. algebra A_0, such that $\mathrm{id}(A_0) \subseteq \mathrm{id}(W)$. Writing the group $\mathbb{Z}/2 = \{u_0, u_1\}$ where u_0 is the identity element, consider the f.d. superalgebra $A = A_0 \otimes F[\mathbb{Z}/2]$, where the elements of $A \otimes u_0$ are even and the elements of $A_0 \otimes u_1$ are odd. We claim that $\mathrm{id}_2(A) \subseteq \mathrm{id}(W)$. Indeed, if $f(y_1, \ldots, y_n, z_1, \ldots, z_m)$ is a multilinear superidentity of A, then

$$0 = f(a_1 \otimes u_0, \ldots, a_n \otimes u_0, a_1' \otimes u_1, \ldots, a_m' \otimes u_m)$$
$$= f(a_1, \ldots, a_n, a_1', \ldots, a_m') \otimes u_{m \cdot 1},$$

where the a and a' are from A_0. Thus f is also an identity of A_0, and thus of W.

Step 2. Given t_0, t_1 we define a superpolynomial f to be s**-fold** (t_0, t_1)**-alternating** if there are sets of homogeneous indeterminates $X_j = Y_j \cup Z_j$, $1 \leq j \leq s$, with $|Y_j| = t_0$, $|Z_j| = t_1$, (Y_j comprised of even indeterminates, and Z_j of odd indeterminates) such that f is (t_0, t_1)-alternating in each X_j.

Given a f.d. superalgebra A and a T_2-ideal Γ, we define

$$\beta(A) = (\beta_1, \beta_2)$$

where (β_1, β_2) is lexicographically greatest such that for any s, there is an s-fold (β_1, β_2)-alternating superpolynomial $f(X_1, \ldots, X_s; Y) \notin \mathrm{id}_2(A)$.

Now, given $\beta(A) = (t_0, t_1)$, we define $\gamma(A)$ to be the minimal s such that for every (s_0, s_1), where $s_0 + s_1 = s$, and for all large enough μ, every superpolynomial f having μ-fold $\beta(A)$-alternating sets, s_0-fold $(t_0 + 1, 0)$-alternating sets, and s_1-fold $(0, t_1 + 1)$-alternating sets, is in $id_2(A)$. Thus, instead of (t_A, s_A), we work with the triple $(\dim_F \bar{A}_0, \dim_F \bar{A}_1; s)$, where s is the nilpotence index of $\mathrm{Jac}(A)$.

Definition 7.2.17. *The* **Kemer superindex** *of a PI-superalgebra A, denoted* $\mathrm{index}_2 A$, *is* $(\beta(A), \gamma(A))$, *which now has three components (two for β and one for γ). We order this triple lexicographically.*

Clearly, if $\mathrm{id}(A_1) \subseteq \mathrm{id}(A_2)$, then $\mathrm{index}_2(A_1) \geq \mathrm{index}_2(A_2)$.

Step 3. Definition 6.3.16 (of full) is once again applicable (requiring the substitutions to be homogeneous), enabling us to prove the super-versions of Kemer's two key lemmas (describing β and γ in terms of the structure of a suitable superalgebra).

Lemma 7.2.18. *[Kemer's First Superlemma] If A is a full f.d. superalgebra, then* $\beta(A) = (\dim_F \bar{A}_0, \dim_F \bar{A}_1)$.

In the proof we need first to construct for every f.d. supersimple superalgebra a multilinear superpolynomial f which is a non-superidentity and having a $(\dim_F \bar{A}_0, \dim_F \bar{A}_1)$-alternating set. The construction of these polynomials is done by tempering the Capelli polynomial. Afterward, using the same technique as in the proof of the ungraded first Kemer's Lemma, we combine them together with a full superpolynomial to obtain a μ-fold $(\dim_F \bar{A}_0, \dim_F \bar{A}_1)$-alternating supernonidentity proving $\beta(A) \geq (\dim_F \bar{A}_0, \dim_F \bar{A}_1)$.

Definition 7.2.19. *A* (s_0, s_1, μ)-**Kemer superpolynomial** *for W is a superpolynomial f not in* $\mathrm{id}_2(W)$, *in* $s_0 + s_1 + \mu$ *sets of indeterminates X_j, which is s_0-fold $(t_0 + 1, 0)$-alternating, s_1-fold $(0, t_1 + 1)$-alternating, and having another μ folds that are (t_0, t_1)-alternating, where $\beta(W) = (t_0, t_1)$. We call this a* **Kemer superpolynomial***, suppressing s_0, s_1, and μ in the notation.*

A multilinear superpolynomial f has **Property super K** with respect to a f.d. superalgebra A having nilpotence degree s, if it vanishes on any (graded) substitution with fewer than $s-1$ radical substitutions, but has a nonvanishing specialization with $s-1$ radical substitutions. Next we need the construction from Definition 6.6.24, but working with

$$A' = \bar{A}\langle y_1, \ldots, y_\nu, z_1, \ldots, z_\nu \rangle$$

instead of Definition 6.6.21.

Definition 7.2.20. *Suppose we are given a T_2-ideal $\Gamma \supset \mathrm{id}_2(A)$. Take $A' = \bar{A}\langle y_1, \ldots, y_\nu, z_1, \ldots, z_\nu \rangle$ of Definition 1.9.3, where ν is some fixed large number, at least $\dim \mathrm{Jac}(A) = \dim A - \dim \bar{A}$. Let $\mathcal{I}_1$ be the ideal generated by all evaluations in A' of superpolynomials in Γ, and $\mathcal{I}_2$ be the ideal generated by all words having total degree u in $y_1, \ldots, y_\nu, z_1, \ldots, z_\nu$ and define*

$$\hat{A}_{u,\nu;\Gamma} = A'/(\mathcal{I}_1 + \mathcal{I}_2), \tag{7.16}$$

*This algebra is called the u-**generic** superalgebra of A*

Lemma 7.2.21. *The u-generic algebra $\hat{A}_{u;\nu}$ of a f.d. superalgebra A is also a superalgebra.*

Proof. Both ideals $\Gamma = \mathrm{id}_2(A)$ and $\mathcal{I}_2$ are superideals. $\square$

The proof of Lemma 6.6.30 now yields:

Lemma 7.2.22. *For any full nonidentity f of A satisfying Property Super-K, and arbitrary μ, the T_2-ideal generated by f contains a full μ-Kemer superpolynomial.*

Definition 7.2.23. *A superalgebra A is* **super-basic** *if A is not PI_2-equivalent to a finite subdirect product A' of f.d. superalgebras A_k, $1 \leq k \leq m$, each of lower super-Kemer index.*

Proposition 7.2.24 (Kemer's Second Superlemma). *Suppose $\mathrm{char}(F) = 0$. If A is a super-basic f.d. superalgebra, then $\gamma(A) = s$ where s is the super-nilpotence index of $J = \mathrm{Jac}(A)$. In this case, A has a (s_0, s_1, μ)-Kemer superpolynomial, for some (s_0, s_1) such that $s = s_0 + s_1 + 1$, where s is the nilpotence degree of A, and for any μ.*

The proof is the same as before: Using f satisfying Property K, we can tack on as many folds of (t_0, t_1)-alternating superpolynomials as we want, and we have s_0 extra even indeterminates and s_1 extra odd indeterminates that we must insert into the folds.

At the same time, just as in Chapter 6, we can verify the Phoenix property for superbasic superalgebras, that any consequence of a Kemer superpolynomial has a consequence that is a Kemer superpolynomial.

Proposition 7.2.25. *Suppose* $A, A_1, \ldots, A_m$ *are f.d. superalgebras, satisfying* $\mathrm{index}_2(A_i) < \mathrm{index}_2(A)$ *for each* i. *Then for all* μ *large enough, any* μ-*Kemer superpolynomial of* A *is a superidentity of* $A_1, \ldots, A_m$.

Proposition 7.2.26. *Any f.d. superalgebra* A *is* PI_2-*equivalent to a direct product of f.d. superalgebras*

$$\tilde{A} \times A_1 \times \cdots \times A_t,$$

where $A_1, \ldots, A_t$ *are super-basic, and where* $\tilde{A}$ *is a direct product of super-basic superalgebras* $\tilde{A}_1, \ldots, \tilde{A}_u$, *such that*

$$
\begin{aligned}
\mathrm{index}_2(\tilde{A}_1) = \mathrm{index}_2(\tilde{A}_2) = \cdots = \mathrm{index}_2(\tilde{A}_u) \\
> \mathrm{index}_2(A_1) \geq \mathrm{index}_2(A_2) \geq \cdots \geq \mathrm{index}_2(A_t).
\end{aligned}
\tag{7.17}
$$

Furthermore, there exists a Kemer superpolynomial f *for each of* $\tilde{A}_1, \ldots, \tilde{A}_u$, *which is a superidentity for* $A_1 \times \cdots \times A_t$.

Proof. Analogous to that of Proposition 6.7.5.   □

Step 4. We need the super-Phoenix-property.

Proposition 7.2.27. *Suppose* Γ *is a* T_2-*ideal properly containing* $\mathrm{id}_2(A)$ *for some f.d. superalgebra* A. *Then there exists a f.d. superalgebra* A' *for which* $\Gamma \supseteq \mathrm{id}_2(A')$, *but* $\mathrm{index}_2(A') = \mathrm{index}_2(\Gamma)$ *and* A' *and* Γ *share the same* μ-*Kemer superpolynomials for large enough* μ.

Proof. The proof is virtually identical to that of Proposition 6.8.4.   □

Likewise we have the analog to Theorem 6.8.5:

Theorem 7.2.28. *Any Kemer superpolynomial of a* T_2-*ideal* Γ *satisfies the Phoenix property.*

Step 5. Now that we have enough Kemer superpolynomials, we need a graded analog of the trace ring. This is done as in the proof of Proposition 6.8.1.

Theorem 7.2.29. *Let* W *be a superalgebra which satisfies the* ℓ-*th Capelli identity* c_ℓ. *Then* $\mathrm{id}_2(W) = \mathrm{id}_2(\mathcal{W})$, *where* $\mathcal{W}$ *is the relatively free superalgebra of* W *generated by* $\ell - 1$ *even indeterminates and* $\ell - 1$ *odd indeterminates.*

Proof. The reduction to the affine case is achieved by throwing in $\ell - 1$ even and odd indeterminates. $\square$

Remark 7.2.30. *Any superalgebra which is finite over an affine algebra C is super-representable, by the super-version of Theorem 1.6.22; the modification of the proof is straightforward, cf. Exercise 6.*

Proposition 7.2.31. *Suppose $\mathcal{A}$ is some affine relatively free superalgebra of a f.d. superalgebra A over an algebraically closed field of characteristic 0 generated by $y_1, \ldots, y_l, z_1, \ldots, z_l$, and $\mathcal{S}$ be a set of Kemer superpolynomials for A. Let $\mathcal{I}$ be the ideal generated by all $\{f(\mathcal{A}) : f \in \mathcal{S}\}$. Then $\mathcal{A}/\mathcal{I}$ is super-representable (i.e., is embeddable as a superalgebra into a f.d. superalgebra).*

Proof. Consider $\mathcal{A}$ as a supersubalgebra of $A \otimes F[\Lambda]$ (as in Example 1.8.15). Suppose $A = A_1 \times \cdots \times A_m$ is the decomposition of A into product of super basic superalgebras. Define $tr(a_1, \ldots, a_m) = (tr(a_1^e), \ldots, tr(a_m^e))$, for $a_i \in A_i$, where a_i^e is the even part of a_i and $tr(a_i)$ is defined as in Remark 6.7.1. We can extend the trace linearly to $\mathcal{A}$, to take on values in $F[\Lambda]^m$. Suppose $w_1, \ldots, w_p$ is a homogenous Shirshov base of $\mathcal{A}$, where the first p' are even and the rest are odd. Letting C denote the commutative affine subalgebra of $F[\Lambda]^m$ consisting of all the characteristic values of $w_1, \ldots, w_{p'}, w_{p'+1}^2, \ldots, w_p^2 \in \mathcal{A}_0$. We let $\hat{\mathcal{A}} = \mathcal{A}C \subseteq \bar{A}[\Lambda]$. By a small modification of Proposition 2.5.14, $\hat{\mathcal{A}}$ is finite over C, and in particular, is a Noetherian algebra. By Remark 6.7.1, $\mathcal{I}$ is also closed under multiplication by elements of C, so in fact, $\mathcal{I}$ is a common ideal of $\mathcal{A}$ and $\hat{\mathcal{A}}$. Thus, $\mathcal{A}/\mathcal{I} \subseteq \hat{\mathcal{A}}/\mathcal{I}$, which is finite over the image of the affine algebra C, so $\mathcal{A}/\mathcal{I}$ is super-representable, by Remark 7.2.30. $\square$

Definition 7.2.32. *Given a f.d.-superalgebra A, with $s := s(A)$, consider the superalgebra $\hat{A}_{s;\nu}$ of Definition 7.2.20, where ν is large enough. $\hat{A}_{s;\nu}$ is f.d. and graded by Remark 6.6.26. We let $\mathcal{A}$ be the affine relatively free superalgebra of A generated by $y_1', \ldots, y_l', z_1', \ldots, z_l'$ elements, which we consider as a sub-superalgebra of $\hat{A}_{s;\nu}[\Lambda = \{\lambda_{i,j} : i = 1 \ldots \ell, \ j = 1, \ldots, \dim_F \bar{A}\}]$, where y_i is sent to $\sum_j \lambda_{i,j} a_{j,0} + \bar{y}_i$ and z_i is sent to $\sum_j \lambda_{i,j} a_{j,1} + \bar{z}_i$. (Here $a_{1,u}, \ldots, a_{\dim_F \bar{A}_u, u}$ is a basis of $\bar{A}_u$. Take $\mathcal{A}_1$ to be the subalgebra of $\hat{A}_{s;\nu}[\Lambda]$ generated by $\mathcal{A}$ and $\{\bar{y}_i, \bar{z}_i : 1 \leq i \leq \mu\}$.*

We define a trace on $A[\Lambda]$ as in Remark 6.7.1(vi) and Proposition 7.2.31, and extend it to $\mathcal{A}_1$ such that the trace of every word including some $\bar{y}_i$ or $\bar{z}_i$ is 0. $\mathcal{A}_2$ is the subalgebra of $\hat{A}_{s;\nu}[\Lambda]$ obtained by adjoining the traces of $\mathcal{A}$ to $\mathcal{A}_1$.

Step 6. Next we introduce representable superalgebras.

Proposition 7.2.33. *Let Γ_2 be the superideal of $\mathcal{A}_2$ obtained by all evaluations of Γ on $\mathcal{A}_2$. Then $\mathcal{A}_2/\Gamma_2$ is finite over its center, and thus is representable.*

Proof. As for Proposition 6.9.4. Let $\mathcal{S}_\mu$ be the set of μ-Kemer polynomials of A. In constructing $\mathcal{A}_1$, recall that we have a set of even and odd nilpotent indeterminates $\bar{y}_j, \bar{z}_j : 1 \leq j \leq \nu$; we took this set large enough to include all of the designated alternating indeterminates in a μ-Kemer polynomial. (Thus, $\nu \geq \nu_\Gamma(\mu)$.) As in Corollary 6.8.3, we can find a specialization of every $f \in \mathcal{S}_\mu$ of $\mathcal{S}_\mu$ to $\mathcal{A}_1$, for which each designated indeterminate y_i, z_i is sent to $\bar{y}_i, \bar{z}_i$, respectively, and $f \notin \Gamma_1$, where Γ_1 is the ideal of $\mathcal{A}_1$ consisting all evaluations of polynomials of Γ on $\mathcal{A}_1$. Denote the set of all such specializations by $\overline{\mathcal{S}_\mu}$.

Claim. We claim that $\Gamma_2 \cap \overline{\mathcal{S}_\mu}$ is empty. Indeed suppose

$$f = \sum_j v_j w_j \in \Gamma_2,$$

where v_j is a product of traces of elements of $\overline{A}[\Lambda]$ and $w_j \in \Gamma_1$. By substituting 0 for $\bar{y}_i$ and $\bar{z}_i$ we may assume that all the monomials of every w_j contain all the $\bar{y}_i$'s and $\bar{z}_i$'s. Thus, every $\bar{y}_i$ and $\bar{z}_i$ appears exactly once in any monomial of every w_j. Applying the alternator (for one of the even folds) to f yields a multiple of f, but on the right side we get $\sum_j v_j w_j'$. Since v_j is a product of traces of elements from $\mathcal{A}_\infty$, iterating Theorem I of Section 1.5 shows that $v_j w_j$ is an element of Γ_1. Hence $f \in \Gamma_1$, so $f \notin \overline{\mathcal{S}_\mu}$.

Let $\mathcal{A}' = \mathcal{A}_2/\Gamma_2$. Thus, $\mathrm{id}_2(\mathcal{A}') \supseteq \Gamma$. The claim shows that the Kemer polynomials $\mathcal{S}_\mu$ remain non-superidentities of $\mathcal{A}'$.

On the other hand, by construction and Shirshov's Theorem, $\mathcal{A}_2$ is finite over its center. Hence $\mathcal{A}'$ is finite over its center and thus is super-representable, by Remark 7.2.30. □

Step 7. We are ready, finally, to conclude the proof of Theorem 7.0.1, analogously to the proof of Theorem 6.9.1:

Proof. Let $\kappa = \mathrm{index}_2(\Gamma)$. $\Gamma \supseteq \mathrm{id}_2(A)$ for some suitable f.d. superalgebra A, by Step 1. By Proposition 7.2.27, we may assume that $\mathrm{index}_2(A) = \kappa$, and furthermore, for μ large enough, Γ and A satisfy precisely the same μ-Kemer superpolynomials.

Let $\mathcal{S}_\mu$ be the set of μ-Kemer superpolynomials of Γ, and $\langle \mathcal{S}_\mu \rangle_{T_2}$ the T_2-ideal generated by $\mathcal{S}_\mu$. and let $\Gamma' = \Gamma + \langle \mathcal{S}_\mu \rangle_{T_2}$. Writing $\kappa = (t, s)$, we see by definition that

$$\mathrm{index}(\Gamma') < \kappa,$$

so by induction there is some f.d. superalgebra A' with $\mathrm{id}_2(A') = \Gamma'$.

We just found a f.d. superalgebra A'' for which all elements of Γ are superidentities, but none of the $\mathcal{S}_\mu$ are superidentities. We claim, that

$$\Gamma = (\Gamma + \langle \mathcal{S}_\mu \rangle_{T_2}) \cap \mathrm{id}_2(A'') = \mathrm{id}_2(A' \times A''),$$

as desired. Indeed, if $f \in \mathrm{id}_2(A'')$ and $f = g + h$ for $g \in \Gamma$ and $h \in \langle \mathcal{S}_\mu \rangle_{T_2}$, then $h = f - g \in \mathrm{id}_2(A'')$. If h is not in Γ, then by the Phoenix property there is some $h' \in \mathcal{S}_\mu \cap \mathrm{id}_2(A'')$, which is absurd. Thus $h \in \Gamma$, yielding $f \in \Gamma$. □

7.3   Kemer's Main Theorem Concluded

Theorem 7.3.1. *[Kemer] The ACC holds for T_2-ideals of affine superalgebras over a field of characteristic 0.*

Proof. We want to show that any chain

$$\mathcal{I}_1 \subseteq \mathcal{I}_2 \subseteq \mathcal{I}_3 \subseteq \ldots \tag{7.18}$$

of T_2-ideals terminates. Since ordered pairs of natural numbers satisfy the descending chain condition, we may start far enough along our chain of T-ideals and assume that all the A_j in (6.12) have the same Kemer superindex κ. We do this for κ minimal possible, and induct on κ.

By Proposition 7.2.26, we can write each

$$A_j = \tilde{A}_j \times A_{j1} \times \cdots \times A_{j,t_j},$$

where each $\text{index}_2(A_{j,i}) < \kappa$. But by the super-version of Remark 6.10.1, we may assume that $\tilde{A}_j = \tilde{A}$ for some superalgebra $\tilde{A}$ that is a direct product of super-basic components of Kemer superindex κ.

Let $\mathcal{I}$ be the T-ideal generated by all Kemer superpolynomials of $\tilde{A}$ that are superidentities of $A_{1,1} \times \cdots \times A_{1,t}$. As we argued in Chapter 6, $\tilde{A}/\mathcal{I}(\tilde{A})$ cannot contain any Kemer superpolynomials (for κ) that are superidentities of $A_{1,1} \times \cdots \times A_{1,t}$, so its Kemer superindex must decrease and the chain

$$\mathcal{I}_1 + \mathcal{I} \subseteq \mathcal{I}_2 + \mathcal{I} \subseteq \mathcal{I}_3 + \mathcal{I} \ldots \tag{7.19}$$

must stabilize by induction. On the other hand, for each j,

$$\mathcal{I} \cap \text{id}_2(\tilde{A}) = \mathcal{I} \cap \text{id}_2(A_1) = \mathcal{I} \cap \mathcal{I}_1 \subseteq \mathcal{I} \cap \mathcal{I}_j = \mathcal{I} \cap \text{id}_2(A_j) \subseteq \mathcal{I} \cap \text{id}_2(\tilde{A}),$$

so equality must hold at each stage, i.e., $\mathcal{I} \cap \mathcal{I}_1 = \mathcal{I} \cap \mathcal{I}_j$ for each j. But coupled with (7.19), this means the chain (7.18) stabilizes, contradiction. $\square$

We are finally ready for Kemer's solution to Specht's problem.

Theorem 7.3.2. *[Kemer] The ACC holds for T-ideals, over any field of characteristic 0.*

Proof. Combine Theorem 7.2.1 with Theorem 7.3.1. $\square$

As we shall see in Chapter 9, Theorem 7.3.2 is true only in characteristic 0, although Theorem 7.3.1 can be proved more generally in arbitrary characteristic.

7.4   Consequences of Kemer's Theory

Kemer's theorem opens the door to a new structure theory of PI-algebras. We shall present the outline here, and give an application of Berele to GK-dimension in Chapter 11. We want to transfer the theory of T-ideals of $F\{X\}$ to an arbitrary algebra A over F, by using the ACC theory of §1.1.7.1.

7.4.1 T-ideals of relatively free algebras

Our starting point is for relatively free algebras. Kemer's theorem implies the ACC for T-ideals of any relatively free algebra $\mathcal{A}$. This means using Theorem 1.1.31. Here is the version for T-ideals.

Definition 7.4.1. *A T-ideal Γ is called **verbally prime** if for every two T-ideals $\mathcal{I}, \mathcal{J}$ such that $\mathcal{I}\mathcal{J} \subseteq \Gamma$, implies $\mathcal{I}$ or $\mathcal{J}$ is contained in Γ.*
*An algebra W is **verbally prime** if $\mathrm{id}(W)$ is verbally prime.*

There is another way to describe verbally prime which also explains their name.

Lemma 7.4.2. *For any (ungraded) polynomials $f = f(x_1,\ldots,x_n)$ and $g = g(y_1,\ldots,y_m)$ (in distinct indeterminates), denote by $\mathcal{I}$ and $\mathcal{J}$ the respective T-ideals generated by f and g, such that $fg \in \Gamma$, either f or g is in Γ. Then $\mathcal{I}\mathcal{J}$ is T–generated by fg.*

Proof. Since the characteristic is 0, we may assume that f and g are multilinear. It is easy to see that $\mathcal{I}\mathcal{J}$ is T- generated by fzg for another indeterminate z. Since

$$x_{\sigma(1)} \cdots x_{\sigma(n)} z = z x_{\sigma(1)} \cdots x_{\sigma(n)} + \sum_{i=1}^{n} x_{\sigma(1)} \cdots x_{\sigma(i-1)} [x_{,\sigma(i)}, z] x_{\sigma(i+1)} \cdots x_{\sigma(n)},$$

we get

$$f(x_1,\ldots,x_n) z = z f(x_1,\ldots,x_n) + \sum_{i=1}^{n} f(x_1,\ldots,x_{i-1},[x_i,z],x_{i+1},\ldots,x_n).$$

Hence $\mathcal{I}\mathcal{J}$ is T-generated by fg.   □

Proposition 7.4.3. *A T-ideal Γ is verbally prime iff for any (ungraded) polynomials $f = f(x_1,\ldots,x_n)$ and $g = g(y_1,\ldots,y_m)$ such that $fg \in \Gamma$, either f or g is in Γ.*

Proof. ($\Rightarrow$) In the notation of the lemma, if $fg \in \Gamma$, then $\mathcal{I} \subseteq \Gamma$ or $\mathcal{J} \subseteq \Gamma$, implying $f \in \Gamma$ or $g \in \Gamma$.
($\Leftarrow$) Suppose $\mathcal{I}, \mathcal{J}$ are T-ideals such that $\mathcal{I}\mathcal{J} \subseteq \Gamma$. If $\mathcal{I}, \mathcal{J}$ were both not contained in Γ, we could find $f(x_1,\ldots,x_n) \in \mathcal{I} \setminus \Gamma$ and $g(y_1,\ldots,y_m) \in \mathcal{J} \setminus \Gamma$, but $fg \notin \Gamma$, contradicting the lemma.   □

Since verbally prime is "prime" for the class of T-ideals, in the sense of Definition 1.1.29, we can apply Theorem 1.1.31 and Corollary 1.1.33 to get:

Theorem 7.4.4. *Any T-ideal $\mathcal{I}$ of $\mathcal{A}$ has a radical $\sqrt{\mathcal{I}}$ that is a finite intersection of verbally prime T-ideals, and $\sqrt{\mathcal{I}}^{k} \subseteq \mathcal{I}$ for some k.*

Viewed another way, for any relatively free algebra $\mathcal{A}$ there is a nilpotent T-deal $\mathcal{I}$, such that $\mathcal{A}/\mathcal{I}$ is the subdirect product of relatively free algebras (whose identities constitute verbally prime T-ideals).

Since verbally prime T-ideals are so fundamental from this point of view, let us describe them explicitly.

Lemma 7.4.5. *Suppose $A = A_1 \times \cdots \times A_m$ is a verbally prime algebra, then $\mathrm{id}(A) = \mathrm{id}(A_j)$ for some $1 \leq j \leq m$.*

Proof. We are done unless $m > 1$. Let $A' = A_2 \times \cdots \times A_m$. By induction on m, $\mathrm{id}(A') = \mathrm{id}(A_j)$ for some j. But $\mathrm{id}(A_1), \mathrm{id}(A') \not\subseteq \mathrm{id}(A)$. If $\mathrm{id}(A_1) \subseteq \mathrm{id}(A)$, then $\mathrm{id}(A) = \mathrm{id}(A_1)$, so we may assume that $\mathrm{id}(A') \subseteq \mathrm{id}(A)$. Thus $\mathrm{id}(A) = \mathrm{id}(A') = \mathrm{id}(A_j)$, as desired. $\qquad\square$

Proposition 7.4.6. *If $A = A_1 \times \cdots \times A_m$ is a product of superalgebras such that $G(A)$ is verbally prime , then $\mathrm{id}(G(A)) = \mathrm{id}(G(A_j))$ for some j.*

Proof. Follows from the lemma, since

$$G(A) = G(A_1) \times \cdots \times G(A_m).$$

$\qquad\square$

Definition 7.4.7. *A finite dimensional superalgebra A is G-**minimal** if for every superalgebra R,*

$$\mathrm{id}(G(A)) = \mathrm{id}(G(R)) \Longrightarrow \dim A \leq \dim R.$$

Lemma 7.4.8. *If A is G-minimal and $G(A)$ is verbally prime, then A is semisimple.*

Proof. We need to show that $J = 0$, where $J = \mathrm{Jac}(A)$. Suppose otherwise, so $\mathrm{id}(G(A)) \subsetneq \mathrm{id}(G(\bar{A}))$, viewing $\bar{A} = A/J$ as a superalgebra since J is a superideal. Since A is G-minimal), there is a multilinear superpolynomial in $f \in \mathrm{id}(G(\bar{A})) \setminus \mathrm{id}(G(A))$. Hence $0 \neq f(G(A)) \subseteq G(J)$. Since $G(J)^t = 0$ for some t, we have $f(X_1)f(X_2) \cdots f(X_t) \in \mathrm{id}(G(A))$ (using t sets of distinct indeterminates), contrary to $G(A)$ being verbally prime. $\qquad\square$

So we obtain the translation:

Theorem 7.4.9. *Suppose that R is a superalgebra which is also an ungraded PI-algebra. Then R is verbally prime if and only if it is graded PI-reducible to the Grassmann envelope of a supersimple f.d. superalgebra.*

Proof. ($\Rightarrow$) By the representability theorem, $\Gamma = \mathrm{id}(R)$ is the T-ideal of $G(A)$, for some f.d. superalgebra A. Assume also that A is G-minimal. By Lemma 7.4.8, A is semisimple. Since every semisimple superalgebra is a direct product of supersimple superalgebras, by Lemma 7.4.5, we may replace A by a supersimple algebra.

($\Leftarrow$) We shall show that $G(A)$ is verbally prime, for any f.d. supersimple superalgebra A. For any T-ideal $\mathcal{I}$ not contained in $\Gamma = \mathrm{id}(G(A))$, denote

$$\mathcal{I}(A) = \mathrm{span}_F \{a \in A : a \otimes w = f(\tilde{x}_1, \ldots, \tilde{x}_n)\}$$

taken over all $0 \neq w \in G$, multilinear $f \in \mathcal{I}$, and

$$\tilde{x}_1 = a_1 \otimes w_1, \ldots, \tilde{x}_n = a_n \otimes w_n \in G(A).$$

It is straightforward to check that $\mathcal{I}(A)$ is a nonzero superideal of R. Thus $\mathcal{I}(A) = R$. For any two T-ideals $\mathcal{I}, \mathcal{J}$ not contained in Γ, we conclude that $\mathcal{I}\mathcal{J}(A) = \mathcal{I}(A)\mathcal{J}(A) = A \neq 0$, so $\mathcal{I}\mathcal{J} \nsubseteq \Gamma$. $\qquad\square$

Putting everything together, we can classify the verbally prime T-ideals in characteristic 0.

Theorem 7.4.10. *The following list is a complete list of the verbally prime T-ideals of $F\{X\}$, for any field F of characteristic 0:*

(i) $\mathrm{id}(M_n(F))$;

(ii) $\mathrm{id}(M_{k,\ell}(F))$;

(iii) $\mathrm{id}(M_n(G))$.

Proof. Suppose that Γ is a verbally prime T-ideal. Writing $\Gamma = \mathrm{id}(G(A))$ for a supersimple, f.d. superalgebra A, we see that $\mathrm{id}(A)$ must be one of $\mathcal{M}_{n,F}$, $\mathcal{M}_{k,\ell}$, and $\mathcal{M}'_n$. It remains to translate these back to R.

Case 1. $A = M_n(F)$ with the trivial grade. Then $R = M_n(F)$.

Case 2. $A = M_{k,\ell}(F)$. Then (for $n = k + \ell$) R is the subalgebra of $M_n(G)$ spanned by $G_0 e_{ij}$ for $1 \leq i, j \leq k$ or $k + 1 \leq i, j \leq n$, and by $G_1 e_{ij}$ for the other i, j ($i > k$ and $j \leq \ell$, or $i \leq k$ and $j > \ell$).

Case 3. $A = M_n(F) \oplus M_n(F)c$, where $c_2 = 1$. Then

$$R = G_0 M_n(F) \oplus G_1 M_n(F)c,$$

which is clearly PI-equivalent to $M_n(G)$. $\qquad\square$

Remark 7.4.11. *Note that whereas (i) is a prime ideal of $F\{X\}$, (ii) and (iii) are only verbally prime (as T-ideals), since the only prime relatively free domains are the algebras of generic matrices.*

Although Theorem 7.4.10 is very explicit and is proved directly by passing to superalgebras, the transition between T-ideals and superalgebras often is quite delicate, involving taking all of the different f_I^* of an identity f, cf. Proposition 7.1.11. This makes it difficult to answer such basic questions as:

"What are the minimal identities of $M_n(G)$?"

Having determined the verbally prime T-ideals in characteristic 0, we are led to try to find the analogs of basic structure theorems related to the classic Nullstellensatz of Hilbert. Any verbally prime T-ideal is the intersection of maximal T-ideals, cf. Exercise 7. Unfortunately, the Jacobson radical of a relatively free algebra, although a T-ideal, need not necessarily be nilpotent, as evidenced by the relatively free algebra of the Grassmann algebra (whose radical is nil but not nilpotent). Nevertheless, one has the following positive result, following immediately from Kemer's Theorem.

Remark 7.4.12. *If the image of a T-ideal $\mathcal{I}$ of a relatively free algebra $\mathcal{A}$ is zero in every verbally prime homomorphic image of $\mathcal{A}$, then $\mathcal{I}$ is nilpotent, by Corollary 1.1.35.*

The situation in characteristic p is much harder, as we shall see in Chapter 9. Even though much goes wrong, the analog to Remark 7.4.12 remains open, and in fact is conjectured by Kemer to be true.

7.4.2   Verbal ideals of algebras

We would like to apply ACC on T-ideals to arbitrary algebras. Unfortunately, unless A is relatively free, there could be too few homomorphisms from A to itself for Definition 1.7.5 to be meaningful. To proceed further, we need another definition.

Definition 7.4.13. *A **verbal ideal** of an F-algebra A is the set of evaluations in A of a T-ideal $\mathcal{I}$ of $F\{X\}$, where X is a countably infinite set of indeterminates.*
 In other words, the verbal ideal of A corresponding to $\mathcal{I}$ is its image

$$\{f(a_1,\ldots,a_m):\ f(x_1,\ldots,x_m)\in\mathcal{I},\ a_i\in A\}.$$

The key to our discussion lies in the following easy observation:

Remark 7.4.14. *Given any $r_j = f_j(a_1,\ldots,a_{m_j})$, there is a homomorphism $\Phi : F\{X\} \to A$ such that*

$$r_1 = \Phi(f_1(x_1,\ldots,x_{m_1})), \quad r_2 = \Phi(f_2(x_{m_1+1},\ldots,x_{m_1+m_2})),\ldots;$$

the point here is that different indeterminates x_i appear in the different positions.

We write $\mathcal{I}(A)$ for the verbal ideal of A corresponding to a T-ideal $\mathcal{I}$.

Lemma 7.4.15.

(i) $\mathcal{I}(F\{X\}) = \mathcal{I}$.

(ii) Any verbal ideal is an ideal.

(iii) $(\mathcal{I}_1 + \mathcal{I}_2)(A) = \mathcal{I}_1(A) + \mathcal{I}_2(A)$.

(iv) $(\mathcal{I}_1\mathcal{I}_2)(A) = \mathcal{I}_1(A)\mathcal{I}_2(A)$.

(v) *If* $\mathcal{I}$ *is a verbal ideal of* A, *then* $\mathcal{I} = \{f \in F\{X\} : f(A) \subseteq \mathcal{I}\} \triangleleft_T F\{X\}$.

(vi) *For* char $F = 0$, *any* F-*algebra satisfies the ACC for verbal ideals.*

Proof. (i) is clear.

(ii) Follows easily from Remark 7.4.14: For example, if $r = f(a_1, \ldots, a_m)$ and $a \in A$, then take a homomorphism Φ sending $x_i \mapsto a_i$ for $1 \le i \le m$ and $x_{m+1} \mapsto a$, and conclude $ar = \Phi(x_{m+1}f(x_1, \ldots, x_m)) \subseteq \mathcal{I}$. The same argument proves (iii) and (iv).

(v) $f(h_1, \ldots, h_m)(a_1, \ldots, a_n) = f(h_1(a_1, \ldots, a_n), \ldots, h_m(a_1, \ldots, a_n)) \in \mathcal{I}$ for all $a_i \in A$.

(vi) Given verbal ideals $\mathcal{I}_1 \subset \mathcal{I}_2 \subset \ldots$, let

$$\widehat{\mathcal{I}}_j = \{f \in F\{X\} : f(A) \subseteq \mathcal{I}_j\},$$

a T-ideal of $F\{X\}$, so we conclude using Kemer's theorem. $\qquad\square$

Thus, the program of Chapter 1 is available:

Theorem 7.4.16. *Every verbal ideal* $\mathcal{I}$ *has only a finite number of verbal primes* $P_1, \ldots, P_t$ *minimal over* $\mathcal{I}$, *and some finite product of the* P_i *is contained in* $\mathcal{I}$.

Proof. By Theorem 1.1.31. $\qquad\square$

We are led to characterize the prime verbal ideals.

Proposition 7.4.17. *Each prime verbal ideal of* A *is the image of a verbally prime* T-*ideal.*

Proof. Suppose $\mathcal{P}$ is a verbally prime verbal ideal of A, and take $\widehat{\mathcal{P}}$ as in Lemma 7.4.15(v). If $\mathcal{I}_1\mathcal{I}_2 \subseteq \mathcal{P}$ for T-ideals $\mathcal{I}_1$ and $\mathcal{I}_2$, then

$$\mathcal{I}_1(A)\mathcal{I}_2(A) \subseteq \mathcal{P}(A) = \mathcal{P},$$

so some $\mathcal{I}_j \subseteq \widehat{\mathcal{P}}$. This proves that $\widehat{\mathcal{P}}$ is a verbally prime T-ideal. $\qquad\square$

Here is a related result, inspired by Kemer's theory.

Definition 7.4.18. *A verbal ideal* $\mathcal{I}$ *is* **verbally irreducible** *if we cannot write* $\mathcal{I} = \mathcal{J}_1 \cap \mathcal{J}_2$ *for verbal ideals* $\mathcal{J}_1, \mathcal{J}_2 \supset \mathcal{I}$.

Proposition 7.4.19. *Every verbal ideal* $\mathcal{I}$ *is a finite intersection of verbally irreducible verbal ideals.*

Proof. By Theorem 1.1.31. $\qquad\square$

7.4.3   Standard identities versus Capelli identities

Since so much of the theory depends on sparse identities, and more specifically Capelli identities, let us conclude this chapter with some results about Capelli identities. The material uses some of Kemer's ideas, although it does not seem to be a direct application of his solution to Specht's problem. In Example 1.8.25 we saw that the Capelli identity c_n implies the standard identity s_n, but s_n does not imply c_n. Furthermore, G does not satisfy any Capelli identity at all. This raises the following question: Does s_n imply c'_n for suitable n'? The solution, due to Kemer [Kem78], although not used in the remainder of this book, is quite pretty, and has historical interest since it is the first time Grassmann techniques were applied to the Capelli polynomial. Our presentation incorporates a slight modification in Berele [Ber85a].

We prove:

Theorem 7.4.20 (Kemer). *Any algebra of characteristic 0 satisfying s_n satisfies $c_{n'}$ for some n' that depends only on n.*

The rest of this subsection is dedicated to the proof of this theorem.

Example 7.4.21. *Let us construct the relatively free superalgebra $\tilde{G}$ with respect (only) to the graded identities of (7.2) of Example 7.1.6. Thus, we consider sets of usual indeterminates $\bar{X} = \{\bar{x}_i : i \in I\}$ and odd indeterminates $\bar{Z} = \{\bar{z}_i : i \in I\}$, where the $\bar{z}_j$ must strictly anticommute with each other. (We do not bother with the even indeterminates.) Let $\tilde{G} = F\{\bar{X}, \bar{Z}\}$. Since all the identities involve the $\bar{z}_i$, we see $F\{\bar{X}\} \cong F\{X\}$, the free algebra. Let $\tilde{G}_m$ denote all expressions in $\tilde{G}$ that are linear in each of $\bar{z}_1, \ldots, \bar{z}_m$. Since the identities are homogeneous, $\tilde{G}_m$ is well-defined, and Remark 1.2.32 has a more precise formulation.*

Remark 7.4.22. *Write $F\{X; Z\}$ for the free algebra in (usual) indeterminates $x_1, x_2, \ldots$ and $z_1, z_2, \ldots$. Let F_m denote the polynomials that are alternating in $z_1, z_2, \ldots, z_m$. There is a 1:1 correspondence $F_m \mapsto \tilde{G}_m$ given by*

$$\sum_{\sigma \in S_n} (\mathrm{sgn}\sigma) h_1 z_{\sigma(1)} \cdots h_m, z_{\sigma(m)} h_{m+1} \mapsto h_1 \bar{z}_1 \cdots h_m \bar{z}_m h_{m+1}.$$

Let W_m denote the set of words of length m in the $\bar{x}_j$. Also let $\langle W_m \rangle$ denote the ideal of $F\{\bar{X}; \bar{Z}\}$ generated by the images of the elements of W_m.

Lemma 7.4.23. *If $w_1, \ldots, w_n \in W_m$, then*

$$w_{\sigma(1)} a_1 \ldots a_{n-1} w_{\sigma(n)} = \epsilon w_1 a_1 \ldots a_{n-1} w_n, \quad \forall a \in \tilde{G},$$

where $\epsilon = 1$ for m even and $\epsilon = \mathrm{sgn}(\sigma)$ for m odd.

Proof. Since every permutation is a product of transpositions, it is enough to check this for $n = 2$, in which case, this follows at once from the definition, since we are making m^2 switches of odd elements, and $m^2 \equiv m \pmod 2$. $\square$

Let $\mathcal{I}_m$ be the T-ideal of s_m, and let $\mathcal{J}_n$ be the T-ideal of c_n. Given a T-ideal $\mathcal{T}$, we let $\mathcal{T}(\tilde{G})$ denote the evaluations of $\mathcal{T}$ on $\tilde{G}$, under $x_i \mapsto \bar{x}_i$ and $z_i \mapsto \bar{z}_i$.

Lemma 7.4.24. $W_m \subseteq \mathcal{I}_m(\tilde{G})$.

Proof. $z_{i_1} \ldots z_{i_m} = \frac{1}{m!} s_n(z_{i_1}, \ldots, z_{i_m})$. $\qquad\square$

Lemma 7.4.25. *For any T-ideal $\mathcal{T}$ of $F\{X\}$, $\mathcal{J}_m \subseteq \mathcal{T}$ iff $\langle W_1 \rangle^m \subseteq \mathcal{T}(\tilde{G})$.*

Proof. We take $a_i \in \tilde{G}$ arbitrary and $w_i \in W_1$.

($\Rightarrow$) $m! a_0 w_1 a_1 \ldots w_m a_{m+1} = a_0 c_m(w_1, \ldots, w_m; a_1, \ldots, a_m) \in \tilde{G}\mathcal{T}(\tilde{G}) \subseteq \mathcal{T}(\tilde{G})$.

($\Leftarrow$) We need to show that $c_m \in \mathcal{I}$. But

$$c_m(\bar{z}_1, \ldots \bar{z}_m; \bar{x}_1, \ldots \bar{x}_m) = m! \bar{z}_1 \bar{x}_1 \ldots \bar{z}_m \bar{x}_m \bar{z}_{m+1} \in \langle W_m \rangle^m \subseteq \mathcal{T}(\tilde{G}).$$

Thus, there is $f \in \mathcal{T}$ such that

$$f(\bar{z}_1, \ldots, \bar{z}_{m+1}; \bar{x}_1, \ldots, \bar{x}_m) = c_m(\bar{z}_1, \ldots, \bar{z}_m; \bar{x}_1, \ldots, \bar{x}_m).$$

Taking the alternator

$$\tilde{f} = \sum_{\sigma \in S_n} \text{sgn}(\sigma) f(z_{\sigma(1)}, \ldots, z_{\sigma(m)}; x_1, \ldots, x_m),$$

which is m-alternating in the z_i, we see that

$$\tilde{f}(\bar{z}_1, \ldots, \bar{z}_m; \bar{x}) = m! c_m((\bar{z}_1, \ldots, \bar{z}_m; \bar{x})),$$

and thus Remark 7.4.22 shows $c_m = \frac{1}{m!} f \in \mathcal{T}$. $\qquad\square$

We recall $\tilde{s}_n(x_1, \ldots, x_n) = \sum_{\pi \in S_n} x_{\pi(1)} \ldots x_{\pi(t)}$, the symmetric polynomial (1.7) of Example 1.2.14.

Proposition 7.4.26. *If $a_1, \ldots, a_{t+1} \in W_m \tilde{G}$, then*

$$\tilde{s}_t(a_1, \ldots, a_t) a_{t+1} \in \begin{cases} \langle W_{2m} \rangle + \mathcal{I}_{2t}(G), & m \text{ odd} \\ \langle W_{2m-1} \rangle + \mathcal{I}_{2t}(G), & m \text{ even.} \end{cases}$$

Proof. For m odd, write $a_i = w_i g_i$ for $w_i \in W_m$ and $g_i \in \tilde{G}$; then Lemma 7.4.23 yields

$$\tilde{s}_t(a_1, \ldots, a_t) a_{t+1} = \sum_\sigma w_{\sigma(1)} g_{\sigma(1)} \cdots w_{\sigma(t)} g_{\sigma(t)} w_{t+1} g_{t+1}$$

$$= \sum_\sigma \text{sgn}(\sigma) w_1 g_{\sigma(1)} \cdots w_t g_{\sigma(t)} w_{t+1} g_{t+1} \qquad (7.20)$$

$$= \frac{1}{t!} \sum_{\sigma, \tau} \text{sgn}(\sigma\tau) w_{\tau(1)} g_{\sigma(1)} \cdots w_{\tau(t)} g_{\sigma(t)} w_{t+1} g_{t+1}.$$

But the last sum differs from $s_{2t}(w_1, g_1, \ldots, w_t, g_t)$ only by terms where two w's are adjacent, i.e., in $\langle W_{2m} \rangle$, so we conclude that the right side must be in $\langle W_{2m} \rangle + \mathcal{I}_{2t}(\tilde{G})$.

For m odd, write $a_i = w_i g_i$ for $w_i \in W_{m-1}$ and $g_i \in W_1 \tilde{G}$, and apply the same reasoning. (We need the subscript of W to be odd, in order for Lemma 7.4.23 to be applicable.) $\qquad \square$

The continuation of the proof of Theorem 7.4.20 has the flavor of the proof we have given of the Amitsur-Levitzki theorem, relying heavily on the Dubnov-Ivanov-Nagata-Higman Theorem 1.8.28. Take β as in Theorem 1.8.28.

Corollary 7.4.27. $\langle W_m \rangle^{\beta(t)} \subseteq \langle W_{2m} \rangle^{2t} + \mathcal{I}_{2t}(G)$ *for* m *odd, and* $\langle W_m \rangle^{\beta(t)} \subseteq \langle W_{2m-1} \rangle^{2t} + \mathcal{I}_{2t}(G)$ *for* m *even.*

Proof. Theorem 1.8.28 says that $(W_m \tilde{G})^{\beta(t)}$ is contained in the T-ideal without 1 generated by $\tilde{s}_t$, which by the Proposition is contained in $\langle W_{2m} \rangle + \mathcal{I}_{2t}(G)$ or $\langle W_{2m-1} \rangle + \mathcal{I}_{2t}(G)$. $\qquad \square$

Now define the function $\gamma : \mathbb{N} \to \mathbb{N}$ inductively by $\gamma(1) = 1$ and

$$\gamma(m+1) = \begin{cases} 2\gamma(m) & \text{for } m \text{ odd;} \\ 2\gamma(m) - 1 & \text{for } m \text{ even.} \end{cases}$$

Corollary 7.4.28. *Suppose* $n = 2t$, *and* m *satisfies* $\gamma(m) > n$. *Then*

$$\langle W_1 \rangle^{\beta(t)^m} \subseteq W_{\gamma(m)} + \mathcal{I}_n(\tilde{G}) \subseteq \mathcal{I}_n(\tilde{G}).$$

Proof. Clear, since $W_{\gamma(m)} \subseteq W_n \subseteq \mathcal{I}_n(\tilde{G})$ by Remark 7.4.24. $\qquad \square$

Confronting this corollary with Lemma 7.4.25, we see for n even and $m' = \beta(m)^t$ that $\mathcal{J}_{m'} \subseteq \mathcal{I}_n$, i.e., s_n implies $c_{m'}$. Since s_n implies s_{n+1}, we may always assume n is even, and thus the corollary is applicable and we have proved Theorem 7.4.20. Furthermore, the proof is constructive, depending on the function β of Theorem 1.8.28, which can be improved, cf. Exercise 8.7.

The characteristic p version is even more decisive, cf. Theorem 5.0.1.

7.4.4   Specht's problem for T-spaces

Recall the definition of T-space from Definition 1.7.12(i). Shchigolev [Shch01] strengthened Kemer's theorem and proved that T-spaces are finitely based in characteristic 0. The proof is based on the following result.

Theorem 7.4.29 (Shchigolev). *Let* S *be a* T-*space in the free algebra* $A = F\{X\}$ *over a field of characteristic 0.*
    (i) If $S \backslash [A, A] \neq \emptyset$, *then* S *includes a nonzero* T-*ideal.*
    (ii) If $S \subseteq [A, A]$, *then there exists a* T-*ideal* Γ *such that* $[\Gamma, A] \subseteq S$.

Using this theorem, Shchigolev then applied Kemer's theory to prove in characteristic 0 that T-spaces are finitely based. In Chapter 9, we consider T-spaces from the opposite point of view; in characteristic $p > 0$ we shall construct T-spaces that are not finitely based, and use them as a launching board to construct T-ideals that are not finitely based.

Exercises for Chapter 7

1. If $A = A_0 \oplus A_1$ is a superalgebra, then for any proper ideal $\mathcal{I}$ of A_0, $\mathcal{I} + A_1\mathcal{I}A_1 \lhd A_0$. If A is supersimple, then $\mathcal{I} \cap A_1\mathcal{I}A_1 = 0$, yielding a Peirce decomposition.

2. Define the Grassmann involution for T-ideals over an arbitrary infinite field F. In other words, extend the $(*)$-action naturally to homogeneous polynomials. Extend the $(*)$-action linearly to any T-ideal by noting that any polynomial can be written canonically as a sum of homogeneous polynomials.

3. Show that $G \otimes G \sim_{\mathrm{PI}} M_{1,1}$; $M_{k,\ell} \otimes G \sim_{\mathrm{PI}} M_{k+\ell,k+\ell}$ for $k, \ell \neq 0$, and

$$M_{k_1,\ell_1} \otimes M_{k_2,\ell_2} \sim_{\mathrm{PI}} M_{k_1 k_2 + \ell_1 \ell_2, k_1 \ell_2 + k_2 \ell_1}.$$

   Conclude that the PI-degree of the tensor product of two algebras can be more than the product of their PI-degrees.

4. (For those who know ultraproducts) Any relatively free verbally prime algebra U_A can be embedded into an ultraproduct of copies of A. (Hint: Take the filter consisting of those components on which the image of some element is 0.)

5. An algebra A is called **Grassmann representable** if A can be embedded into $M_n(G)$ for suitable n and for a suitable Grassmann algebra over a suitable commutative ring. Show that the Grassmann envelope of a f.d. superalgebra is Grassmann representable. Conclude that any relatively free PI-algebra U in characteristic 0 is Grassmann representable. (Hint: $U = U_A$ where A is the Grassmann envelope of a f.d. superalgebra, and construct U by means of generic linear combinations of a base of A.)

6. Suppose C is a commutative affine superalgebra over a field F, and $R \supset C$ is a PI-superalgebra that is f.g. as a C-module. Then R is super-representable. (Hint: R is spanned by homogeneous elements, so is left super-Noetherian; proceed as in the proof of Theorem 1.6.22.)

7. For any relatively free algebra $\mathcal{A}$ in an infinite number of indeterminates, $\mathrm{Jac}(\mathcal{A})$ is contained in every maximal T-ideal, and thus is the intersection of the maximal T-ideals of $\mathcal{A}$. (Hint: Any maximal T-ideal M is contained in a maximal ideal, which clearly contains $\mathrm{Jac}(\mathcal{A})$, so $M + \mathrm{Jac}(\mathcal{A})$ is a T-ideal containing M, and, thus, equals M.)

PI-superalgebras

To exploit Theorem 7.4.9 further, we develop more structure theory for PI-superalgebras, along the lines of Lemma 7.2.5. We always assume our algebras contain the element $\frac{1}{2}$.

8. If $M_n(F)$ is graded, it has a central superpolynomial. (Hint: Any multilinear central polynomial must have some nonzero evaluation of homogeneous elements; designate the corresponding indeterminates as even or odd according to this substitution.)

9. Recall Lemma 7.2.4, which attaches to any superalgebra $R = R_0 \oplus R_1$ the automorphism $\sigma : r_0 + r_1 \mapsto r_0 - r_1$. Then $P \cap \sigma(P) \lhd_2 R$ for any $P \lhd R$.

10. Suppose R is a superalgebra, notation as in Exercise 9.

    (i) If P is a prime ideal of R, then $P \cap \sigma(P)$ is a prime superideal, in the sense that if $I_1, I_2 \lhd_2 R$ with $I_1 I_2 \supseteq P \cap \sigma(P)$, then I_1 or I_2 is contained in $P \cap \sigma(P)$.

    (ii) Conversely, given a prime superideal P, take $P' \lhd R$ maximal with respect to $P' \cap \sigma(P') \subseteq P$, which exists by Zorn's lemma; then P' is a prime ideal of R.

    (Hint for (i): $I_1 I_2 \subseteq P$, so we may assume $I_1 \subseteq P$; hence, $I_1 = I_1 \cap \sigma(I_1) \subseteq P \cap \sigma(P)$.)

11. If R is superprime PI with center $C = C_0 \oplus C_1$, and $S = C_0 \setminus \{0\}$, then the localization $S^{-1}R$ is supersimple and finite dimensional over $F = S^{-1}C_0$, the field of fractions of C_0. If $S^{-1}R$ is not simple, then it is the direct product of two isomorphic simple algebras. (Hint: R must be torsion-free over C_0. Indeed, more generally, no homogeneous central element c can be a zero divisor. Otherwise, if $rc = 0$ for $r \neq 0$, then the homogeneous parts of r also annihilate c, implying c is annihilated by a superideal I, and then $I(Rc) = 0$, contrary to R superprime. Hence one can construct $S^{-1}R$, which is a superalgebra whose even component is the field F. Replacing R by $S^{-1}R$, one may assume $C_0 = F$ and needs to prove that R is supersimple. Any proper superideal I intersects the field C_0 at 0, implying I must have a nonzero odd element c, and by the previous paragraph $0 \neq c^2 \in I \cap F$, contradiction. Taking a maximal ideal P yields $P \cap \sigma(P) = 0$, implying $R \cong R/P \times R/\sigma(P)$.)

12. Any superprime superalgebra R is PI_2-equivalent to one of the algebras in Proposition 7.2.8.

13. Here is the super-version of Definition 7.4.1. A T_2-ideal Γ is called **verbally prime** if for every two T_2-ideals $\mathcal{I}, \mathcal{J}$ such that $\mathcal{I}\mathcal{J} \subseteq \Gamma$, implies $\mathcal{I}$ or $\mathcal{J}$ is contained in Γ.

    Show that any verbally prime T_2-ideal $\text{id}_2(A)$ can be written as $\text{id}_2(A)$ for a supersimple algebra A. (Hint: One may assume that A is the relatively free superalgebra of a f.d. superalgebra, and thus is superrepresentable inside some f.d. superalgebra $\hat{A}$. But $\text{Jac}(A)$ is a nilpotent T_2-ideal, cf. Lemma 7.2.5 and Proposition 1.8.21, reducing us to the case where $\text{Jac}(A) = 0$.

    Hence $\text{Jac}(\hat{A}) \cap A = 0$, implying A is superrepresentable inside the super-semisimple superalgebra $\hat{A}/\text{Jac}(\hat{A})$. It follows that $\text{id}_2(A) = \text{id}_2(\hat{A})$, which one decomposes.)

14. The verbally prime T_2-ideals are precisely $\mathcal{M}_n$, $\mathcal{M}_{k,\ell}$, and $\mathcal{M}'_n$.

15. Suppose Γ is a verbally prime T-ideal. Writing $\Gamma = \text{id}(G(R))$ for A supersimple PI-algebra R over an algebraically closed field, show that $\text{id}_2(R)$ must be a verbally prime T_2-ideal, and thus one of $\mathcal{M}_n$, $\mathcal{M}_{k,\ell}$, and $\mathcal{M}'_n$.

Chapter 8

Trace Identities

Several times in this book we used results about traces. Now we shall develop these *trace identities* formally, showing in one crucial respect that they behave better than polynomial identities. Razmyslov, Procesi, and Helling proved that all trace identities of $M_n(\mathbb{Q})$ formally are consequences of a single trace identity, which Razmyslov identified with the Cayley-Hamilton equation (rewritten as a trace identity via Newton's formulas.) We prove this basic result in this chapter, and also provide tools for further investigation. In the process, following Regev [Reg84], we see that the trace identities have a cocharacter theory analogous to that of the PI-theory, and the two can be tied together quite neatly.

8.1 Trace Polynomials and Identities

In line with Definition 1.1.4, we define a formal *trace symbol* tr, linear over C, which will be stipulated to satisfy

(i) $\mathrm{tr}(x_{i_1} x_{i_2} \ldots x_{i_n}) = \mathrm{tr}(x_{i_n} x_{i_1} \ldots x_{i_{n-1}})$;

(ii) $\mathrm{tr}(f)x_i = x_i \mathrm{tr}(f)$;

(iii) $\mathrm{tr}(f\mathrm{tr}(g)) = \mathrm{tr}(f)\mathrm{tr}(g)$ for all $f, g \in C\{X\}$.

Note that (iii) enables us to eliminate tr^2, since

$$\mathrm{tr}^2(f) = \mathrm{tr}(1\mathrm{tr}(f)) = \mathrm{tr}(1)\mathrm{tr}(f).$$

Definition 8.1.1. *A* *simple trace monomial* *is an expression of the form*

$$\text{tr}(x_{i_1} \ldots x_{i_t}).$$

A **pure trace monomial** *is a product of simple trace monomials, i.e., of the form*

$$\text{tr}(x_{i_{t_1+1}} \cdots x_{i_{t_1+2}} \cdots x_{i_{t_2}})\text{tr}(x_{i_{t_2+1}} \cdots x_{i_{t_3}}) \cdots ,$$

In general, a **(mixed) trace monomial** *is the product of a regular monomial with an arbitrary number of pure trace monomials, i.e., of the form*

$$h = x_{i_1} \ldots x_{i_{t_1}} \text{tr}(x_{i_{t_1+1}} \cdots x_{i_{t_1+2}} \cdots x_{i_{t_2}})\text{tr}(x_{i_{t_2+1}} \cdots x_{i_{t_3}}) \cdots .$$

Thus, the trace monomial is a pure trace monomial iff $t_1 = 0$.

The **degree** $\deg_i h$ *of a trace monomial is computed by counting each* x_i, *regardless of whether or not it appears inside a* tr *symbol.*

Definition 8.1.2. *The* **free algebra with traces** $T'\{X\}$ *is spanned by the trace monomials, with the obvious multiplication (juxtaposition), satisfying the above conditions (i), (ii), and (iii); its elements are called* **trace polynomials**. *The* **algebra of pure trace polynomials** $T = T\{X\}$ *is the subalgebra of* $T'\{X\}$ *generated by the simple trace monomials.*

Thus, any element of $T'\{X\}$ is written as a linear combination of

$$w_1 \text{tr}(w_2) \cdots \text{tr}(w_k),$$

where $w_1, \ldots, w_k$ are words in the x_i, permitted to be 1. It is easy to see that $T'\{X\}$ is an associative algebra, on which tr acts linearly via

$$\text{tr}(w_1 \text{tr}(w_2) \cdots \text{tr}(w_k)) = \text{tr}(w_1)\text{tr}(w_2) \cdots \text{tr}(w_k);$$

in particular, $\text{tr}(\text{tr}(w_2) \cdots \text{tr}(w_k)) = \text{tr}(1)\text{tr}(w_2) \cdots \text{tr}(w_k)$. Note that T is a commutative subalgebra of $T'\{X\}$.

We define **homogeneous** and **multilinear** trace polynomials, according to the degrees of their monomials obtained by erasing the tr. For example,

$$\text{tr}(x_1 x_2)\text{tr}(x_1 x_3^2) - \text{tr}(x_1^2 x_2 x_3^2)$$

is a homogeneous pure trace polynomial. Also

$$\text{tr}(x_1 x_2)\text{tr}(x_3) + \text{tr}(x_1 x_2 x_3)$$

is a multilinear pure trace polynomial.

Definition 8.1.3. *A* **trace identity** *of an algebra* A *with trace function* tr *is a trace polynomial which vanishes identically on* A, *whenever we substitute elements of* A *for the* x_i, *and traces for* tr.

Remark 8.1.4. *As with ordinary identities, one can multilinearize trace identities by applying the operators* Δ_i *of Definition 1.2.12 or of §3.5; when the field F has characteristic zero, the trace identities can be recovered from the multilinear trace identities.*

Our interest in this chapter is in $A = M_n(F)$, for a field F, where tr is the usual trace function; hence $\text{tr}(1) = n$. A trace identity is *trivial* if it holds for $M_n(F)$ for all n; examples were given in Remark 1.4.10.

Example 8.1.5. *Assume* $\text{char}(F) \neq 2$. *In* $M_2(F)$,

$$\det(a) = \frac{\text{tr}(a)^2 - \text{tr}(a^2)}{2},$$

(seen most easily by considering diagonal matrices and applying Zariski density), so the Cayley-Hamilton Theorem yields the trace identity

$$x^2 - \text{tr}(x)x + \frac{\text{tr}(x)^2 - \text{tr}(x^2)}{2},$$

which multilinearizes (via the usual technique of substituting $x_1 + x_2$ for x) to

$$x_1 x_2 + x_2 x_1 - \text{tr}(x_1)x_2 - \text{tr}(x_2)x_1 + \text{tr}(x_1)\,\text{tr}(x_2) - \text{tr}(x_1 x_2).$$

Let us formulate this example in general.

Example 8.1.6. *By the Cayley-Hamilton Theorem, any matrix a annihilates its characteristic polynomial $\chi_a(x)$. Newton's formulas (Remark 1.4.5) imply that in characteristic 0, $M_k(F)$ satisfies the* **Cayley-Hamilton** *trace identity*

$$g_k = x^k + \sum_{j=1}^{k} \left(\sum_{j_1 + \cdots + j_u = j} \alpha_{(j_1,\ldots,j_u)} \, \text{tr}(x^{j_1}) \ldots \text{tr}(x^{j_u}) x^{k-j} \right)$$

where the $\alpha_{(j_1,\ldots,j_u)} \in \mathbb{Q}$ can be computed explicitly, as to be elaborated in Example 8.1.20.

Of course g_k is a mixed trace polynomial that is homogeneous but not multilinear. Since the Cayley-Hamilton trace identity g_k depends on k, it is nontrivial, and our object here is to prove that every trace identity of $M_k(\mathbb{Q})$ is a consequence of g_k. This is a considerable improvement (and much easier to prove!) than Kemer's theorem, since the minimal identity s_{2n} of $M_n(\mathbb{Q})$ does not generate $\text{id}(M_n(\mathbb{Q}))$ as a T-ideal, cf. Exercise 6.2. However, it still remains a mystery as to how to prove in general that $\text{id}(M_n(\mathbb{Q}))$ is finitely based as a corollary of the Hellman-Procesi-Razmyslov Theorem 8.2.1.

Here is the basic setup, which we shall use throughout:

Remark 8.1.7. *Let $V = F^{(k)}$ and $V^* = \text{Hom}(V, F)$ denote its dual space. Let $e_1, \ldots, e_k \in V$ be a base of V and $\theta_1, \ldots, \theta_k \in V^*$ its dual base.*

There is a bilinear form $V \times V^* \to F$ denoted by $\langle v, w \rangle$ for $v \in V$ and $w \in V^*$. Thus, we can identify $V \otimes V^*$ with $\mathrm{End}(V) = M_n(F)$, where $v \otimes w$ corresponds to the transformation on V given by $(v \otimes w)(y) = \langle y, w \rangle v$.

For example,

$$(e_i \otimes \theta_j)(e_\ell) = \langle e_\ell, \theta_j \rangle e_i = \delta_{j\ell} e_i,$$

where $\delta_{j\ell}$ is the Kronecker delta, so $e_i \otimes \theta_j$ is identified with the matrix unit e_{ij}.

Remark 8.1.8. *Define $\mathrm{tr}(v \otimes w) = \langle v, w \rangle$. This matches the usual definition of trace on matrices as seen by checking base elements:*

$$\mathrm{tr}(e_i \otimes \theta_j) = \langle e_i, \theta_j \rangle = \delta_{ij} = \mathrm{tr}(e_{ij}).$$

Now

$$
\begin{aligned}
(v_1 \otimes w_1)(v_2 \otimes w_2)(v) &= (v_1 \otimes w_1)(\langle v, w_2 \rangle v_2) \\
&= \langle v, w_2 \rangle (v_1 \otimes w_1)(v_2) \\
&= \langle v, w_2 \rangle \langle v_2, w_1 \rangle v_1 \\
&= \langle v_2, w_1 \rangle \langle v, w_2 \rangle v_1 \\
&= \langle v_2, w_1 \rangle (v_1 \otimes w_2)(v).
\end{aligned}
\tag{8.1}
$$

Since this holds for all $v \in V$, we have $(v_1 \otimes w_1)(v_2 \otimes w_2) = \langle v_2, w_1 \rangle (v_1 \otimes w_2)$. By induction on s, it follows that

$$(v_{j_1} \otimes w_{j_1}) \cdots (v_{j_s} \otimes w_{j_s}) = \langle v_{j_2}, w_{j_1} \rangle \cdots \langle v_{j_s}, w_{j_{s-1}} \rangle (v_{j_1} \otimes w_{j_s}). \tag{8.2}$$

8.1.1   The Kostant-Schur trace formula

The theory of trace identities of matrices over a field is based on two main facts. The first is the description of the kernel of the map

$$\varphi_{n,k} : F[S_n] \to \mathrm{End}(T^n(V)),$$

defined in Definition 3.3.1, where $\dim V = k$.

Theorem 8.1.9. *If $F[S_n] = \oplus_{\lambda \vdash n} I_\lambda$, the direct sum of its minimal (two-sided) ideals, then*

$$\ker \varphi_{n,k} = \bigoplus_{\lambda \vdash n,\, \lambda_1' \geq k+1} I_\lambda.$$

This classical theorem of Schur and Weyl, a reformulation of Theorem 3.3.3(i), can be found in more general "hook" form in Theorem 3.20 of [BerRe87].

If $\bar{\mu} = (1^{k+1})$, then the corresponding ideal $I_{\bar{\mu}} = F e_{\bar{\mu}}$, where

$$e_{\bar{\mu}} = \sum_{\pi \in S_{k+1}} \mathrm{sgn}(\pi) \pi.$$

Corollary 8.1.10. *Suppose* $\bar{\mu} = (1^{k+1})$ *and let* $n \geq k+1$. *Then by the Branching Theorem,* $\ker \varphi_{n,k}$ *is the ideal generated in* $F[S_n]$ *by* $I_{\bar{\mu}}$. *Thus,* $\ker \varphi_{n,k}$ *is spanned over* F *by the elements*

$$\tau \left(\sum_{\pi \in S_{k+1}} \mathrm{sgn}(\pi)\pi \right) \zeta, \quad \tau, \zeta \in S_n.$$

The second fact is a trace formula due to Kostant [Kos58] which is a generalization of a formula of I. Schur.

Remark 8.1.11. *We have the identification* $T^n(V)^* = T^n(V^*)$ *via*

$$\langle v_1 \otimes \cdots \otimes v_n, w_1 \otimes \cdots \otimes w_n \rangle = \prod_{i=1}^{n} \langle v_i, w_i \rangle.$$

Theorem 8.1.12. *Write* $\sigma \in S_n$ *as a product of disjoint cycles*

$$\sigma = (i_1, \ldots, i_a)(j_1, \ldots, j_b) \cdots (k_1, \ldots, k_c),$$

and let $A_1, \ldots, A_n \in \mathrm{End}(V)$, *so that*

$$(A_1 \otimes \cdots \otimes A_n) \circ \varphi_{n,k}(\sigma) \in \mathrm{End}(T^n(V)).$$

Then

$$\mathrm{tr}((A_1 \otimes \cdots \otimes A_n) \circ \varphi_{n,k}(\sigma)) = \mathrm{tr}(A_{i_1} \cdots A_{i_a}) \mathrm{tr}(A_{j_1} \cdots A_{j_b}) \cdots \mathrm{tr}(A_{k_1} \cdots A_{k_c}). \tag{8.3}$$

Here the trace on the left is applied to $\mathrm{End}(T^n(V))$, *while the traces on the right are applied to* $\mathrm{End}(V)$.

Proof. The proof is by employing rather elementary linear algebra to the set-up of Remark 8.1.7.

Step 1. First, applying tr to both sides of (8.2), we obtain

$$\mathrm{tr}((v_{j_1} \otimes w_{j_1}) \cdots (v_{j_s} \otimes w_{j_s})) = \langle v_{j_2}, w_{j_1} \rangle \cdots \langle v_{j_s}, w_{j_{s-1}} \rangle \langle v_{j_1}, w_{j_s} \rangle$$
$$= \prod_{i \in \{j_1, \ldots, j_s\}} \langle v_{\sigma(i)}, w_i \rangle, \tag{8.4}$$

for any cycle $\sigma = (j_1, \ldots, j_s)$.

Step 2. If $c \in \mathrm{End}(V)$, then

$$\mathrm{tr}(c) = \sum_{\ell=1}^{k} \langle ce_\ell, \theta_\ell \rangle.$$

This is clear when $c = e_i \otimes \theta_j$, and then we extend it by linearity to all $c \in \mathrm{End}(V)$.

Step 3. Note that $\{e_{i_1} \otimes \cdots \otimes e_{i_n} \mid 1 \leq i_1, \ldots, i_n \leq k\}$ is a basis of $T^n(V)$, and $\{\theta_{i_1} \otimes \cdots \otimes \theta_{i_n} \mid 1 \leq i_1, \ldots, i_n \leq k\}$ its dual basis.

Thus, if $c \in \operatorname{End}(T^n(V))$, it follows that

$$\operatorname{tr}(c) = \sum_{1 \leq i_1, \ldots, i_n \leq k} \langle c(e_{i_1} \otimes \cdots \otimes e_{i_n}), \theta_{i_1} \otimes \cdots \otimes \theta_{i_n} \rangle.$$

Step 4. We need a lemma.

Lemma 8.1.13. *Let* $\sigma \in S_n$, $A_i = v_i \otimes w_i \in V \otimes V^*$, $i = 1, \ldots, n$, *and let* $c = (A_1 \otimes \cdots \otimes A_n) \circ \varphi_{n,k}(\sigma) \in \operatorname{End}(T^n(V))$. *Then*

$$\operatorname{tr}(c) = \prod_{i=1}^{n} \langle v_{\sigma(i)}, w_i \rangle.$$

Proof. By direct calculation,

$$c(e_{j_1} \otimes \cdots \otimes e_{j_n}) = \left(\prod_{i=1}^{n} \langle e_{j_{\sigma(i)}}, w_i \rangle \right) (v_1 \otimes \cdots \otimes v_n),$$

and therefore

$$\langle c(e_{j_1} \otimes \cdots \otimes e_{j_n}), \theta_{j_1} \otimes \cdots \otimes \theta_{j_n} \rangle = \prod_{i=1}^{n} (\langle e_{j_{\sigma(i)}}, w_i \rangle \langle v_i, \theta_{j_i} \rangle).$$

Summing over all the k^n tuples $\{(j_1, \ldots, j_n) \mid 1 \leq j_i \leq k\}$ yields $\operatorname{tr}(c)$ on the left. We claim that the right hand side $\sum_{1 \leq j_i \leq k} \prod_{i=1}^{n} (\langle e_{j_{\sigma(i)}}, w_i \rangle \langle v_i, \theta_{j_i} \rangle)$ equals $\prod_{i=1}^{n} \langle v_{\sigma(i)}, w_i \rangle$. By linearity, we may assume that $v_i = e_{r_i}$ and $w_i = \theta_{s_i}$, for $i = 1, \ldots, n$. Since $\langle e_a, \theta_b \rangle = \delta_{a,b}$, it follows that

$$\prod_{i=1}^{n} (\langle e_{j_{\sigma(i)}}, w_i \rangle \langle v_i, \theta_{j_i} \rangle) = \prod_{i=1}^{n} (\langle e_{j_{\sigma(i)}}, \theta_{s_i} \rangle \langle e_{r_i}, \theta_{j_i} \rangle) = \prod_{i=1}^{n} \delta_{j_{\sigma(i)}, s_i} \delta_{r_i, j_i}.$$

The only nonzero summands occur when all $j_i = r_i$, in which case

$$\prod_{i=1}^{n} \delta_{j_{\sigma(i)}, s_i} \delta_{r_i, j_i} = \prod_{i=1}^{n} \delta_{r_{\sigma(i)}, s_i} = \prod_{i=1}^{n} \langle e_{r_{\sigma(i)}}, \theta_{s_i} \rangle = \prod_{i=1}^{n} \langle v_{\sigma(i)}, w_i \rangle.$$

Step 5. Completion of the proof of Theorem 8.1.12. By linearity, we may assume $A_i = v_i \otimes w_i$, $v_i \in V$, $w_i \in V^*$, $i = 1, \ldots, n$. Then, by Lemma 8.1.13,

$$\operatorname{tr}(c) = \prod_{i=1}^{n} \langle v_{\sigma(i)}, w_i \rangle.$$

Now $\sigma = (i_1, \ldots, i_a) \cdots (k_1, \ldots, k_c)$, hence

$$\prod_{i=1}^{n} \langle v_{\sigma(i)}, w_i \rangle = \left(\prod_{i \in \{i_1, \ldots, i_a\}} \langle v_{\sigma(i)}, w_i \rangle \right) \cdots \left(\prod_{i \in \{k_1, \ldots, k_c\}} \langle v_{\sigma(i)}, w_i \rangle \right),$$

which by (8.4) equals $\text{tr}(A_{i_1} \cdots A_{i_a}) \cdots \text{tr}(A_{k_1} \cdots A_{k_c})$, as desired. $\qquad\square$

8.1.2 Pure trace polynomials

Let us consider the pure trace polynomials, and then give, in Theorem 8.1.18, a complete description of the multilinear pure trace identities of the $k \times k$ matrices. The multilinear pure trace polynomials can be described as elements of the group algebras $F[S_n]$ as follows.

Definition 8.1.14. *Write $\sigma \in S_n$ as a product of disjoint cycles*

$$\sigma = (i_1, \ldots, i_a)(j_1, \ldots, j_b) \cdots (k_1, \ldots, k_c)$$

and denote

$$M_\sigma(x_1, \ldots, x_n) = \text{tr}(x_{i_1} \cdots x_{i_a})\text{tr}(x_{j_1} \cdots x_{j_b}) \cdots \text{tr}(x_{k_1} \cdots x_{k_c}),$$

the corresponding pure trace monomial. Given $a = \sum_\sigma \alpha_\sigma \sigma \in F[S_n]$, denote the corresponding formal trace polynomial

$$f_a(x_1, \ldots, x_n) = \sum_\sigma \alpha_\sigma M_\sigma(x_1, \ldots, x_n).$$

Remark 8.1.15. *Given a multilinear pure trace monomial $M(x_1, \ldots, x_n)$, there exists $\sigma \in S_n$ such that $M(x_1, \ldots, x_n) = M_\sigma(x_1, \ldots, x_n)$. Given a multilinear pure trace polynomial $f(x_1, \ldots, x_n)$, there exists $a \in F[S_n]$ such that $f(x_1, \ldots, x_n) = f_a(x_1, \ldots, x_n)$.*

This vector-space map $a \mapsto f_a$ has the following property, which will be applied later.

Lemma 8.1.16. *Let $a \in F[S_n]$ with corresponding pure trace polynomial f_a, and let $\eta \in S_n$. Then*

$$f_{\eta a \eta^{-1}}(x_1, \ldots, x_n) = f_a(x_{\eta(1)}, \ldots, x_{\eta(n)}).$$

Proof. It is enough to check pure trace monomials. Let

$$\sigma = (i_1, \ldots, i_a) \cdots (k_1, \ldots, k_c),$$

so

$$M_\sigma(x_1, \ldots, x_n) = \text{tr}(x_{i_1} \cdots x_{i_a}) \cdots \text{tr}(x_{k_1} \cdots x_{k_c}).$$

Now $\eta\sigma\eta^{-1} = (\eta(i_1), \ldots, \eta(i_a)) \cdots (\eta(k_1), \ldots, \eta(k_c))$ and therefore,

$$M_{\eta\sigma\eta^{-1}}(x_1, \ldots, x_n) = \mathrm{tr}(x_{\eta(i_1)} \cdots x_{\eta(i_a)}) \cdots \mathrm{tr}(x_{\eta(k_1)} \cdots x_{\eta(k_c)})$$
$$= M_\sigma(x_{\eta(1)}, \ldots, x_{\eta(n)}),$$

as desired.   □

Corollary 8.1.17. *Let $a \in F[S_n]$ with $f_a(x)$ the corresponding trace polynomial, and let $A_1, \ldots, A_n \in \mathrm{End}(V)$. Then*

$$f_a(A_1, \ldots, A_n) = \mathrm{tr}((A_1 \otimes \cdots \otimes A_n) \circ \varphi_{n,k}(a)).$$

Proof. Follows at once from Theorem 8.1.12.   □

We now give a complete description of the multilinear trace identities of $M_k(F)$.

Theorem 8.1.18. *Let $f(x) = f_a(x) = f_a(x_1, \ldots, x_n)$ be a multilinear trace polynomial, where $a \in F[S_n]$. Then $f_a(x)$ is a trace identity of $M_k(F)$ if and only if*

$$a \in \bigoplus_{\lambda \vdash n,\ \lambda_1' \geq k+1} I_\lambda.$$

Proof. Follows easily from the fact that

$$\ker \varphi_{n,k} = \bigoplus_{\lambda \vdash n,\ \lambda_1' \geq k+1} I_\lambda.$$

Indeed, let $\bar{A} = A_1 \otimes \cdots \otimes A_n \in T^n(\mathrm{End}(V))$ ($\dim V = k$). Now, $f_a(x)$ is a trace identity for $M_k(F) = \mathrm{End}(V)$ if and only if for all such $\bar{A}$, $f_a(\bar{A}) = 0$, and thus, by Corollary 8.1.17, if and only if $\mathrm{tr}(\bar{A} \circ \varphi_{n,k}(a)) = 0$ for all $\bar{A} \in T^n(\mathrm{End}(V)) \equiv \mathrm{End}(T^n(V))$.

By the non-degeneracy of the trace form, the above happens if and only if $\varphi_{n,k}(a) = 0$, namely $a \in \ker \varphi_{n,k}$. The assertion now follows from the above description of $\ker \varphi_{n,k}$.   □

8.1.3   Mixed trace polynomials

There is an obvious transition between pure and mixed trace identities, as follows. If $p(x_1, \ldots, x_n)$ is a mixed trace identity of an algebra A with trace, then trivially $\mathrm{tr}(x_0 \cdot p(x_1, \ldots, x_n))$ is a pure trace identity of A. When $A = M_k(F)$, the multilinear converse also holds: Let $f(x_0, x_1, \ldots, x_n)$ be a multilinear pure trace polynomial. Each monomial of f can be written as

$$\mathrm{tr}(g_0)\mathrm{tr}(g_1) \cdots \mathrm{tr}(g_t)$$

for monomials $g_0, \ldots, g_t$, which we can rearrange so that x_0 appears in g_0. Furthermore, performing a cyclic permutation on g_0, we may assume that g_0 starts with x_0. Writing $g_0 = x_0 g_0'$, we have

$$\text{tr}(g_0)\text{tr}(g_1)\ldots\text{tr}(g) = \text{tr}(x_0 g_0'\text{tr}(g_1)\cdots\text{tr}(g_t)).$$

Letting p be the sum of these $g_0'\text{tr}(g_1)\cdots\text{tr}(g_t)$, a mixed trace polynomial, we see that $f = \text{tr}(x_0 p)$. For example,

$$\text{tr}(x_2 x_0 x_3)\text{tr}(x_1 x_4) = \text{tr}(x_0 x_3 x_2)\text{tr}(x_1 x_4) = \text{tr}(x_0 x_3 x_2\text{tr}(x_1 x_4)).$$

The non-degeneracy of the trace on $M_k(F)$ implies the following result.

Proposition 8.1.19. *Suppose* $f(x_0, x_1, \ldots, x_n) = \text{tr}(x_0 p(x_1, \ldots, x_n))$. *Then* $f(x_0, \ldots, x_n)$ *is a pure trace identity of* $M_k(F)$ *iff* $p(x_1, \ldots, x_n)$ *is a mixed trace identity of* $M_k(F)$.

We are ready for the main example of a trace identity of matrices.

Example 8.1.20.

(i) *Let* $\bar{\mu} = (1^{k+1})$ *and*

$$e_{\bar{\mu}} = \sum_{\pi \in S_{k+1}} \text{sgn}(\pi)\pi,$$

with the corresponding pure trace polynomial

$$f_{e_{\bar{\mu}}} = f_{e_{\bar{\mu}}}(x_0, x_1, \ldots, x_k).$$

By Theorem 8.1.18, $f_{e_{\bar{\mu}}}$ is a trace identity of $M_k(F)$. Moreover, it is the multilinear pure trace identity of the lowest degree, and any other such identity of degree $k+1$ is a scalar multiple of $f_{e_{\bar{\mu}}}$. Thus, $f_{e_{\bar{\mu}}}$ is called the **fundamental pure trace identity**.

(ii) *Let $g_k(x)$ denote the trace identity of $M_k(F)$ given in Example 8.1.6, and take its multilinearization*

$$\tilde{g}_k = \tilde{g}_k(x_1, \ldots, x_k).$$

Then $\tilde{g}_k$ is a multilinear mixed trace identity of $M_k(F)$. Hence,

$$\text{tr}(x_0 \tilde{g}_k(x_1, \ldots, x_k))$$

is a multilinear pure trace identity of $M_k(F)$; by (i), it is a scalar multiple of $f_{e_{\bar{\mu}}}$.

8.2   Finite Generation of Trace T-Ideals

In this section, we prove the fundamental theorem of trace identities, due independently to Razmyslov [Raz74b], Helling [Hel74], and Procesi [Pro76]. Note that the variables $x_0, \ldots, x_k$ are replaced by $x_1, \ldots, x_{k+1}$, to enable us to notate permutations more easily.

Theorem 8.2.1.

(i) *The T-ideal of the pure trace identities of $k \times k$ matrices is generated by the single trace polynomial $f_{e_{\bar{\mu}}}$, where $\bar{\mu} = (1^{k+1})$ and*

$$e_{\bar{\mu}} = \sum_{\pi \in S_{k+1}} \operatorname{sgn}(\pi)\pi.$$

(ii) *The T-ideal of the mixed trace identities of $k \times k$ matrices is generated by the two elements $f_{e_{\bar{\mu}}}$ and $\tilde{g}_k$ (cf. Example 8.1.20(iii)).*

Proof. We follow [Pro76].

(i) In view of multilinearization, it suffices to prove that any multilinear pure trace identity of $M_k(F)$ is a consequence of $f_{e_{\bar{\mu}}}$. By Corollary 8.1.10, it suffices to prove this assertion for the pure trace polynomials that correspond to the elements

$$\tau e_{\bar{\mu}} \zeta = \tau \left(\sum_{\pi \in S_{k+1}} \operatorname{sgn}(\pi)\pi \right) \zeta, \quad \tau, \zeta \in S_n,$$

where $n \geq k+1$. Now $\tau e_{\bar{\mu}} \zeta = \tau(e_{\bar{\mu}}\zeta\tau)\tau^{-1} = \tau(e_{\bar{\mu}}\eta)\tau^{-1}$ where $\eta = \zeta\tau$. Hence, we have the following reduction.

Reduction. It suffices to prove the assertion for the trace polynomials that correspond to the elements $e_{\bar{\mu}}\eta$, where $\eta \in S_n$ and $n \geq k+1$.

The essence of the argument is the following lemma.

Lemma 8.2.2. *Let $n \geq k+1$ and let $\eta \in S_n$. Then there exist permutations $\sigma \in S_{k+1}$ and $\gamma \in S_n$ such that $\eta = \sigma \cdot \gamma$, and each cycle of γ contains at most one of the elements $1, 2, \ldots, k+1$.*

Proof. On the contrary, assume for example that both 1 and 2 appear in the same cycle of η:

$$\eta = (1, i_1, \ldots, i_r, 2, j_1, \ldots, j_s) \cdots .$$

Then

$$(1,2) \cdot \eta = (1, i_1, \ldots, i_r)(2, j_1, \ldots, j_s) \cdots .$$

Continue by applying the same argument to $(1,2) \cdot \eta$, and the proof of the lemma follows by induction. $\qquad\square$

Let us prove the theorem. For any $\sigma \in S_{k+1}$,

$$e_{\bar{\mu}}\sigma = \sum_{\pi \in S_{k+1}} \text{sgn}(\pi)\pi\sigma = \text{sgn}(\sigma) \sum_{\rho \in S_{k+1}} \text{sgn}(\rho)\rho = \pm e_{\bar{\mu}}.$$

In view of the lemma, it follows that $e_{\bar{\mu}} \cdot \eta = \pm e_{\bar{\mu}} \cdot \gamma$. Thus, it suffices to show that the trace polynomial corresponding to $e_{\bar{\mu}} \cdot \gamma$ is obtained from $f_{e_{\bar{\mu}}}$.

Consider $\pi \in S_{k+1}$ and $\pi\gamma$, with their corresponding trace monomials $M_\pi(x_1, \ldots, x_{k+1})$ and $M_{\pi\gamma}(x_1, \ldots, x_n)$, given that the cycle decomposition of γ is:

$$(1, i_2, \ldots, i_r)(2, j_2, \ldots, j_s) \cdots (k+1, \ell_2, \ldots, \ell_t)(a_1, \ldots, a_u) \cdots (b_1, \ldots, b_v).$$

A straightforward calculation shows that the cycle decomposition of $\pi\gamma$ is obtained from the cycle decomposition of π as follows:

First write the cycle decomposition of π; then replace 1 by the string $1, i_1, \ldots, i_r$, replace 2 by the string $2, j_1, \ldots, j_s$, $\cdots$, replace $k+1$ by the string $k+1, \ell_1, \ldots, \ell_t$, and finally multiply (from the right) by the remaining cycles $(a_1, \ldots, a_u) \cdots (b_1, \ldots, b_v)$.

It follows that

$$M_{\pi\gamma}(x_1, \ldots, x_n) = M_\pi(h_1, \ldots, h_{k+1}) \cdot \text{tr}(x_{a_1} \cdots x_{a_u}) \cdots \text{tr}(x_{b_1} \cdots x_{b_v})$$

where $h_1 = x_1 x_{i_2} \cdots x_{i_r}$, $h_2 = x_2 x_{j_2} \cdots x_{j_s}$, $\ldots$, $h_{k+1} = x_{k+1} x_{\ell_2} \cdots x_{\ell_t}$.

As a consequence, we deduce that the trace polynomial that corresponds to $e_{\bar{\mu}} \cdot \gamma$ is

$$F(h_1, \ldots, h_{k+1}) \cdot \text{tr}(x_{a_1} \cdots x_{a_u}) \cdots \text{tr}(x_{b_1} \cdots x_{b_v}),$$

and the proof of (i) of the theorem is now complete.

(ii) Let $h = h(x_1, \ldots, x_n)$ be a mixed trace identity of $M_k(F)$. Then $\text{tr}(hx_{n+1})$ is a pure trace identity of $M_k(F)$. By (i), $\text{tr}(hx_{n+1})$ is a linear combination of terms of the form $af_{e_{\bar{\mu}}}(h_1, \ldots, h_{k+1})$, where the h_i are ordinary monomials, and a is a pure trace monomial which behaves like scalars. Since $\text{tr}(hx_{n+1})$ is linear in x_{n+1}, we may assume x_{n+1} is of degree 1 in each such summand. Thus, we have the following two cases.

Case 1. x_{n+1} appears in a; then $a = \text{tr}(hx_{n+1})$, so

$$af_{e_{\bar{\mu}}}(h_1, \ldots, h_{k+1}) = \text{tr}(hf_{e_{\bar{\mu}}}(h_1, \ldots, h_{k+1})x_{n+1}).$$

Case 2. x_{n+1} appears in, say, h_{k+1}: $\quad h_{k+1} = q_1 x_{n+1} q_2$. By applying cyclic permutations in the monomials of $\tilde{g}_k(h_1, \ldots, h_{k+1})$, it follows that

$$a\tilde{g}_k(h_1, \ldots, h_{k+1}) = \text{tr}(aq_2\tilde{g}_k(h_1, \ldots, h_k)q_1 x_{n+1}).$$

Consequently,

$$\text{tr}(h(x_1, \ldots, x_n)x_{n+1}) =$$
$$\text{tr}\left(\left[\sum a'\tilde{g}_k(h_1, \ldots, h_{k+1}) + \sum aq_2\tilde{g}_k(h_1, \ldots, h_k)q_1\right]x_{n+1})\right).$$

Part (ii) now follows, because of the nondegeneracy of the trace. $\qquad\square$

Corollary 8.2.3. *Every trace identity of $M_k(F)$, and in particular every polynomial identity, is a formal consequence of the Cayley-Hamilton polynomial.*

Proof. $f_{e_{\tilde{n}}}$ is a consequence of $\tilde{g}_k$, by Example 8.1.20(ii).   □

See Exercise 2 for a direct proof of the Amitsur-Levitzki theorem from the Cayley-Hamilton polynomial.

Recently, Bresar, M., Procesi, C., and Spenko [BrPS12] have shown that Corollary 8.2.3 can be extended to a theorem about trace superidentities of the Grassmann algebra, starting with Remark 7.1.6 (which explains Rosset's proof of the Amitsur-Levitzki Theorem).

8.3   Trace Codimensions

Having classified trace identities, it makes sense to incorporate them into the S_n-representation theory. Suppose X is infinite; let $T_n \subseteq T = T\{X\}$ denote the subspace of the multilinear pure trace polynomials in $x_1, \ldots, x_n$, and similarly let $T'_n \subseteq T' = T'\{X\}$ denote the subspace of the multilinear mixed trace polynomials.

Remark 8.3.1. *The map $p(x_1, \ldots, x_n) \mapsto \operatorname{tr}(x_0 p(x_1, \ldots, x_n))$, yields an isomorphism $T'_n \cong T_{n+1}$, as we saw above.*

The pure trace identities of $M_k(F)$ form a submodule $\operatorname{tr-id}_{k,n} \subseteq T_n$. By Theorem 8.1.18, we have the $F[S_n]$-module isomorphism

$$T_n / \operatorname{tr-id}_{k,n} \cong \bigoplus_{\lambda \vdash n, \, \lambda'_1 \leq k} I_\lambda.$$

Call the S_n-character of $T_n / \operatorname{tr-id}_{k,n}$ the *n-th pure trace cocharacter of $M_k(F)$* and denote it $\chi_n^{Tr}(M_k(F))$. The preceding arguments show that $\chi_n^{Tr}(M_k(F))$ equals the $F[S_n]$-module $\bigoplus_{\lambda \vdash n, \, \lambda'_1 \leq k} I_\lambda$ (under conjugation). It is well known that the character of I_λ under conjugation equals the character $\chi^\lambda \otimes \chi^\lambda$, where $\otimes$ indicates here the Kronecker (inner) product of characters. This proves the following result.

Theorem 8.3.2. *Notation as above,*

$$\chi_n^{Tr}(M_k(F)) = \bigoplus_{\lambda \vdash n, \, \lambda'_1 \leq k} \chi^\lambda \otimes \chi^\lambda.$$

Definition 8.3.3. $t_n(M_k(F)) = \dim(T_n / \operatorname{tr-id}_{k,n})$ *is the* (**pure**) **trace codimension** *of $M_k(F)$.*

Corollary 8.3.4. *Notation as above, we have*

$$t_n(M_k(F)) = \sum_{\lambda \vdash n,\ \lambda'_1 \leq k} (f^\lambda)^2,$$

where $f^\lambda = \deg \chi^\lambda$.

We can now prove a codimension result.

Lemma 8.3.5.

$$c_{n-1}(M_k(F)) \leq \sum_{\lambda \vdash n,\ \lambda'_1 \leq k} (f^\lambda)^2.$$

Proof. The assertion follows from the nondegeneracy of the trace form. Define $g : V_{n-1} \to T_n$ by

$$g(p(x_1, \ldots, x_{n-1})) = \operatorname{tr}(p(x_1, \ldots, x_{n-1})x_n).$$

Clearly g is one-to-one. Now $p(x_1, \ldots, x_{n-1})$ is a (non) identity of $M_k(F)$ if and only if $g(p(x_1, \ldots, x_{n-1}))$ is a (non) trace identity of $M_k(F)$, so the assertion follows. $\qquad\square$

Remark 8.3.6.

(i) *Applying Regev's comparison of the cocharacter theories to some deep results of Formanek [For84], it could be shown that when n goes to infinity, $c_{n-1}(M_k(F))$ is asymptotically equal to $t_n(M_k(F))$:*

$$c_{n-1}(M_k(F)) \sim t_n(M_k(F)).$$

(ii) *It can be shown that as n goes to infinity,*

$$t_n(M_k(F)) \sim c \left(\frac{1}{n}\right)^{(k^2-1)/2} k^{2n},$$

where c is the constant

$$c = \left(\frac{1}{\sqrt{2\pi}}\right)^{k-1} \left(\frac{1}{2}\right)^{(k^2-1)/2} 1!2! \cdots (k-1)! k^{(k^2+4)/2}.$$

Combining (i) and (ii) yields the precise asymptotics of the codimensions $c_n(M_k(F))$.

Remark 8.3.7. *This theory has been formulated and proved in exactly the same way for algebras without 1, cf. Exercise 1.*

8.4   Kemer's Matrix Identity Theorem in Characteristic p

Although the theory has been described here in characteristic 0, the characteristic p version has been developed by Donkin [Don94] and Zubkov [Zu96]. A much stronger version of Kemer's Capelli Theorem holds in characteristic p:

Theorem 8.4.1 ([Kem95], [Bel99]). *Any PI-algebra A in characteristic $p > 0$ satisfies the identities of $M_n(\mathbb{Z}/p)$ for suitable n.*

Although the proof is beyond the scope of our book, let us give an overview. First, Kemer [Kem93] showed (cf. Exercise 5.10) that every PI-algebra in characteristic > 0 satisfies a standard identity, although the bounds are very large. Applying the representation theory in characteristic p described briefly at the end of Chapter 3 to [Kem80], Kemer proved in [Kem95]:

Any PI-algebra over a field of characteristic p satisfies all *multilinear* identities of some matrix algebra, and in particular, the appropriate Capelli identity.

This is already a striking result which is clearly false for characteristic 0, when we recall that the Grassmann algebra of an infinite dimensional vector space over $\mathbb{Q}$ does not satisfy a Capelli identity.

As a consequence, the Razmyslov-Zubrilin theory expounded in Chapter 4 (and Exercise 6.17) becomes available to all PI-algebras of characteristic p. Furthermore, Donkin [Don92] proved the following generalization of Theorem 8.2.1:

Theorem 8.4.2. *Every GL-invariant in several matrices is a consequence of the coefficients of the characteristic polynomials.*

Zubkov [Zu96] showed that relations among these invariants follow from the Cayley-Hamilton polynomial. Putting the pieces together yields the proof of Theorem 8.4.1.

Let us conclude this chapter by noting that Kemer [Kem96] also has linked the codimension theory to the n-good words in the sense of Definition 3.2.18. Namely, let A be the algebra of m generic $n \times n$ matrices, and let c_m denote the codimension of the space of multilinear trace polynomials of length m. Kemer proved c_m is also the number of n-good words among all words of length m. Hence, using Formanek's methods in [For84], one could obtain explicit formulas for the n-good words. This ties Shirshov's theory in with the representation theory.

Exercises for Chapter 8

1. The trace identity theory goes through for algebras without 1.

2. Prove the Amitsur-Levitzki Theorem for $M_n(F)$ as a formal consequence of the Cayley-Hamilton trace identity. (Hint: Linearize and then substitute commutators. Then take the $2n$-alternator, cf. Definition 1.2.31, and dispose of the superfluous terms using Remark 1.4.10.)

3. Reformulate the Razmyslov-Zubrilin theory of Chapter 4 in terms of trace identities.

4. Define a theory of Young diagrams analogous to that of Chapter 2, by means of Definition 8.1.14 and Theorem 8.1.18. Use the trick of passing from mixed trace identities to pure trace identities by adding an indeterminate x_0.

5. Define an operation φ on $F[S_n]$ by $\varphi(\pi) = \operatorname{sgn}(\pi)\pi$. Under the correspondence of Definition 8.1.14, show that

$$\varphi(\operatorname{tr}(x_{i_1} \cdots x_{i_{m_1}})\operatorname{tr}(x_{j_1} \cdots x_{j_{m_2}}) \cdots \operatorname{tr}(x_{k_1} \cdots x_{k_{m_t}})) \\ = (-1)^t \operatorname{tr}(x_{i_1} \cdots x_{i_{m_1}})\operatorname{tr}(x_{j_1} \cdots x_{j_{m_2}}) \cdots \operatorname{tr}(x_{k_1} \cdots x_{k_{m_t}}). \tag{8.5}$$

6. Using the correspondence of Exercise 4, show that φ of Exercise 5 sends a Young tableau T_λ to its transpose.

The Nagata-Higman Theorem

Recall the symmetric polynomials $\tilde{s}_n$, cf. Equation (1.7).

7. (Improved in Exercise 8.) If A is an algebra without 1 satisfying the identity x^n, then A is nilpotent of index $\leq 2n^2 - 3n + 2$. (Hint: A satisfies the Cayley-Hamilton trace identity, where formally we take $\operatorname{tr}(a) = 0$ for all $a \in A$. Hence, A satisfies all the identities of $n \times n$ matrices, including s_{2n}, whose Young diagram is a $1 \times 2n$ column. But A also satisfies the $\tilde{s}_n$, whose Young diagram is an $n \times 1$ row. By the Branching Theorem, any Young diagram containing a row of length n or a column of length $2n$ must be an identity of A. This means the largest possible non-identity has $(n-1)(2n-1) = 2n^2 - 3n + 1$ boxes.)

8. (Razmyslov's improved version of Theorem 1.8.28.) If A is an algebra without 1 satisfying the identity x^n, then A is nilpotent of index $\leq n^2$. (Hint: Use the Young theory with traces of Exercise 4, adding one extra variable x_0. Any non-identity must not contain a column of length $n+1$, nor, by Exercise 6, a row of length $n+1$, so has length at most n^2.)

9. (Kuzmin's lower bound.) The T-ideal (without 1) $\mathcal{I}$ of $\mathbb{Q}\{x,y\}$ generated by X^n does not imply nilpotence of degree $\frac{n^2+n-2}{2}$. (Hint: It is enough to show that

$$yx^1yx^2y\cdots yx^{n-1} \notin \mathcal{I}.$$

Note that $\mathcal{I}$ is generated by values of the symmetric polynomials $\tilde{s}_n$. Given a function $\pi : \{0,\ldots,k\} \to \mathbb{N}$, identify π with the monomial

$$w_\pi = x^{\pi(0)}yx^{\pi(1)}yx^{\pi(2)}y\ldots yx^{\pi(n-1)};$$

it suffices to show that $w_{(1)} \notin \mathcal{I}$. Let V be the T-space generated by $\{w_\pi : \pi \in S_n\}$; let $\mathcal{J}$ be the T-space generated by all terms of the following form:

(i) x^j for all $j \geq n$.

(ii) w_π unless π is $1:1$.

(iii) $w_\pi - w_{\pi\tau}$ when π is $1:1$, for all transpositions τ.

The intuition for $n = k$: viewing these as reduction procedures, one may eliminate any monomial w_π unless π is a permutation of $\{0,\ldots,n-1\}$, and then use (iii) to transpose the powers of x. However, one permits $k \leq n$ to facilitate the induction procedure.

Since $w_{(1)} \notin \mathcal{J}$, it suffices to show that $\mathcal{I} \cap V \subseteq \mathcal{J}$.

Take $a = \tilde{s}_n(u_1,\ldots,u_n) \in \mathcal{I} \cap V$. Rearranging the u_i in decreasing order according to $\deg_y u_i$, assume that $\deg_y u_i = 0$ for all $i \geq k$. It suffices to prove $a \in \mathcal{J}$; $k \leq n-1$ since $\deg_y a = n-1$.

$\mathcal{J}$ is invariant under the derivations $x^i\frac{\partial}{\partial x}$, so one can build a from terms of the form $\tilde{s}_n(v_1,\ldots,v_{k-1},x,x,\ldots,x)$; thus, assume that a has this form. Prove the stronger assertion:

$$\sum_{j_1+\cdots+j_k=n+1-k} x^{j_1}v_1x^{j_2}v_2\ldots v_{k-1}x^{j_k} \in \mathcal{J}.$$

This is seen by induction on k, using (iii) above to rearrange the terms in ascending order, and then breaking off the initial term and applying induction on k.)

Chapter 9

PI-Counterexamples in Characteristic p

The Grassmann envelope developed in Chapter 7, to pass from the affine case to the general case, relied on the assumption of characteristic 0. Accordingly, the theory in characteristic $p \neq 0$ requires a separate discussion. The most striking feature is that without the characteristic 0 assumption, Specht's problem now has a negative solution, and this chapter is devoted to various counterexamples. In Volume II, we actually get a positive solution for affine PI-algebras.

9.1 De-multilinearization

The major obstacle in characteristic p is how to reverse the multilinearization process. Here is our main technique.

Remark 9.1.1. *Suppose we want to verify that a given homogeneous polynomial $f(x_1, \ldots, x_d)$ is an identity of A. Let $n_i = \deg_i f$. If each $n_i < p$, i.e., p does not divide $n_i!$ for any i, then it is enough to check the multilinearization of f, cf. Remark 1.2.13(ii).*

In general, we define a **word specialization** *to be a specialization to words in the generators of A. To verify f is an identity, it is enough to check that every partial multilinearization of f is an identity, and also that f vanishes under all word specializations, seen by working backwards from the definition of Δ in Definition 1.2.12, e.g.,*

$$f(x_1 + x_2; y_1, \ldots, y_m) = f(x_1; y_1, \ldots, y_m) + f(x_2; y_1, \ldots, y_m) \\ + \Delta f(x_1, x_2; y_1, \ldots, y_m). \tag{9.1}$$

But then continuing inductively, it is enough to check that the partial multilinearizations (as well as f) vanish under all word specializations.

This remark enables us to employ the combinatorics of the previous chapters even in characteristic p. We shall see several instances of this in the sequel, and start now with a trivial example, which is just a restatement of the Frobenius automorphism.

Remark 9.1.2. *Suppose $\mathrm{char}(F) = p$, and $A = F\{a_1, \ldots, a_\ell\}$ is a commutative algebra without 1, and $a_i^p = 0$ for $1 \le i \le \ell$. Then $x^p \in \mathrm{id}(A)$. Indeed, this is true for products of the generators, whereas all coefficients of $(a+b)^p - a^p - b^p$ are multiples of p and thus 0.*

Similarly, we shall see in Exercise 4 that the Grassmann algebra in characteristic $p > 2$ satisfies the identity $(x + y)^p - x^p - y^p$. (For $p = 2$, one gets $(x + y)^4 - x^4 - y^4$, cf. Exercise 9.)

Remark 9.1.3. *We continue the logic of Remark 9.1.1. Suppose that $\deg_i f = n_i = p$. We claim that if the linearization $\tilde{f}_i$ of f in x_i is an identity of A, then every partial multilinearization g of f in x_i is also an identity of A.*

Indeed, by induction it is enough to take $g = \Delta_i f$, notation as in (1.3) of Chapter 1. Write $g = \sum_{j=1}^{n_i-1} f_j$ where f_j has degree j in x_i (and thus degree $n_i - j$ in x_i'). As in (1.5) of Chapter 1, we can recover $j!(n_i - j)!f_j$ from $\tilde{f}_i$. But p does not divide $j!(n_i - j)!$, so we see that $\tilde{f}_i \in \mathrm{id}(A)$ iff each $f_j \in \mathrm{id}(A)$, which clearly implies $g \in \mathrm{id}(A)$.

Repeating this argument for each original indeterminate x_i, we see that if $\deg_i f \le p$ for each p and the total multilinearization of f is an identity of A, then each partial multilineariation is also an identity of A.

9.2 The Extended Grassmann Algebra

As in characteristic 0, the key to much of our theory is a close study of the Grassmann algebra G, whose identities were seen in Exercise 6.1 to be consequences of the Grassmann identity $[[x_1, x_2], x_3]$. Now assume for the remainder of this section that $\mathrm{char}(F) = 2$. (Thus, $[a, b] = ab + ba$.) Unfortunately, the Grassmann algebra is now commutative, so we need a replacement algebra for G. Surprisingly, its identities are also are generated by the Grassmann identity. Furthermore, this is generalized in arbitrary characteristic in [Dor12, DorKV15], cf. Exercises 10ff.

Definition 9.2.1. *Define $F[\varepsilon_i : i \in \mathbb{N}]$ to be the commutative polynomial algebra modulo the relations $\varepsilon_i^2 = 0$. We denote this algebra as $F[\varepsilon]$.*

Thus, any element $c \in F[\varepsilon]$ inherits the natural degree function, and if c has constant term 0, then $c^2 = 0$. (This is seen at once from Remark 9.1.2, since $F[\varepsilon]$ is commutative.) This provides us the perfect setting for our analog of the Grassmann algebra in characteristic 2.

Definition 9.2.2. *The* **extended Grassmann algebra** G^+ *is defined over $F[\varepsilon]$ by means of the quadratic relations*

$$[e_i, e_j] = \varepsilon_i \varepsilon_j e_i e_j, \ \forall i \neq j.$$

Thus, G^+ is a superalgebra, and in contrast with G, we now have $e_i^2 \neq 0$. Nevertheless we do have useful positive information. G^+ has the base B consisting of words $e_{i_1} \cdots e_{i_t}$ with $i_1 \leq i_2 \leq \cdots \leq i_t$. Given a word $b = e_{i_1} \cdots e_{i_t} \in B$ let us write $\varepsilon_b = \sum_{j=1}^{t} \varepsilon_{j_u}$.

Remark 9.2.3. $\varepsilon_b^2 = 0$ *for any word b, since*

$$\left(\sum_{u=1}^{t} \varepsilon_{j_u} \right)^2 = \sum_{u=1}^{t} \varepsilon_{j_u}^2 = \sum_{u=1}^{t} 0 = 0.$$

Definition 9.2.4. *There is a natural F-linear map $\varphi : F[\varepsilon] \to F$ given by each $\varepsilon_i \mapsto 1$. We call an element of $F[\varepsilon]$* **even** *if it is in $\ker \varphi$, and* **odd** *otherwise. Likewise, an element of G^+ is* **even** *if, written as a linear combination of elements of B, each coefficient is even.*

Lemma 9.2.5. G^+ *satisfies the following properties, for $b = e_{j_1} \cdots e_{j_t}$:*
 (i) $[e_i, b] = \varepsilon_i \varepsilon_b e_i b$;
 (ii) $\varepsilon_b b \in \mathrm{Cent}(G^+)$.

Proof. First we show that $\varepsilon_i e_i \in \mathrm{Cent}(G^+)$, by proving it commutes with each e_j :

$$[\varepsilon_i e_i, e_j] = \varepsilon_i [e_i, e_j] = \varepsilon_i^2 \varepsilon_j e_i e_j = 0.$$

Now

$$[e_i, b] = \sum_{u=1}^{t} e_{j_1} \cdots e_{j_{u-1}} [e_i, e_{j_u}] e_{j_{u+1}} \cdots e_{j_t}$$

$$= \sum_{u=1}^{t} \varepsilon_i \varepsilon_{j_u} e_{j_1} \cdots e_{j_{u-1}} e_i e_{j_u} e_{j_{u+1}} \cdots e_{j_t}$$

$$= \sum_{u=1}^{t} \varepsilon_{j_u} e_{j_1} \cdots e_{j_{u-1}} (\varepsilon_i e_i) e_{j_u} e_{j_{u+1}} \cdots e_{j_t}$$

$$= \varepsilon_i e_i \sum_{u=1}^{t} \varepsilon_{j_u} e_{j_1} \cdots e_{j_{u-1}} e_{j_u} e_{j_{u+1}} \cdots e_{j_t}$$

$$= \varepsilon_i (\sum_{u=1}^{t} \varepsilon_{j_u}) e_i e_{j_1} \cdots e_{j_{u-1}} e_{j_u} e_{j_{u+1}} \cdots e_{j_t},$$

proving (i).

To prove (ii), it is enough to commute with any e_i:

$$\left[e_i, \left(\sum_{u=1}^{t} \varepsilon_{j_u} \right) b \right] = \sum_{u=1}^{t} \varepsilon_{j_u} [e_i, b]$$

$$= \sum_{u=1}^{t} \varepsilon_{j_u} \varepsilon_i \left(\sum_{u=1}^{t} \varepsilon_{j_u} \right) e_i e_{j_1} \cdots e_{j_t}$$

$$= \varepsilon_i \left(\sum_{u=1}^{t} \varepsilon_{j_u} \right)^2 e_i e_{j_1} \cdots e_{j_t} = 0.$$

$\square$

This is generalized in Exercise 7. However we take another tack. We say a sum is **homogeneous** if each summand has the same degree in each indeterminate. For example, the sum in Remark 6.2.4 is homogeneous.

Remark 9.2.6. *In the free associative algebra $C\{X\}$, any commutator $[w_1, w_2]$ of words in the x_i can be rewritten as a homogeneous sum $\sum_i [w_{3i}, x_i]$ for suitable words w_{3i}. This is seen immediately by applying induction to Remark 6.2.4.*

Proposition 9.2.7. G^+ *satisfies the Grassmann identity. (Equivalently, $[x_1, x_2]$ is G^+-central.)*

Proof. It suffices to prove $[w_1, w_2]$ is central for any words w_1, w_2 in G^+; by Remark 9.2.6 we may assume that w_2 is a letter e_i, so we are done by Lemma 9.2.5. $\square$

Corollary 9.2.8. $[x_1^2, x_2] \in \mathrm{id}(G^+)$. *(Equivalently, x_1^2 is G^+-central.)*

Proof. For any $a, b \in G^+$, we have

$$[a^2, b] = a[a, b] + [a, b]a = a[a, b] + a[a, b] = 2a[a, b] = 0.$$

$\square$

We also are interested in squares in G^+.

Corollary 9.2.9. $[x_1, x_2]^2 \in \mathrm{id}(G^+)$.

Proof. This is by Corollary 6.2.6. $\square$

Proposition 9.2.10. *Suppose $a \in G^+$ is homogeneous of degree 1 in each of in $e_1, \ldots, e_t$. Then for some $\gamma \in F$,*

$$a^2 = \gamma^2 (e_1 \ldots e_t)^2.$$

Proof. Noting that $e_j e_i = e_i e_j + [e_j, e_i]$, we can rewrite a as

$$ce_1 \ldots e_t \quad + \quad \text{commutators},$$

where $c \in F[\varepsilon]$. But each commutator is central, so

$$a^2 = c^2 (e_1 \ldots e_t)^2 \quad + \quad \text{sums of squares of commutators}.$$

But each commutator squared is 0, by Corollary 9.2.9, so

$$a^2 = c^2 (e_1 \ldots e_t)^2;$$

writing $c = \gamma + c'$ where $\gamma \in F$ and $c' \in F[\varepsilon]$ has constant term 0, we have

$$c^2 = \gamma^2 + (c')^2 = \gamma^2 + 0 = \gamma^2,$$

as desired. $\square$

It remains to compute squares of arbitrary words:

Proposition 9.2.11. *Let*

$$\bar{\varepsilon}_t = \sum_{j=0}^{[t/2]} \sum_{i_1 < i_2 \cdots < i_{2j}} \varepsilon_{i_1} \ldots \varepsilon_{i_{2j}},$$

i.e., the sum of all even products of the ε_i, where the term for $j = 0$ is 1. Then

$$(e_1 \ldots e_t)^2 = \bar{\varepsilon}_t e_1^2 \ldots e_t^2.$$

Proof. By induction on t, the assertion being obvious for $t = 1$. But then by induction

$$(e_1 \ldots e_{t-1})^2 = \bar{\varepsilon}_{t-1} e_1^2 \ldots e_{t-1}^2,$$

and letting $b = e_1 \ldots e_{t-1}$ we have

$$(e_1 \ldots e_t)^2 = (be_t)^2 = be_t be_t = b(be_t + [b, e_t])e_t = b(1 + \varepsilon_t \varepsilon_b)be_t e_t$$
$$= (1 + \varepsilon_t \varepsilon_b)b^2 e_t^2 = (1 + \varepsilon_t \varepsilon_b)\bar{\varepsilon}_{t-1} e_1^2 \ldots e_{t-1}^2 e_t^2$$

so it remains to show that $(1 + \varepsilon_t \varepsilon_b)\bar{\varepsilon}_{t-1} = \bar{\varepsilon}_t$, i.e., we need

$$\varepsilon_t \varepsilon_b \bar{\varepsilon}_{t-1} = \bar{\varepsilon}_t - \bar{\varepsilon}_{t-1},$$

which is clearly $\sum_j \sum_{i_1 < \cdots < i_{2j-1}} \varepsilon_{i_1} \ldots \varepsilon_{i_{2j-1}} \varepsilon_t$. In other words we want to show that

$$\varepsilon_b \bar{\varepsilon}_{t-1} = \sum_j \sum_{i_1 < \cdots < i_{2j-1}} \varepsilon_{i_1} \ldots \varepsilon_{i_{2j-1}}.$$

But there are $2j - 1$ ways to get a given nonzero product $\varepsilon_{i_1} \cdots \varepsilon_{i_{2j-1}}$ of length $2j - 1$ from some nonzero product of length $2t - 2$. Namely, multiply one of the ε_i by the product of the others. Hence

$$\varepsilon_b \bar{\varepsilon}_{t-1} = \sum_j \sum_{i_1 < \cdots < i_{2j-1}} (2j - 1)\varepsilon_{i_1} \ldots \varepsilon_{i_{2j-1}}$$
$$= \sum_j \sum_{i_1 < \cdots < i_{2j-1}} \varepsilon_{i_1} \cdots \varepsilon_{i_{2j-1}},$$

since $2j - 2 \equiv 0 \pmod 2$. $\square$

The important thing for us will be that the number of terms is

$$1 + \binom{t}{2} + \binom{t}{4} + \cdots = 2^{t-1},$$

which is even.

Corollary 9.2.12. *The product of squares of words involving distinct letters is even, provided that at least one of the factors has length > 1.*

Proof. The products have coefficients whose number of terms are a power of 2, without cancelation, and thus are even. $\square$

For example, $(e_1 e_2)^2 (e_3 e_4)^2 = (1 + \varepsilon_1 \varepsilon_2)(1 + \varepsilon_3 \varepsilon_4)e_1^2 e_2^2 e_3^2 e_4^2$. Note on the other hand that

$$(e_1 e_2)^2 (e_1 e_3)^2 = (1 + \varepsilon_1 \varepsilon_2)(1 + \varepsilon_1 \varepsilon_3)e_1^2 e_2^2 e_1^2 e_3^2 = (1 + \varepsilon_1 \varepsilon_2 + \varepsilon_1 \varepsilon_3)e_1^4 e_2^2 e_3^2,$$

which is odd.

9.2.1 Computing in the Grassmann and extended Grassmann algebras

We now turn to the heart of our examples — some of the combinatorial properties of the Grassmann identity $[[x_1, x_2], x_3]$ which make it possible for us to perform complicated computations involving polynomials. The key role is played by commutators, which by hypothesis are central. We start with some formal (and very useful) consequences of Latyshev's Lemma 6.2.5:

Lemma 9.2.13. *The following identities are consequences of the Grassmann identity:*

(i) $[x_1, x_2]u[x_3, x_4] + [x_1, x_4]u[x_2, x_3]$,

(ii) $[x_1, x_2]u[x_1, x_3]$,

(iii) *The identity* $[x_1, x_2]w[x_3, x_4]$ *holds in any 2-generated subalgebra.*

(iv) *In a 2-generated subalgebra, any single expression (formed by taking regular products and Lie products only) involving two sets of commutators is* 0. *(For example,* $[a_1 a_2, [a_2, a_1 a_2^2] a_1] = 0$.)

Proof. (i) follows immediately from Lemma 6.2.5, since u commutes with all commutators. One gets (ii) by specializing x_4 to x_1.

To prove (iii), one may assume that x_i are words in the e_i, and proceed by an easy induction on their lengths: If all have length 1, then two of them are equal, and the only nontrivial case is $[e_i, e_j]u[e_i, e_k]$ which is 0 by (ii). Finally, the inductive step: If $x_1 = rs$, then

$$[rs, x_2]w[x_3, x_4] = r[s, x_2]w[x_3, x_4] + [r, x_2]sw[x_3, x_4],$$

both of which are 0 by induction.

(iv) Any single commutator is central, and so could be extracted from the expression; thus, we are done by (iii). □

We say a product of m commutators in G or G^+ is in **standard** form if it can be written as a homogeneous sum of the form

$$[e_{j_1}, e_{j_2}][e_{j_3}, e_{j_4}] \ldots [e_{j_{2m-3}}, e_{j_{2m-2}}][v_m, e_{j_{2m}}], \tag{9.2}$$

where v_m is an arbitrary word.

Lemma 9.2.14. *If f is a product of m commutators, then any specialization of f in G^+ can be put in standard form.*

Proof. We need to rewrite $[v_1, w_2] \cdots [v_m, w_m]$ (for any words v_i, w_i) as a sum of the form (9.2).

Step 1. By Remark 9.2.6, we may reduce to the case $w_i = e_{j_i}$ (since our expression is sums of expressions of this form).

Step 2. In view of Lemma 9.2.13(ii), we may assume that the e_{j_i} are distinct.

Step 3. $[v_1, e_{j_1}][v_2, e_{j_2}] = -[e_{j_1}, v_1][v_2, e_{j_2}] = [e_{j_1}, e_{j_2}][v_2, v_1]$, by Lemma

9.2.13(i). Thus, if $m \geq 2$, we may assume that the first commutator is $[e_{j_1}, e_{j_2}]$, say $[e_1, e_2]$. By Step 1 again, we may replace v_1 by some new e_{j_2}. So now we have

$$[e_1, e_2][v_2, e_{j_2}] \cdots [v_m, e_{j_m}]$$

Step 4. Repeating the argument of Step 3, we reduce to

$$[e_1, e_2][e_3, e_4] \cdots [e_{2m-3}, e_{2m-2}][v_m, e_{j_m}],$$

which is the desired form. $\qquad\square$

9.3 Non-Finitely Based T-Ideals in Characteristic 2

indexsubjectindexideal!T-ideal

Here is a rather simple strategy for constructing a non-finitely based T-ideal: Construct a series of polynomials $f_1, f_2, \ldots$ and algebras $A_1, A_2, \ldots$ such that for each n, $f_1, \ldots, f_n \in \mathrm{id}(A_n)$ but $f_m \notin \mathrm{id}(A_n)$ for some $m > n$. This means that the T-ideal generated by all $f_1, f_2, \ldots$ cannot be a consequence of a finite number of f_i, and thus cannot be finitely based. Although somehow this approach is (perhaps surprisingly) defeated in characteristic 0 by the Kemer polynomials, there are enough new identities in characteristic p to carry it through.

We start with the simplest counterexample, which is in characteristic 2. The example is due to Belov and Grishin, with proof shortened by using a construction of Gupta and Krasilnikov. We utilize the extended Grassmann algebra G^+, cf. Definition 9.2.1.

Lemma 9.3.1. *Let I_n be the $\mathbb{Z}/2$-subspace of G^+ spanned by all $\bar{x}_{i_1}^2 \ldots \bar{x}_{i_m}^2$, where $m < n$, for $\bar{x}_i \in G^+$. Then*

$$e_{i_1}^2 e_{i_2}^2 \ldots e_{i_n}^2 \notin I_n,$$

for any n. (We do not require $i_1, i_2, \ldots i_n$ to be distinct.)

Proof. Taking homogeneous parts in $e_{i_1}, e_{i_2}, \ldots e_{i_n}$, it suffices to show that

$$e_{i_1}^2 e_{i_2}^2 \ldots e_{i_n}^2$$

cannot be written as a sum of products of fewer than n squares. But each such product is even, by Corollary 9.2.12, so the sum must be even, whereas $e_{i_1}^2 e_{i_2}^2 \ldots e_{i_n}^2$ has coefficient 1. $\qquad\square$

Example 9.3.2. *Define*

$$f_n = [y_1^2, y_2]x_1^2 x_2^2 \ldots x_n^2 [y_1^2, y_2]^3.$$

Over a field F of characteristic 2, the T-ideal generated by $\{f_n : n \in \mathbb{N}\}$ is not finitely based.

Proof. We work in the matrix algebra $M_5(G^+)$ with its standard set of matrix units e_{ij}, not to be confused with the $e_i \in G^+$ used above. Let R be the subalgebra of $M_5(G^+)$ consisting of upper triangular matrices, and let A be the F-subalgebra of R generated by 1, $v = \sum_{i=1}^{4} e_{i,i+1}$ and all $\{b(e_{22} + e_{44}) : b \in G^+\}$. Thus, A consists of elements of the form

$$
w_j = \begin{pmatrix}
\alpha_j & b'_j & * & * & * \\
0 & b_j & b''_j & * & * \\
0 & 0 & \alpha_j & b'_j & * \\
0 & 0 & 0 & b_j & b''_j \\
0 & 0 & 0 & 0 & \alpha_j
\end{pmatrix}
$$

where $\alpha_j, \beta_j \in F$ and $b_j, b'_j, b''_j \in G^+$. Note that $G^+ e_{15} \subseteq A$ since

$$
be_{15} = vb(e_{22} + e_{44})v^3.
$$

On the other hand, $b(e_{22} + e_{44})$ and v both annihilate $G^+ e_{15}$ from each side, so any F-subspace of $G^+ e_{15}$ is an ideal of A; in particular, $I_n e_{15} \triangleleft A$, taking I_n as in Lemma 9.3.1. The continuation is an easy exercise in computing upper triangular matrices.

Let $A_n = A/I_n e_{15}$. In view of Corollary 9.2.8, $b_j^2 \in \mathrm{Cent}(G^+)$, and certainly $\alpha_j^2 \in \mathrm{Cent}(G^+)$, so any evaluation of $[w_1^2, w_2]$ in A_n has diagonal 0 thus is of the form

$$
s = \begin{pmatrix}
0 & b' & * & * & * \\
0 & 0 & b'' & * & * \\
0 & 0 & 0 & b' & * \\
0 & 0 & 0 & 0 & b'' \\
0 & 0 & 0 & 0 & 0
\end{pmatrix}.
$$

If $r = w_1^2 \dots w_m^2 \in A$, the 2,2 entry of r is $b_1^2 \dots b_m^2$. Then, taking s as above, the only possible nonzero entry of srs^3 in A comes from

$$
(b'e_{12})re_{22}(b''e_{23})(b'e_{34})(b''e_{45}),
$$

so is $(b'b'')^2 b_1^2 \dots b_m^2 e_{15}$. This is in $I_n e_{15}$ whenever $m < n - 2$, implying $f_1, \dots, f_{n-3} \in \mathrm{id}(A_n)$.

On the other hand,

$$
[(e_{22} + e_{44})^2, s] = [e_{22} + e_{44}, s] = (b'e_{23} + b'e_{45}) - (b''e_{12} + e_{34}) = s
$$

since $+1 = -1$, so taking $\bar{x}_i = e_i$ and $b' = 1$ and $b'' = e_n$ in G^+, we see that f_{n-1} takes on the value $e_1^2 \dots e_{n-1}^2 e_n^2$, which is not in I_n, so $f_{n-1} \notin \mathrm{id}(A_n)$.

Thus, the chain of T-ideals generated by $\{f_1, \dots, f_n : n \in \mathbb{N}\}$ never stabilizes, as desired. $\qquad\square$

9.3.1 T-spaces evaluated on the extended Grassmann algebra

Unfortunately, Shchigolev [Shch00] showed that the natural analog to Lemma 9.3.1 in odd characteristic p is finitely based; the T-space generated by the polynomials $Sh_n = x_1^p \cdots x_n^p$ is finitely based (even in the free algebra!), by the set $Sh_1, \ldots, Sh_{(p+1)/2}$.

Accordingly, we shall take a different approach to constructing infinitely based T-ideals, which although more intricate, leads naturally to examples in all positive characteristics. The idea is to build a T-space (cf. Definition 1.7.12) that is not finitely based, and then modify the construction for T-ideals. Thus, we divide the problem into easier parts, but the down side is that we can only use one test algebra at a time. In characteristic 2, it is natural to try the test algebra G^+.

Finding a non-finitely based T-space means we want to find a sequence $f_1, f_2, \ldots$ of polynomials such that, for each n, some value of f_n cannot be spanned over F by specializations of $f_m, m < n$. It turns out that these polynomials are not linear, so we have to worry about linearizing the f_m.

Toward this end, by Remark 9.1.1, it suffices to check this for word specializations and for partial linearizations. It turns out in all our examples that the contribution of the partial linearizations *must be* 0! Let us make the idea explicit by means of a definition.

Definition 9.3.3. *A p-monomial is a monomial h for which $\deg_i h \in \{0, p\}$ for each i. A polynomial is called p-**admissible** if it is a linear combination of p-monomials. A specialization of a homogeneous polynomial f in G (resp. G^+ for $p = 2$) to words v_i in the e_j is p-**admissible** if each e_j occurs of degree a multiple of p, for $1 \le i \le 2n$.*

We shall choose $f_n(x_1, \ldots, x_{2n})$ with $\deg_i f_n = p$ for all i. Assuming that $\text{char}(F) = p$, we then see by Remark 9.1.3 that to check the partial linearizations it is enough to show in the multilinearization of f_m that the homogeneous parts of degree p in each e_i of any specialization of f_m is 0. Thus, we shall show for each $m < n$ that each p-admissible word specialization of a multilinearization of f_m is 0, and also show that the p-admissible word specializations of the f_m cannot cancel f_n.

Let us see how this works for $p = 2$.

Lemma 9.3.4. *If f is a product of m commutators, then any 2-admissible specialization of f in G^+ is 0.*

Proof. In view of Lemma 9.2.14, we need show that (9.2) is 0 if it is 2-admissible. We argue by induction on the degree ℓ of v_m, and then on m. If $\ell = 1$ and $m = 1$, then our specialization is $[e_1, e_1] = 0$.

If $\ell = 1$ and $m > 1$, then v_m is some e_i, say $v_m = e_i$. We consider the general situation after dealing with two important cases.

Special Case. Suppose e_i appears in v_m with degree 1, where $i \ne m$.

By definition of 2-admissibility, e_i must appear somewhere else; let us assume for notational convenience that $i = 1$. Then we write $v_m = e_1 v'$ and, since commutators are central,

$$
\begin{aligned}
[e_1, e_2][e_3, e_4] \ldots [v_m, e_{j_m}] &= \ldots [e_1 v', e_{j_m}] \\
&= [e_3, e_4] \ldots [e_1[e_1, e_2]v', e_{j_m}] \\
&= [e_3, e_4] \ldots [\varepsilon_1 \varepsilon_2 e_1^2 e_2 v', e_{j_m}] \\
&= [e_3, e_4] \ldots [e_2 v', e_{j_m}] \varepsilon_1 \varepsilon_2 e_1^2.
\end{aligned}
$$

But, by induction, $[e_3, e_4] \ldots [e_2 v', e_{j_m}] = 0$ since it is a 2-admissible specialization of a product of commutators with smaller m and the same ℓ, implying (9.2) is 0.

General Case. If $v_m = v' e_i e_j v''$, we claim that we can replace v_m by $v' e_j e_i v''$. Indeed, it suffices to prove the difference,

$$[e_1, e_2][e_3, e_4] \ldots [e_{2m-3}, e_{2m-2}][v'[e_i, e_j]v'', e_{j_m}],$$

is 0. But $[e_i, e_j]$ is central, so we get

$$[e_1, e_2][e_3, e_4] \ldots [e_{2m-3}, e_{2m-2}][e_i, e_j][v'v'', e_{j_m}].$$

But the length of $v'v''$ is $\ell - 2$, so this is 0 by induction, as desired.

Now we handle (9.2) in general. By the previous paragraph, we can rearrange the e_i in v_m in ascending order. But any e_i^2 is central, so can be removed from the commutator. Disregarding this e_i^2 still produces a 2-admissible specialization of f, which by induction on ℓ is 0 (since we have decreased ℓ by 2). Hence we are done unless all the e_i appearing in v_m are distinct, but then we are done by Case 1 unless $v_m = e_m$, in which case we have $\ldots [e_m, e_m] = 0$. □

9.3.2 Another counterexample in characteristic 2

Define

$$P_n = x_1[x_1, x_2]x_2 x_3[x_3, x_4]x_4 \ldots x_{2n-1}[x_{2n-1}, x_{2n}]x_{2n}. \tag{9.3}$$

This definition is only valid for characteristic 2; we give the version for characteristic p below.

Lemma 9.3.5. *Any 2-admissible specialization of any partial linearization P_n in G^+ is 0.*

Proof. In view of Remark 9.1.3, it suffices to check the multilinearization. Let us linearize first at x_1. Then $x_1[x_1, x_2]x_2$ becomes

$$x_1[x_1', x_2]x_2 + x_1'[x_1, x_2]x_2$$

which (recalling that we are in characteristic 2) is

$$[x_1, x_1']x_2^2 + [x_1x_2, x_1'x_2].$$

Now linearizing at x_2 yields

$$[x_1, x_1'][x_2, x_2'] + [x_1x_2, x_1'x_2'] + [x_1x_2', x_1'x_2].$$

Applying this argument to each pair gives us a sum of products of commutators, each of which is 0 by Lemma 9.3.4. $\qquad\square$

Let us now check word specializations.

Lemma 9.3.6. *For any words $v_1, \ldots, v_{2m}$ such that $\deg_i v_1 \ldots v_{2m} \le 2$ for each i, the term*

$$v_1[v_1, v_2]v_2v_3[v_3, v_4]v_4 \cdots v_{2m-1}[v_{2m-1}, v_{2m}]v_{2m}$$

is even, cf. Definition 9.2.4, unless all the v_i are single letters.

Proof. Suppose

$$q_j = v_{2j-1}[v_{2j-1}, v_{2j}]v_{2j} \ne 0.$$

If e_i occurs in v_1 it already occurs twice and cannot occur in any other v_j. Likewise with v_2, so the ε_i occurring in the calculation of q_1 cannot occur in the calculation of any other q_j. It follows at once that the parity of $q_1 \ldots q_m$ is the product of the parities of $q_1, q_2, \ldots, q_m$ individually. So it suffices to prove that some q_j is even. Now

$$v_{2j-1}[v_{2j-1}, v_{2j}]v_{2j} = v_{2j-1}^2 v_{2j}^2 - (v_{2j-1}v_{2j})^2.$$

These are even, by Corollary 9.2.12, unless both v_{2j-1} and v_{2j} are single letters. So we are done unless $v_1, \ldots, v_{2m}$ are all single letters, as claimed. $\quad\square$

Theorem 9.3.7. *The T-space generated by $\{P_n : n \in \mathbb{N}\}$ is not finitely based.*

Proof. $P_n(e_1, \ldots, e_n) = \varepsilon_1 \ldots \varepsilon_{2n} e_1^2 \ldots e_{2n}^2$ which is 2-admissible and odd (nonzero). Lemma 9.3.5 shows that any multilinearization of any P_m cannot cancel this term, so it suffices to check word specializations.

But Lemma 9.3.6 says for $m < n$ that any word specialization of P_m provides an even contribution to $e_1^2 \cdots e_{2n}^2$, so we do not obtain 0 by adding word specializations. $\qquad\square$

9.4 Non-Finitely Based T-Ideals in Odd Characteristic

Our next task is to construct a non-finitely based T-ideal in characteristic $p > 2$. It turns out that the natural analog of the polynomials P_n in characteristic p plays the key role, but the analysis is rather intricate. First we shall find a non-finitely based T-space containing the consequences of the Grassmann identity. Then we modify this to a non-finitely based T-ideal. However, we must be careful, for we already saw in Exercise 6.1 that any T-ideal containing the Grassmann identity is finitely based!

Since the Grassmann algebra is 2-graded, we are led to utilize superidentities in our treatment. We shall use w_i, x_i to denote ordinary indeterminates, y_i for even indeterminates, and z_i for odd indeterminates.

Throughout, we work over an infinite field F of characteristic $p > 2$. We start with the Grassmann algebra G over a countably infinite dimensional F-vector space V. As usual, we write $\{e_i : i \in \mathbb{N}\}$ for the standard base of V.

9.4.1 Superidentities of the Grassmann algebras

Let us obtain some superidentities of G that lie at the core of our investigation.

Remark 9.4.1. *We recall Remark 7.1.6, which says $zxz \in \mathrm{id}_2(G)$, for all z odd.*

Lemma 9.4.2. *Suppose x_i are indeterminates, y_i are even indeterminates, and z_i are odd indeterminates. Then G satisfies the following superidentities:*

(i) $(y + z)^n = y^n + ny^{n-1}z$;

(ii) $x_0(y + z)x_1(y + z) \cdots x_{n-1}(y + z)x_n$

$$= y^n x_0 \cdots x_n + y^{n-1} \sum_{k=0}^{n-1} x_0 \cdots x_k z x_{k+1} \cdots x_n;$$

(iii) $zx(y + z)^n = zxy^n$;

(iv) $[y_1 + z_1, y_2 + z_2] = 2z_1z_2$.

Proof. (i) In the expansion of $(y + z)^n$, any monomial containing two occurrences of z vanishes. But this only leaves the monomials of degree at most 1 in z, and since y is central, we get (i) by the binomial expansion.

(ii) The same argument as in (i), although here we note that z does not necessarily commute with the x_i.

(iii) The same idea as before, noting that z already occurs at the beginning so the only monomial that survives in the expansion of $(y + z)^n$ is y^n.

(iv) The y_i are central, so the only nonzero part of the commutator is $[z_1, z_2]$, which is $2z_1z_2$ since the z_i anticommute. $\square$

This already lays the groundwork for an example by Shchigolev [Shch00] of a non-finitely based T-space. However, since Belov proved Specht's conjecture for affine algebras in characteristic $p > 0$, to be presented in Volume II, we proceed directly toward a counterexample to Specht's problem in the non-affine case.

9.4.2   The test algebra A

We assume henceforth that $\mathrm{char}(F) = p > 2$. In order to get a nonfinitely based T-ideal, we need an example that intrinsically requires an infinite number of indeterminates and we use a natural generalization of the P_n used earlier in characteristic 2. The ensuing computations require a sophisticated analysis that occupies the remainder of this chapter. Our choice of test algebra is crucial, since the Grassmann algebra G cannot do the trick.

Rather than working over G itself, we take the relatively free superalgebra $\mathcal{G}$ of G, which has generic even elements $\bar{y}_i$ and generic odd elements $\bar{z}_i$ satisfying the various superidentities of G. Let $A = \mathcal{G}/I$, where I is the ideal generated by all words h on the $\bar{y}_i, \bar{z}_i$, for which $\deg_i h \geq p$ for some i. Since the $\bar{y}_i$ and $\bar{z}_j$ either commute or anticommute, clearly the p-th power of any word in A is 0. Also, A is a superalgebra, where a word is odd or even depending on how many $\bar{z}_i$ appear in it. Furthermore, A inherits the natural degree from $\mathcal{G}$.

Lemma 9.4.3. $\bar{y}^p = 0$ *for any even element* $\bar{y} \in A$ *without a constant term.*

Proof. By Remark 9.1.2, since y is spanned by even elements whose p power is 0. $\square$

Proposition 9.4.4. *A satisfies the identity*

$$w x_1 w x_2 \cdots x_{p+1} w$$

for indeterminates w, x_i.

Proof. Consider a specialization $x_i \mapsto \bar{x}_i$, and $w \mapsto \bar{y} + \bar{z}$ where $\bar{y}$ is even (and thus central) and $\bar{z}$ is odd. Then $w x_1 w x_2 \cdots x_{p+1} w$ specializes to a sum of terms where each w is replaced by $\bar{y}$ or $\bar{z}$. But if $\bar{z}$ appears twice, then the term is 0 by Remark 7.1.6. Thus, we only are concerned when at most one of the w are replaced by $\bar{z}$. Then $\bar{y}$ appears p times and is central, so our term equals $\bar{y}^p$ times the remaining part. But $\bar{y}^p = 0$ by Lemma 9.4.3. $\square$

The key to our analysis is the polynomial

$$P(x_1, x_2) = x_1^{p-1}[x_1, x_2]x_2^{p-1}. \tag{9.4}$$

Clearly, $P(x_1, x_2)$ is a p-admissible polynomial, cf. Definition 9.3.3.

Lemma 9.4.5 (Shchigolev).

(i) *Given any monomial h, write h_i for the monomial obtained by substituting 1 for x_i. Then*

$$[w, h] = \sum d_i[w, x_i]x_i^{d_i}h_i \tag{9.5}$$

holds identically in A, where $d_i = \deg_i h$.

(ii) *Any p-admissible polynomial is central or an identity of A.*

(iii) *The polynomial $P(x_1, x_2)$ is A-central.*

Proof. (i) One can show easily by induction that

$$[w, x_{i_1} \cdots x_{i_n}] = \sum_{j=1}^{n} x_{i_1} \cdots x_{i_j - 1}[w, x_{i_j}]x_{i_j + 1} \cdots x_{i_n}. \tag{9.6}$$

Let $h = x_{i_1} \cdots x_{i_n}$. Since commutators in A are central, we can move the commutator in (9.6) to the left, and get

$$[w, h] = \sum_j [w, x_{i_j}]x_{i_1} \cdots x_{i_j - 1}x_{i_j + 1} \cdots x_{i_n}.$$

Lemma 9.2.13(ii) yields the identity

$$[w, x_i]vx_i = [w, x_i]x_iv - [w, x_i][x_i, v] = [w, x_i]x_iv,$$

which enables us to move each unbracketed occurrence of the x_i to the left, yielding

$$[w, x_i]h = [w, x_i]x_i^{d_i}h_i.$$

Noting that each x_i occurs exactly d_i times, we combine repetitions and get (9.5).

(ii) By (i), $[w, h] \in \mathrm{id}(A)$ if h is a p-monomial.

(iii) Substituting

$$e_{ip+1} + e_{ip+2} + \cdots + e_{(i+1)p}$$

for x_i gives a nonzero value in A. $\qquad\square$

Recall Definition 2.6.3, of *cyclic shift* and *cyclically conjugate*.

Proposition 9.4.6.

(i) *Suppose h is a p-monomial, and h' is cyclically conjugate to h. Then h and h' have the same value in A under any word specialization.*

(ii) *Suppose g is any central polynomial in A, and h, h' are two cyclically conjugate words. If gh is p-admissible, then $gh - gh'$ always specializes to 0 under word specializations.*

Proof. (i) It suffices to consider the case of a cyclic shift. Write $d_j = \deg_j h$. We need to prove that $[x_j, h] \in \mathrm{id}(A)$ whenever $x_j h$ is a p-monomial, i.e., $d_j = p-1$ and $d_i \equiv 0 \pmod{p}$ for all $i \neq j$. But this follows from Lemma 9.4.5(i), since $[x_j, h]$ can be replaced by

$$\sum_j d_i [x_i, x_j] x_i^{d_i} h_i;$$

each summand is 0, because $d_i \equiv 0$ for $j \neq i$, and because $[x_i, x_i] = 0$ for $j = i$.

(ii) Immediate from (i). If $h = uv$ and $h' = vu$, then gh takes on the same value as ugv, which is cyclically conjugate to gvu. □

Corollary 9.4.7. *Suppose g is a central polynomial of A, and words $v = v_1 v_2$, $v' = v_3 v_4$ are cyclically conjugate, with gv p-admissible. Then*

$$v_1 g v_2 = gv = gv' = v_3 g v_4$$

holds identically under all word specializations in A.

9.4.3   Shchigolev's non-finitely based T-space

Here is the correct version of our polynomial P_n in arbitrary characteristic $p > 0$, cf. (9.3):

$$P_n = \prod_{i=1}^{n} P(x_{2i-1}, x_{2i}) = x_1^{p-1}[x_1, x_2]x_2^{p-1} \cdots x_{2n-1}^{p-1}[x_{2n-1}, x_{2n}]x_{2n}^{p-1}.$$

We formally write the indeterminates $x_i = y_i + z_i$ for y_i even and z_i odd. The bulk of the proof will involve considering in turn all the even or odd substitutions for the x_i.

Lemma 9.4.8. *The polynomials $x_1^{p-1}[x_1, x_2]x_2^{p-1}$ and $2z_1 z_2 y_1^{p^n-1} y_2^{p^n-1}$ are equivalent on G.*

Proof. By Lemma 9.4.2 (iv) and then (iii),

$$x_1^{p^n-1}[x_1, x_2]x_2^{p^n-1} = 2x_1^{p^n-1}z_1 z_2 x_2^{p^n-1} = 2z_1 z_2 y_1^{p^n-1} y_2^{p^n-1}.$$

□

Remark 9.4.9. *By Lemma 9.4.8, P_n is equivalent over G to*

$$2z_1 z_2 y_1^{p-1} y_2^{p-1} \cdots 2z_{2n-1} z_{2n} y_{2n-1}^{p-1} y_{2n}^{p-1} = 2^n z_1 y_1^{p-1} \cdots z_{2n} y_{2n}^{p-1},$$

since the y_i are central.

The supermonomial on the RHS is linear in the z_i and has degree $p-1$ in each y_i. For notational convenience, we drop the coefficient 2^n since it is prime to p.

The first step of our argument will be to show that the T-space generated by the P_n, $n \in \mathbb{N}$, is not finitely based. Intuitively, one might try to demonstrate this on A, but the assertion is then false! Indeed,

$$
\begin{aligned}
P_1(y_1 y_2 y_3 + z_1 z_2 z_3, y_4 + z_4) &= 2z_1 z_2 z_3 z_4 (y_1 y_2 y_3)^p y_4^p \\
&= 2z_1 y_1^p z_2 y_2^p z_3 y_3^p z_4 y_4^p \\
&= P_2(y_1 + z_1, y_2 + z_2, y_3 + z_3, y_4 + z_4).
\end{aligned}
$$

So we must modify our test space,

Definition 9.4.10. $\tilde{A}$ *is the subalgebra of A generated by all $\bar{x}_i = \bar{y}_i + \bar{z}_i$.*

Since $\bar{y}_i$ is not an element of $\tilde{A}$, the above counterexample is no longer valid. We shall need to deal with the multilinearization of $P(x_1, x_2)$, which is

$$
\sum_{\sigma_1, \sigma_2 \in S_p} x_{1,\sigma_1(2)} \cdots x_{1,\sigma_1(p)} [x_{1,\sigma_1(1)}, x_{2,\sigma_2(1)}] x_{2,\sigma_2(2)} \cdots x_{2,\sigma_2(p)}
$$

and clearly equivalent to

$$
\sum_{\sigma_1, \sigma_2 \in S_p} x_{1,\sigma_1(2)} \cdots x_{1,\sigma_1(p)} z_{1,\sigma_1(1)} z_{2,\sigma_2(1)} x_{2,\sigma_2(2)} \cdots x_{2,\sigma_2(p)}.
$$

Thus, writing

$$
\hat{P}' = \sum_{\sigma \in S_p} x_{1,\sigma(2)} \cdots x_{1,\sigma(p)} z_{1,\sigma(1)}, \qquad \hat{P}'' = \sum_{\sigma \in S_p} z_{2,\sigma(1)} x_{2,\sigma(2)} \cdots x_{2,\sigma(p)},
$$

we see that $\hat{P}' \hat{P}''$ is the multilinearization of P. We need a more subtle argument to move the other occurrences of x past y as in Lemma 9.4.8, since the odd parts do not repeat so conveniently (and thus Lemma 9.4.2 is not available).

Lemma 9.4.11. *Writing $x_{1,i} = y_{1,i} + z_{1,i}$ for $2 \le i \le p$, we have*

$$
\hat{P}' = \sum_{\sigma \in S_p} y_{1,\sigma(2)} \cdots y_{1,\sigma(p)} z_{1,\sigma(1)}.
$$

Proof. We could see this by multilinearizing Lemma 9.4.8, but let us prove it directly. Substituting $y_{1,i} + z_{1,i}$ for $x_{1,i}$ and opening $\hat{P}'$ accordingly we consider terms involving some $z_{1,i}$ other than $z_{1,\sigma(1)}$. Suppose $i = \sigma(j)$. Switching $z_{1,\sigma(1)}$ and $z_{1,\sigma(j)}$ yields

$$
\cdots z_{1,\sigma(j)} \cdots z_{1,\sigma(1)} = -\cdots z_{1,\sigma(1)} \cdots z_{1,\sigma(j)},
$$

which cancels the term corresponding to $\sigma \tau$ where τ is the transposition $(1j)$. Thus, all these terms pair off with opposite signs and cancel, leaving only the terms in the assertion. $\square$

Since the $y_{1,i}$ are central, we can rewrite

$$\hat{P}'(x_{11}, \ldots, x_{1p}) = \sum_{\sigma_1 \in S_p} z_{1,\sigma(1)} y_{1,\sigma(2)} \cdots y_{1,\sigma(p)}, \tag{9.7}$$

which has the same form as $\hat{P}''$. Writing $\hat{P}$ for $\hat{P}'$, we see that the multilinearization of P_m is

$$\hat{P}(x_{1,1}, \ldots, x_{1,p}) \hat{P}(x_{2,1}, \ldots, x_{2,p}) \ldots \hat{P}(x_{2m,1}, \ldots, x_{2m,p}). \tag{9.8}$$

Theorem 9.4.12. *(char $F = p > 2$.) P_n does not belong to the T-space generated by the P_m, $\forall m < n$.*

Proof. Since $P_n(\bar{x}_1, \bar{x}_2, \ldots \bar{x}_{2n}) = 2^n \bar{y}_1^{p-1} \bar{z}_1 \bar{y}_2^{p-1} \bar{z}_2 \cdots \bar{y}_{2n}^{p-1} \bar{z}_{2n}$, we shall show that for any $m < n$, the part of any evaluation $P_m(u_1, \ldots, u_{2m})$ of P_m which has degree precisely $p-1$ in each $\bar{y}_i$ and degree 1 in each $\bar{z}_i$, $1 \le i \le n$, must be 0. This will show that P_n cannot be a consequence of P_m for $m < n$, thereby yielding the desired result.

As in Remark 9.1.1, this can be checked in two steps: First when the u_i is a word in $\tilde{A}$, and second to show each partial linearization of P_m lacks the desired terms. For any word v, we write v' for the odd part and v'' for the even part. Thus, $\bar{x}_i' = \bar{z}_i$ and $\bar{x}_i'' = \bar{y}_i$.

Step I. We check the assertion for any substitutions of words. Suppose some u_j is a word of degree > 1, i.e., $u_j = \bar{x}_i v_j$. We claim that P_m then has value 0. For notational simplicity, we assume that j is odd, i.e., $j = 2k - 1$, and have

$$P(u_j, u_{j+1}) = P(\bar{x}_i v_j, u_{j+1}) = 2(\bar{x}_i v_j)' u_{j+1}'((\bar{x}_i v_j)'')^{p-1}(u_{j+1}'')^{p-1}.$$

Then $(\bar{x}_i v_j)' = \bar{x}_i' v_j'' + \bar{x}_i'' v_j' = \bar{z}_i v_j'' + \bar{y}_i v_j'$, and $(\bar{x}_i v_j)'' = \bar{y}_i v_j'' + \bar{z}_i v_j'$, so we get

$$2(\bar{z}_i v_j'' + \bar{y}_i v_j') u_{j+1}'(\bar{y}_i v_j'' + \bar{z}_i v_j')^{p-1}(u_{j+1}'')^{p-1}.$$

But

$$(\bar{z}_i v_j'' + \bar{y}_i v_j') u_{j+1}' \bar{z}_i v_j' = 0$$

since it is a sum of terms in which either the odd element $\bar{z}_i$ or v_j' repeats, so we are left with

$$2(\bar{z}_i v_j'' + \bar{y}_i v_j') u_{j+1}'(\bar{y}_i v_j'')^{p-1}(u_{j+1}'')^{p-1},$$

or

$$2(\bar{z}_i \bar{y}_i^{p-1} v_j''^p + \bar{y}_i^p v_j''^{p-1} v_j') u_{j+1}'(u_{j+1}'')^{p-1},$$

each term of which is 0 by Lemma 9.4.3. (This argument could be streamlined a bit using Lemma 9.2.14.)

In other words, for word specializations, the only way to get a nonzero value in P_m is each u_j to be linear in the $\bar{z}$, i.e., $\bar{z}_i \bar{y}_i^{p-1}$. In this case, the degree is $2m < 2n$, and we cannot cancel P_n. Then we can delete any linear substitution and continue by induction. Thus, we may assume that there are no word specializations.

So it remains to prove:

Step 2. Every p-admissible word specialization in $\tilde{A}$ of any partial linearization (cf. Definition 1.2.12) of P_m is 0. Since we have introduced the 2-graded structure into our considerations, we consider a partial linearization f of P_m in the super-sense, i.e., take f to be a homogeneous component of

$$\ldots (z_{i,1} + z_{i,2})(y_{i,1} + y_{i,2})^{p-1} - z_{i,1}y_{i,1}^{p-1} - z_{i,2}y_{i,2}^{p-1} \ldots,$$

i.e., the total i, 1-degree (in y_i and z_i together) is some k_i and the i, 2-degree is $p - k_i$, where $0 < k_i < p$, and $1 \leq i \leq 2m$. We consider substitutions (from $\tilde{A}$) in f in total degree p in each $\bar{x}_i$, and need to show, in any such substitution the part of degree $p-1$ in each $\bar{y}_i, i \leq n$, and degree 1 in each $\bar{z}_i, i \leq n$, must be 0. In order to avoid restating this condition, we call this the "prescribed part" of the substitution.

In view of Remark 1.2.13(iv), since all indeterminates have degree $\leq p$, it is enough to check all (total) linearizations of f in the indeterminates in which we do not make word specializations. But we already ruled out word specializations, so we may assume that f is the multilinearization of P_m; in other words, we may assume that f has the form (9.7), which we write more concisely as

$$\prod_{j=1}^{2n} \hat{P}(x_{j1}, \ldots, x_{jp}).$$

It suffices to check that all word specializations of each $x_{j1}, \ldots, x_{jp}$ have prescribed part 0. We shall prove this for all m (even including $m \geq n$), and argue by induction on n.

So we consider the various substitutions

$$x_{jk} \mapsto w_{jk} = \bar{x}_{jk1} \cdots \bar{x}_{jkt},$$

where $t = t(j, k)$ depends on j and k. Now each $\bar{x}_{jk\ell} = \bar{y}_{jk\ell} + \bar{z}_{jk\ell}$, the sum of its even and odd parts, and we have to consider all terms arising from one or the other. Toward this end, we define the **choice vector** $\mathbf{m} = (m_{jk\ell})$, where $m_{jk\ell}$ is 0 if we are to choose $\bar{y}_{jk\ell}$ (the even part) and $m_{jk\ell}$ is 1 if we are to choose $\bar{z}_{jk\ell}$ (the odd part) taken in v_{jk}.

Thus, we define $\bar{x}_{jk\ell}^{\mathbf{m}}$ to be $y_{jk\ell}$ if $m_{jk\ell} = 0$, and to be $z_{jk\ell}$ if $m_{jk\ell} = 1$. Also we write $w_{jk}^{\mathbf{m}}$ for

$$\bar{x}_{jk1}^{\mathbf{m}} \cdots \bar{x}_{jkt}^{\mathbf{m}}.$$

Thus, we have

$$w_{jk} = \sum_{\mathbf{m}} w_{jk}^{\mathbf{m}},$$

summed over the 2^t possibilities for $\mathbf{m}$. Of course, many of these automatically give 0 in f. We consider only those substitutions for which

$$\sum_{sg} w^{\mathbf{m}}_{j,\sigma(1)} w^{\mathbf{m}}_{j,\sigma(2)} \cdots w^{\mathbf{m}}_{j,\sigma(p)}$$

could be nonzero.

For example, by Lemma 9.4.11, when $k = \sigma(1)$, we must take $\bar{z}_{jk\ell}$ an odd number of times, and for all other k, we must take $\bar{z}_{jk\ell}$ an even number of times. In other words, to evaluate, we first designate the special k_0 which will go to the first position of (9.7), pick $\bar{z}_{jk_0\ell}$ an odd number of times, pick $\bar{z}_{jk\ell}$ an even number of times for all other k, sum over all such choices, and sum over those σ such that $\sigma(1) = k_0$.

We need to evaluate

$$w_j = \sum_{sg \in S_p} \sum_{\mathbf{m}} w^{\mathbf{m}}_{j,\sigma(1)} w^{\mathbf{m}}_{j,\sigma(2)} \cdots w^{\mathbf{m}}_{j,\sigma(p)}, \qquad (9.9)$$

and show that $w_1 w_2 \ldots w_n = 0$. Note that each summand of $w_1 w_2 \ldots w_n$ is p-admissible; we should always bear this in mind, and this is why we have to take the product (since an individual w_j may not have length a multiple of p, and then the proof would fail.)

We subdivide (9.9) as $S_1 + S_2$, where S_1 is taken over all $\mathbf{m}$ such that $\sum_{\ell=1}^{t(j,k)} m_{jk\ell} \leq 1$ for all j, k, and S_2 is taken over all other $\mathbf{m}$. In other words, when we evaluate our specialization of P_m, which is

$$\hat{P}(w_{1,1}, \ldots, w_{1,p})\hat{P}(w_{2,1}, \ldots, w_{2,p}) \ldots \hat{P}(w_{2m,1}, \ldots, w_{2m,p}), \qquad (9.10)$$

S_1 consists of all terms where from *each* $\hat{P}(w_{j,1}, \ldots, w_{j,p})$, we only take those summands with at most one z, and S_2 consists of all terms containing a summand of *some* $\hat{P}(w_{j,1}, \ldots, w_{j,p})$ that has at most two zs.

We conclude by showing that each of S_1 and S_2 are 0.

Proof that S_1 is 0.

Induction on n. We may assume that each $w^{\mathbf{m}}_{jk}$ has fewer than two odd terms, and thus, as shown above, $w^{\mathbf{m}}_{jk}$ has no odd terms unless $k = \sigma(1)$. Thus, the only nonzero contribution to S_1 comes from

$$x_{jk} \mapsto \bar{y}_{jk1} \ldots \bar{y}_{jkt}, \quad k \neq \sigma(1);$$

$$x_{jk} \mapsto \bar{z}_{jk1}\bar{y}_{jk2} \ldots \bar{y}_{jkt} + \bar{y}_{jk1}\bar{z}_{jk2} \ldots \bar{y}_{jkt} + \cdots + \bar{y}_{jk1}\bar{y}_{jk2} \ldots \bar{z}_{jkt}, \quad k = \sigma(1).$$

Since all $\bar{y}_{jk\ell}$ are central, we can rearrange the letters. Writing d_{ijk} for the degree of $\bar{x}_i$ in the substitution of x_{jk}, and $d_{ij} = \sum_k d_{ijk}$, we see $\hat{P}(x_{j1}, \ldots, x_{jp})$

specializes to

$$\sum_{i=1}^{2n}\sum_{k=1}^{p}\sum_{\sigma\in S_p} d_{ijk}\bar{z}_i\bar{y}_1^{d_{1,j}}\cdots\bar{y}_i^{d_{i,j}-1}\cdots\bar{y}_{2n}^{d_{2n,j}}$$

$$= (p-1)!\sum_{i=1}^{2n}\sum_{k=1}^{p} d_{ijk}\bar{z}_i\bar{y}_1^{d_{1,j}}\cdots\bar{y}_i^{d_{i,j}-1}\cdots\bar{y}_{2n}^{d_{2n,j}}$$

$$= -\sum_{i=1}^{2n} d_{i,j}\bar{z}_i\bar{y}_1^{d_{1,j}}\cdots\bar{y}_i^{d_{i,j}-1}\cdots\bar{y}_{2n}^{d_{2n,j}}.$$

(We have the coefficient $(p-1)!$ because we sum over the $(p-1)!$ permutations σ for which $\sigma(1) = k$, but $(p-1)! \equiv -1 \pmod{p}$.)

At this stage, if $m < n$, then in multiplying these to get our specialization of f, we see that in each summand a suitable $\bar{z}_i$ does not appear, so we could factor out $\bar{y}_i^p$. Writing $S_1(\hat{x}_i)$ to denote the sum corresponding to the further specialization of the w_{jk} where we erased all occurrences of $\bar{x}_i$ (i.e., substitute 1 for $\bar{x}_i$), we see that

$$S_1 = \sum_{i_1=1}^{n}\bar{y}_{i_1}^p S_1(\hat{x}_{i_1}) - \sum_{i_1,i_2=1}^{n}\bar{y}_{i_1}^p\bar{y}_{i_2}^p S_1(\hat{x}_{i_1},\hat{x}_{i_2})$$

$$+\cdots\pm\sum_{i_1,i_2,\ldots i_{n-m}=1}^{n}\bar{y}_{i_1}^p\bar{y}_{i_2}^p\cdots\bar{y}_{i_{n-m}}^p S_1(\hat{x}_{i_1},\ldots,\hat{x}_{i_{n-m}}).$$

Each of the summands is 0 by induction on n, so it remains to prove $S_1 = 0$ when $m = n$. Taking the product over j we see that f specializes to

$$-\sum d_{i_1,1}\cdots d_{i_{2n},2n}\bar{z}_{i_1}\cdots\bar{z}_{i_{2n}}\bar{y}_1^{p-1}\cdots\bar{y}_{2n}^{p-1}. \tag{9.11}$$

If i repeats, then the same $\bar{z}_i$ occurs twice and the term is 0. Thus, $i_1,\ldots,i_{2n}$ are distinct, and

$$\bar{z}_{i_1}\cdots\bar{z}_{i_{2n}} = \operatorname{sgn}(\pi)\bar{z}_1\cdots\bar{z}_{2n}$$

where $\pi \in S_{2n}$ is the permutation sending $\pi(j) = i_j$. Thus, (9.11) equals

$$-\sum \operatorname{sgn}(\pi)d_{i_1,1}\ldots d_{i_{2n},2n}\bar{z}_1\ldots\bar{z}_{2n}\bar{y}_1^{p-1}\cdots\bar{y}_{2n}^{p-1}$$

$$= -\det(d_{i,j})\bar{z}_1\ldots\bar{z}_{2n}\bar{y}_1^{p-1}\cdots\bar{y}_{2n}^{p-1}.$$

But $\sum d_{i,j} = 0 \pmod{p}$, so the determinant is 0 (since $\operatorname{char}(F) = p$), yielding the desired result.

In summary, the proof of this case boils down to showing that a certain determinant that calculates the coefficients of S_1 is always 0 modulo p.

Proof that S_2 is 0.

The idea here is to pull out double occurrences of z, and thus reduce to the previous case via induction. Unfortunately, the argument is extremely delicate, since our reduction involves counting certain terms more than once; we need to show that the extraneous terms we counted also total 0, by means of a separate argument.

By assumption, we are only counting terms where some w_{jk} involves at least two odd substitutions. Fixing this j, k, we make the following claim.

Claim 1. *For any $t' \geq 2$, the sum of terms in which*

$$m_{j,k,\ell_1} = m_{j,k,\ell_1+1} = 1$$

for specified ℓ_1 (i.e., in which v_{jk} has odd terms in the two specified positions $(\ell_1, \ell_1 + 1)$) is 0.

Indeed, written explicitly,

$$w_{jk} = \bar{x}_{jk1}^{\mathbf{m}} \cdots \bar{x}_{jk\ell_1-1}^{\mathbf{m}} \bar{z}_{jk\ell_1} \bar{z}_{jk\ell_1+1} \bar{x}_{jk\ell_1+2}^{\mathbf{m}} \cdots \bar{x}_{jk\ell_t}^{\mathbf{m}}.$$

But $\bar{z}_{jk\ell_1} \bar{z}_{jk\ell_1+1}$ is even and thus central so we can write

$$w_{jk} = \bar{z}_{jk\ell_1} \bar{z}_{jk\ell_1+1} w_{jk}^{\mathbf{m}'}$$

where

$$w_{jk}^{\mathbf{m}'} = \bar{x}_{jk1}^{\mathbf{m}} \cdots \bar{x}_{jk\ell_1-1}^{\mathbf{m}} \cdots \bar{x}_{jk\ell_2-1}^{\mathbf{m}} \bar{x}_{jk\ell_2}^{\mathbf{m}} \cdots \bar{x}_{jk\ell_{t'}}^{\mathbf{m}}, \ldots.$$

Since $\bar{z}_{jk\ell_1} \bar{z}_{jk\ell_{t'}}$ is central, so we could pull it out of the evaluation of f, i.e.,

$$\hat{P}(w_{j1}^{\mathbf{m}}, \ldots, w_{jp}^{\mathbf{m}}) = \bar{z}_{jk\ell_1} \bar{z}_{jk\ell_2} \hat{P}(w_{j1}^{\mathbf{m}}, \ldots, w_{jk}^{\mathbf{m}'}, \ldots, w_{jp}^{\mathbf{m}}).$$

But $\bar{x}_{j,k,\ell_1}$ is some $\bar{x}_{i_1}$, $\bar{z}_{j,k,\ell_1+1}$ is some x_{i_2}. Having used up the odd occurrences of $\bar{x}_{i_1}$, $\bar{x}_{i_2}$, we only have even occurrences left in the prescribed part, which can also be pulled out of those v's in which they appear. Writing v_{jk} for that part of $w_{jk}^{\mathbf{m}}$ obtained by deleting all occurrences of $\bar{x}_{i_1}$, $\bar{x}_{i_2}$, we see that

$$\hat{P}(w_{j1}^{\mathbf{m}}, \ldots, w_{jp}^{\mathbf{m}}) = \bar{z}_{i_1} \bar{y}_{i_1}^{p-1} \bar{z}_{i_2} \bar{y}_{i_2}^{p-1} \hat{P}(v_{j1}, \ldots, v_{jp}), \qquad (9.12)$$

which involves two fewer indeterminates. Thus, the prescribed part in (9.12) is 0, proving Claim 1.

Next, recall that $[x_i, x_j] = 2z_i z_j$, so Claim 1 implies that we get 0 when we take the sum in (9.9) over all terms containing a commutator $[\bar{x}_{i,j,\ell}^{\mathbf{m}}, \bar{x}_{i,j,\ell+1}^{\mathbf{m}}]$. But

$$\bar{x}_{i,j,\ell}^{\mathbf{m}} \bar{x}_{i,j,\ell+1}^{\mathbf{m}} = [\bar{x}_{i,j,\ell}^{\mathbf{m}}, \bar{x}_{i,j,\ell+1}^{\mathbf{m}}] + \bar{x}_{i,j,\ell+1}^{\mathbf{m}} \bar{x}_{i,j,\ell}^{\mathbf{m}},$$

so this proves that the sum in (9.9) is still the same if we interchange two positions. This proves:

Claim 2. *For any ℓ_1, ℓ_2, the sum of terms in which $m_{j,k,\ell_1} = m_{j,k,\ell_2} = 1$ for specified ℓ_1, ℓ_2 (i.e., in which $w_{jk}^{\mathbf{m}}$ has odd letters in two specified positions ℓ_1, ℓ_2) is 0.*

Since any term in S_2 has some $m_{j,k,\ell_1} = m_{j,k,\ell_2} = 1$, by definition of S_2, we would like to conclude the proof by summing over all such possible pairs (ℓ_1, ℓ_2). Unfortunately, letting

$$\bar{S}_2 = \sum_{\sigma \in S_p} \sum_{m_{j,k,\ell_1} = m_{j,k,\ell_2} = 1} w^{\mathbf{m}}_{j\sigma(1)} \cdots w^{\mathbf{m}}_{j,\sigma(p)},$$

we see that $\bar{S}_2 - S_2$ consists of terms which are doubly counted, and we want to show that this discrepancy also is 0. Toward this end, for any $(\ell_1, \ldots, \ell_q)$, we define $S_{j,k,(\ell_1,\ldots,\ell_q)}$ to be the sum of terms in S_2 for which

$$m_{j,k,\ell_1} = \cdots = m_{j,k,\ell_q} = 1,$$

since this is the source of our double counting.

Claim 3. $S_{j,k,(\ell_1,\ldots,\ell_q)} = 0$ *whenever $q \geq 2$ is even.*

Claim 3 is proved analogously to Claim 1, this time drawing out $\frac{q}{2}$ pairs of odd letters at a time from $w^{\mathbf{m}}_{jk}$ instead of a single pair.

Now define $\bar{S}_{j,k,q} = \sum_{(\ell_1,\ldots,\ell_q)} S_{j,k,(\ell_1,\ldots,\ell_q)}$. Clearly $S_2 = \sum c_{j,k,q} \bar{S}_{j,k,q}$ for suitable integers $c_{j,k,q}$. Thus, it remains to prove:

Claim 4. $c_{j,k,q} = 0$ *for q odd.*

Given Claim 4, we see that S_2 is summed over $\sum c_{j,k,q} \bar{S}_{j,k,q}$ for q even, so is $\sum c_{j,k,q} 0 = 0$, proving the theorem. So it remains to prove Claim 4, which we do by showing that for q even, the computation of

$$c_{j,k,2} \bar{S}_{j,k,2} + c_{j,k,4} \bar{S}_{j,k,4} + \cdots + c_{j,k,q} \bar{S}_{j,k,q}$$

counts each $(2m+1)$-tuple exactly once.

For example, consider $\bar{S}_{j,k,3}$, i.e., suppose $w^{\mathbf{m}}_{jk}$ has (at least) three odd letters, which we assume are in positions $\ell = 1, 2, 3$. Then we counted $z_{j,k,1} z_{j,k,2} z_{j,k,3}$ three times (the number of times we can choose two places of three). But when we moved $z_{j,k,1} z_{j,k,3}$ out first, we had to move $z_{j,k,3}$ past $z_{j,k,2}$, which reverses the sign, so in counting the duplication, we have twice with sign $+1$ and once with sign -1, so the total is $1 + 1 - 1 = 1$, and there is no extra duplication.

When we try to apply this argument in case $w^{\mathbf{m}}_{jk}$ has (at least) four odd letters, there is extra duplication, as seen in considering $z_{j,k,1} z_{j,k,2} z_{j,k,3} z_{j,k,4}$. Here we choose two of four, yielding six possibilities, four of which can be pulled out of $w^{\mathbf{m}}_{jk}$ without changing sign (odd positions in $(1,2)$, $(2,3)$, $(3,4)$, or $(1,4)$) and two of which reverse the sign (odd positions in $(1,3)$ or $(2,4)$). Hence, the sums are counted twice.

Next consider $q = 5$. Now we have to worry about the duplication arising from counting $\bar{S}_{j,k,2}$ and counting $\bar{S}_{j,k,4}$. We need to prove that the duplication for each is the same as for $t(j,k) = 4$, for all the duplications to cancel. This is not too difficult to see directly. The contribution from counting $\bar{S}_{j,k,2}$ is the number of pairs of the form $(i, i + 1)$ or $(i, i + 3)$ minus the number of pairs of the form $(i, i + 2)$ or $(i, i + 4)$, which is

$$(4 + 2) - (3 + 1) = 6 - 4 = 2,$$

the same as the contribution from counting $\overline{S_{j,k,2}}$ for $q = 4$. Counting $\overline{S_4}$ for $q = 5$ gives $+$ for (1,2,3,4), (1,2,4,5), and (2,3,4,5) and $-$ for (1,2,3,5) and (1,3,4,5), for a total of $3 - 2 = 1$, the same as for $q = 4$.

To conclude the proof of Claim 4 in general we need to show for any numbers t, u that the contribution arising in counting $\overline{S_{2u}}$ for $q = 2t$ is the same as the contribution arising in counting $\overline{S_{2u}}$ for $q = 2t + 1$.

As before, the contribution arises in the number of switches we need to make to move all the odd terms to the left side; we multiply by (-1) raised to the number of such switches. Given odd positions in $\ell_1, \ell_2, \ldots, \ell_{2u}$, listed in ascending order, we could compute the number of such switches (to arrive at position $(1, 2, \ldots, 2u)$) to be

$$\ell_1 - 1 + \ell_2 - 2 + \cdots + \ell_{2u} - 2u = \sum_{i=1}^{2u} \ell_i - u(2u + 1).$$

Since the crucial matter is whether this is even or odd, we can replace $u(2u+1)$ by u and define:

$\gamma(u, q, 0)$ is the number of $(\ell_1, \ell_2, \ldots, \ell_{2u})$ (written in ascending order) with $\ell_{2u} \leq q$, such that $(\sum_{i=1}^{2u} \ell_i) - u$ is even.

$\gamma(u, q, 1)$ is the number of $(\ell_1, \ell_2, \ldots, \ell_{2u})$ (written in ascending order) with $\ell_{2u} \leq q$, such that $(\sum_{i=1}^{2u} \ell_i) - u$ is odd.

$\delta(u, q) = \gamma(u, q, 0) - \gamma(u, q, 1)$.

Then $\delta(u, q)$ is the number of times we are counting terms in $\overline{S_{2u}}$ with q odd letters. We showed above that $\delta(2, 2) = \delta(2, 3)$, $\delta(2, 4) = \delta(2, 5)$, and $\delta(4, 4) = \delta(4, 5)$; we need to prove that $\delta(u, 2m) = \delta(u, 2m + 1)$ for each m.

Note that the difference between $\gamma(u, 2m, 0)$ and $\gamma(u, 2m + 1, 0)$ arises precisely from those sequences $(\ell_1, \ell_2, \ldots, \ell_{2u})$ for which $\ell_{2u} = 2m+1$; likewise for $\gamma(u, 2m, 1)$ and $\gamma(u, 2m + 1, 1)$.

Thus, we need only consider sequences $(\ell_1, \ell_2, \ldots, \ell_{2u-1}, \ell_{2m+1})$ used in the computation of $\delta(u, 2m + 1)$. Since $2m + 1 - u$ is a constant that appears in all the sums, we need to prove:

Claim 5. *Let $\beta(u, m, 0) =$ the number of sequences $(\ell_1, \ell_2, \ldots, \ell_{2u-1})$ with $\ell_{2u-1} \leq 2m$ and $\sum \ell_i$ even, and*

$\beta(u, m, 1) =$ *the number of sequences $(\ell_1, \ell_2, \ldots, \ell_{2u-1})$ with $\ell_{2u-1} \leq 2m$ and $\sum \ell_i$ odd.*

Then $\beta(u, m, 0) = \beta(u, m, 1)$.

(Note that for this claim to be non-vacuous we must have $u \leq m$.) We prove this by induction on m. For $m = 1$ it is trivial, since then $u = 1$ and there are two numbers to choose from, namely 2, which is even, and 1, which is odd.

We say $\ell_1, \ell_2, \ldots, \ell_{2u-1}$ has:

type 1 if $\ell_{2u-1} \leq 2(m-1)$;

type 2 if $\ell_{2u-2} \leq 2(m-1)$ but $\ell_{2u-1} \geq 2m-1$;

type 3 if $\ell_{2u-2} = 2m-1$ and so $\ell_{2u-1} = 2m$.

In type 1 we are counting $\beta(u, m-1, 0)$ and $\beta(u, m-1, 1)$, which are equal by induction on m.

In type 3 we are counting $\beta(u-1, m-1, 0)$ and $\beta(u-1, m-1, 1)$, which also are equal by induction on m.

So it remains to consider type 2. In this case there is a 1:1 correspondence between the sequences with $\ell_{2u-1} = 2m-1$ and the sequences with $\ell_{2u-1} = 2m$ (obtained by switching ℓ_{2u-1} from $2m-1$ to $2m$), which obviously have opposite parities, and so their contributions cancel. This concludes the proof of Claim 4 and thus of Theorem 9.4.12. $\qquad\square$

Corollary 9.4.13. *The T-space generated by $\{P_n : n \in \mathbb{N}\}$ is not finitely based.*

9.4.4   The next test algebra

Unfortunately, as noted earlier, in order to obtain a non-finitely based T-ideal, we need to pass to another test algebra that does not satisfy the Grassmann identity. For convenience we introduce notation for $\tilde{A}$ of Definition 9.4.10, which we shall carry to the end of this section. We write the image of x_{2i-1} as a_i and the image of x_{2i} as b_i. Thus, $\tilde{A}$ is generated by $\{a_i, b_i : i \in \mathbb{N}\}$, and any word product in which either a_i or b_i repeats p times is 0.

We extend the algebra $\tilde{A}$ to an algebra $\widehat{A}$ by adjoining new elements e, f, t to A, satisfying the following relations where $q = p^2$:

$$t^{4q+1} = 0, \quad [t, a_i] = [t, b_i] = 0.$$
$$[t, fe] = [fe, a_i] = [fe, b_i] = 0,$$
$$a_i e = b_i e = te = e^2 = 0,$$
$$fa_i = fb_i = ft = f^2 = 0,$$

and any word with more than q occurrences of e or more than q occurrences of f is 0.

As before, we shall exhibit a countably infinite set of polynomials whose T-ideal in $\widehat{A}$ is not finitely based, and so the T-ideal itself is not finitely based. Before defining these polynomials, let us list some properties of $\widehat{A}$, especially in connection with the Grassmann polynomial.

Remark 9.4.14.

(i) *The elements t and fe centralize $\tilde{A}$, and e, f, t are nilpotent.*

(ii) For any $c \in \{a_i, b_j, e, f, t\}$, $ce = 0$ unless $c = f$, and $fc = 0$ unless $c = e$.

(iii) Each nonzero word in $\widehat{A}$ can be written as

$$w = e^{\varepsilon_1} s v f^{\varepsilon_2}$$

where $\varepsilon_1, \varepsilon_2 \in \{0, 1\}$, $s = t^{k_1}(fe)^{k_2}$ for suitable k_1, k_2, and v is a product of various a_i and b_j. (This follows at once from (i) and (ii).)

(iv) In (iii), if w has respective degree d_1, d_2 in e, f, then letting $d = \min\{d_1, d_2\}$, we see that $k_2 = d$ and $\varepsilon_i = d_i - d$ for $i = 1, 2$.

Definition 9.4.15. *The word w in $\widehat{A}$ is called a p-**word** if v is a p-monomial in $\tilde{A}$. A p-**element** of $\widehat{A}$ is a sum of p-words.*

Remark 9.4.16. *In view of (ii) and (iii) of Remark 9.4.14, every commutator of elements of $\widehat{A}$ has one of the forms (up to reversing the order of the commutator, which means multiplying by -1):*

$$[es_1 v_1, es_2 v_2] = 0;$$

$$[s_1 v_1 f, s_2 v_2 f] = 0;$$

$$[es_1 v_1, s_2 v_2] = es_1 s_2 v_1 v_2;$$

$$[s_1 v_1 f, s_2 v_2] = -s_1 s_2 v_2 v_1 f;$$

$$[es_1 v_1 f, es_2 v_2 f] = es_1 v_1 f es_2 v_2 f - es_2 v_2 f es_1 v_1 f = efes_1 s_2 [v_1, v_2] f;$$

$$[es_1 v_1, s_2 v_2 f] = es_1 v_1 s_2 v_2 f - s_2 v_2 f es_1 v_1;$$

$$[es_1 v_1, es_2 v_2 f] = -es_2 v_2 f es_1 v_1 = es_1 s_2 fev_2 v_1;$$

$$[s_1 v_1 f, es_2 v_2 f] = s_1 v_1 f es_2 v_2 f = s_1 s_2 fev_1 v_2 f.$$

We shall see that the last four equations are irrelevant, since their degrees in e and f are too high. Take new indeterminates $\tilde{e}, \tilde{f}, \tilde{t}$. We want to see when $[[\tilde{e}, \tilde{t}], \tilde{t}]$ can have a nonzero specialization on $\widehat{A}$. Let us linearize this to

$$[[\tilde{e}, \tilde{t}_1], \tilde{t}_2] + [[\tilde{e}, \tilde{t}_2], \tilde{t}_1].$$

Lemma 9.4.17. *Any nonzero word specialization of*

$$[[\tilde{e}, \tilde{t}_1], \tilde{t}_2]$$

to $\widehat{A}$ must have some occurrence of e or f.

Proof. $[[x_1, x_2], x_3]$ is an identity of $\tilde{A}$ and thus of $\tilde{A}[t]$, which is the subalgebra of $\widehat{A}$ without e or f. $\qquad\square$

9.4.5 The counterexample

Definition 9.4.18. *Let* $q = p^2$, *and*

$$Q_n = [[\tilde{e}, \tilde{t}], \tilde{t}] \prod_{i=1}^{n} P_n(x_1, \ldots, x_{2n}) \left([\tilde{t}, [\tilde{t}, \tilde{f}]] [[\tilde{e}, \tilde{t}], \tilde{t}] \right)^{q-1} [\tilde{t}, [\tilde{t}, \tilde{f}]],$$

where $P_n = P(x_1, x_2) \cdots P(x_{2n-1}, x_{2n})$, *with* $P(x_1, x_2) = x_1^{p-1}[x_1, x_2]x_2^{p-1}$.

Theorem 9.4.19. *The T-ideal generated by the polynomials $\{Q_n : n \in \mathbb{N}\}$ is not finitely based on $\widehat{A}$ (and so is not a finitely based T-ideal).*

Lacking a shortcut generalizing the proof of Example 9.3.2, we return to the generic approach based on Remark 9.1.1, which is to verify it first for words and then for partial multilinearizations.

9.4.5.1 Specializations to words

Let us fix arbitrary specializations $\hat{e}, \hat{f}, \hat{t}_i$ in $\widehat{A}$ of $\tilde{e}, \tilde{f}$, and $\tilde{t}_i$.

Lemma 9.4.20. *(i) In any nonzero word specialization in $\widehat{A}$ from partial linearizations of Q_n, the specializations $\bar{x}_i$ of x_i (in P_n) are in $\tilde{A}[t]$, and the evaluation c of P_n is a sum of terms each of which contains suitable a_j or b_j. Furthermore the following conditions must hold:*
e has degree 1 in $[[\hat{e}, \hat{t}_1], \hat{t}_2] + [[\hat{e}, \hat{t}_2], \hat{t}_1]$.
f has degree 1 in $[\hat{t}_1, [\hat{t}_2, \hat{f}]] + [\hat{t}_2, [\hat{t}_1, \hat{f}]]$.

(ii) In each substitution, any word containing a_i or b_i has multiplicity $< p$.

Proof. (i) By Lemma 9.4.17, e and f must have a total of at least $2q$ occurrences, and by the last line defining the relations of $\widehat{A}$, this must mean e and f each appear exactly q times, i.e., each $[[\hat{e}, \hat{t}], \hat{t}]$ has degree exactly 1 in e (or f) in any term contributing to the evaluation.

This does not leave any room for e, f in the $\bar{x}_i$, so these are in $\tilde{A}[t]$. If a_j or b_j does not appear in $\bar{x}_{2i-1}$, then $\bar{x}_{2i-1} \in F[t]$, so $[\bar{x}_{2i-1}, \bar{x}_{2i}] = 0$, yielding the first assertion.

We also need $[\hat{t}_{i_1}, [\hat{t}_{i_2}, \hat{f}]][[\hat{e}, \hat{t}_{i_3}], \hat{t}_{i_4}] \neq 0$. This means $[\hat{t}_{i_1}, [\hat{t}_{i_2}, \hat{f}]]$ either starts with e or ends with f. But if it starts with e, it is annihilated by the evaluation of c that precedes it. Thus, the only relevant contribution of $[\hat{t}_{i_1}, [\hat{t}_{i_2}, \hat{f}]]$ ends with f. The same argument shows that the only relevant contribution of $[[\hat{e}, \hat{t}_{i_3}], \hat{t}_{i_4}]$ starts with e.

(ii) By Remark 9.4.14(iii), all the appearances come together in some subword $v \in \tilde{A}$, so we are done by the defining relations of $\tilde{A}$. $\qquad \square$

The specialization $\tilde{e} \mapsto \hat{e}$, $\tilde{f} \mapsto \hat{f}$, $\tilde{t}_i \mapsto \hat{t}_i$ is called *good* if no a_i, b_i appear in $\hat{e}, \hat{f}$ or $\hat{t}_i$, and the specialization is *bad* otherwise.

Lemma 9.4.21. *Any bad word specialization in Q_n is 0.*

Proof. Since $\tilde{A}$ satisfies the identity $zv_1zv_2\cdot\ldots\cdot v_qz = 0$, cf. Proposition 9.4.4, any term vanishes if it is obtained by a specialization of $\tilde{t}$ or $\tilde{e}$ or $\tilde{f}$ containing a_i or b_i in q. □

Thus, we may consider only the good word specializations.

Remark 9.4.22. *In any good word specialization, if $[[\hat{e},\hat{t}_1],\hat{t}_2]] \neq 0$, then two of these terms are powers of t and the third is et^j for some j. If $\hat{e}$ starts with e, then $[[\hat{e},\hat{t}_1],\hat{t}_2] = \hat{e}\hat{t}_1\hat{t}_2$. (Similarly if $\hat{t}_1$ starts with $\hat{e}$, since then we switch $\hat{e}$ and $\hat{t}_1$ in the notation.) If $\hat{t}_2$ starts with e, then we get 0, since both $\hat{e}$ and $\hat{f}_1$ would have to be polynomials in t, and thus commute.)*

Thus, we may assume that $\hat{e}$ starts with e, and $\hat{t}_i = t^{j_i}$ for $j = 1,2$ and $\hat{e} = et^{j_3}$. Then

$$[[\hat{e},\hat{t}_1],\hat{t}_2] = et^{j_1+j_2+j_3} = t^{j_1+j_2+j_3-2}[[e,t],t].$$

Likewise, if $\hat{f} = t^{j_4}f$, then

$$[\hat{t}_1,[\hat{t}_2,\hat{f}]] = t^{j_1+j_2+j_4}f.$$

Consequently, any good word specialization from $Q_n(x_i,y_i,\tilde{e},\tilde{f},\tilde{t})$ to $\hat{A}$ will be a product of $Q_n(\bar{x}_i,\bar{y}_i,e,f,t)$ and a power of t, where $\bar{x}_i,\bar{y}_i$ are the images of x_i and y_i.

Remark 9.4.23. *Multiplying any good word specialization of Q_n (or of any of its partial linearizations) by any element of $\hat{A}$ yields zero. Indeed, the evaluation starts with e, so the only question for left multiplication is by f, and we saw this is 0 by the last line of the defining relations for $\hat{A}$.*

Combining Remarks 9.4.22 and 9.4.23, we see that any word specialization in the T-ideal of Q_m has the form

$$t^k e P_m(\bar{x}_1,\ldots,\bar{x}_{2m})(fe)^{q-1}f.$$

These cannot cancel $t^k e P_m(\bar{x}_1,\ldots,\bar{x}_{2m})(fe)^{q-1}f$ in view of Theorem 9.4.12. In particular, the T-ideal generated by $\{Q_n : n \in \mathbb{N}\}$ cannot be finitely generated via good word specializations.

9.4.5.2   Partial linearizations

Having disposed of word specializations, we turn to the partial linearizations.

Lemma 9.4.24. *Any p-admissible specialization of any partial linearization of Q_n is 0.*

Proof. Recall the cyclic shift δ from Remark 2.6.3. We define an analogous operation σ that acts on the qn appearances of the term $[\tilde{t}, [\tilde{t}, \tilde{f}]][\tilde{e}, \tilde{t}], \tilde{t}]$ in

$$Q_n = [[\tilde{e}, \tilde{t}], \tilde{t}] \prod_{i=1}^n P_n(x_1, \ldots, x_n)([\tilde{t}, [\tilde{t}, \tilde{f}]][[\tilde{e}, \tilde{t}], \tilde{t}])^{q-1}[\tilde{t}, [\tilde{t}, \tilde{f}]].$$

The last commutator $[\tilde{t}, [\tilde{t}, \tilde{f}]]$ is paired with the first $[[\hat{e}, \hat{t}], \hat{t}]$. More precisely, we write

$$Q_n = \tilde{u} \prod_{i=1}^n P_n(x_1, \ldots, x_n)(\tilde{v}\tilde{u})^{q-1}\tilde{v},$$

where $\tilde{u} = [\tilde{e}, [\tilde{t}, \tilde{t}]]$ and $\tilde{v} = [\tilde{t}, [\tilde{t}, \tilde{f}]]$, and let $\sigma = \delta^2$ applied to any partial linearization, where δ now is the usual cyclic shift with respect to the alphabet $\{\tilde{u}, \tilde{v}\}$.

The terms of the specialization are partitioned into orbits of the operator σ. Since $q = p^2$, the orbit of a specialization is of order p^j for $0 \leq j \leq 2$.

Case 1. $j = 0$, i.e., the specialization is invariant. This is a word specialization, so is 0 by Lemma 9.4.21.

Case 2. $j \neq 0$. The length of the orbit is a multiple of p. By Lemma 9.4.5 combined with Corollary 9.4.7, applied to the cyclic shift, if any two p-admissible specializations belong to the same orbit, then they are equal in $\hat{A}$; hence their sum is a multiple of p and is thus zero. $\qquad\square$

9.4.5.3 Verification of the counterexample

Having proved the word specialization case and the partial multilinearization case, we can now finish the proof of Theorem 9.4.19.

Proof of Theorem 9.4.19. To show that Q_n is not in the T-ideal generated by the $\{Q_i\}_{i<n}$, we consider the ideal generated by its specializations in $\hat{A}$; since Q_n is admissible, and we can take homogeneous parts, we only consider admissible specializations.

By Lemma 9.4.24, any specialization of $\tilde{e}, \tilde{f}$, or $\tilde{t}$ in a partial linearization of Q_n on $\tilde{A}$ is zero. Thus, we may consider good specializations of Q_n. In view of Remarks 9.4.23 and 9.4.22, we may take the specialization to be $\hat{t} \mapsto t$, $\hat{e} \mapsto e$, and $\hat{f} \mapsto f$ (since other good specializations are proportional) and substitute words for x_i and y_i. This reduces us to substitutions as a T-space, which is covered by Theorem 9.4.12. $\qquad\square$

Note 9.4.25.

(i) *We can unify our examples by noting that the same proof holds for $p = 2$, virtually word for word, if we go back to Section 9.4.2 and define the test algebra $A = \mathcal{G}/I$, where I is the ideal generated by all words h on the $\bar{y}_i, \bar{z}_i$ for which $\deg_i h \geq 4$ for some i.*

(*ii*) *In the example of $p > 2$, we could lower the degrees, and actually take $q = p$. Of course, bad nonzero invariant specializations could arise but, as Shchigolev has noted, their action produces exactly one pth power v^p. So we could replace each polynomial by its product with v^p. On the other hand, the T-ideals could be constructed using products of polynomials Q_m for each $m \geq 1$. The double commutators $[[e, t], t]$ and $[t, [t, f]]$ could be replaced by other polynomials implied by the Grassmann identity. The key was to use a test algebra in which the polynomials from Lemma 9.4.5 are central.*

Exercises for Chapter 9

Identities of Grassmann algebras

1. In any algebra satisfying the Grassmann identity, the terms $[v_1, v_2]wv_1$ and $v_1[v_1, v_2]w$ are equal. (Hint: $[v_1, v_2][w, v_1] = 0$.)

2. If a, b are odd elements of G, then $(a + b)^m = 0$ for any $m > 2$. (Hint: In each word of $(a + b)^m$, either a or b repeats.)

3. In characteristic p, G satisfies the superidentity $(y + z)^p = y^p$ for y even and z odd. (Hint: Lemma 9.4.2.)

4. Any Grassmann algebra in characteristic $p > 2$ satisfies the *Frobenius identity* $(x + y)^p - x^p - y^p$. (Hint: Writing $a = a_0 + a_1$ and $b = b_0 + b_1$, for $a, b \in G$, use Exercise 3 to show

$$(a + b)^p = (a_0 + b_0)^p = a_0^p + b_0^p = a^p + b^p.)$$

5. Derive Exercise 4 directly from the Grassmann identity.

6. The Frobenius identity implies $(x + y)^{p^k} - x^{p^k} - y^{p^k}$ for all $k \geq 1$.

   The extended and generalized Grassmann algebras (for char(F) = 2)

7. $[a, b] = \varepsilon_a \varepsilon_b ab$ for all words a, b in G^+. (Hint: Write $a = e_i a_1$ and proceed by induction, since

$$[a, b] = [e_i a_1, b] = e_i[a_1, b] + [e_i, b]a_1.) \tag{9.13}$$

8. Conclude from Exercise 7 that if v is a word in G^+ and v' is a permutation of the letters of v, then $v^2 = (v')^2$.

9. Every identity of G^+ is a consequence of the Grassmann identity; one such identity is $(x+y)^4 - x^4 - y^4$. Conclude the identity $(x+y)^{2^k} - x^{2^k} - y^{2^k}$ for all $k \geq 2$. (Compare with Exercises 4, 5, 13.)

Generalized Grassmann algebras (for arbitrary characteristic), [DorKV15]

10. Take the commutative ring $C[\varepsilon] = C[\theta, \varepsilon_1, \varepsilon_2, \ldots]$, subject to the relations $\varepsilon_i^2 = \theta\varepsilon_i$ and $\theta^2 = 2$. The **generalized Grassmann algebra** $\mathcal{G}$ is the $C[\varepsilon]$ algebra generated by elements $e_1, e_2, \ldots$, subject to the relations

$$[e_i, e_j] = \varepsilon_i\varepsilon_j e_i e_j$$

for every i, j. (In particular, $\theta\varepsilon_i e_i^2 = \varepsilon_i^2 e_i^2 = 0$.) Equivalently,

$$e_j e_i = (1 - \varepsilon_i\varepsilon_j)e_i e_j.$$

Show that the elements e_j^2 are central, and $\mathcal{G}/\theta\mathcal{G}$ is the extended Grassmann algebra G^+ over $C/2C$.

11. $[e_i, [e_j, e_k]] = 0$ for all i, j, k. $[e_i, e_j][e_m, e_k] + [e_j, e_k][e_i, e_m] = 0$ for all i, j, k, m.

12. $[e_i, e_j][u, e_k] + [e_j, e_k][e_i, u] = 0$ for all $u \in \mathcal{G}$.

13. (Generalizing Exercise 9) Every identity of $\mathcal{G}$ is a consequence of the Grassmann identity. (Hint: First show that all commutators are central, to verify the Grassmann identity, and then reduce an arbitrary identity first to its homogeneous parts and then to 0. Use Exercise 1.)

Non-finitely based T-ideals

14. In characteristic 2, use the extended Grassmann algebra G^+, together with the relations $x_i^3 = 0$. Then x^2 is central, and define

$$\tilde{Q}_n = [[E, T], T] \prod_{i=1}^n x_i^2 ([T, [T, F]][[E, T], T])^3 [T, [T, F]].$$

Carry out the analog to the proof that was given in characteristic $p > 2$, to prove $\{\tilde{Q}_n : n \in \mathbb{N}\}$ generates a non-finitely based T-ideal in characteristic 2.

15. The construction given in the text can be generalized to non-finitely generated T-ideals for varieties of Lie and special Jordan algebras, with virtually the same proofs. Here are the modifications needed for the Jordan case in characteristic $p > 2$.

Let $Q(x, y) = x^{p-1}y^{p-1}[x, y]$, let $\mathfrak{r}[x]$ denote right multiplication by x, and let $\{x, y, z\} = (xy)z - x(yz)$ be the associator. The polynomials are considered as polynomials in operators. The relation $t^{4q+1} = 0$ is replaced by $t^{8q+1} = 0$.

$$Q_n = \{E, T^2, T^2\} \prod_{i=1}^{n} Q(\mathfrak{r}[x_i], \mathfrak{r}[y_i]) \mathfrak{r}[x_i]^2 \cdot$$

$$\cdot \left(\mathfrak{r}(\{E, T^2, T^2\}) \mathfrak{r}[\{T^2, T^2, F\}] \right)^{q-1} \mathfrak{r}[\{T^2, T^2, F\}]$$

(9.14)

Chapter 10

Recent Structural Results

This short chapter is devoted to some recent structural results of a combinatorial nature, obtained by applying Corollary 5.2.6 in combination with "pumping" and techniques from Kemer's theory of Chapter 6.

10.1   Left Noetherian PI-Algebras

Rather than review the full theory of left Noetherian PI-algebras, cf. [Ro80, Chapter 5] and [GoWa89] for example, we jump straight to a recent theorem of Belov that answers a question of Irving, Bergman, and others. It is a lovely example of Kemer's theory developed in Chapter 2, applied to a question which, at the outset, seems totally unrelated to this theory.

Definition 10.1.1. *An algebra A is* **finitely presented** *(as an algebra) if A can be written in the form $F\{X\}/I$ where I is finitely generated as a two-sided ideal.*

Clearly there are only countably many such I when F is a countable field, so the same argument as quoted in Example 1.6.19 shows that there are uncountably many non-finitely presented affine algebras. In fact, uncountably many are weakly Noetherian, cf. Exercise 2.

Our aim in this section is to prove the following theorem:

Theorem 10.1.2. *Every affine left Noetherian PI-algebra A (over an arbitrary commutative ring) is finitely presented.*

(Note that the theorem is false if R is not affine. For example, the field of rational functions $F(\lambda)$ in one indeterminate is commutative Artinian but is not finitely presented. On the other hand, Resco and Small [ReS93] produced an example to show that the theorem is not true without the PI hypothesis, at least in positive characteristic.)

We introduce the following notation. Write $\mathcal{F} = F\{x_1, \ldots, x_\ell\}$. Consider a left Noetherian affine algebra

$$A = F\{a_1, \ldots, a_\ell\} = \mathcal{F}/\mathcal{I} \tag{10.1}$$

for suitable $\mathcal{I} \lhd \mathcal{F}$. We need to show that $\mathcal{I}$ is f.g. as an ideal. Let $\mathcal{I}_n$ be the ideal generated by elements of $\mathcal{I}$ of degree $\leq n$, which is certainly f.g. (by at most the number of words of length $\leq n$), and let $A_n = \mathcal{F}/\mathcal{I}_n$. Thus, there is a natural surjection $\psi_n : A_n \to A$, given by $a + \mathcal{I}_n \mapsto a + \mathcal{I}$. We aim to show that ψ_n is an isomorphism for all large enough n.

Toward this end, we define $\psi_{m,n}$ to be the natural surjection $A_m \to A_n$, for $m < n$, and let

$$\mathcal{K}_{m,n} = \ker \psi_{m,n}. \tag{10.2}$$

If $n_1 \leq n_2$, then $\mathcal{K}_{m,n_1} \subseteq \mathcal{K}_{m,n_2}$, and

$$\bigcup_{n \in \mathbb{N}} \mathcal{K}_{m,n} = \ker \psi_n.$$

Recall the definition $\mathcal{I}^{\mathrm{cl}}$ of Shirshov closure, from Definition 4.7.3. By Proposition 4.7.4, the chain $\{\mathcal{K}_{m,n}^{\mathrm{cl}} : n \in \mathbb{N}\}$ stabilizes, so given m, we have n such that $\mathcal{K}_{m,n}^{\mathrm{cl}} = \ker \psi_n^{\mathrm{cl}} = \ker \psi_n$.

The main technique is a way to use the Noetherian property to encode information into a finite set of relations.

Lemma 10.1.3. *Suppose we are given $b \in A$ arbitrarily. There is some $t \in \mathbb{N}$ (depending on b) and words $w_1, \ldots, w_t$ in the a_i satisfying the property that for any $a \in A$, there are $r_j \in A$ (depending on a as well as b) such that*

$$ba = \sum_{j=1}^{t} r_j b w_j. \tag{10.3}$$

Proof. $M = AbA \subset A$ is a Noetherian A-module, so we can write $M = \sum_{j=1}^{t} Abw_j$ for finitely many words $w_1, \ldots, w_t$ in the a_i, for suitable t. This means for any $a \in A$ that $ba = \sum r_j b w_j$ for suitable $r_j \in A$, yielding (10.3). □

We can push this farther, using the same proof.

Lemma 10.1.4. *For any k, there is some t (depending on k) and words $w_1, \ldots, w_t$ in the a_i satisfying the property that for any $a \in A$, there are $r_j \in A$ such that*

$$va = \sum_{j=1}^{t} r_j v w_j \tag{10.4}$$

for each word v of length $\leq k$, where the $r_j = r_j(a)$ depend on a, but are independent of the choice of v. Furthermore, the relations (10.4) are all consequences of finitely many relations (viewing $A = \mathcal{F}/\mathcal{I}$).

Proof. Consider the direct sum $M = \oplus_v AvA$, where v runs over all words in $a_1, \ldots, a_\ell$ of length $\leq k$. There are finitely many ($\leq (\ell + 1)^k$) of these, and each AvA is an ideal of A and thus f.g. as A-module, implying M is a f.g. A-module. Let $\hat{v}$ be the vector whose v-component is v. Then $M = \sum_w A\hat{v}w$, summed over all words w in $a_1, \ldots, a_\ell$, and since M is Noetherian we can write $M = \sum_{j=1}^t A\hat{v}w_j$ for finitely many words $w_1, \ldots, w_t$ in the a_i, for suitable t. This means for any $a \in A$ that $\hat{v}a = \sum r_j\hat{v}w_j$ for suitable $r_j \in A$, and checking components we get (10.4), with the r_j independent of v.

It remains to show that (10.4) is the consequence of a finite number of relations. Let k' be the maximal length of the w_j. We have a finite number of relations (10.4) where a runs over words in the a_i of length $\leq k' + 1$, since there are only finitely many such words a. From now on, we assume only these relations, but claim they suffice to prove (10.4) for all $a \in A$.

Indeed, by linearity, we need check the assertion only for words a in the a_i, and we proceed by induction on $t = |a|$. The assertion is given for $t \leq k' + 1$. In general, suppose $t > k' + 1$ and write $a = a'a''$ where $|a'| = k' + 1$ and thus, $|a''| = t - (k' + 1)$. By hypothesis (10.4) holds for a', so we can write

$$va' = \sum_{k=1}^t r_k(a')vw_k,$$

and thus,

$$va = va'a'' = \sum_{k=1}^t r_k(a')vw_ka''.$$

But $|w_ka''| \leq k' + (t - (k' + 1)) \leq t - 1$, so by induction

$$r_k(a')v(w_ka'') = \sum_j r_k(a')r_j(w_ka'')vw_j.$$

Thus,

$$va = \sum_{j=1}^t \left(\sum_{k=1}^t r_k(a')r_j(w_ka'') \right) vw_j,$$

as desired, proving the claim, taking

$$r_j(a) = \left(\sum_k r_k(a')r_j(w_ka'') \right),$$

which is independent of v. $\qquad\square$

Recall the definition of "folds" of a polynomial, from Definition 6.4.4, and the designated indeterminates x_i (which alternate in the folds). Also recall Definition 5.2.2, of d-identity, and from Remark 6.6.14 that a polynomial is "monotonic" if the undesignated indeterminates y_j occur in a fixed order, and

Lemma 10.1.5. *Suppose f is a monotonic identity of a Noetherian algebra A. For any $d > 0$, there is n such that f is a d-identity of A_n. Furthermore, if A_n is spanned over its center by words of length $\leq d$, then f is an identity of A_n.*

Proof. Consider $f(\bar{x}_1, \ldots, \bar{x}_m; \bar{y}_1, \ldots, \bar{y}_q)$ for suitable m, q. Taking $\hat{v}$ to be a vector of words $v_1, \ldots, v_m$ in $a_1, \ldots, a_\ell$ of length $\leq d$, we write each $\bar{x}_i$ as some v_{j_i} and apply Lemma 10.1.4 (where $a = \bar{y}_m$) to replace $\hat{v}\bar{y}_m$ by terms ending in the w_j, and next apply the claim to replace the $\hat{v}\bar{y}_{m-1}r_k(\bar{y}_m)$ by terms ending in the w_j, and so forth. After passing through each $\bar{y}_i$, we get a sum of terms of the form

$$r f(\bar{x}_1, \ldots, \bar{x}_m; w_{j_1}, \ldots, w_{j_m}).$$

Applying Lemma 10.1.4 requires only finitely many relations. Since the $\bar{x}_i$ are words in the generators of bounded length, and there are a finite number of w_j, we require only finitely many extra relations declaring that $f(\bar{x}_1, \ldots, \bar{x}_m; w_{j_1}, \ldots, w_{j_m}) = 0$, so for n_0 suitably large (the maximum length of these relations) we see that f is a k-identity of A_n whenever $n \geq n_0$.

The last assertion is clear. $\square$

10.1.1 Proof of Theorem 10.1.2

Our goal is to show that $A_n = A$ for suitably large n. We shall rely heavily on Lemma 10.1.5.

Suppose $\mathcal{J}$ is any ideal of $\mathcal{F}$ not contained in $\mathcal{I}$. Then

$$0 \neq (\mathcal{J} + \mathcal{I})/\mathcal{I} \triangleleft A = \mathcal{F}/\mathcal{I};$$

write $\bar{A} = A/\left((\mathcal{J} + \mathcal{I})/\mathcal{I} \right)$. By Noetherian induction we may assume that the theorem holds for

$$\bar{A} \cong \mathcal{F}/(\mathcal{J} + \mathcal{I}_n).$$

Thus, taking $\mathcal{J}_n$ to be the ideal generated by the elements of $\mathcal{J}$ of degree $\leq n$, and

$$\bar{A}_n = \mathcal{F}/(\mathcal{J}_n + \mathcal{I}_n),$$

we have $\bar{A}_n = \bar{A}$ for large enough n, i.e., $\mathcal{J} + \mathcal{I} = \mathcal{J}_n + \mathcal{I}_n$.

Our goal is to find such an ideal $\mathcal{J}$ such that $\mathcal{J}_n \cap \ker \psi_n = 0$ for all large enough n. Indeed, then we could consider the exact sequences

$$0 \to \mathcal{J}_n \to A_n \to \bar{A}_n \to 0; \tag{10.5}$$

$$0 \to \mathcal{J} \to A \to \bar{A} \to 0. \tag{10.6}$$

Applying ψ_n to (10.5) would yield isomorphisms $\mathcal{J}_n \cong \mathcal{J}$ and $\bar{A}_n \cong \bar{A}$ at the two ends, so ψ_n also would be an isomorphism at the middle, i.e., $A_n = A$, as desired.

So we look for such an $\mathcal{J}$. If $\mathcal{J} = \mathcal{F}h\mathcal{F}$ for some $h \in \mathcal{F}$, then $\mathcal{J}_n = A_n \bar{h} A_n$ for $\bar{h} = h + \mathcal{I}_n$, which is a finite left ideal by Lemma 10.1.3.

Recall by Kemer's Capelli Theorem (Chapter 5) that A satisfies some Capelli identity c_m. Hence, Lemma 10.1.5 shows thats c_m is an m-identity for A_n for all large enough n. In view of Theorem 5.2.7, A_n is a PI-algebra (satisfying c_{2m-1}), for all suitably large n. Thus,

$$\mathrm{id}(A_n) \subseteq \mathrm{id}(A_{n+1}) \subseteq \mathrm{id}(A_{n+2})\dots,$$

so letting $\kappa_n = (t_n, s_n)$ be the Kemer index of A_n, we have $\kappa_n \geq \kappa_{n+1} \geq \dots$. This sequence must reach a minimum at some n, i.e., at this stage

$$\kappa_n = \kappa_{n+1} = \dots = \kappa$$

for some $\kappa = (t, s)$.

Claim. $\kappa = \mathrm{index}(A)$.

This requires some machinery from Chapter 6. (Strictly speaking, that material was formulated over algebraically closed fields of characteristic 0, but as was noted in Remark 6.6.4, the basic definitions as well as the arguments used here are not dependent on the base ring.) Otherwise, in view of Proposition 6.6.11, for any large enough μ, A_n has a Kemer μ-polynomial f such that all of its monotonic components are identities of A. In view of Proposition 6.11.2, taking $d = t$, we may assume that f is not a d-identity of A_n, for some n. But, by Lemma 10.1.5, f is a d-identity of A_n for large enough n. This contradiction shows that $\mathrm{index}(A) = \kappa$, as desired.

Now take a Kemer polynomial f of A which is also a Kemer polynomial of A_n. Thus, there is $h \in f(\mathcal{F})$ whose image in A is not 0. Let $\mathcal{J} = \langle a \rangle$. Write

$$\bar{\mathcal{J}} = (\mathcal{J} + \mathcal{I})/\mathcal{I} \triangleleft A.$$

Note that if $s > 1$, then we may choose $\mathcal{J}$ such that $\bar{\mathcal{J}}_n^2 = 0$, in view of Proposition 6.9.5, and we conclude easily by the above argument. Thus, we may assume that $s = 1$, and again we could use Proposition 6.9.5 to get central polynomials and conclude fairly easily. However, this argument has been laid out only in characteristic 0, so we provide a characteristic-free argument based on the obstruction to integrality (Section 4.4).

Let $\bar{L}_m$ denote the obstruction to integrality for A_m of degree t (recalling that $\kappa = (t, s)$, cf. Definition 4.7.1), and likewise $\bar{L}$ denotes the obstruction to integrality for A of degree t. For large enough m, $\bar{\mathcal{J}} = (\mathcal{J} + \mathcal{I}_m)/\mathcal{I}_m$ is f.g. as a left ideal of A_m, in view of Lemma 10.1.3 (which enables us to express the relations with respect to a finite number of terms). By Theorem I of §1.5, $\bar{I}$ annihilates the obstruction to integrality (to degree t) of A_m, and thus, recalling Equation (10.2), of $A_n = A_m/\mathcal{K}_{m,n}$ for all $n > m$. Hence $\bar{I}$ is a module over $A_n/\bar{L}_n$ for all $n > m$. Taking n large enough, these stabilize, i.e., $A_n/\bar{L}_n \cong A/\bar{L}$. But now $\overline{\mathcal{J}}$ is f.g. as a module over the Noetherian ring $A/\bar{L}$, and thus is a Noetherian module.

Let $\mathcal{K}'_n = \overline{\mathcal{J}} \cap \mathcal{K}_{m,n}$. Then $\{\mathcal{K}'_n : n \in \mathbb{N}\}$ is an ascending chain of submodules of the Noetherian module $\overline{\mathcal{J}}$, so has a maximal member $\mathcal{K}'_n$; now for any $n' \geq n$ we have $\overline{\mathcal{J}} \cap \mathcal{K}_{n,n'} = 0$ since otherwise its preimage in $\mathcal{F}/\mathcal{I}_n$ would be more than L_n, contrary to choice of n. It follows that $\overline{\mathcal{J}} \cap \ker \psi_n = 0$, which is what we needed to show.

10.2   Identities of Group Algebras

Representability plays a special role in the PI-theory of group algebras.

Example 10.2.1. *Suppose G is any group that contains an Abelian group A of finite index. Then $F[G]$ is finite free over the commutative algebra $F[A]$, so is weakly representable in the sense of Definition 1.6.1.*

This easy observation led group theorists to wonder if there are other examples of group algebras satisfying a PI. Isaacs and Passman [IsPa64] proved the striking result that these are the only examples in characteristic 0, although the situation in characteristic $p > 0$ is somewhat more complicated. Their original proof was long, but in characteristic 0 there is a short, simple proof found in Passman [Pas89], which we present here. Let us start by quoting some well-known facts from the theory of group algebras (to be found in [Pas77] or [Ro88b]), which we shall use repeatedly without citation. We start with some characteristic free observations.

One of the key tools is $\Delta = \Delta(G)$, the subgroup of G consisting of elements having only finitely many conjugates. Note that an element $g \in G$ is in Δ iff its centralizer $C_G(g)$ is of finite index in G.

Fact I. *Given $g \in \Delta$, the (finite) sum $\sum h$ of the conjugates of g is an element of $\mathrm{Cent}(F[G])$, and conversely $\mathrm{Cent}(F[G])$ has a base consisting of such elements. Furthermore, if $\sum_{i=1}^{t} h_i \in \mathrm{Cent}(F[G])$, then the set $\{h_1, \ldots, h_t\}$ includes all conjugates of h_1, implying $h_1 \in \Delta$. In particular, $\mathrm{Cent}(F[G]) \subseteq F[\Delta]$.*

Fact II ([Pas77, Lemma 4.1.6]). *For any f.g. subgroup H of Δ, the commutator subgroup H' is torsion.*

Fact III ([Pas77, Theorem 4.2.12], [Ro88b]). *$F[G]$ is semiprime, for any group G and any field F of characteristic 0.*

Lemma 10.2.2. *Suppose $R = F[G]$ has PI-class n, with $\mathrm{char}(F) = 0$. Then $[G : \Delta] \leq n^2$.*

Proof. Take an n^2-alternating multilinear central polynomial h. By Remark G of §1.5, any n^2 elements of R are dependent over $h(R)$. But if h is a sum of say m monomials, any element of $h(G)$ is a sum of at most m elements, each of which thus lies in Δ. Hence, $h(R) \subseteq F[\Delta]$. Thus, $n^2 \geq [G : \Delta]$. $\quad\square$

Lemma 10.2.3. *Suppose $G = \Delta(G)$ and $F[G]$ satisfies a multilinear identity $f = f(x_1, \ldots, x_d)$ with coefficients in $F[Z]$, where Z is a torsion subgroup of $Z(G)$. Then G has a subgroup A of finite index with A' finite.*

Proof. By induction on the number of monomials of the identity f. Write $f = f_1 + f_2$ where f_1 is the sum of monomials in which x_i appears before x_j, and f_2 is the sum of the monomials where x_i appears after x_j. Clearly we could choose i, j such that f_i and f_j are both nonzero, and we assume for convenience that $i = 1$ and $j = 2$. Let Z_0 be the subgroup of Z generated by the support of the coefficients of f. Then Z_0 is a f.g. torsion Abelian group and thus is finite. If $G' \subseteq Z_0$ is Abelian we are done, so we take $v, w \in G$ with $vw \notin Z_0 wv$. Then $H = C_G(v) \cap C_G(w)$ has finite index in G, and for all $h_i \in H$, we have

$$vw f_1(h_1, \ldots, h_d) + wv f_2(h_1, \ldots, h_d) = f(vh_1, wh_2, h_3, \ldots, h_d) = 0,$$

but also

$$vw(f_1(h_1, \ldots, h_d) + f_2(h_1, \ldots, h_d)) = vw f(h_1, \ldots, h_d) = 0,$$

yielding $(vw - wv)f_2(h_1, \ldots, h_d) = 0$. Thus, letting $s = vwv^{-1}w^{-1} \notin Z_0$,

$$(s - 1)f_2(h_1, \ldots, h_d)wv = (s - 1)wv f_2(h_1, \ldots, h_d) = 0, \quad \forall h_i \in H,$$

implying $(s - 1)f_2(h_1, \ldots, h_d) = 0, \forall h_i \in H$.

If $s \notin H$, then matching components in $F[H]$ shows that

$$f_2(h_1, \ldots, h_d) = 0, \quad \forall h_i \in H,$$

i.e., f_2 is an identity of $F[H]$, and we are done by induction (using H instead of G).

Thus, we may assume that $s \in H$, and so $s \in Z(H)$ since v, w both centralize H. But $s \in H$ is torsion, so we replace Z by $Z\langle s \rangle$, a torsion central subgroup of H, and again are done by induction, since $(s - 1)f_2$ is an identity of $F[H]$. (Note that $(s - 1)f_2 \neq 0$ since $s \notin Z_0$.) $\quad\square$

We are ready for the characteristic 0 case.

Theorem 10.2.4. *If $\mathrm{char}(F) = 0$ and $F[G]$ satisfies a PI, then G has a normal Abelian subgroup of finite index.*

Proof. By the Lemma, G has a subgroup $A \subseteq \Delta(G)$ of finite index, such that the commutator subgroup A' is finite; we take A such that $|A'|$ is minimal. We shall prove that $A' = \{1\}$, i.e., that A is Abelian.

Since $F[A]$ is semiprime, we need to show that the image of each element of A' in any prime homomorphic image $R = F[A]/P$ of $F[A]$ is 1. By Corollary D of §1.5, taking $C = \text{Cent}(R)$ and $S = C \setminus \{0\}$, the central localization $S^{-1}R$ is finite over $S^{-1}C$, and thus, is generated by the images of finitely many elements of A; since these each have only finitely many conjugates, their centralizer in A is a subgroup H of finite index, whose image in R/P is central. Clearly $A' \supseteq H'$ so, by hypothesis, $A' = H'$, whose image in R/P is 1.

Any subgroup A of index n contains a normal subgroup of index $\leq n!$, cf. Exercise 3. □

10.3 Identities of Enveloping Algebras

The situation for the enveloping algebra $U(L)$ of a Lie algebra L turns out to be trivial when $\text{char}(F) = 0$. In this case, we define

$$\Delta = \Delta(L) = \{a \in L : [L : C_L(a)] < \infty\},$$

and have the following information in characteristic 0:

Fact IV. *$U(L)$ is a domain (as seen at once from the Poincaré-Birkhoff-Witt Theorem).*

Fact V ([BergP92]). $\text{Cent}(U(L)) \subseteq U(\Delta)$.

Lemma 10.3.1. *Suppose the enveloping algebra $U(L)$ of a Lie algebra L satisfies a PI. Then*

(i) *L has an Abelian Lie ideal of finite codimension.*

(ii) *Any Abelian Lie ideal A of L is central.*

Proof. Remark G of §1.5 implies $[L : \Delta] < \infty$, and also, letting $C = \text{Cent}(U(L)$ and $S = C \setminus \{0\}$, $S^{-1}U(\Delta)$ is finite dimensional over $S^{-1}C$, spanned, say, by $a_1, \ldots, a_m$ in $U(\Delta)$.

(i) Since each element of Δ has centralizer in L of finite codimension, this implies Δ has a subspace A of finite codimension centralizing all a_i, and thus, A is in the center of Δ, and is certainly Abelian.

(ii) $K = S^{-1}U(A)$ is commutative, and thus a finite field extension of $S^{-1}C$. But for any $b \in L$, ad_b induces a derivation δ on K over F, which must be 0 since K/F is separable [Jac80]. This means A is central. □

Theorem 10.3.2 (cf. Lichtman [Lic89]). *The enveloping algebra $U(L)$ of a Lie algebra L over a field F of characteristic 0 satisfies a PI iff L is Abelian.*

Proof. ($\Leftarrow$) $U(L)$ is commutative, so certainly is PI.

($\Rightarrow$) Tensoring by the algebraic closure of F, we may assume that F is algebraically closed. Take an Abelian Lie ideal A of L, of minimal (finite) codimension. We claim that L/A is Abelian. Indeed, otherwise there is some $a \in L/A$, such that ad_a has a nontrivial eigenvector $b \in L/A$; thus, $Fa + Fb$ is a 2-dimensional Lie algebra whose Lie ideal Fb is noncentral, contrary to Lemma 10.3.1(ii).

Now for any $b \in L$, $A + Lb \lhd L$ since L/A is Abelian, and is Abelian by Lemma 10.3.1 (ii), contrary to the maximality of A unless L is Abelian. $\quad\Box$

A similar argument gives a result of Bakhturin, cf. Exercise 5.

Exercises for Chapter 10

1. Any prime PI-ring R with Noetherian center C is a f.g. C-module. (Hint: Take any evaluation $c \neq 0$ of an n^2-alternating central polynomial and note $R \cong Rc$, which is contained in a f.g. C-module, by Theorem 1.5.2.)

2. An uncountable class of non-finitely presented monomial algebras satisfying ACC on ideals: Consider all monomial algebras generated by three elements a, b, c, supported by ba^m, $a^n c$, $ba^q c$, where $m, n \in \mathbb{N}$ are arbitrary, but q is in a specified infinite subset of $\mathbb{N}$ whose complement is also infinite. There are uncountably many such sets, and an uncountable number of such algebras, each of which satisfies ACC on ideals. But there are only countably many finitely presented algebras, so we are left with uncountably many non-finitely presented ones. (Details are as in Example 1.6.19.)

3. If H is a subgroup of G of index n, then H contains a normal subgroup of index $\leq n!$. (Hint: Consider the homomorphism $G \to S_n$ by identifying S_n with the permutations of the cosets gH of H, sending $a \in G$ to the permutation given by $gH \mapsto agH$.)

4. Prove the analog of Lemma 10.3.1 for prime enveloping algebras of restricted Lie algebras.

5. Prove Bachturin's theorem, that the (unrestricted) enveloping algebra of a Lie algebra L in characteristic p satisfies a PI iff L has an Abelian ideal of finite codimension and the adjoint representation of L is algebraic. (Hint: Define $L^p = \{a^p : a \in L\}$, and take $H = \sum_{i \in \mathbb{N}} L^{p^i} \subseteq U(L)$, a restricted Lie algebra, whose restricted enveloping algebra $u(H) = U(L)$ is prime. By Exercise 4, H has an Abelian ideal of finite codimension, which translates to the required conditions on L.)

Chapter 11

Poincaré-Hilbert Series and Gel'fand-Kirillov Dimension

Our aim in this chapter is to study growth of PI-algebras by means of the techniques developed so far. To do this, we shall define the **Hilbert series** of an algebra, otherwise called the Poincaré-Hilbert series, and the important invariant known as the **Gel'fand-Kirillov** dimension, abbreviated as GK dim. The GK dim of an affine PI-algebra A always exists (and is bounded by the Shirshov height), and is an integer when A is representable. (However, there are examples of affine PI-algebras with non-integral GK dim.) These reasonably decisive results motivate us to return to the Hilbert series, especially in determining when it is a rational function. We shall discuss (in brief) some important ties of the Hilbert series to the codimension sequence.

11.1   The Hilbert Series of an Algebra

We extend the definition of graded algebras from §1.3.

Definition 11.1.1. *An algebra A has a* **filtration** *with respect to $\mathbb{N}$ if A has f.d. subspaces $\tilde{A}_0 \subseteq \tilde{A}_1 \subseteq \ldots$ with $\cup_n \tilde{A}_n = A$ and $\tilde{A}_m \tilde{A}_n \subseteq \tilde{A}_{m+n}$. Given an* *algebra A with a filtration, define the* **Hilbert series** *or* **Poincaré series** *to be*

$$H_A = 1 + \sum_{n \geq 1} d_n \lambda^n, \tag{11.1}$$

where $d_n = \dim_F(\tilde{A}_n / \tilde{A}_{n-1})$ and λ is an indeterminate. When A is ambiguous, we write $d_n(A)$ for d_n.

Likewise, an A-module M is **filtered** *if M has f.d. subspaces $\widetilde{M}_n$ with $\cup \widetilde{M}_n = M$ and $\tilde{A}_n \widetilde{M}_m \subseteq \widetilde{M}_{m+n}$. We define*

$$H_M = \sum_{n \geq 0} d_n \lambda^n,$$

where $d_0 = 1$ and $d_n = \dim_F(\widetilde{M}^n / \widetilde{M}^{n-1})$. (Note we are assuming $d_n < \infty$.)

Any finite module $M = \sum A w_j$ over a filtered algebra becomes filtered, by putting $\widetilde{M}_n = \sum_j \tilde{A}_n w_j$. However, at times, it is preferable to choose different filtrations. For example, if $M \lhd A$, one might prefer to take the filtration of M given by $\widetilde{M}_n = M \cap \tilde{A}_n$.

Remark 11.1.2. *A is finite dimensional, iff $\tilde{A}_n = A$ for some n, iff H_A is a polynomial of degree $\geq n$. In this case, $d_n = \dim_F(A_n)$.*

Thus, the theory is of interest mainly for infinite dimensional algebras.

Remark 11.1.3.

(i) *Any $\mathbb{N}$-graded algebra $A = \oplus A_n$ is filtered, where $\tilde{A}_n = \sum_{m \leq n} A_m$. Then*

$$H_A = 1 + \sum_{n \geq 1} \dim A_n \lambda^n. \tag{11.2}$$

(ii) *If A, B are graded algebras over a field F, then under the tensor product grading of Remark 1.3.3,*

$$d_n(A \otimes B) = \sum_{j=0}^{n} d_j(A) d_{n-j}(B),$$

so

$$H_{A \otimes B} = \sum_{j=0}^{n} d_j(A) d_{n-j}(B) \lambda^n = \sum_j d_j(A) \lambda^j \sum_k d_k(B) \lambda^k = H_A H_B.$$

This gives rise to several standard examples.

Example 11.1.4.

(i) $A = F[\lambda]$. *Then* $A_n = F\lambda^n$, *so each* $d_n = 1$, *and*

$$H_A = \sum_{i \geq 0} \lambda^i = \frac{1}{1 - \lambda},$$

a rational function.

(ii) $A = F[\lambda_1, \ldots, \lambda_n] = \otimes_{i=1}^n F[\lambda_i]$. *Then* $H_A = \frac{1}{(1-\lambda)^n}$.

(iii) $A = F\{x_1, \ldots, x_\ell\}$, *the free algebra. Then* A_n *has a base comprised of words of length* n, *so* $d_n(A) = \ell^n$, *which is exponential in* n, *and its Hilbert series*

$$H_A = 1 + \sum_{n \geq 1} \ell^n \lambda^n \tag{11.3}$$

is rational.

Fortunately, the imposition of a polynomial identity makes the growth behavior more like (ii) than like (iii).

Definition 11.1.5. *The* **associated graded algebra** $\operatorname{Gr}(A)$ *of a filtered algebra* A *is defined as the vector space* $\oplus(\tilde{A}^n/\tilde{A}^{n-1})$ *under multiplication*

$$(r_n + \tilde{A}^{n-1})(s_m + \tilde{A}^{m-1}) = r_n s_m + \tilde{A}^{n+m-1},$$

for $r_n \in \tilde{A}^n$ *and* $s_m \in \tilde{A}^m$.

Unless otherwise indicated, $A = F\{a_1, \ldots, a_\ell\}$ will denote an affine algebra, which we shall study in terms of the vector space $V = \sum_{i=1}^\ell Fa_i$. V is called the *generating space* of A, and clearly depends on the choice of generators $a_1, \ldots, a_\ell$. Then A is endowed with the **natural filtration** given by

$$\tilde{A}_n = F + V + V^2 \cdots + V^n. \tag{11.4}$$

(Here V^n is spanned by all $a_{i_1} \cdots a_{i_n}$.) Clearly, V depends on the choice of generators; when V is ambiguous, we write $H_{A;V}$ for H_A. Note

$$\tilde{A}_n/\tilde{A}_{n-1} \cong V^n/(V^n \cap \tilde{A}_{n-1}),$$

so its dimension d_n measures the growth of A in terms of its generators. This will be our point of view throughout, and we are interested in the Hilbert series as a way of determining the asymptotic behavior of the d_n.

Remark 11.1.6 ([KrLe00, Proposition 7.1 and Lemma 6.7]). $\operatorname{Gr}(A)$ *is affine, and every* a_i *is homogeneous of degree 1, by definition.* $H_A = H_{\operatorname{Gr}(A)}$, *seen immediately from the definition.*

11.1.1   Monomial algebras

Definition 11.1.7. *A* **monomial algebra** *is an algebra defined by genera-tors and relations, i.e., $F\{X\}/\mathcal{I}$, where the relation ideal $\mathcal{I}$ is generated by monomials.*

Since monomials are homogeneous in the free algebra under the natural grade given by degree (Example 1.3.14), any monomial algebra inherits this grading. Let us strengthen Remark 11.1.6.

Proposition 11.1.8. *Every affine algebra $A = F\{a_1, \dots, a_\ell\}$ has the same Hilbert series as an appropriate monomial algebra.*

Proof. We need a total order on words, so we order the words first by length and then by the lexicographic order of Chapter 2. (We saw there that any two words of the same length are comparable.) Consider the set S of words in the free algebra $F\{x_1, \dots, x_\ell\}$, which when specialized to A under $x_i \mapsto a_i$ can be written as a linear combination of words of smaller order. Clearly, any word containing a word in S also is in S, so the linear span of S comprises an ideal $\mathcal{I}$ in $F\{x_1, \dots, x_\ell\}$, and $\tilde{A} = F\{x_1, \dots, x_\ell\}/\mathcal{I}$ is by definition a monomial algebra that has the same growth as A. (In the above notation, V^n/V^{n-1} is spanned by elements corresponding to elements of S; conversely, any depen-dence of elements of A can be expressed as writing the largest word as a linear combination of the others.)   □

Unfortunately, the monomial algebra $\tilde{A}$ formed in the proof need not be finitely presented, even when A is finitely presented. Likewise a given PI of A need not pass to $\tilde{A}$. The theory of monomial PI-algebras was studied carefully in [BelBL97], and we shall see interesting examples of monomial PI-algebras below and in Exercise 18.

11.2   The Gel'fand-Kirillov Dimension

As above, we work with an affine algebra $A = F\{a_1, \dots, a_\ell\}$ under the natural filtration of (11.4).

Definition 11.2.1. *The* **Gel'fand-Kirillov** *dimension*

$$\operatorname{GKdim}(A) = \varlimsup_{n \to \infty} \log_n \tilde{d}_n, \tag{11.5}$$

where $\tilde{d}_n = \dim_F \tilde{A}_n$, and $\tilde{A}_n = \sum_{i=0}^n V^i$ is as in (11.4).

GK dim also can be defined for non-affine algebras, but we stick to the affine case.

GKdim(A) can be computed at once from its Hilbert series with respect to a generating set. The Gel'fand-Kirillov dimension was originally introduced by Gel'fand and Kirillov to study growth of f.d. Lie algebras, but quickly became an important invariant of any affine algebra A, being independent of the choice of its generating set (in contrast to the Hilbert series). In particular, we may assume that $1 \in V$, in which case $V^i \subseteq V^j$ for all $j \geq i$, and thus $\tilde{A}_n = V^n$. At times it is convenient to make this assumption; at other times we want to work with a specific generating set which may not include 1.

The standard reference on GK dim is the book by Krause and Lenagan [KrLe00], which also contains the main facts about GKdim for PI-algebras. Our goal here is to enrich that account with some observations based on the theory of Chapters 2 and 6; we shall also study Hilbert series of PI-algebras. Let us recall some basic facts.

Lemma 11.2.2. *The GK dim of the commutative polynomial algebra* $A = F[\lambda_1, \ldots, \lambda_\ell]$ *is* ℓ.

Proof. The space $\tilde{A}_n$ of polynomials of degree $\leq n$ is spanned by the monomials $\lambda_1^{n_1} \cdots \lambda_\ell^{n_\ell}$, where $n_1 + \cdots + n_\ell \leq n$; hence its dimension is at least $(n/\ell)^\ell$ (choosing only those exponents up to $\frac{n}{\ell}$) and at most n^ℓ. Thus

$$\ell(1 - \log_n \ell) \leq \log_n(\tilde{d}_n) \leq \ell,$$

and the limit as $n \to \infty$ is ℓ on both sides. $\square$

Corollary 11.2.3. *For* A *a (commutative) affine integral domain,* $\mathrm{GKdim}(A) = \mathrm{trdeg}_F A$, *an integer.*

For A noncommutative the situation becomes much more complex— $\mathrm{GKdim}(F\{X\})$ is infinite, for example. (There are different levels of infinite growth, which we shall not discuss here, since we shall see that Shirshov's Height Theorem implies that all affine PI-algebras have finite GKdim.)

11.2.1 Bergman's gap

Obviously $\mathrm{GKdim}(A) = 0$ if $[A : F] < \infty$. But if $[A : F]$ is infinite, then $V^{n+1} \supset V^n$ for each n, so $\tilde{d}_n \geq n$ and $\mathrm{GKdim}(A) \geq \varlimsup\limits_{n \to \infty} \log_n n = 1$. Since no affine algebra has GK dim properly between 0 and 1, we are ready for an interesting combinatoric result due to Bergman.

Theorem 11.2.4 (Bergman's gap). *If* $\mathrm{GKdim}(A) > 1$, *then* $\mathrm{GKdim}(A) \geq 2$.

Proof. There are several proofs in the literature, including [KrLe00, Lemma 2.4], but here is a quick proof using the ideas from Chapter 2.

By Proposition 11.1.8, we may assume that $A = F\{a_1, \ldots, a_\ell\}$ is a monomial algebra, so $\tilde{d}_k$ now is the number of nonzero words of length $\leq k$ in $a_1, \ldots, a_\ell$.

Assume $\mathrm{GKdim}(A) > 1$. We claim that there is some hyperword in the a_i

that is not quasiperiodic. Otherwise, if all hyperwords were quasiperiodic, then by the König Graph Theorem (since we have a finite number of a_i), there would also be a bound k on the preperiodicity. But then starting after the k position and applying the König Graph Theorem to the periods, we also would have a bound m on the periodicity. Hence, the number of periods would be at most the number of words of length $\leq m$, which is $(\ell+1)^m$, so any word of length $\geq k$ on the a_i could be extended in at most $(\ell+1)^m$ possible ways, implying

$$\tilde{d}_{k+n} = \dim \tilde{A}_{k+n} \leq n(\ell+1)^m \tilde{d}_k, \tag{11.6}$$

so $\log_n \tilde{d}_{k+n} \leq \log_n n + m\log_n(\ell+1) + \log_n \tilde{d}_k$, whose limit as $n \to \infty$ is 1, contrary to GKdim(A) > 1.

Since there is some hyperword h that is not quasiperiodic, Lemma 2.3.9 implies $\nu_n(h) < \nu_{n+1}(h)$ for every n. But $\nu_1(h) > 1$, so this means that each V^n has dimension $> n$, and

$$\tilde{d}_n \geq 2 + \cdots + n + 1 = \frac{n^2 + 3n}{2};$$

thus GKdim(A) ≥ 2. $\square$

Remark 11.2.5.

(i) *We proved a technically stronger result, that when GKdim(A) > 1, the $\tilde{d}_n$ grow at least at the rate $\frac{n^2+3n}{2}$.*

(ii) *We can turn the argument around to obtain more information about an algebra A with GKdim 1. Namely, A cannot have a hyperword that is not periodic, since that would imply GKdim(A) ≥ 2. Hence, the inequality (11.6) holds, implying the $\tilde{d}_n$ grow at most linearly in n, i.e., there is a constant N such that*

$$\tilde{d}_n - \tilde{d}_{n-1} \leq N, \quad \forall n.$$

*(For example take $N = (\ell+1)^m \tilde{d}_k$.) The smallest such constant N, which we call the **linear growth rate**, plays an important role, cf. Exercise 6.*

Unfortunately, Bergman's gap does not extend to higher GK-dim, and there is an example of a PI-algebra whose GK-dim is any given real number ≥ 2, cf. Example 11.2.9 and Exercise 18. Nevertheless, we shall see soon that GKdim(A) is an integer if A is representable.

Small, Stafford, and Warfield proved that if GKdim(A) = 1, then A is a PI-algebra. Pappacena, Small, and Wald found a nice proof of this theorem for A semiprime, using the linear growth rate, which we sketch in Exercise 6ff. It follows for A prime of GKdim 1 that A is finite over a central polynomial algebra.

There are well-known examples of algebras of GKdim 2 which are not

PI, for example. the Weyl algebra and the enveloping algebra of the nontrivial 2-dimensional Abelian Lie algebra. However, both of these examples are primitive rings. Recently a prime, non-PI example has been found [SmoVi03] that is not primitive.

Braun and Vonessen [BrVo92] extended the Small-Stafford-Warfield theorem to $\mathrm{GKdim}(A) = 2$, in the sense that any prime affine PI-algebra of $\mathrm{GKdim}\,2$ is "Schelter integral" over a commutative subalgebra K, by which we mean that any element satisfies an equation

$$a^n = f(a)$$

where $f \in A\{x\}$ is a generalized polynomial (cf. Definition 1.9.3) and each generalized monomial of f has total degree $< n$ in x. This form of integrality had been studied by Schelter [Sch76] in connection with Shirshov's Theorem, and will be described in Volume II; Braun-Vonessen [BrVo92] shows that this concept is very useful in the study of extensions of arbitrary PI-algebras.

11.2.2   Examples of affine algebras and their GK dimensions

The structure theory of affine PI-algebras is so nice that one begins to expect all of the theory of commutative algebras to carry over to affine algebras. Lewin showed this is not the case, by producing nonrepresentable affine PI-algebras, although, as we have noted, Lewin's argument is based on countability, and does not display an explicit example. Our aim here is to present some general sources of examples: Algebras obtained via affinization, and monomial algebras.

Example 11.2.6. *Suppose $A = F\{a_1, \dots, a_\ell\}$ and write A_j for $A/\mathcal{I}_j$ for $j = 1, 2$, which are affine algebras, generated by some sets V_j. Suppose A is as in (1.24) of Theorem H of §1.5, i.e.,*

$$A \subseteq W = \begin{pmatrix} A_1 & M \\ 0 & A_2 \end{pmatrix}.$$

Then

$$\mathrm{GKdim}(A) \leq \mathrm{GKdim}(A_1) + \mathrm{GKdim}(A_2).$$

(Indeed, writing

$$a_i = \begin{pmatrix} a_{1i} & a_i' \\ 0 & a_{2i} \end{pmatrix},$$

i.e., $a_{ji} = e_{jj} a_i e_{jj}$ and $a_i' = e_{11} a_i e_{22}$, we see that the image of A is contained in the affine subalgebra of W generated by A_1, A_2, and $a_1', \dots, a_\ell'$. Any word of length n has to start with, say, k elements from A_1, then (at most) one element from $\{a_1', \dots, a_\ell'\}$, followed from elements from A_2, so the dimension is at most

$$\sum_{k=0}^{n} (\dim V_1^k)\ell(\dim V_2^{n-k-1});$$

taking logarithms and suprema gives our result.)

Clearly, equality holds in the case of a block triangular algebra, cf. Definition 1.6.11. This is an important example that plays a prominent role in the codimension theory of Giambruno and Zaicev.

11.2.2.1   Affinization

One of the ideas behind Lewin's example shows how to take a "bad" algebra and make it affine. Recall Example 1.6.20 from Chapter 1. Beidar and Small found a way of refining Examples 1.6.20 and 1.6.21, used to advantage by Bell [Bell03].

Example 11.2.7. *Take $T = F\{x,y\}$, the free algebra in indeterminates x and y, and $L = Ty$, a principal left ideal. Then $F + L$ is a free algebra on the infinite set of "indeterminates" $\{x^i y : i \in \mathbb{N}\}$. Define R as in Example 1.6.20. Now suppose W is any prime, countably generated algebra, presumably not affine. We have a natural surjection*

$$\phi : F + L \to W,$$

sending the $x^i y$ to the respective generators of W. Let $P = \ker \phi$, a prime ideal of $F + L$, and $Q_0 = RPe_{11}R$, an ideal of R. Clearly, $e_{11}Q_0 e_{11} = Pe_{11}$, so Zorn's Lemma yields an ideal $Q \supseteq Q_0$ of S maximal with respect to $e_{11}Qe_{11} = Pe_{11}$.

Let us consider some properties of Q.

First, Q is a prime ideal of R, by the maximality hypothesis. Furthermore, if $\mathcal{I} \triangleleft R$ satisfies $e_{11}\mathcal{I}e_{11} \subseteq Pe_{11}$, then $e_{11}(Q+\mathcal{I})e_{11} \subseteq Pe_{11}$ implies $Q+\mathcal{I} = Q$ by maximality of Q, implying $\mathcal{I} \subseteq Q$. Also

$$e_{11}(R/Q)e_{11} \cong (F + L)/\ker\phi = W.$$

We call $A = R/Q$ the **affinization** of W with respect to ϕ.

Proposition 11.2.8. *[Bell03] The algebra A defined in Example 11.2.7 has GK dim equal to $\mathrm{GKdim}(W) + 2$.*

This proposition provides a host of interesting examples, but unfortunately (from our point of view) these need not be PI, and may even have properties which are impossible in affine PI-algebras. More examples are given in the exercises.

11.2.2.2   GK dimension of monomial algebras

In view of Proposition 11.1.8, we should be able to display whatever GK dimensions that exist by means of a monomial algebra (although one could lose the PI in the process).

There is a way of constructing monomial algebras that provides a rather easy example of a PI-algebra with non-integral GK dim. The basic idea goes back to Warfield, cf. [KrLe00, pp. 19–20], and is carried out to fruition in [BelBL97] and [BelIv03].

Example 11.2.9. *(A PI-algebra with non-integral GK dim.) Given* $k_1, \ldots, k_4$, *let A be the monomial algebra $F\{a, b\}$ whose nonzero monomials are all subwords of*

$$a_1^{k_1} b_1 a_2^{k_2} b_2 a_3^{k_3} b_3 a_4^{k_4}.$$

We claim that A satisfies the PI

$$[x_1, x_2][x_3, x_4][x_5, x_6][x_7, x_8].$$

To see this, suppose on the contrary there were a nonzero word substitution

$$[v_1, v_2][v_3, v_4][v_5, v_6][v_7, v_8]$$

in A. Rearranging the indices, we may assume that $v_1 v_2 \ldots v_8 \neq 0$. Hence $v_1 v_2 \ldots v_8$ is a subword of $a_1^{k_1} b_1 a_2^{k_2} b_2 a_3^{k_3} b_3 a_4^{k_4}$. Let $m_i = \deg(v_i)$. If $m_1 + m_2 \leq k_1$, then $v_1 = x_1^{m_1}$ and $v_2 = x_1^{m_2}$, so $[v_1, v_2] = 0$. Hence, $m_1 + m_2 > k_1$, and v_3 must start with x_j for $j \geq 2$. The same argument applied to $[v_3, v_4]$ shows that $m_1 + m_2 + m_3 + m_4 > k_1 + 1 + k_2$, and so v_5 must start with x_j for $j \geq 3$, and likewise v_7 must start with x_4. But then v_7, v_8 both are powers of x_4, so $[v_7, v_8] = 0$. Thus, one of the factors must make the word substitution zero.

We can control GKdim(A) *by adding stipulations on the k_i. For example, we shall now construct an affine PI-algebra A with* GKdim$(A) = 3\frac{1}{2}$. *Toward this end, we require $k_3^2 \geq k_2$. Any nonzero monomial is determined by the sequence (k_1, k_2, k_3, k_4), so arguing as in Lemma 11.2.2, the number of nonzero words of degree $\leq n$ is less than $n^{3 + \frac{1}{2}} = n^{7/2}$. On the other hand, we can bound the number of nonzero words of degree $\leq n$ from below, by letting k_1, k_4 run from 0 to $\frac{n}{4}$ and k_3 from 0 to $\sqrt{n}$, and thus get*

$$(n/4)^2 \sum_{k_3=0}^{\sqrt{n}} \sum_{k_2=0}^{k_3^2} 1 = (n/4)^2 \sum_{k_3=0}^{\sqrt{n}} k_3^2 \geq \frac{\sqrt{n^7}}{96},$$

so as $n \to \infty$ we see $\frac{7}{2} \leq$ GKdim$(A) \leq \frac{7}{2}$, *and thus* GKdim$(A) = \frac{7}{2}$.

We shall see below (Corollary 11.2.24) that this example is nonrepresentable. It not difficult to generalize this to arbitrary GKdim for affine PI-algebras, and furthermore permits us to produce affine PI-algebras having arbitrary upper and lower Gel'fand-Kirillov dimensions, cf. Exercise 19. A finitely presented PI-algebra with non-integral GKdim is given in [BelIv03].

11.2.3   Affine algebras that have integer GK dimension

Having seen that any GK dim ≥ 2 can appear as the GK dim of an affine PI-algebra, we are led to ask whether it is an integer for a "nice" class of affine PI-algebras.

Lemma 11.2.10. *Suppose $R = A[s^{-1}]$, the localization of A by a regular central element s. Then* GKdim$(R) =$ GKdim(A).

Proof. Given generators $a_1, \ldots, a_\ell$ of A, we have generators $a_1 s^{-1}, \ldots, a_\ell s^{-1}$, $s^{-1}, 1, s$ of R, so $W = Vs^{-1} + Fs^{-1} + F + Fs$ is a generating subspace of R. Then letting $V' = V + F + Fs + Fs^2$, we have $(V')^n = (sW)^n = s^n W^n$. In other words, left multiplication by s^n gives us an isomorphism from W^n to $(V')^n$, so $\dim_F (V')^n = \dim_F W^n$ implying GKdim$(A) =$ GKdim(R). $\square$

Proposition 11.2.11.

   (i) GKdim$(S^{-1}A) \leq$ GKdim(A), *for any finite submonoid S of* Cent(A).

  (ii) GKdim$(S^{-1}A) \leq$ GKdim(A), *for any finite regular submonoid S of* Cent(R).

Proof. (i) Pass from A to its homomorphic image in $S^{-1}A$. (ii) Induction applied to Lemma 11.2.10. $\square$

Theorem 11.2.12. *For any prime affine PI-algebra A,* GKdim(A) *is an integer, equal to the transcendence degree of the center of A.*

Proof. In view of Theorem J of §1.5, Shirshov's Theorem, and Proposition 11.2.11, localizing at finitely many values of a central polynomial, we may assume that A is finite over its center, which is commutative affine, so we are done by Corollary 11.2.3. $\square$

Lemma 11.2.13 ([KrLe00, Corollary 3.3]). *If A is a subdirect product of $A_1, \ldots, A_q$, then*

$$\text{GKdim}(A) = \max\{\text{GKdim}(A_j) : 1 \leq j \leq q\}.$$

Proof. GKdim$(A) \leq \max\{$GKdim$(A_j) : 1 \leq j \leq q\}$ since $A \subseteq \prod A_j$. The reverse inequality holds since A projects to each component. $\square$

Corollary 11.2.14. GKdim(A) *is an integer, for any semiprime affine PI-algebra A.*

Proof. A is a finite subdirect product of prime affine PI-algebras. $\square$

In the next section (Theorem 11.2.21), we also see that representable PI-algebras have integer GK dim. Lorenz and Small [LoSm82] proved when A is two-sided Noetherian PI that GKdim$(A) = \max\{$GKdim$(A/P)\}$, P taken over the minimal prime ideals of A. On the other hand, [BoKr76] already has examples of an affine PI-algebra A with a nilpotent ideal N for which GKdim$(A/N) <$ GKdim(A).

11.2.4   The Shirshov height and GK dimension

As indicated earlier, Shirshov's Height Theorem 2.2.2 is an important tool in studying GK dim. Our first observation, due to Amitsur and Drensky, is that Shirshov's height yields a bound on the GK dim of any PI-algebra.

Theorem 11.2.15. *If $\mu = \mu(\ell, d)$ is as in Shirshov's Height Theorem, then*

$$\mathrm{GKdim}(A) \leq \mu$$

for any affine algebra $A = F\{a_1 \ldots, a_\ell\}$ of PI-degree d.

Proof. By Shirshov's Height Theorem we could take our generating vector space V to be a Shirshov base. Furthermore, by Remark 2.2.7, the reductions used in its proof do not change the degree of a word. So for $n \geq \mu$, $\dim V^n$ is at most the number of formal products $w_1^{k_1} \cdots w_\mu^{k_u}$ for which $\sum k_j = n$. This is $m^\mu t_{n,\mu}$ where $m = \ell^d + \ell^{d-1} + \cdots + 1$ is the number of words of length at most d in the a_i, and $t_{n,\mu}$ is the number of ways of picking $k_1, \ldots, k_\mu$ such that $\sum k_j = n$. This is at most n^μ, so

$$\dim V^n \leq m^\mu n^\mu = (mn)^\mu.$$

Hence, $\mathrm{GKdim}(A) \leq \displaystyle\lim_{n \to \infty} \log_n((mn)^\mu) = \lim_{n \to \infty} (\mu(1 + \log_n m)) = \mu.$ □

(Note: It is well-known from combinatorics that $t_{n,\mu} = \binom{n+\mu-1}{n}$, but our estimate is good enough, as seen in the discussion of Lemma 11.2.2.) In particular, this gives Berele's result that any affine PI-algebra has finite GK dim, and has immediate consequences:

Corollary 11.2.16. *If A is an affine PI-algebra over a field, then any chain of prime ideals has bounded length $\leq \mu$. In other words, A satisfies the ascending and descending chain conditions on prime ideals.*

Proof. (This is a well-known fact about rings with finite GK dim whose prime factors are right Goldie, [KrLe00, Corollary 3.16].)   □

The formula $\mathrm{GKdim}(A) \leq \mu$ of Theorem 11.2.15 cannot be an equality in general, since we saw examples of affine PI-algebras whose GK dim is not an integer. But equality holds in a certain sense for representable PI-algebras. To see this, we extend the notion of Shirshov base. (We also modify the notation a bit, writing W where in Chapter 2 we would have used $\bar{W}$.)

Recall the essential Shirshov height from Definition 2.3.20. If W and W' are given as in Definition 2.3.20, then the number of elements of (2.1) such that $\sum k_i \leq n$, grows at most at the order of $|W|^\nu n^\nu$, where $\nu = \nu(A)$. Thus, we have the following bound.

Proposition 11.2.17. $\mathrm{GKdim}(A) \leq \varlimsup\limits_{n \to \infty} \log_n(|W|n)^\nu \leq \nu.$

We aim to prove that GKdim$(A) = \nu(A)$, for A representable. Toward this end, we need some preliminaries.

Proposition 11.2.18. *Suppose* $A \subseteq M_t(C)$, *where* C *is a commutative* F-*algebra. Suppose for suitable elements* $v_0, \cdots, v_j, w_1, \ldots, w_j \in A$,

$$\sum_{k_1 + \cdots + k_j \leq t} \alpha_I v_0 w_1^{m_1 + k_1} v_1 \ldots w_j^{m_j + k_j} v_j = 0 \qquad (11.7)$$

holds for all $m_1, \ldots, m_j$ *satisfying* $0 \leq m_i \leq t - 1$ *for all* i *(the coefficients* $\alpha_I \in F$ *not depending on the* m_i*). Then we also have*

$$\sum_{k_1 + \cdots + k_j \leq t} \alpha_I v_0 w_1^{q_1 + k_1} v_1 \ldots w_j^{q_j + k_j} v_j = 0 \qquad (11.8)$$

for all $q_1, \ldots, q_j \in \mathbb{N}$.

Proof. Induction on each q_i in turn. Indeed, if each $q_i \leq t - 1$, then (11.8) holds by (11.7). So assume some $q_i \geq t$. Then the Cayley-Hamilton Theorem lets us rewrite $w_i^{q_i + k_i}$ as a C-linear combination of $w_i^{q_i - n + k_i}, \ldots, w_i^{q_i - 1 + k_i}$. But by induction Equation (11.8) holds for each of these smaller values, so therefore Equation (11.8) holds with the original q_i. $\qquad\square$

Proposition 11.2.19. *Let* U *be the* F-*linear span of all elements of the form*

$$a = v_0 w_1^{q_1} v_1 \cdots w_j^{q_j} v_j, \qquad q_1, \ldots, q_j \in \mathbb{N}.$$

Then the F-*linear span of those elements for which one of the* q_i *is smaller than* t, *is the same as* U.

Proof. One might want to repeat the Cayley-Hamilton argument used above, but a difficulty now arises insofar as we want to take the F-linear span and not the C-linear span. (In the previous result, we only needed to take the linear span of 0, which remains 0.) Thus, we need a slightly more intricate argument, and turn to an easy version of pumping.

Given a as above, we call $(q_1, \ldots, q_j)$ the *power vector* of a. We say that a is *good* if $q_i < t$ for some i. It is enough to show that any element $a \in U$ is a linear combination of good elements.

We order the power vectors $(q_1, \ldots, q_j)$ lexicographically, and proceed by induction on the power vector. Suppose a is "bad," i.e., each $q_i > t$. In (11.7) take that term with $(\hat{k}_1, \ldots, \hat{k}_j)$ maximal among the $(k_1, \ldots, k_t)$. Let $m_i = q_i - \hat{k}_i$. Since $q_i \geq t$, we see $m_i \geq 0$ and $y_i^{q_i} = y_i^{m_i} \hat{y}_i^{\hat{k}_i}$. Now Equation (11.8) enables us to write $v_0 w_1^{q_1} v_1 \ldots w_j^{q_j} v_j$ as a linear combination of terms

$$v_0' w_1^{q_1 + k_1 - \hat{k}_1} v_1' \ldots w_j^{q_j + k_j - \hat{k}_j} v_j'.$$

Write $v_j' = w^{\hat{k}_j - k_j} v_j$, i.e., having smaller power vectors

$$(q_1 + k_1 - \hat{k}_1, \ldots, q_j + k_j - \hat{k}_j),$$

which by induction yield linear combinations of good elements. Thus, a also is a linear combination of good elements, as desired. $\square$

Corollary 11.2.20. *Write $S(q)$ for the dimension of the linear span of all elements of the form $v_0 w_1^{q_1} v_1 \ldots w_j^{q_j} v_j$, such that each $q_i \leq q$. If Equation (11.7) holds, then there exists a constant κ, such that $S(q) < \kappa q^{j-1}$.*

Proof. In view of Proposition 11.2.19, we may take a base of "good" elements; i.e., we have degree at most t in one of the positions, and we have q^{j-1} choices for the powers $w_i^{q_i}$, in the other $j-1$ positions. Thus, we have at most $|W'|^j q^{j-1}$ possibilities for our base elements, and we may take $\kappa \leq |W'|^j$. $\square$

Theorem 11.2.21. *Suppose $A \subseteq M_n(C)$ is affine, and $\mathrm{GKdim}(A) > 0$. Then $\mathrm{GKdim}(A) = \nu(A)$.*

Proof. Let ν be the essential height of $A = F\{a_1, \ldots, a_\ell\}$, and we take the v_j and w_i as in Definition 2.3.20. We already know that $\mathrm{GKdim}(A) \leq \nu$, so our aim is to show that if $\mathrm{GKdim}(A) < \nu$, then we can reduce ν, contrary to choice of ν.

Let $\mathcal{V}$ denote the (at most) t^j-dimensional vector space with base elements of the form
$$v_0 w_1^{m_1+k_1} v_1 \ldots w_j^{m_j+k_j} v_j, \quad 0 \leq m_i < t.$$
(This base is determined by the vectors $(m_1, \ldots, m_j)$ of nonnegative integers, $m_i < t$, since the v_i and w_i are given.) We use this as our generating space for A, since we may choose whichever one we like in calculating $\mathrm{GKdim}(A)$.

If $\mathrm{GKdim}(A) < j$, then given ε, $\dim_F \mathcal{V} < n^{j-\varepsilon}$ for large enough n, thereby yielding Equation (11.7). Define a new (finite) set
$$W'' = \{v_i w_i^{k_i} v_{i+1} : 1 \leq i \leq \nu, 1 \leq k_i \leq t\}.$$

But each "good" word can be written in length $\nu - 1$, since the small power of one of the w_i has been absorbed in W''. Hence, we have reduced ν, as desired. $\square$

We can push this argument a bit further.

Remark 11.2.22. *A slight modification of the proof shows*
$$\nu \leq \liminf_{n \to \infty} \log_n (\dim(\mathcal{V}^n)).$$
This also proves that $\mathrm{GKdim}(A) = \liminf_{n \to \infty} \log_n (\dim(\mathcal{V}^n))$.

Corollary 11.2.23. *The essential height of a representable algebra A is independent of the choice of the set W, and $\mathrm{GKdim}(A)$ is an integer.*

Corollary 11.2.24. *Any affine PI-algebra with non-integral GK-dimension is nonrepresentable. (Compare with Example 11.2.9.)*

V.T. Markov first observed that the GK dim of a representable algebra is an integer, cf. [BelBL97]. It would be interesting to construct a (non-representable) PI-algebra in which the essential height depended on the choice of essential Shirshov base. Soon we shall obtain a polynomial bound on Gel'fand-Kirillov dimension of a PI-algebra (and hence on the essential height), first noted by Berele.

11.2.5 Other upper bounds for GK dimension

Suppose $A = F\{a_1, \ldots, a_\ell\}$ has PI-degree d. In Theorem 11.2.15, we saw that GKdim(A) is bounded by the essential height (i.e., the lowest possible Shirshov height) and we also saw in Theorem 11.2.21 that equality holds for representable algebras. On the other hand, every relatively free algebra is representable, by Theorem 11.3.13 in characteristic 0, and to be proved in Volume II in characteristic p. So, given any T-ideal $\mathcal{I}$ of the free algebra $F\{x_1, \ldots, x_\ell\}$, we have an effective bound for the GK dim of any PI-algebra A for which id(A) $= \mathcal{I}$, namely the essential height of the relatively free algebra.

This motivates us even more to compute a formula for GKdim(A), which would also give a lower bound for the essential height. One special case has been known for a long time — Procesi showed in 1967 ([Pro73]) for A of PI-class n that GKdim(A) $\leq (\ell - 1)n^2 + 1$, equality holding iff A is isomorphic to the algebra of generic $n \times n$ matrices. Once Kemer's theory became available, Berele [Ber94] was able to generalize this formula for all PI-algebras, using a simple structure-theoretical argument.

Proposition 11.2.25. *If* char(F) $= 0$ *and* $A = F\{a_1, \ldots, a_\ell\}$ *is PI, then* GKdim(A) $\leq s((\ell - 1)t + 1)$, *where* index(A) $= (t, s)$ *is the Kemer index, cf. Definitions 6.4.7, 6.6.2, and 6.6.3.*

Proof. Since A is a homomorphic image of its relatively free algebra, we may assume that A is relatively free; in view of Proposition 7.4.19, we may assume that A is the relatively free algebra of a basic algebra (Definition 6.3.9) of some Kemer index $\kappa = (t, s)$. But then $J^s = 0$ and A/J is relatively free semiprime and thus prime PI, of PI-class n where $t = n^2$. [KrLe00, Corollary 5.10] shows that

$$\text{GKdim}(A) \leq s\,\text{GKdim}\,A/J,$$

so we conclude with Procesi's result.   □

Berele's original proof did not require the Kemer index, but did involve some more computation, cf. Exercise 32.

11.3   Rationality of Certain Hilbert Series

By definition, the GKdim is determined by the Hilbert series under the natural filtration; since the GK dim of representable PI-algebras is so well behaved, this leads us to look more closely at the Hilbert series. Unfortunately, unlike the GKdim, the Hilbert series depends on the choice of generating space V, so we should be careful in selecting V. To emphasize this point, we write $H_{A;V}$ (resp. $H_{M;V}$) instead of H_A (resp. H_M), when appropriate.

The next question is whether H_A is a rational function, i.e., the quotient $\frac{p(\lambda)}{q(\lambda)}$ of two polynomials.

We start with a classical result, given in [Jac80, p. 442]:

Lemma 11.3.1. *If C is an $\mathbb{N}$-graded affine commutative algebra, then any homogeneous generating set is rational, and also the corresponding Hilbert series of any finite graded C-module is rational.*

This yields an ungraded result, by means of Remark 11.1.6.

Proposition 11.3.2.

(i) *Any generating set of a commutative affine algebra R is rational.*

(ii) *Any finitely generated module M over a commutative affine algebra A has a rational Hilbert series.*

Proof. Reduce to the graded case by passing to $\mathrm{Gr}(A)$ and $\mathrm{Gr}(M)$. $\qquad\square$

Presumably one could have an affine PI-algebra which has a rational Hilbert series with respect to one set of generators, but which is nonrational with respect to a different set of generators. So our goal is to find a suitable finite generating set for which the Hilbert series is rational. We shall call this a *rational generating set*.

Proposition 11.3.3. *Suppose R is an affine algebra that is finite over a commutative subalgebra A. Then any finite subset of R can be expanded to a rational generating set of R (as an affine algebra).*

Proof. A is affine by the Artin-Tate lemma (Proposition 1.1.21). Taking a generating space V_0 of A, we write $R = \sum_{j=1}^{t} A r_j$ and write $r_j r_k = \sum_{m=1}^{t} a_{jkm} r_m$. Then letting $V = V_0 + \sum_{j,k,m} F a_{jkm}$, we see that R now is a filtered module over A, whose Hilbert series as a module with respect to V is the same as the Hilbert series as an algebra with respect to $V + \sum F r_j$. $\qquad\square$

From now on, we assume that our generating space V contains 1 (since we can clearly expand V by adjoining F, spanned by the element 1), so that the natural filtration of A is now given by $\tilde{A}_n = V^n$.

Remark 11.3.4.
$$H_{R;W} = H_{A;V'},$$
in the setup of Proposition 11.2.10(ii).

We strengthen slightly a theorem of Bell [Bell04].

Theorem 11.3.5. *Suppose A is a prime affine PI-algebra. Then any generating set of A can be expanded to a rational generating set.*

Proof. Notation as in Theorem I(ii) of §1.5, we could invert any single element c in $h_n(A)$ to get an algebra that is f.d. free over its center, so we are done by Proposition 11.3.2 and Remark 11.3.4.   □

Having proved that, at times, the Hilbert series of a PI-algebra is rational, we want to know if it must have certain form. To this end we quote [KrLe00, p. 175]:

Theorem 11.3.6. *If A is an affine PI-algebra and $H_A = 1 + \sum \tilde{d}_n \lambda^n$ is rational, then there are numbers m, d such that*

$$H_A = \frac{p(\lambda)}{(1 - \lambda^m)^d},$$

with $p(1) \neq 0$. In this case, there are polynomials $p_1, \ldots, p_{m-1}$ such that $\tilde{d}_n = p_j(n)$ for all large enough $n \equiv j \pmod{m}$, and $\max \deg p_j = d - 1$; also, GKdim$(A) = d$.

Thus, a non-rational Hilbert series implies non-integral GK dim. On the other hand, even if a PI-algebra has integer GK dim, the growth of the subspaces might be sufficiently grotesque as to keep it from having a rational Hilbert series, cf. Exercise 18. Furthermore, although the GK dim of a representable algebra is finite, there are representable algebras whose Hilbert series are non-rational. One way of seeing this is via monomial algebras, cf. We quote a fact from [BelBL97]:

Proposition 11.3.7. *A monomial algebra is representable iff the following two conditions hold:*

(i) *A is PI (which for monomial algebras is equivalent to Shirshov's Theorem holding);*

(ii) *the set of all defining relations of A is a finite union of words and relations of the form "$v_1^{k_1} v_2^{k_2} \ldots v_s^{k_s} = 0$ iff $p(k_1, \ldots, k_s) = 0$." where p is an exponential diophantine polynomial, e.g., $p = \lambda^{k_1} + k_2 + k_3^4 3^{k_2}$.*

The proof is sketched in Exercise 20. Using this result, we have:

Example 11.3.8. *A representable affine PI-algebra A that does not have a rational Hilbert series. Consider the monomial algebra supported by monomials of the form*

$$h = x_1 x_2^m x_3 x_4^n x_5$$

and also satisfying the extra relation $h = 0$ if $m^2 - 2n^2 = 1$. This is representable by Proposition 11.3.7 but the dimensions u_n grow at the approximate rate of polynomial minus $c \ln n$, so H_A cannot be rational, in view of Theorem 11.3.6.

11.3.1 Hilbert series of relatively free algebras

Our other result along these lines has a slightly different flavor, since it deals with a generating set that comes naturally. Namely, any relatively free algebra is graded naturally by degree, and our generating set is composed of the elements of degree 1, i.e., the images of the letters. Let us record an observation that enables us to apply induction of sorts.

Lemma 11.3.9. *Suppose $\mathcal{I} \triangleleft A$ with the filtration $\mathcal{I}_n = \mathcal{I} \cap \tilde{A}_n$.*

(i) $H_A = H_{A/\mathcal{I}} + H_{\mathcal{I}}$.

(ii) If $\mathcal{I}, \mathcal{I}' \triangleleft A$ are graded, then

$$H_{A/\mathcal{I} \cap \mathcal{I}'} = H_{A/\mathcal{I}} + H_{A/\mathcal{I}'} - H_{A/(\mathcal{I}+\mathcal{I}')}.$$

Proof. (i) Immediate from the definition.

(ii) $A/(\mathcal{I} + \mathcal{I}') \cong (A/\mathcal{I}')/((\mathcal{I} + \mathcal{I}')/\mathcal{I}')$ and $(\mathcal{I} + \mathcal{I}')/\mathcal{I}' \cong \mathcal{I}/(\mathcal{I} \cap \mathcal{I}')$, so $H_{A/\mathcal{I}'} = H_{A/(\mathcal{I}+\mathcal{I}')} + H_{(\mathcal{I}+\mathcal{I}')/\mathcal{I}'} = H_{A/(\mathcal{I}+\mathcal{I}')} + H_{\mathcal{I}/(\mathcal{I}\cap\mathcal{I}')}$, implying

$$H_{A/\mathcal{I} \cap \mathcal{I}'} = H_{A/\mathcal{I}} + H_{\mathcal{I}/(\mathcal{I}\cap\mathcal{I}')} = H_{A/\mathcal{I}} + H_{A/\mathcal{I}'} - H_{A/(\mathcal{I}+\mathcal{I}')}.$$

$\square$

Another general technique that is available is the trace ring of Proposition 2.5.14. We shall continue the characteristic 0 assumption here. Let $\mathcal{I}$ be as in Remark 2.5.10. Since $\hat{A}$ is Noetherian and $\mathcal{I} \subseteq \hat{A}$, we see that $\mathcal{I}$ is a finite $\hat{A}$-module. On the other hand, the natural filtration of $\hat{A}$ might not produce the same Hilbert series as the natural filtration on A.

Definition 11.3.10. *Suppose $A = F\{a_1, \ldots, a_\ell\}$ is representable, and let $\mathcal{I}$ be as in Remark 2.5.10. We say the trace ring $\hat{A}$ has a filtration **compatible with A** if*

$$A \cap \hat{A}_n = \tilde{A}_n.$$

Proposition 11.3.11. *Notation as in Definition 11.3.10, suppose A has a compatible filtration with $\hat{A}$, with respect to which $A/\mathcal{I}$ has a rational Hilbert series. Then A has a rational Hilbert series.*

Proof. Take the filtration $\mathcal{I}_n = \mathcal{I} \cap \hat{A}_n$ of Lemma 11.3.9(i). Since $\hat{A}_n$ is finite over its affine center, $H_{\mathcal{I}}$ is rational, and we can apply Lemma 11.3.9.   □

Corollary 11.3.12. *If A is representable and graded, having a set of homogeneous generators with respect to which $A/\mathcal{I}$ has a rational Hilbert series ($\mathcal{I}$ as in Lemma 2.5.10), then A has a rational Hilbert series.*

Proof. We can grade $\hat{A}$ by giving $\mathrm{tr}(w)$ the same grade as w. In view of Theorem I of §1.5, this grade is compatible with that of A, so we can apply the proposition. (Note that $\mathcal{I}$ is a graded ideal, yielding a grade on $A/\mathcal{I}$.)   □

Using these facts, we have a nice result for relatively free PI-algebras:

Theorem 11.3.13. *In characteristic 0, every relatively free, affine PI-algebra A has a rational Hilbert series.*

Proof. In view of the ACC on T-ideals, we may assume by Noetherian induction (cf. Remark 1.1.20) that $A/\mathcal{I}$ has a rational Hilbert series for every T-ideal $\mathcal{I} \neq 0$ of A. In view of Proposition 7.4.19, we may assume that A is PI-irreducible; we take the generators to be the generic elements $\bar{x}_1, \ldots, \bar{x}_\ell$, i.e., $V = \sum F\bar{x}_i$. Then A has a Kemer polynomial f, and let $\mathcal{I}$ be the T-ideal of A generated by $\langle f(A) \rangle$. $\mathcal{I}$ is also an ideal of the trace algebra $\hat{A}$, defined in the proof of Proposition 6.8.1.

We grade $\hat{A}$ by degree, where $\deg \mathrm{tr}(f) = \deg(f)$. For our generating set of $\hat{A}$, we take the generating set $\hat{V}$ of $\hat{A}$ consisting of V together with the homogeneous parts of the traces we added to A (which are finite in number). Since our algebras are relatively free, they are all graded, so clearly the compatibility hypothesis in Proposition 11.3.11 is satisfied. We thus can apply induction to Lemma 11.3.9(i) in conjunction with Proposition 11.3.2(ii).   □

The graded version of this result is proved by Aljadeff and Belov [AB12]. The same general idea holds in nonzero characteristic.

Proposition 11.3.14. *Any relatively free, representable affine algebra has a rational Hilbert series.*

Proof. In characteristic 0, this has already been proven. Thus, we assume $\mathrm{char}(F) = p > 0$. We define the characteristic closure $\hat{A}$ as in Definition 2.5.13, i.e., using suitably high powers $\bar{w}^q$ of the words $\bar{w}$. Now Proposition 2.5.15 is available, so we just take $\mathcal{I}$ to be the ideal generated by evaluations in A of the appropriate Capelli polynomial, and conclude with the last paragraph of the previous proof.   □

Note that we bypassed difficulties in the last assertion by assuming the algebra is representable. In Volume II we prove that any relatively free affine algebra of arbitrary characteristic is representable.

Here is another result connected to Noetherian algebras.

Theorem 11.3.15. *If A is affine and its graded algebra is Noetherian PI, then A has a rational generating set.*

Proof. Since A has the same Hilbert series as its graded algebra, we may assume that A is graded Noetherian. But now A is representable, so we take $\mathcal{I}$ as above. By Noetherian induction, $A/\mathcal{I}$ has a generating set for which the Hilbert series with respect to any larger set is rational. We lift this to A and apply Corollary 11.3.12. $\qquad\square$

Remark 11.3.16. *It is still open whether the Hilbert series of a left Noetherian PI-algebra A is rational. In view of Anan'in's theorem, A is representable, so Proposition 2.5.14 is available, and a Noetherian induction argument would enable one to assume $J^{s-1}\mathcal{I} \neq 0$, and thus is a common ideal with A and $\hat{A}$. In order to conclude using Corollary 11.3.12, it would suffice to show $\hat{A}$ has a filtration compatible with the natural filtration on A. The natural candidate would be for us to declare that $\mathrm{tr}(a_n) \in \hat{A}_n$ for each $a \in A_n$, but it is not clear without some extra hypothesis that this filtration is compatible with that of A.*

11.4   The Multivariate Poincaré-Hilbert Series

The concept of Hilbert series can be refined even further, cf. [KrLe00, pp. 180–182]. We consider such a situation. Suppose A is filtered, generated by F and $V = A_1$, and we assume V has a base $a_1, \ldots, a_\ell$. For example, if $A = F\{x_1, \ldots, x_\ell\}$, we could take $V = \sum_{i=1}^{\ell} Fx_i$.

Given an ℓ-tuple $\mathbf{k} = (k_1, \ldots, k_\ell)$, define the **length** $|\mathbf{k}| = k_1 + \cdots + k_\ell$; when $|\mathbf{k}| = n$ define

$$B_{\mathbf{k}} = \{b = a_{i_1} \ldots a_{i_n} : \deg_i(x_{i_1} \ldots x_{i_n}) = k_i, \forall i\}.$$

In other words, $b = a_{i_1} \cdots a_{i_n} \in B_{\mathbf{k}}$ iff a_i appears exactly k_i times for each i.

$A_n = V^n$ is spanned by the $b = a_{i_1} \cdots a_{i_n}$. We write $\deg_i b$ for the number of times a_i occurs in this product. Thus, $(\deg_1 b, \ldots, \deg_\ell b)$ is the multi-degree. (Note, however, that a certain b could occur in $B_{\mathbf{k}}$ for several different $\mathbf{k}$ of the same length, so the multi-degree is not well-defined without a further assumption, such as A being graded.)

Definition 11.4.1. *Ordering the $\mathbf{k}$ first by length and then by lexicographic order, define $\tilde{V}_{\mathbf{k}}$ to be the vector space spanned by all $B_{\mathbf{k}'}$, $\mathbf{k}' \leq \mathbf{k}$, and $\tilde{V}'_{\mathbf{k}}$ to be the vector space spanned by all $B_{\mathbf{k}'}$, $\mathbf{k}' < \mathbf{k}$. Define $d_{\mathbf{k}} = \dim_F(V_{\mathbf{k}}/V_{\mathbf{k}'})$, intuitively the number of base elements not occurring in "lower" components, and the **multivariate** Poincaré-Hilbert series*

$$H(A) = \sum_{\mathbf{k} \in \mathbb{N}^{(\ell)}} d_{\mathbf{k}} \lambda_1^{k_1} \ldots \lambda_\ell^{k_\ell}.$$

Remark 11.4.2. *Specializing all $\lambda_i \mapsto \lambda$ in the multivariate $H(A)$ gives the original Hilbert series.*

As in Example 11.1.4, it is easy to see that

$$H(F[\lambda_1, \ldots, \lambda_\ell]) = \prod_{i=1}^{\ell} \frac{1}{1 - \lambda_i},$$

and likewise Remark 11.1.3 carries over. Accordingly, one can also obtain the results of Section 11.3.1, and thus generalize Belov's theorem to prove the multivariate Poincaré-Hilbert series of any relatively free PI-algebra is rational, cf. Exercise 33.

This generalization is not merely a formal exercise. There is another, completely different, approach, to the Hilbert series of relatively free algebras, namely, by using the representation theory described briefly in Chapter 3 (and to be elaborated in Chapter 12.) Formanek describes this approach in detail in [For84, Theorem 7], which identifies the multivariate Hilbert series with the series of codimensions. This enables him to compute the Hilbert series for the trace algebra of generic $n \times n$ matrices (cf. Chapter 8), and in the case of $n = 2$, he derives the Hilbert series for generic 2×2 matrices (due to Formanek-Halpin-Li for two matrices, and due to Drensky for an arbitrary number of generic 2×2 matrices). The formula for generic $n \times n$ matrices remains unknown, for $n > 2$, but the technique is powerful, mentioned at the end of Chapter 8. We shall discuss this issue further in Chapter 12.

Exercises for Chapter 11

1. Define the following relation on the set of all growth functions:

   $f \succ g$, if there exist c and k such that $cf(kn) \geq g(n)$ for all n. This gives rise to an equivalence relation: $f \equiv g$, if $f \succ g$ and $g \succ f$. This equivalence relation does not depend on the choice of generators; this is a general fact about GK dim.

2. $\text{GKdim}(A) = \text{GKdim}(M_n(A))$.

3. $\text{GKdim}(A) = \text{GKdim}(\text{End}_A M)$, for any finite A-module M.

4. The GK dim of any algebra finite over A equals $\text{GKdim}(A)$.

5. To present the proof of Bergman's gap as quickly as possible from start to finish, avoiding hyperwords and the König graph theorem, define a word w to be **unconfined** if it occurs infinitely many times as a nonterminal subword in the various $\tilde{A}_n$, and $\nu'(n)$ to be the number of

unconfined subwords. As in the text, $\text{GKdim}(A) > 1$ implies $\nu'(1) > 1$ and $\nu'(n+1) > \nu'(n)$ for each n; hence $\tilde{d}_n \geq \sum_{m=1}^{n} \nu'(m) \geq \frac{n^2+3n}{2}$.

Algebras having linear growth

The object of the next few exercises is to prove highlights of the Small-Stafford-Warfield theorem, that any algebra of GKdim 1 is PI. We follow Pappacena-Small-Wald [PaSmWa].

6. (Pappacena–Small–Wald.) If $A = F\{a_1, \dots, a_\ell\}$ with $\text{GKdim}(A) = 1$, then any simple A-module M is f.d. over F. (Hint, using some structure theory: Factoring A by the annihilator of M, one may assume that M is also a faithful A-module, i.e., A is a primitive algebra. Write $M = Aw$, and let $D = \text{Hom}_A(M, M)$. If M has m linearly independent elements $b_1, \dots, b_m$ over D, using the density theorem, take $s_i \in A$, $1 \leq i \leq m$, such that $s_i b_j = 0$ for $j \leq i$ and $s_i b_i = b_1$. Take n such that each $b_i \in V^n$, where $V = \sum_{i=1}^{\ell} F a_i$, and let L_k be the left annihilator of $b_1, \dots, b_k$ for each $k \leq n$. Note that $L_k w_i = M$ and the L_k are independent over F. If M is infinite dimensional over F, then comparing growth rates, show that $m \leq n$. But this means M has dimension at most n over D, i.e., $A = M_n(D)$ is simple Artinian. By the Artin-Tate lemma, the center Z of A is affine over F and thus, f.d. over F; hence, one may assume that $Z = F$. One shows that A has PI-class n by splitting A in the same way as in the standard proofs of Kaplansky's theorem.)

7. Suppose P is a maximal ideal of $A = F\{a_1, \dots, a_\ell\}$, and A/P has linear growth rate N. Let n be the PI-class of A/P. Then any word of length $(n+1)N$ in the a_i is a linear combination of smaller words.

8. Notation as in Exercise 7, show that $n \leq \frac{1}{2}(N^2 + \sqrt{N^4 + 4N^2 - 4N})$. (Hint: The words of length $(n+1)N - 1$ span R/P, which has dimension n^2, so $(n+1)N^2 - N \geq n^2$.)

9. If A is semiprimitive affine of GKdim 1, then A is PI, and its PI-class does not exceed the function of Exercise 8. (Hint: Apply Exercises 6 and 8 to each primitive homomorphic image.)

10. If $\text{GKdim}(A) = 1$ and A is affine with no nonzero nil ideals, then A is PI, of PI-class not exceeding the function of Exercise 8. (Hint: Expand the base field to an uncountable field; then one may assume that A is semiprimitive and apply Exercise 9.)

11. (Small-Warfield [SmWa84].) If A is prime with $\text{GKdim}(A) = 1$, then every chain of principal left annihilators of A has length bounded by the linear growth rate of A. (Hint: Given principal left annihilators $L_1 \supset L_2$, show that $L_1 a = 0$ but $L_2 a \neq 0$ for some $a \in A$; $L_2 a$ is a faithful A-module, and thus has linear growth rate at least 1; continuing in this way, show that the linear growth rate of A is at least the length of any chain of principal left annihilators.)

12. If A is semiprime affine with $\mathrm{GKdim}(A) = 1$, then A is PI. (Hint: One may assume that A is prime. By Exercise 11, A satisfies the ACC on principal left annihilators, so, as in [Ro88b, Proposition 2.6.24], A satisfies the conditions of Exercise 10.)

13. (Small-Warfield.) If A is prime of GK dim 1, then A is finite over a central subring isomorphic to the polynomial ring $F[\lambda]$, and in particular, A is Noetherian. (Hint: Let $C = \mathrm{Cent}(A)$. A is PI, by Exercise 12. If C is a field, then A is simple, so assume that C is not a field. Then C has GK dimension 1, implying C is integral over $F[\lambda]$, and thus, finite over $F[\lambda]$, by Exercise 10.1.)

Examples of PI-algebras having unusual growth

14. The affine PI-algebra $A = \begin{pmatrix} F[\lambda_1] & F[\lambda_1, \lambda_2] \\ 0 & F[\lambda_2] \end{pmatrix}$ has GK dim 2, although modulo its radical has GK dim 1.

15. Define the algebra $A^\infty = F\{x, y\}/\langle y \rangle^2$, i.e., y occurs at most once in each nonzero word. Then $V^n = n(n+3)/2$.

16. Define the algebra $A^n = F\{x, y\}/\mathcal{I}_y^n$, where $\mathcal{I}_y^n$ is the ideal generated by monomials that do not contain subwords of the form $yx^k y$, $k < n$. Then $2^{[k/n]} \le d_k(A^n) = d_k(A^\infty) = k + 1$. (Hint: There are $2^{[k/n]}$ words having y in the positions not divisible by n.)

17. A prime affine F-algebra of GK dim 3, having non-nil Jacobson radical. Notation as in Example 11.2.7, take T to be the localization of $F[\lambda]$ at the prime ideal $F[\lambda]\lambda$. Then $\mathrm{GKdim}(A) = 3$. Also $e_{11}J(A)e_{11} \subseteq J(T) = T\lambda$. But A is not PI.

Monomial algebras having unusual growth

18. Suppose ψ, φ are functions satisfying

$$\lim_{n\to\infty} \frac{\ln \psi(n)}{n} = 0, \qquad \lim_{n\to\infty} \left(\varphi(n) - \frac{n(n+3)}{2} \right) = \infty.$$

Then there exist an algebra A and two infinite subsets $\mathcal{K}$ and $\mathcal{L}$ of $\mathbb{N}$, such that:

(i) $d_n > \psi(n), \forall n \in \mathcal{K}$.

(ii) $d_n < \varphi(n), \forall n \in \mathcal{L}$.

If $\psi(n) < Cn^k$, then there exists a PI-algebra A satisfying these conditions.

(Hint: Let $\alpha : \ell_1 < k_1 < \ell_2 < k_2 < \ldots < \ell_i < k_i < \ell_{i+1} < \ldots$ be a sequence of positive integers, where $\mathcal{I}(\alpha)$ denotes the ideal generated by the sets $\mathcal{W}_1$ and $\mathcal{W}_2$ defined as follows:

$$\mathcal{W}_1 = \{w : \ell_i \leq |w| < k_i \quad \text{and} \quad \exists k < \ell_i : yx^k y \subset w\},$$
$$\mathcal{W}_2 = \{w : k_i \leq |w| < \ell_{i+1} \quad \text{and} \quad \exists k < \ell_{i+1} : yx^k y \subset w\}.$$

Define $A(\alpha) = F\{x,y\}/\mathcal{I}(\alpha)$. If $\ell_i \leq n < k_i$, then $d_n(A(\alpha)) = d_n(A^{\ell_i})$, and, if $k_i \leq n < \ell_{i+1}$, $d_n(A(\alpha)) = d_n(A^\infty) = n+1$. Exercise 16 allows one to choose ℓ_{i+1} so that

$$d_{\ell_{i+1}-1}(A(\alpha)) < \varphi(\ell_{i+1} - 1).$$

A similar argument allows one to choose k_i so that

$$d_{k_i-1}(A(\alpha)) = d_{k_i-1}(A^{\ell_i}) > \psi(k_i - 1).$$

Hence, $A(\alpha)$ satisfies (i) and (ii).

The proof of the last assertion is analogous. Note that $F\{x,y\}/\langle y\rangle^k$ is a PI-algebra; the number of words with y occurring k times, such that the distance between each two occurrences of y is greater than some given large enough number, has growth n^k.)

19. In Exercise 18, let $\psi(n) = e^{\sqrt{n}}$, and $\varphi(n) = n(n+3)/2 + \ln(n)$. Then the growth function of A starts slowly and lags behind φ, but then begins to grow very fast and outstrips ψ.

20. Prove Proposition 11.3.7. (Hint: If A is representable, then one can examine elements in Jordan form, and by computing their k power obtain any given set of exponential diophantine polynomials with respect to k; the converse is by an inductive construction.)

Essential Shirshov bases

21. Suppose u is a non-periodic word, of length n. Let A_u be the monomial F-algebra whose generators are letters of u, with relations $v = 0$ iff v is not a subword of the hyperword $h = u^\infty$. Prove that $A_u \hookrightarrow M_n(F[\lambda])$, and $A_u \sim_{\mathrm{PI}} M_n(F[\lambda])$ when F is infinite. (Hint: Take the cyclic graph of length n whose edges are labeled with the letters in u. Identify the edge between i and $i+1$ with λ times the matrix unit $e_{i,i+1}$. For example, if the letter x_1 appears on the edges connecting 1 and 2, 3 and 4, and 7 and 8, then x_1 is sent to $\lambda(e_{12} + e_{34} + e_{78})$. This gives the injection of A_u into $M_n(F[\lambda])$. Hence, id$(A) \supseteq$ id$(F[\lambda])$. If F is infinite, conclude the reverse inclusion by using the surjection $A_u \to M_n(F)$ sending $\lambda \mapsto 1$.)

22. Let A_u be as in Exercise 21. The set of non-nilpotent words in A_u is the set of words that are cyclically conjugate to powers of u.

23. Every prime monomial PI-algebra A is isomorphic to some A_u as defined in Exercise 21. (Hint: Use Shirshov's Height Theorem, and show that A has bounded height over some word.)

24. Suppose A is an algebra with generators $a_s \succ \cdots \succ a_1$, and M is a finite right A-module spanned by $w_k \succ \cdots \succ w_1$. Put $w_1 \succ a_s$, and $a_1 \succ x$. Make $\tilde{A} = A \oplus M$ into an algebra via the "trivial multiplication"

$$0 = w_i w_j = a_i w_j \quad \forall i, j.$$

Let A'' denote $\tilde{A} * F\{x\}/\mathcal{I}$, where the ideal $\mathcal{I}$ is generated by the elements xw_i. If $MA^k \neq 0$ for all k, then the minimal hyperword in the algebra A'' is pre-periodic of pre-periodicity $\leq n + 1$. (Hint: By Theorem 2.7.3.)

25. The set of lexicographically minimal words in a PI-algebra A of PI-class n has bounded height over the set of words of length $\leq n$. (Hint: Let $d = 2n$. As A has bounded height over the set of words of degree not greater than d, it is enough to prove that if $|u|$ is a nonperiodic word of length not greater than n, then the word u^k is a linear combination of lexicographically smaller words, for k sufficiently big. This is done in several steps.

(i) Consider the right A-module N defined by the generator v and relations $vw = 0$ whenever $w \prec u^\infty$ (i.e., when w is smaller than some power of u). The correspondence $t : vs \to vus$ is a well-defined endomorphism of N; hence N is an $A[t]$-module. The goal is to prove that $Nt^k = 0$ for some k.

(ii) If $Mt^k \in MJ$, where $\operatorname{Ann} M$ is the annihilator of M in A and $J := \operatorname{Jac}(\operatorname{Ann} M)$ (as an algebra without 1), then $Mt^{\ell k} \in MJ^\ell$ and $Nt^{\ell k} = 0$, for ℓ sufficiently large. Hence, viewing N as a module over $A[t]/J$, one may assume that $J = 0$.

(iii) Reduce the proof to the case when M is a faithful module over a semiprime ring B.

(iv) The elements from $\operatorname{Cent}(B)\setminus\{0\}$ all have annihilator 0. Localize with respect to them, and, by considering an algebraic extension of $\operatorname{Cent}(B)$, reduce to the case that B is a matrix algebra over a field.

(v) Applying Exercise 24, get a minimal nonzero right hyperword vu^∞ and hence, a contradiction to Corollary 2.7.4 of the Independence Theorem.)

26. A set of words S in a relatively free algebra $\mathcal{A}$, is an essential Shirshov base iff, for each word u of length not greater than the PI-degree of $\mathcal{A}$, S contains a word that is cyclically conjugate to some power of u. (Hint: Since $(uv)^n = u(vu)^{n-1}v$, if S contains two cyclically conjugate words, then one of them can be deleted from the base. Hence, sufficiency

is a consequence of Exercise 25. To prove necessity, suppose that u is non-cyclic of length $\leq n$. By Exercise 21, $A_u \sim_{\mathrm{PI}} A$, implying A_u is a quotient algebra of $\mathcal{A}$. The image of S must contain a non-nilpotent element. Conclude by Exercise 22.)

27. (Converse to Shirshov's Height Theorem.) Suppose A is a C-algebra. We make $a \in A$ **generically integral** of degree m if we replace A by

$$A \otimes_C C[\lambda_1, \ldots, \lambda_{m-1}] \Big/ \langle a^m - \sum_{i=0}^{m-1} \lambda_i a^i \rangle,$$

where $\lambda_1, \ldots, \lambda_{m-1}$ are new commuting indeterminates over C. Repeating this process for each a in a given subset W of A, we say W has been made **generically integral** of degree m. Also we say W is a **Kurosch set** for A if A becomes finite over C when we make W generically integral. Show that W is an essential Shirshov base iff W is a Kurosch set. (Hint: ($\Rightarrow$) is easy. Conversely, first note that if an ideal $\mathcal{I}$ of A has bounded essential height over W and $A/\mathcal{I}$ has bounded essential height over W, then A also has bounded essential height over W. Take $\mathcal{I}$ to be the evaluations of μ-Kemer polynomials for large enough μ; apply Zubrilin traces in view of Exercise 6.18. Proposition 6.11.3 guarantees that the induction terminates.)

28. Reformulate Exercise 27 as follows: W is an essential Shirshov base of A, iff $\varphi(A)$ is integral for any algebra homomorphism $\varphi : A \to B$ such that $\varphi(W)$ is integral (over the center of B).

29. Suppose A is affine PI, $\mathcal{I} \lhd A$, and $W \subset A$ such that $(W + \mathcal{I})/\mathcal{I}$ is an essential Shirshov base in $A/\mathcal{I}$. Then there exists a finite set of elements $S \subseteq \mathcal{I}$, such that $W \cup S$ is an essential Shirshov base in A. (Hint: Let A_k be the module spanned by words from W of degree $\leq k$. A_k is finite-dimensional for all k. By the definition of the essential height, there exists $\mu \in \mathbb{N}$ and a finite set S_k such that A_k is spanned by elements of essential height μ with respect to $S_k \cup W$, modulo a finite linear combination (which depends on k) of elements from $\mathcal{I}$. It remains to choose k sufficiently large.)

30. If $\mathcal{I} \lhd A$ and $A/\mathcal{I}$ respectively have bounded essential height over W and its image, then A also has bounded essential height over W.

31. (Essential height in the graded case.) Let A be a finitely generated graded associative algebra without 1 of PI-degree d, and $B \subset A$ be a finite set of homogeneous elements that generate A as an algebra. If $\bar{A} = A/\langle B^m \rangle$ is nilpotent, then A has bounded essential height over B. (Hint: Use Exercise 5.3 to show that A is spanned by elements of the form $v_0 b_0^{j_0} v_1 b_1^{j_1} \ldots b_{s-1}^{j_{s-1}} v_s$ where, for all i, $b_i \in M$, $|v_i| < m$, and $j_i \geq d$, and not more than $d - 1$ words v_i have length $\geq d$, and not more than $d - 1$ of the v_i are equal.)

32. (Berele's proof of Proposition 11.2.25.) By Theorem 7.4.4, there is a nilpotent ideal $\mathcal{J}$ such that $A/\mathcal{J}$ is the subdirect product of relatively free T-prime algebras $A_1, \ldots, A_q$ of T-prime ideals. Then $\mathcal{J}^s = 0$ by Lemma 7.2.24. It remains to compute GKdim(A) when A is relatively free T-prime. But these algebras were determined in Corollary 14, so it remains to compute the GKdim for the three kinds of prime T-ideals, cf. Definition 7.2.14. Note that $n \leq [d/2]$.

Case 1. If $\mathcal{I} = \mathcal{M}$, the GK dim is $(\ell - 1)n^2 + 1$, by Procesi's result.

Case 2. If $\mathcal{I} = \mathcal{M}'$, the GK dim is $(\ell - 1)n^2 + 1$. To see this, one displays the relatively free algebra in terms of generic matrices over the Grassmann algebra.

Case 3. If $\mathcal{I} = \mathcal{M}_{j,k}$ where $j + k = n$, the relatively free algebra is contained in that of Case 2, so the GK dim is bounded by that number. (Note: Berele [Ber93] computed the GK dim for $\mathcal{I} = \mathcal{M}_{j,k}$ to be precisely $(\ell - 1)(j^2 + k^2) + 2$.)

The multivariate Poincaré-Hilbert series

33. (Suggested by Berele.) Show that the multivariate Poincaré–Hilbert series satisfies the properties described in Lemma 11.3.9 and the ensuing theory, and conclude that Belov's theorem about the rationality of the Hilbert series of a relatively free PI-algebra holds in this more general context. (Hint: Over an infinite field, every T-ideal is homogeneous with respect to each variable, so the relatively free algebra is multi-graded. Every Kemer polynomial generates a homogeneous module with respect to formal traces, in view of Theorem J of §1.5.)

Chapter 12

More Representation Theory

In this chapter we introduce finer tools from representation theory (in characteristic 0), in order to gain even more precise information about the nature of polynomial identities. In particular, we refine some of the results of Chapter 3 on $F[S_n]$-modules (also called S_n-**modules**). Our main objective is to bring in GL-representations and use them to investigate homogeneous identities, in parallel to our earlier study of multilinear identities via S_n-representations.

We appeal again to the theory of Young diagrams, which describes the irreducible S_n-representations in terms of partitions λ of $\{1, \ldots, n\}$. Recall that $V_n = V_n(x_1, \ldots, x_n)$ is the F-vector space of the multilinear polynomials in $x_1, \ldots, x_n$, viewed naturally as an S_n-module, and, fixing a PI-algebra A, we define

$$\Gamma_n = \Gamma_n(A) = \mathrm{id}(A) \cap V_n. \tag{12.1}$$

The main innovations here are the more systematic use of the character of the module V_n/Γ_n itself, rather than just its dimension (the n-**codimension** c_n of A), and the use of GL-representations to study homogeneous identities. Much of this theory (in its PI-formulation) is due to Regev. Perhaps surprisingly, the same theory can be applied to trace identities, to be discussed in Volume II.

12.1 Cocharacters

Definition 12.1.1. *The S_n-character afforded by the quotient S_n-module V_n/Γ_n is called the n-th **cocharacter** of the algebra A, and is denoted by*

$$\chi_n(A) = \chi_{S_n}(V_n/\Gamma_n).$$

Obviously $c_n(A) = \deg \chi_n(A)$.

Remark 12.1.2. *In this way we have an action of S_n on V_n/Γ_n, i.e., the nonidentities. As usual, we assume that $\mathrm{char}(F) = 0$; hence, S_n-characters are completely reducible. This implies that*

$$\chi_n(A) = \sum_{\lambda \in \mathrm{Par}(n)} m_\lambda(A)\chi^\lambda$$

*for suitable **multiplicities** $m_\lambda(A)$. Determining the multiplicities would essentially determine the structure of $\mathrm{id}(A)$. Accordingly, the problem of computing the $m_\lambda(A)$ for a given PI-algebra A is perhaps the major open question in PI-theory. Recently, using the theory of Formanek [For84], Berele [Ber05] has shown for $M_k(\mathbb{Q})$ that $\sum_{\lambda \in \mathrm{Par}(n)} m_\lambda(A)$ asymptotically is $\alpha n^{\binom{k^2}{2}}$ for some constant α, and computed them for $k = 2$.*

Remark 12.1.3. *We can view the standard identity*

$$s_n = \sum_{\sigma \in S_n} \mathrm{sgn}(\sigma)\sigma$$

as the (central) semi-idempotent in the two-sided ideal J_λ, where $\lambda = (1^n)$.

As an example of the usefulness of cocharacters, let us show how the Capelli identity c_m is characterized in terms of strips of Young diagrams, following Regev [Reg78].

Proposition 12.1.4. *The algebra A satisfies the Capelli identity c_m iff its cocharacters are supported on a strip of Young diagrams of height $< m$:*

$$\chi_n(A) = \sum_{\lambda \in H(m,0;n)} m_\lambda(A)\chi^\lambda$$

(cf. Definition 3.3.3); namely, $m_\lambda(A) = 0$ if the length λ_1' of the first column of λ is $\geq m$.

Proof. ($\Rightarrow$) Assume that A satisfies c_m. Let T_λ be a tableau of shape λ with corresponding semi-idempotent $e_{T_\lambda} = R_{T_\lambda}^+ C_{T_\lambda}^- \in F[S_n] \equiv V_n$. By construction, $C_{T_\lambda}^-$ is alternating in λ_1' variables, so e_{T_λ} is a linear combination of such polynomials.

In general, suppose $f = f(x_1, \ldots, x_n) \in V_n$ is t-alternating. If f contains the monomial $\alpha \cdot w_0 x_1 w_1 x_2 \cdots w_{t-1} x_t w_t$ with coefficient $\alpha \neq 0$, where $w_0, \ldots, w_t$ are words in $x_{t+1}, \ldots, x_n$, some possibly equal to 1, then f contains

$$\alpha \cdot \sum_{\sigma \in S_t} \operatorname{sgn}(\sigma) w_0 x_{\sigma(1)} w_1 x_{\sigma(2)} \cdots w_{t-1} x_{\sigma(t)} w_t.$$

This is a consequence of c_m, provided $t \geq m$. Thus, if $\lambda_1' \geq m$, all the elements of the two-sided ideal $I_\lambda \subseteq F[S_n]$ are identities of A, which implies that $m_\lambda(A) = 0$.

($\Leftarrow$) Assume that $\chi_n(A) = \sum_{\lambda \in H(m,0;n)} m_\lambda(A) \chi^\lambda$. We claim that A satisfies c_m. Indeed, given any n, let λ be a partition of n with the corresponding two-sided ideal I_λ in $F[S_n]$. By assumption, the elements of I_λ are identities of A if $\lambda_1' > m$. Let $\mu = (1^m)$, and let J denote the right ideal $J = s_m F[S_{2m}]$ in $F[S_{2m}]$. Then Corollary 3.2.15 implies

$$J \subseteq \bigoplus_{\mu \leq \lambda} I_\lambda.$$

But $\mu \subseteq \lambda$ implies that $\lambda_1' \geq m$, and it follows that the elements of J are identities of A. In particular, $e_m \pi$ is an identity of A for any $\pi \in S_{2m}$.

Choose $\pi \in S_{2m}$ such that

$$x_1 \cdots x_{2m} \pi = x_1 x_{m+1} x_2 x_{m+2} \cdots x_m x_{2m},$$

cf. Lemma 3.1.10. Then $c_m = e_m \pi \in \operatorname{id}(A)$. $\qquad\square$

See Exercise 1 for an application of this result. Also, in Exercise 3.4, we saw how Amitsur and Regev [AmRe82] used the polynomials p^* from Definition 3.1.9 to create an explicit sparse identity with very nice properties.

12.1.1 A hook theorem for the cocharacters

In this subsection we elaborate Proposition 3.4.20. Here e always denotes the number $2.718 \cdots$.

Theorem 12.1.5. *Suppose A is a PI-algebra over a field F of characteristic 0. Then there exist positive integers k and ℓ such that the cocharacters $\chi_n(A)$ all are supported on the "(k, ℓ) hook" of Young diagrams; that is,*

$$\chi_n(A) = \sum_{\lambda \in H(k,\ell;n)} m_\lambda(A) \chi^\lambda.$$

Equivalently, $E_\mu^* \in \operatorname{id}(A)$ where μ is the $(k+1) \times (\ell+1)$ rectangle.
Such k and ℓ can be found explicitly: If A satisfies an identity of degree d, then we can choose any k and ℓ satisfying

$$\ell > \frac{e}{2} \cdot (d-1)^4 = a \qquad \text{and} \qquad k > \frac{a\ell}{\ell - a};$$

or in particular, we can take $k = \ell > e(d-1)^4$.

We need a preliminary for the proof. Let s^λ denote the number of standard tableaux of shape λ.

Remark 12.1.6. *Although in general Γ_n is not a two-sided ideal, Theorem 3.4.3 and Lemma 3.4.11 indicate that when n is large, Γ_n contains a large two-sided ideal of V_n, since for most $\lambda \vdash n$, $s^\lambda \geq (d-1)^2 > c_n(A)$.*

Proof of Theorem 12.1.5. Assume that A satisfies an identity of degree d. By Theorem 3.4.3, $c_n(A) \leq (d-1)^{2n}$. Let $\alpha = (d-1)^4$ and choose integers u, v such that $uv/(u+v) > e\alpha/2 = e(d-1)^4/2$. For example, choose (any) $u = v > e(d-1)^4$. Let $\lambda = (v^u) \vdash n$, where $n = uv$. Then, by Lemma 3.4.13,

$$s^\lambda > \left(\frac{2}{e} \cdot \frac{uv}{v+v}\right)^n > \alpha^n = (d-1)^{4n} \geq (d-1)^{2n} > c_n(A).$$

Hence by Lemma 3.4.11, $I_\lambda \subseteq \mathrm{id}(A)$, and in particular, $E_\lambda \in \mathrm{id}(A)$.

Moreover, if $\mu \vdash m$, $n \leq m \leq 2n-1$ and $\mu \geq \lambda$ then

$$s^\mu \geq s^\lambda \geq (d-1)^{4n} \geq (d-1)^{2m};$$

hence, also $I_\mu \subseteq \mathrm{id}(A)$. By Corollary 3.2.15(i) $F[S_m]I_\lambda F[S_m] \subseteq \mathrm{id}(A)$, and by Theorem 3.4.16 it follows that $E_\lambda^* \in \mathrm{id}(A)$. The proof now follows by Proposition 3.4.20, with $k = u-1$ and $\ell = v-1$. $\qquad\square$

12.2 $GL(V)$-Representation Theory

For the remainder of this chapter, fixing k, we let $V = \sum_{i=1}^k Fx_i$. Recall $T^n(V)$ from Definition 1.3.4, and the identification

$$F\{x_1, \ldots, x_k\} = T(V) = \oplus T^n(V).$$

Also $E_{n,k}$ denotes $\mathrm{End}_F(T^n(V))$. Let $GL(V)$ be the corresponding general linear group. The **diagonal** action of $GL(V)$ on $T^n(V)$ is given by:

$$g \mapsto \bar{g}, \quad \bar{g}(v_1 \otimes \cdots \otimes v_n) := g(v_1) \otimes \cdots \otimes g(v_n),$$

which, extended by linearity, clearly makes $T^n(V)$ into a left $GL(V)$-module. The $GL(V)$-representations obtained in this way are called **polynomial representations**.

Analogously, the Lie algebra $gl(V)$ acts on $T^n(V)$ by derivations, as follows:

$$\tilde{g}(v_1 \otimes \cdots \otimes v_n) = \sum_{i=1}^n (v_1 \otimes \cdots \otimes v_{i-1} \otimes gv_i \otimes v_{i+1} \otimes \cdots \otimes v_n).$$

We saw in Definition 3.3.1 that the group S_n also acts naturally on $T^n(V)$ via

$$\pi \mapsto \hat{\pi}, \quad \hat{\pi}(v_1 \otimes \cdots \otimes v_n) := v_{\pi(1)} \otimes \cdots \otimes v_{\pi(n)},$$

reversing the order of multiplication. (This action does not preserve identities.) Thus, $T^n(V)$ is a representation for $GL(V)$ as well as for S_n, and hence, for $GL(V) \times S_n$. We use the notation $\bar{g}$ and $\hat{\pi}$ to avoid ambiguity concerning the actions.

Remark 12.2.1.

(i) If $p(x) = p(x_1, \ldots, x_k) \in F\{X\}$ and $g \in GL(V)$, then

$$\bar{g}p(x_1, \ldots, x_k) = p(g(x_1), \ldots, g(x_k))$$

In particular, if $p(x) \in \mathrm{id}(A)$, then also $\bar{g}p(x) \in \mathrm{id}(A)$.

(ii) *Clearly, $\bar{g} \in E_{n,k}$ and the map $\psi_{n,k} : GL(V) \to E_{n,k}$, given by $\psi_{n,k} : g \to \bar{g}$, is multiplicative: $\overline{g_1 g_2} = \bar{g}_1 \bar{g}_2$. Similarly, the map $\Psi_{n,k} : g \to \tilde{g}$ is a Lie algebra homomorphism.*

12.2.1 Applying the Double Centralizer Theorem

Definition 12.2.2. *Denote by $\bar{B}(n,k)$ the subalgebra generated by the image $\psi_{n,k}(GL(V))$ in $E_{n,k}$.*

Note that for $\sigma \in S_n$,

$$(\bar{g}\hat{\sigma})(v_1 \otimes \cdots \otimes v_n) = (\hat{\sigma}\bar{g})(v_1 \otimes \cdots \otimes v_n).$$

Schur's celebrated "Double Centralizer Theorem" connects the S_n and GL actions:

Theorem 12.2.3 (Schur). *Let S_n and $GL(V)$ act on $T^n(V)$ as above, with corresponding subalgebras $A(n,k)$ and $\bar{B}(n,k)$ in $E_{n,k}$. Then $\bar{B}(n,k)$ is the centralizer of $A(n,k)$ in $E_{n,k}$, and $A(n,k)$ is the centralizer of $\bar{B}(n,k)$ in $E_{n,k}$.*

Also $\bar{B}(n,k) = \langle \Psi_{n,k}(gl(V)) \rangle$, as explained in [BerRe87]. This algebra is called the **enveloping algebra** of $GL(V)$ — or $gl(V)$ — in $\mathrm{End}_F(T^n(V))$.

Part of Schur's Double Centralizer Theorem can be obtained through easy considerations about modules, so we pause for a moment to collect these results.

Proposition 12.2.4. *Assume F is algebraically closed. Let M be a finite-dimensional F vector-space and let $A \subseteq \mathrm{End}_F(M)$ be a semisimple subalgebra. View M naturally as a left A-module. Write $A = \oplus_{i=1}^r A_i$ where $A_i \cong M_{d_i}(F)$, and let $J_i \subseteq A_i$ be minimal left ideals. Let $B = \mathrm{End}_A(M)$. Then*

(i) B is semisimple and has an isotypic decomposition similar to that of A (in Remark 3.2.2):

$$B = \bigoplus_{i=1}^{r} B_i \quad \text{where} \quad B_i \cong \operatorname{End}_{A_i}(J_i^{\oplus m_i}) \cong M_{m_i}(F).$$

(ii) Notation as in Remark 3.2.2, $M_i = B_i M$. If $L_i \subseteq B_i$ are minimal left ideals, then $\dim L_i = m_i$, and as a left B-module,

$$M \cong \bigoplus_{i=1}^{r} L_i^{\oplus d_i}.$$

Thus, the decomposition in Remark 3.2.2 is also the isotypic decomposition of M as a B-module.

(iii) $A = \operatorname{End}_B(M)$ and $A_i \cong \operatorname{End}_{B_i}(M_i)$. Also $\operatorname{End}_F(M_i) \cong A_i \otimes_F B_i$.

Remark 12.2.5. If M is a left A-module and $B = \operatorname{End}_A(M)$, then aM is a B-submodule of M, $\forall a \in A$. Similarly, bM is an A-submodule of M, $\forall b \in B$.

The following lemma will be useful in the sequel.

Lemma 12.2.6. As in Proposition 12.2.4, let $A, B \subseteq \operatorname{End}_F(M)$ be two semisimple algebras satisfying $B = \operatorname{End}_A(M)$ and $A = \operatorname{End}_B(M)$.

(i) Let $N \subseteq M$ be a left B-submodule of M. Then $N = eM$ for some idempotent $e \in A$.

(ii) Let $J_1, J_2 \subseteq A$ be left ideals with $J_1 \cap J_2 = 0$. Then $J_1 M \cap J_2 M = 0$. Similarly for any direct sum of left ideals.

(iii) If J is a minimal right ideal of A, then JM is a simple left B-submodule of M. Thus, if $e \in A$ is a primitive idempotent, then eM is B-simple.

Proof. (i) By Proposition 12.2.4 B is semisimple, so there is a B-submodule $M_2 \subseteq M$ with $M = N \oplus N'$. Let $e : M \to N$ be the projection map, so $e \in \operatorname{End}_B(M) = A$. Clearly, $N = eM$ and also, since e is a projection, $e = e^2$.

(ii) By Proposition 3.2.3, there are idempotents $e_1, e_2 \in A$ with $J_i = e_i A$ and thus, $J_i M = e_i M$, $i = 1, 2$. Suppose $a = e_1 w_1 = e_2 w_2 \in J_1 M \cap J_2 M$. Then $0 = e_1 e_2 w_2 = e_1^2 w_1 = e_1 w_1 = a$.

(iii) J is in some simple component A_i of A. Replacing A by A_i and B by B_i, we may assume A, B are simple, say $A \cong M_s(F)$ and $B \cong M_t(F)$. Then $\dim M = st$. Decompose $A = \bigoplus_{j=1}^{i} J_i$ with $J_1 = J$ and all $J_i \cong J$. By (ii), $M = \oplus_i J_i M$ and each $J_i M \neq 0$. Since $J_i M$ is a left B-module, $\dim J_i M \geq t$. Hence $st = \dim M = \sum \dim J_i M \geq st$, so equality holds and $\dim J_i M = t$ for each $1 \leq i \leq s$, proving that JM is a simple module. $\qquad\square$

12.2.2 Weyl modules

Remark 12.2.5 allows the construction of irreducible representations of $GL(V)$ and $gl(V)$ from those of S_n. This follows from the obvious fact that a subspace $M \subseteq T^n(V)$ is a $GL(V)$-module ($gl(V)$-module) exactly when it is a $\bar{B}(n,k)$-module. If $a \in F[S_n]$ (or $a \in A(n,k)$), then $aT^n(V)$ is a $B(n,k)$-module, hence a $GL(V)$-module ($gl(V)$-module). This leads us to another notion.

Definition 12.2.7. *Given a tableau T_λ, recall from Section 3.2 that e_{T_λ} is its corresponding semi-idempotent. Define the **Weyl module***

$$V_k^\lambda = e_{T_\lambda} T^n(V).$$

(Our notation emphasizes that $V = \sum_{i=1}^k Fx_i$.) These are parametrized by the strips $H(k,0;n)$, cf. Definition 3.3.3, in the next result.

Theorem 12.2.8. *Let $\lambda \in H(k,0;n)$, with T_λ a tableau of shape λ. Then*

(i) *The Weyl module V_k^λ is a simple $GL(V)$ (or $gl(V)$)-module. Moreover, if $\mu \vdash n$ and $\mu \notin H(k,0;n)$, then $I_\mu T^n(V) = 0$.*

(ii) *All simple $GL(V)$- (or $gl(V)$)-submodules of $W_\lambda = I_\lambda T^n(V)$ are isomorphic to V_k^λ.*

(iii) *As a $GL(V)$ (or $gl(V)$)-module, $W_\lambda \cong (V_k^\lambda)^{\oplus s^\lambda}$. Moreover,*

$$W_\lambda = \bigoplus_{T_\lambda} e_{T_\lambda} T^n(V),$$

where the sum is over all standard tableaux of shape λ.

(iv) *Let S^λ denote the corresponding Specht module. Then, as an S_n-module, $W_\lambda \cong (S^\lambda)^{\oplus s_k^\lambda}$, where $s_k^\lambda = \dim V_k^\lambda$. (This clearly follows from Proposition 12.2.4.)*

(v) *If the λ are tableaux of different shapes λ, then*

$$\sum_\lambda V_k^\lambda = \bigoplus_\lambda V_k^\lambda.$$

A straightforward application of Proposition 12.2.4 yields the following analog to Theorem 3.3.4:

Proposition 12.2.9.
$$B(n,k) = \bigoplus_{\lambda \in H(k,0;n)} B_\lambda.$$

(i) *Also $W_\lambda = B_\lambda T^n(V)$, and the decomposition in Theorem 3.3.4(iii) is also the isotypic decomposition of $T^n(V)$ as a $B(n,k)$ module.*

(ii) B_λ is the centralizer of A_λ in $\mathrm{End}_F(W_\lambda)$ and also is a simple algebra.

(iii) $\mathrm{End}_F(W_\lambda) \cong A_\lambda \otimes_F B_\lambda$.

Let us adjust Lemma 12.2.6 to the anti-homomorphism $\varphi_{n,k}$.

Lemma 12.2.10. *Let* $A = A(n,k) \subseteq E_{n,k} = \mathrm{End}_F(T^n(V))$, *and let* $B = B(n,k)$ *be its centralizer in* $E_{n,k}$, *as in Theorem 12.2.3.*

(i) *If* $J_1 \cap J_2 = 0$ *for left ideals* $J_1, J_2 \subseteq A(n,k)$, *then*

$$J_1 T^n(V) \cap J_2 T^n(V) = 0.$$

(ii) *If* $L \subseteq A$ *is a minimal right ideal in* $A(n,k)$, *then* $LT^n(V)$ *is a simple right* $B(n,k)$ *submodule of* $T^n(V)$. *Thus, if* $e \in A(n,k)$ *is a primitive idempotent, then* $eT^n(V)$ *is* $B(n,k)$*-simple. In particular, the Weyl modules are simple.*

Putting everything together yields the following result:

Theorem 12.2.11. (i) *The* $GL(V) \times S_n$ *decomposition of* $T^n(V)$ *is given by*

$$T^n(V) \cong \bigoplus_{\lambda \in H(k,0;n)} V_k^\lambda \otimes S^\lambda,$$

where S^λ *is the corresponding Specht module.*

(ii) *Thus, as a* $GL(V)$*-module,*

$$T^n(V) \cong \bigoplus_{\lambda \in H(k,0;n)} (V_k^\lambda)^{\oplus s^\lambda},$$

and as S_n*-modules,*

$$T^n(V) \cong \bigoplus_{\lambda \in H(k,0;n)} (S^\lambda)^{\oplus s_k^\lambda},$$

where $s_k^\lambda = \dim V_k^\lambda$.

The dimension of the Weyl module V_k^λ is given by tableaux as follows.

Definition 12.2.12. *Let* T_λ *be a tableau* λ *filled with integers* ≥ 0 *(repetitions allowed!).* T_λ *is called* **semi-standard** *if its rows are weakly increasing from left to right, and its columns are strictly increasing from top to bottom. Such a* T_λ *is a* **k-tableau** *if it is filled with the integers from the set* $\{1, \ldots, k\}$.

For example, $\begin{array}{|c|c|} \hline 1 & 1 \\ \hline 3 \\ \cline{1-1} \end{array}$, is a 3- (but not 2-) semi-standard tableau.

Theorem 12.2.13. $\dim(V_k^\lambda) = \dim(e_{T_\lambda} W_{n,k}) = s_k^\lambda$, *where s_k^λ is the number of k-semi-standard tableaux of shape λ.*

A k-tableau is a tableau filled, possibly with repetitions, with some of the integers from $\{1, \ldots, k\}$.

There is also a hook formula for $s_k^\lambda = \dim V_k^\lambda$; cf. [MacD95].

Theorem 12.2.14 (The GL hook formula). *For any partition λ,*

$$s_k^\lambda = \frac{\prod_{(i,j)\in\lambda}(k+j-i)}{\prod_{(i,j)\in\lambda} h_{i,j}(\lambda)}.$$

Consequently, if $\lambda_{k+1} > 0$, then $s_k^\lambda = 0$. Thus, $s_k^\lambda \neq 0$ if and only if $\lambda \in H(k,0;n)$.

12.2.3 Homogeneous identities

Let A be a PI-algebra. Recall that $V = \sum F_{i=1}^k x_i$. Let $\widehat{\Gamma}_n$ denote $T^n(V) \cap \mathrm{id}(A)$, the homogenous identities of degree n in $x_1, \ldots, x_k$. Clearly, in view of Remark 3.5.1, $\widehat{\Gamma}_n$ is a $GL(V)$-submodule of $T^n(V)$. As in the case of multilinear identities, we are led to consider the quotient space $T^n(V)/\widehat{\Gamma}_n$.

Definition 12.2.15. *The $GL(V)$-character of*

$$T^n(V)/\widehat{\Gamma}_n,$$

*which designates its isomorphism type as a $GL(V)$-module, is called the **homogeneous cocharacter** of the PI-algebra A, and is denoted by*

$$\chi_{n,k}(A) = \chi_{GL(V)}(T^n(V)/\widehat{\Gamma}_n).$$

By Theorem 12.2.8(3), for suitable multiplicities $m_\lambda'(A) \leq s^\lambda$,

$$T^n(V)/(T^n(V) \cap \mathrm{id}(A)) \cong \bigoplus_{\lambda\in H(k,0;n)} (V_k^\lambda)^{\oplus m_\lambda'(A)} = \bigoplus_{\lambda\in H(k,0;n)} m_\lambda'(A) \cdot V_k^\lambda$$

and therefore

$$\chi_{n,k}(A) = \bigoplus_{\lambda\in H(k,0;n)} m_\lambda'(A) \cdot V_k^\lambda.$$

In this way we have an action of GL_n on the homogeneous nonidentities.

12.2.4 Multilinear and homogeneous multiplicities and cocharacters

Having seen that the PI-theory studies multilinear identities via S_n-representations, and homogeneous identities via GL-representations. It behooves us to indicate how the two approaches coalesce.

Theorem 12.2.16 (Berele [Ber82], Drensky [Dr81b]). *The multiplicities m_λ for the homogeneous cocharacters are the same as for the multilinear cocharacters. More precisely:*

Let A be a PI-algebra and let

$$\chi_n(A) = \sum_{\lambda \vdash n} m_\lambda(A) \cdot \chi^\lambda$$

be its multilinear cocharacter. Let

$$\chi_{n,k}(A) = \sum_{\lambda \in H(k,0;n)} m'_\lambda(A) \cdot V_k^\lambda$$

denote the corresponding homogeneous cocharacter of A. Then $m'_\lambda(A) = m_\lambda(A)$ for every $\lambda \in H(k,0;n)$.

Note that the homogeneous cocharacters "capture" only the multiplicities m_λ where $\ell(\lambda) \leq k$.

Proof. The argument here is due to Berele [Ber82]. See [Dr81b] for a proof based on the Hilbert series.

Since

$$\chi_n(A) = \sum_{\lambda \vdash n} m_\lambda(A) \cdot \chi^\lambda,$$

therefore, up to isomorphism,

$$V_n \cong \Gamma_n \bigoplus \left(\bigoplus_{\lambda \vdash n} J_\lambda^{\oplus m_\lambda(A)} \right),$$

where J_λ is a minimal left ideal in I_λ, cf. Section 3.2. By Lemma 12.2.10, we have

$$T^n(V) = V_n T^n(V) = \Gamma_n T^n(V) \bigoplus \left(\bigoplus_{\lambda \in H(k,0;n)} (J_\lambda T^n(V))^{\oplus m_\lambda(A)} \right).$$

The proof now follows from the definition of $\chi_{n,k}(A)$ since, by Proposition 3.5.6, $\Gamma_n \cap T^n(V) = \widehat{\Gamma}_n$, and by Lemma 12.2.10(ii), $J_\lambda T^n(V)$ is isomorphic to the simple $GL(V)$-module V_k^λ, provided $\lambda \in H(k,0;n)$. $\qquad\square$

Exercises for Chapter 12

1. (Regev) If A satisfies the Capelli identity c_{m+1} and B satisfies c_{n+1}, then $A \otimes B$ satisfies c_{mn+1}. (Hint: Use Proposition 12.1.4.)

2. In the group algebra $F[S_n] = \oplus I_\lambda$, the element

$$E_\lambda = \sum_{\sigma \in S_n} \chi^\lambda(\sigma)\sigma \in I_\lambda$$

   is the (central) unit in I_λ.

Part IV

Supplementary Material

Part IV

Supplementary Material

List of Theorems

Chapter 1

(Theorem 1.1.31). *Suppose that a collection S of ideals satisfies the ACC. Then for any $\mathcal{I} \in S$, there are only finitely many primes $P_1, \ldots, P_n$ in S minimal over $\mathcal{I}$, and some finite product of primes is contained in $\mathcal{I}$.*

(Theorem 1.3.25). *For F algebraically closed, if $A = M_n(F)$ is G-graded, then $n = n_1 n_2$, where G has a subgroup H of order n_1^2, and elements $\{g_1, \ldots, g_{n_2}\}$, such that there is a graded isomorphism $A \cong A' \otimes A''$, where $A' = M_{n_1}(F)$ has a fine H-grading and $A'' = M_{n_2}(F)$ has an elementary grading.*

(Theorem 1.4.11, Amitsur-Levitzki). *s_{2n} is an identity of $M_n(C)$, for every commutative ring C.*

(Theorem 1.4.17, Razmyslov). *The $1 : 1$ correspondence between the multilinear central polynomials of $M_n(C)$ and the multilinear 1-weak identities that are not identities.*

Review of Major Structure Theorems in PI Theory

(Theorem A, Kaplansky). *Any primitive algebra A satisfying a PI of degree d has the form $M_t(D)$ where D is a division algebra of dimension m^2 over $\mathrm{Cent}(A)$, with $mt \leq \left[\frac{d}{2}\right]$.*

(Theorem B, Amitsur). *Any semiprime PI-algebra A has no nonzero left or right nil ideals.*

(Theorem C, Rowen). *Every ideal of a semiprime PI-algebra A intersects $\mathrm{Cent}(A)$ nontrivially.*

(Corollary E). *Any semiprime PI-algebra has suitable PI-class n (and PI-degree $2n$) for suitable n.*

(Theorem H). *Suppose A has PI-class n and $h(x_1, \ldots, x_t, x_{t+1}; \vec{y})$ is the t-alternating central polynomial of Remark F. If $h_n(a_1, \ldots, a_{n^2}, r_1, \ldots, r_m) = 1$, then A is a free $\mathrm{Cent}(A)$-module with base $a_1, \ldots, a_{n^2}$.*

More generally, if $h_n(a_1, \ldots, a_{n^2}, r_1, \ldots, r_m) = c$, then Ac is contained in a f.g. free C-submodule of A.

(Theorem I, Lewin).

(i) *Suppose* $I_1, I_2 \lhd A$. *Then there is an* $A/I_1, A/I_2$ *bimodule* M, *such that* $A/I_1 I_2$ *can be embedded into* $\begin{pmatrix} A/I_1 & M \\ 0 & A/I_2 \end{pmatrix}$ *via* $a \mapsto$ $\begin{pmatrix} a+I_1 & \delta(a) \\ 0 & a+I_2 \end{pmatrix}$, *where* $\delta : A \to M$ *is a derivation.*

(ii) *If the algebras* $F\{X\}/I_i$ *are embeddable in* $M_{n_i}(C)$, *then* $F\{X\}/I_1 I_2$ *is embeddable in an* (n_1, n_2)-*block triangular algebra.*

(Theorem J). *If A spans $M_n(K)$ over a field K, so that $t = n^2$, and if $\mathcal{I}$ denotes the additive subgroup of A generated by $h_n(A)$, then $\alpha \mathcal{I} \subseteq \mathcal{I}$ for every characteristic coefficient α of every element of A (viewed as a matrix).*

(Example 1.6.19, Example 1.6.21). *Uncountably many non-representable affine PI-algebras exist.*

(Theorem 1.6.22, Wehrfritz, Beidar). *Any finite algebra over a commutative Noetherian algebra over a field F, is representable.*

(Theorem 1.8.14). *For every field F and every n, $F\{\bar{Y}\}_n$ is a (noncommutative) domain, whose ring of central fractions, denoted $\mathrm{UD}(n, F)$, is a division algebra D of dimension n^2 over its center.*

(Theorem 1.8.26, Amitsur). *Every algebra A of PI-degree d satisfies a power s_{2n}^k of the standard identity, for suitable $n \leq \lceil \frac{d}{2} \rceil$ and for suitable k.*

(Theorem 1.8.28, Dubnov-Ivanov-Nagata-Higman). *Any $\mathbb{Q}$-algebra A without 1 which is nil of bounded index, is nilpotent.*

Chapter 2

(Theorem 2.2.2, Shirshov's Height Theorem).

(i) *Suppose $A = C\{a_1, \dots, a_\ell\}$ is an affine PI-algebra of PI-degree d, and let W be the set of words of length $\leq d$. Then $\{\bar{w}_1^{k_1} \dots \bar{w}_\mu^{k_\mu} : w_i \in W, \ k_i \geq 0\}$ span A.*

(ii) *If in addition $\bar{w}$ is integral over C for each word w in W, then A is finite over C.*

(Shirshov's Lemma 2.3.1). *For any ℓ, k, d, there is $\beta = \beta(\ell, k, d)$ such that any d-indecomposable word w of length $\geq \beta$ in ℓ letters must contain a nonempty subword of the form u^k, with $|u| \leq d$.*

(Theorem 2.3.11). *Any hyperword h is either d-strongly decomposable or is preperiodic of periodicity $< d$.*

(Theorem 2.5.8, Wedderburn's Principal Theorem and Mal'cev's Inertia Theorem). *For any f.d. algebra $\tilde{A}$ over an algebraically closed field, there is a vector space isomorphism*

$$\tilde{A} = \bar{A} \oplus \tilde{J}$$

where $\tilde{J} = \mathrm{Jac}(\tilde{A})$ is nilpotent and $\bar{A}$ is a semisimple subalgebra of $\tilde{A}$ isomorphic to $\tilde{A}/\tilde{J}$.

Furthermore, if there is another decomposition $\tilde{A} = \bar{A}' \oplus \tilde{J}$, then there is some invertible $a \in \tilde{A}$ such that $\bar{A}' = a\bar{A}a^{-1}$.

(Theorem 2.6.16). *Shirshov's function $\beta(\ell, k, d) \leq d^2 k \ell^d$.*

(Theorem 2.7.3, Independence of hyperwords). *Let h be a minimal non-zero hyperword on A, and w_i denote its initial subword of length i. Then, for any n, one of the following two conditions holds:*

(i) $\bar{w}_1, \ldots, \bar{w}_n$ are linearly independent.

(ii) h is pre-periodic of pre-periodicity $\leq n$, and its period is a tail of w_n.

(Theorem 2.8.2, Dilworth). *Let d be the maximal number of elements of an antichain in the sequence w. Then w can be subdivided into d disjoint chains.*

(Theorem 2.8.3). *$\beta(d, k, \ell) < 2^{27} \ell(kd)^{3 \log_3(kd) + 9 \log_3 \log_3(kd) + 36}$.*

(Theorem 2.8.4). *The essential height of an ℓ-generated PI-algebra of PI-degree d over the set of words of length $\leq d$ is less than $2d^{3^{\lceil \log_3 d \rceil} + 4} \ell$.*

(Theorem 2.8.5). *$\mu(\ell, d) \leq 2^{96} d^{12 \log_3 d + 36 \log_3 \log_3 d + 91} \ell$.*

Chapter 3

(Theorem 3.2.1, isotypic decomposition). *Let $A = F[G]$, where G is a finite group and F is a field of characteristic 0. Then A is semisimple, implying that any module M over A is semisimple; in particular, any submodule N of M has a **complement** N' such that $N \oplus N' = M$. Furthermore,*

$$A = \bigoplus_{i=1}^{r} I_i$$

where the $\mathcal{I}_i$'s are the minimal two-sided ideals of A.

(Theorem 3.2.11).

(i)

$$F[S_n] = \bigoplus_{\lambda \vdash n} I_\lambda = \bigoplus_{\lambda \vdash n} \left(\bigoplus_{T_\lambda \text{ is standard}} F[S_n] e_{T_\lambda} \right).$$

(ii)

$$I_\lambda \cong (S^\lambda)^{\oplus s^\lambda},$$

and hence,

$$F[S_n] \cong \bigoplus_{\lambda \vdash n} (J_\lambda)^{\oplus s^\lambda}$$

as S_n-modules.

(Branching Theorem 3.2.14).

(i) (Branching-up) If $\lambda \vdash n$, then

$$\chi^\lambda \uparrow^{S_{n+1}} = \sum_{\mu \in \lambda^+} \chi^\mu.$$

(ii) (Branching-down) If $\mu \vdash n+1$, then

$$\chi^\mu \downarrow_{S_n} = \sum_{\lambda \in \mu^-} \chi^\lambda.$$

(Theorem 3.2.16). *Let $\sigma \in S_n$ and*

$$\sigma \mapsto (P_\lambda, Q_\lambda)$$

in the RSK correspondence, and let $d = \ell(\lambda)$. Then d is the length of a maximal chain $1 \leq i_1 < \cdots < i_d \leq n$ such that $\sigma(i_1) > \cdots > \sigma(i_d)$.

(Theorem 3.2.20, Young-Frobenius formula). *Given*

$$\lambda = (\lambda_1, \ldots, \lambda_k) \vdash n,$$

let $t_i = \lambda_i + k - i$, $1 \leq i \leq k$. Then

$$s^\lambda = \frac{n!}{t_1! \cdots t_k!} \operatorname{Disc}_k(t_1, \ldots, t_k).$$

(Theorem 3.2.22, S_n hook formula). *If $\lambda \vdash n$, then $s^\lambda = \dfrac{n!}{\prod_{(i,j) \in \lambda} h_{(i,j)}(\lambda)}$.*

(Theorem 3.3.4).

(i) $\varphi_{n,k}(I_\lambda) \cong I_\lambda$ if $\lambda \in H(k,0;n)$, and $\varphi_{n,k}(I_\lambda) = 0$ otherwise.

(ii) $A(n,k) = \varphi_{n,k}(F[S_n]) = \bigoplus_{\lambda \in H(k,0;n)} A_\lambda \cong \bigoplus_{\lambda \in H(k,0;n)} I_\lambda$, *where for each* $\lambda \in H(k,0;n)$, $A_\lambda = \varphi_{n,k}(I_\lambda) \cong I_\lambda$.

(iii) $T^n(V) = \bigoplus_{\lambda \in H(k,0;n)} W_\lambda$, *where* $W_\lambda = I_\lambda T^n(V) = A_\lambda T^n(V)$, *and this is the isotypic decomposition of* $T^n(V)$ *as an* $A(n,k)$-*module, hence, as an* $F[S_n]$-*module.*

(Theorem 3.4.3, Regev-Latyshev). $c_n(A) \le (d-1)^{2n}$.

(Theorem 3.4.7, Regev). *If* A *and* B *are two PI-algebras over a field* F, *then* $A \otimes B$ *is also PI.*

(Theorem 3.4.16, Amitsur). *Let* A *be a PI-algebra and let* $I \subseteq F[S_n]$ *be a two-sided ideal in* $F[S_n] = V_n$.

(i) *Let* $f = f(x) \in I$ *and assume for* $n \le m \le 2n-1$ *that the elements of the two-sided ideal generated by* f *in* $F[S_m]$ *are identities of* A. *Then* $f^* \in \mathrm{id}(A)$.

(ii) *Assume* $I \subseteq \mathrm{id}(A)$ *and that for each* $f \in I$, $f^* \in \mathrm{id}(A)$.

Then for all $m \ge n$, *all the elements of the two-sided ideal generated by* I *in* $F[S_m]$ *are identities of* A: $F[S_m]IF[S_m] \subseteq \mathrm{id}(A)$.

(Theorem 3.4.17).

(i) $F[S_m]e_{T_{\lambda_0}}F[S_m] \subseteq \Gamma_m$, *for any tableau* T_{λ_0} *of* $\lambda_0 = (v^u)$.

(ii) Γ_m *contains every two-sided ideal of* S_m *corresponding to a shape containing the rectangle* (v^u).

Chapter 4

(Theorem 4.0.1, Braun-Kemer-Razmyslov). *The Jacobson radical* Jac(A) *of any affine PI-algebra* A *over a commutative Jacobson ring is nilpotent.*

(Theorem 4.0.2, Razmyslov). *If an affine algebra* A *over a field satisfies a Capelli identity, then its Jacobson radical* Jac(A) *is nilpotent.*

(Theorem 4.0.3, Razmyslov). *Let* A *be an affine algebra over a commutative Noetherian ring* C. *If* A *satisfies a Capelli identity, then any nil subring (without 1) of* A *is nilpotent.*

(Theorem 4.0.4, Kemer). *In characteristic zero, any affine PI-algebra satisfies some Capelli identity.*

(Theorem 4.0.6, Braun). *Any nil ideal of an affine PI-algebra over an arbitrary commutative Noetherian ring is nilpotent.*

(Theorem 4.0.7, Kemer). *If A is a PI-algebra (not necessarily affine) over a field F of characteristic $p > 0$, then A satisfies some Capelli identity.*

(Proposition 4.2.9, Zubrilin). *Let $f(x_1,\ldots,x_n; y_1,\ldots,y_n,\vec{z})$ be doubly alternating in $x_1,\ldots,x_n$ and in $y_1,\ldots,y_n$. Then for any polynomial h,*

$$\delta_{k,h}^{(\vec{x},n)}(f) \equiv \delta_{k,h}^{(\vec{y},n)}(f) \quad modulo \; \mathcal{CAP}_{n+1}; \tag{12.2}$$

namely,

$$\sum_{1 \le i_1 < \cdots < i_k \le n} f\,|_{x_{i_j} \to hx_{i_j}} \equiv \sum_{1 \le i_1 < \cdots < i_k \le n} f\,|_{y_{i_j} \to hy_{i_j}} \quad modulo \; \mathcal{CAP}_{n+1}.$$

(Proposition 4.3.9, Zubrilin). *Let $\overline{\mathcal{M}}$ be the module given by Definition 4.3.1. Then $\mathcal{I}_{n,\overline{C\{t\}}} \cdot \overline{\mathcal{M}} = 0$.*

(Theorem 4.6.6). *$\mathrm{Obst}_n(A) \cdot (\mathcal{CAP}_n(A))^2 = 0$, for any PI-algebra $A = C\{a_1,\ldots,a_\ell\}$ satisfying the Capelli identity c_{n+1}.*

Chapter 5

(Theorem 5.0.1, Kemer's Capelli Theorem). *Any PI-algebra over a field F of characteristic $p > 0$ satisfies a Capelli identity c_n, for large enough n.*

(Corollary 5.1.4). *If A satisfies a multilinear PI of degree n, then the identity of algebraicity $0 \ne D^t(y_1 \cdots y_n) \in \mathrm{id}(A)$.*

(Theorem 5.1.8). *Let $A = C[a_1,\ldots,a_\ell]$ be an affine PI-algebra satisfying a multilinear PI f of degree n. Let W be the set of words in a_i of degree $< n$, and $D^t(f)$ be the identity of algebraicity (Lemma 5.1.5) holding in A. Then A satisfies the Capelli identity c_q for q given in the proof.*

(Theorem 5.2.7). *If c_d is a d-identity of A (with respect to the d alternating variables $x_1,\ldots,x_d$), then $c_{2d-1} \in \mathrm{id}(A)$.*

(Theorem 5.2.8). *Every affine algebra $A = C\{a_1,\ldots,a_\ell\}$ satisfying a sparse system of degree d satisfies the Capelli identity c_n, where $n = \ell^d + d$.*

(Theorem 5.2.9). *Every PI-algebra satisfies a sparse system.*

(Theorem 5.2.14). *Every PI-algebra over a field of characteristic 0 satisfies non-trivial strong identities, described explicitly.*

(Theorem 5.2.15). *Every affine PI-algebra over a field of characteristic* 0 *satisfies some Capelli identity, described explicitly.*

(Theorem 5.2.21). *Every affine PI-algebra over a field of characteristic* 0 *satisfies some Capelli identity, described explicitly.*

(Example 5.2.18). *The* **wide staircase,** *a Young tableau satisfying Fayer's criterion.*

(Theorem 5.2.23). *Any affine PI-algebra over a commutative Noetherian base ring C satisfies some Capelli identity.*

Chapter 6

(Specht's Lemma 6.2.2). *Every multilinear identity f of an algebra A (with* 1*) is a consequence of Spechtian identities.*

(Theorem 6.2.7). *Every Spechtian polynomial f can be reduced modulo the Grassmann identity to a sum of polynomials of the form* $[x_{i_1}, x_{i_2}] \cdots [x_{i_{2k-1}}, x_{i_{2k}}]$*, where* $i_1 < i_2 < \cdots < i_{2k}$.

(Theorem 6.2.9, Regev). *In characteristic* 0*, the T-ideal of G is a consequence of the Grassmann identity.*

(Theorem 6.3.1, Kemer's PI-representability Theorem). *Any affine PI-algebra W over a field F of characteristic* 0 *is PI-equivalent to some f.d. algebra over a purely transcendental extension of F.*

(Proposition 6.5.2, Kemer's First Lemma). *If F is algebraically closed and A is f.d. full, then* $\beta(A) = t_A$.

(Proposition 6.6.31, Kemer's Second Lemma). *If A is a basic f.d. algebra with char*$(F) = 0$*, then* $\gamma(A) = s_A$*. In this case, A has a μ-Kemer polynomial for every μ. In fact, for any full polynomial f satisfying property K, the T-ideal generated by f contains a μ-Kemer polynomial for A whose degree is* $\deg f + \nu_A(\mu)$.

(Theorem 6.8.2). *Let W be an algebra satisfying the ℓ-th Capelli identity c_ℓ. Then* $\mathrm{Id}(\mathcal{W}) = \mathrm{Id}(W)$ *where $\mathcal{W}$ is the relatively free algebra of W generated by $\ell - 1$ indeterminates.*

(Theorem 6.8.5). *Any Kemer polynomial f of a T-ideal Γ satisfies the Phoenix property.*

(Theorem 6.10.2, Kemer). *Any T-ideal of $F\{x_1, \ldots, x_\ell\}$ is finitely based (assuming* char$(F) = 0$).

Chapter 7

(Theorem 7.0.1). *Every affine ungraded-PI superalgebra is PI_2-equivalent to a finite dimensional superalgebra.*

(Theorem 7.1.11). *Let A_0 be a PI-algebra, and let*

$$p = p(x_1, \ldots, x_n) = \sum_{\sigma \in S_n} \alpha_\sigma x_{\sigma(1)} \cdots x_{\sigma(n)}$$

be a multilinear polynomial. Then the following assertions are equivalent:

(i) $p \in \mathrm{id}(A_0)$.

(ii) $\tilde{p}_I^*$ *is a superidentity of $A = A_0 \otimes G$ for some subset $I \subseteq \{1, \ldots, n\}$.*

(iii) $\tilde{p}_I^*$ *is a superidentity of $A = A_0 \otimes G$ for every subset $I \subseteq \{1, \ldots, n\}$.*

(Theorem 7.1.14). *Let A be a PI superalgebra, and let $p = p(x_1, \ldots, x_n) = \sum_{\sigma \in S_n} \alpha_\sigma x_{\sigma(1)} \cdots x_{\sigma(n)}$ be a multilinear polynomial. Then $p \in \mathrm{id}(G(A))$ if and only if $\tilde{p}_I^* \in \mathrm{id}_2(A)$ for every subset $I \subseteq \{1, \ldots, n\}$.*

(Theorem 7.1.15). *There is a correspondence Ψ from $\{$varieties of superalgebras$\}$ to $\{$varieties of algebras$\}$ given in the sense that*

$$A_0 \sim_{\mathrm{PI}} G(A_0 \otimes G), \qquad A_0 \text{ any algebra}$$

and

$$A \sim_{\mathrm{PI}_2} G(A) \otimes G, \qquad A \text{ any superalgebra.}$$

(Theorem 7.2.1, Kemer). *Let A be a PI-algebra over a field of characteristic 0. Then $\mathrm{id}_2(A \otimes G) = \mathrm{id}_2(\tilde{A})$, for a suitable affine superalgebra $\tilde{A}$.*

(Proposition 7.2.8). *Suppose F is an algebraically closed field.*

(i) The only nontrivial $\mathbb{Z}/2$-gradings on $M_n(F)$ yield superalgebras isomorphic to those in Example 1.3.14.

(ii) Any f.d. supersimple superalgebra is either as in (i) or as in Example 7.2.7.

(iii) Any f.d. semisimple superalgebra A is isomorphic (as superalgebra) to a direct product $\prod_{i=1}^k S_i$ of supersimple superalgebras of form (i) or (ii).

(Theorem 7.2.11). *Any f.d. superalgebra A over an algebraically closed field has a semisimple super-subalgebra $\bar{A}$ such that $A = \bar{A} \oplus \mathrm{Jac}(A)$.*

(Theorem 7.2.12). *Suppose that A is a f.d. superalgebra of superdimension (t_0, t_1). Consider $V = A_0$ as a vector space. For any C-linear homogeneous map $T : V \to V$, letting*

$$\lambda^{t_0} + \sum_{k=0}^{t_0-1} (-1)^k \alpha_k \lambda^{t_0-k}$$

denote the characteristic polynomial of T (as a $t_0 \times t_0$ matrix), then for any (t_0, t_1)-alternating superpolynomial f, and any even homogeneous $a_1, \ldots, a_{t_0}$ in V_0, we have

$$\alpha_k f(a_1, \ldots, a_{t_0}, r_1, \ldots, r_m) = \sum f(T^{k_1} a_1, \ldots, T^{k_t} a_{t_0}, r_1, \ldots, r_m),$$

where $r_1, \ldots, r_m \in A$ and summed over all vectors $(k_1, \ldots, k_{t_0})$ for which $k_1 + \cdots + k_{t_0} = k$.

(Theorem 7.2.13, Kemer). *Any affine PI-superalgebra W over a field F of characteristic 0 is **super-PI representable**, i.e., $\sim_{\mathrm{PI}_2}$-equivalent to some f.d. superalgebra over a purely transcendental extension of F.*

(Theorem 7.2.28). *Any Kemer superpolynomial of a T_2-ideal Γ satisfies the Phoenix property.*

(Theorem 7.2.29). *Let W be a superalgebra which satisfies the ℓ-th Capelli identity c_ℓ. Then $\mathrm{id}_2(W) = \mathrm{id}_2(\mathcal{W})$, where $\mathcal{W}$ is the relatively free superalgebra of W generated by $\ell - 1$ even indeterminates and $\ell - 1$ odd indeterminates.*

(Theorem 7.3.1, Kemer). *The ACC holds for T_2-ideals of affine superalgebras over a field of characteristic 0.*

(Theorem 7.3.2, Kemer). *The ACC holds for T-ideals, over any field of characteristic 0.*

(Theorem 7.4.4, Kemer). *Any T-ideal $\mathcal{I}$ of $\mathcal{A}$ has a radical $\sqrt{\mathcal{I}}$ that is a finite intersection of verbally prime T-ideals, and $\sqrt{\mathcal{I}}^k \subseteq \mathcal{I}$ for some k.*

(Theorem 7.4.9, Kemer). *Suppose that R is a superalgebra which is also an ungraded PI-algebra. Then R is verbally prime if and only if it is graded PI-reducible to the Grassmann envelope of a supersimple f.d. superalgebra.*

(Theorem 7.4.10, Kemer). *The following list is a complete list of the verbally prime T-ideals of $F\{X\}$, for any field F of characteristic 0 :*

(i) $\mathrm{id}(M_n(F))$;

(ii) $\mathrm{id}(M_{k,\ell}(F))$;

(iii) $\mathrm{id}(M_n(G))$.

(Theorem 7.4.16, Kemer). *Every verbal ideal $\mathcal{I}$ has only a finite number of verbal primes $P_1, \ldots, P_t$ minimal over $\mathcal{I}$, and some finite product of the P_i is contained in $\mathcal{I}$.*

(Theorem 7.4.20, Kemer). *Any algebra of characteristic 0 satisfying s_n satisfies $c_{n'}$ for some n' that depends only on n.*

(Theorem 7.4.29, Shchigolev). *Let S be a T-space in the free algebra $A = F\{X\}$ over a field of characteristic 0.*
  (i) If $S\backslash[A, A] \neq \emptyset$, then S includes a nonzero T-ideal.
  (ii) If $S \subseteq [A, A]$, then there exists a T-ideal Γ such that $[\Gamma, A] \subseteq S$.

Chapter 8

(Theorem 8.1.9). *If $F[S_n] = \oplus_{\lambda \vdash n} I_\lambda$, the direct sum of its minimal (two-sided) ideals, then*

$$\ker \varphi_{n,k} = \bigoplus_{\lambda \vdash n, \, \lambda_1' \geq k+1} I_\lambda.$$

(Theorem 8.1.12). *Write $\sigma \in S_n$ as a product of disjoint cycles*

$$\sigma = (i_1, \ldots, i_a)(j_1, \ldots, j_b) \cdots (k_1, \ldots, k_c),$$

and let $A_1, \ldots, A_n \in \text{End}(V)$, so that

$$(A_1 \otimes \cdots \otimes A_n) \circ \varphi_{n,k}(\sigma) \in \text{End}(T^n(V)).$$

Then

$$\text{tr}((A_1 \otimes \cdots \otimes A_n) \circ \varphi_{n,k}(\sigma))$$
$$= \text{tr}(A_{i_1} \cdots A_{i_a}) \text{tr}(A_{j_1} \cdots A_{j_b}) \cdots \text{tr}(A_{k_1} \cdots A_{k_c}).$$

(Theorem 8.1.18). *Let $f(x) = f_a(x) = f_a(x_1, \ldots, x_n)$ be a multilinear trace polynomial, where $a \in F[S_n]$. Then $f_a(x)$ is a trace identity of $M_k(F)$ if and only if*

$$a \in \bigoplus_{\lambda \vdash n, \, \lambda_1' \geq k+1} I_\lambda.$$

(Theorem 8.2.1).

  (i) *The T-ideal of the pure trace identities of $k \times k$ matrices is generated by the single trace polynomial $f_{e_{\bar\mu}}$, where $\bar\mu = (1^{k+1})$ and*

$$e_{\bar\mu} = \sum_{\pi \in S_{k+1}} \text{sgn}(\pi)\pi.$$

(ii) The T-ideal of the mixed trace identities of $k \times k$ matrices is generated by the two elements $f_{e_{\bar{\mu}}}$ and $\tilde{g}_k$.

(Theorem 8.3.2). $\chi_n^{Tr}(M_k(F)) = \bigoplus_{\lambda \vdash n, \, \lambda_1' \leq k} \chi^\lambda \otimes \chi^\lambda.$

(Theorem 8.4.1, Kemer). *Any PI-algebra A in characteristic $p > 0$ satisfies the identities of $M_m(\mathbb{Z}/p)$ for suitable m.*

Chapter 9

(Example 9.3.2). *Define*

$$f_n = [y_1^2, y_2] x_1^2 x_2^2 \ldots x_n^2 [y_1^2, y_2]^3.$$

Over a field F of characteristic 2, the T-ideal generated by $\{f_n : n \in \mathbb{N}\}$ is not finitely based.

(Theorem 9.3.7). *The T-space generated by $\{P_n : n \in \mathbb{N}\}$ is not finitely based.*

(Theorem 9.4.12). *(char $F = p > 2$.) P_n does not belong to the T-space generated by the P_m, $\forall m < n$.*

(Theorem 9.4.19). *The T-ideal generated by the polynomials $\{Q_n : n \in \mathbb{N}\}$ is not finitely based.*

Chapter 10

(Theorem 10.1.2, Belov). *Every affine left Noetherian PI-algebra (over an arbitrary commutative ring) is finitely presented.*

(Theorem 10.2.4, Isaacs-Passman). *If $\text{char}(F) = 0$ and $F[G]$ satisfies a PI, then G has a normal Abelian subgroup of finite index.*

(Theorem 10.3.2, Lichtman). *The enveloping algebra $U(L)$ of a Lie algebra L over a field F of characteristic 0 satisfies a PI iff L is Abelian.*

Chapter 11

(Theorem 11.2.4, Bergman's gap). *If* $\mathrm{GKdim}(A) > 1$, *then* $\mathrm{GKdim}(A) \geq 2$.

(Example 11.2.9). *A PI-algebra with non-integral GK dim.*

(Theorem 11.2.12). *For any prime affine PI-algebra* A, $\mathrm{GKdim}(A)$ *is an integer, equal to the transcendence degree of the center of* A.

(Theorem 11.2.15). *If* $\mu = \mu(\ell, d)$ *is as in Shirshov's Height Theorem, then* $\mathrm{GKdim}(A) \leq \mu$ *for any affine algebra* $A = F\{a_1 \ldots, a_\ell\}$ *of PI-degree* d.

(Theorem 11.2.21). *Suppose* $A \subseteq M_n(C)$ *is affine, and* $\mathrm{GKdim}(A) > 0$. *Then* $\mathrm{GKdim}(A) = \nu(A)$.

(Theorem 11.3.5, Bell). *Suppose* A *is a prime affine PI-algebra. Then any generating set of* A *can be expanded to a rational generating set.*

(Theorem 11.3.6). *If* A *is an affine PI-algebra and its Hilbert series* $H_A = 1 + \sum \tilde{d}_n \lambda^n$ *is rational, then there are numbers* m, d *such that*

$$H_A = \frac{p(\lambda)}{(1 - \lambda^m)^d},$$

with $p(1) \neq 0$. *In this case,* $\mathrm{GKdim}(A) = d$.

(Theorem 11.3.13). *In characteristic* 0, *every relatively free, affine PI-algebra has a rational Hilbert series.*

(Theorem 11.3.15). *If* A *is affine and its graded algebra is Noetherian PI, then* A *has a rational generating set.*

Chapter 12

(Theorem 12.1.5). *Suppose* A *is a PI-algebra over a field* F *of characteristic* 0. *Then there exist positive integers* k *and* ℓ *such that the cocharacters* $\chi_n(A)$ *all are supported on the "(k, ℓ) hook" of Young diagrams; that is,*

$$\chi_n(A) = \sum_{\lambda \in H(k, \ell; n)} m_\lambda(A) \chi^\lambda.$$

(Theorem 12.2.3, Schur's Double Centralizer Theorem). *Let* S_n *and* $GL(V)$ *act on* $T^n(V)$ *as above, with corresponding subalgebras* $A(n, k)$ *and* $\bar{B}(n, k)$ *in* $E_{n,k}$. *Then* $\bar{B}(n, k)$ *is the centralizer of* $A(n, k)$ *in* $E_{n,k}$, *and* $A(n, k)$ *is the centralizer of* $\bar{B}(n, k)$ *in* $E_{n,k}$.

(Theorem 12.2.8). *Let $\lambda \in H(k,0;n)$, with T_λ a tableau of shape λ. Then*

(i) *The Weyl module V_k^λ is a simple $GL(V)$ (or $gl(V)$)-module. Moreover, if $\mu \vdash n$ and $\mu \notin H(k,0;n)$, then $I_\mu T^n(V) = 0$.*

(ii) *All simple $GL(V)$- (or $gl(V)$)-submodules of $W_\lambda = I_\lambda T^n(V)$ are isomorphic to V_k^λ.*

(iii) *As a $GL(V)$ (or $gl(V)$)-module, $W_\lambda \cong (V_k^\lambda)^{\oplus s^\lambda}$, where s^λ equals the number of standard tableaux of shape λ. Moreover,*

$$W_\lambda = \bigoplus_{T_\lambda} e_{T_\lambda} T^n(V),$$

where the sum is over all standard tableaux of shape λ.

(iv) *Let S^λ denote the corresponding Specht module. Then, as an S_n-module, $W_\lambda \cong (S^\lambda)^{\oplus s_k^\lambda}$, where $s_k^\lambda = \dim V_k^\lambda$.*

(v) *If the λ are tableaux of different shapes λ, then $\sum_\lambda V_k^\lambda = \bigoplus_\lambda V_k^\lambda$.*

(Theorem 12.2.11).

(i) *The $GL(V) \times S_n$ decomposition of $T^n(V)$ is given by*

$$T^n(V) \cong \bigoplus_{\lambda \in H(k,0;n)} V_k^\lambda \otimes S^\lambda,$$

where S^λ is the corresponding Specht module.

(ii) *Thus, as a $GL(V)$-module,*

$$T^n(V) \cong \bigoplus_{\lambda \in H(k,0;n)} (V_k^\lambda)^{\oplus f^\lambda},$$

and as S_n-modules, $T^n(V) \cong \bigoplus_{\lambda \in H(k,0;n)} (S^\lambda)^{\oplus s_k^\lambda}$, where $s_k^\lambda = \dim V_k^\lambda$.

(Theorem 12.2.13). $\dim(V_k^\lambda) = \dim(e_{T_\lambda} W_{n,k})$ *is the number of k-semi-standard tableaux of shape λ.*

(Theorem 12.2.14, The GL hook formula). *For any partition λ,*

$$s_k^\lambda = \frac{\prod_{(i,j)\in\lambda}(k+j-i)}{\prod_{(i,j)\in\lambda} h_{i,j}(\lambda)}.$$

(Theorem 12.2.16, Berele, Drensky). *The multiplicities m_λ for the homogeneous cocharacters are the same as for the multilinear cocharacters.*

Some Open Questions

Throughout, G denotes the Grassmann algebra.

Central Polynomials

1. Describe the identities of minimal degree of $M_n(G)$, as well as for the other verbally prime T-ideals. This is more complicated than $M_n(\mathbb{Q})$; the case $n = 2$ by Vishne [Vi02, Vi11] required the computer.

2. What are the central polynomials of minimal degree for $M_n(\mathbb{Q})$?

   Although Razmyslov's method in principle gives all the central polynomials for matrices in terms of 1-weak identities, this question stubbornly remains. Likewise Razmyslov [Raz87] found central polynomials for the superalgebras $M_{k,\ell}(F)$.

3. What are the central polynomials of minimal degree for $M_n(G)$? For the other verbally prime T-ideals?

4. For p prime, does $M_p(\mathbb{Q})$ have a noncentral polynomial f such that f^p is central? This easily formulated question is equivalent to division algebras of degree p being cyclic, a long-standing open question. Thus, it is known to be true for $p \leq 3$, but is open for all other p. It is easy to see for $p \geq 2$ that f cannot be multilinear.

Representability

1. Is every algebra over a field, that is finite over its center, weakly representable?

2. Can an affine PI-algebra R be embedded as a subalgebra of the form $\mathrm{End}_H M$, where M is finite over a commutative algebra H?

   These questions remain open even with extra assumptions.

3. If an algebra R over a field is finite over its center and is a subring of a Noetherian ring W, then is R representable (or just weakly representable)?

4. Suppose an associative algebra R over a field is finite over its center, satisfies the ACC on right annihilators, is local and irreducible, and every prime ideal is maximal. Then is R weakly representable?

5. Is an Artinian PI-algebra over a field F weakly representable? (This holds with radical squared 0, by [RoSm15, Theorem F].)

T-Ideals

The interplay between trace identities and identities remains tantalizing. Although the Cayley-Hamilton polynomial generates all trace identities of matrices, and thus $\mathcal{M}_n$, by restriction, we do not know how to describe these explicitly. This is discussed in Averyanov and Kemer [KemA08], which contains some questions.

1. Find a set of identities that generate the T-ideal $\mathcal{M}_n$. (This is only known for $n = 2$.)

2. Find a set of identities that generate the T-ideal $\mathrm{id}(M_n(G))$.

3. What is the minimal m such that $s_m \in \mathrm{id}(M_n(G))$? in characteristic p? Is $m = (2n-1)p + 1$? Kemer [Kem95] proved that $(2n-3)p - 1 < m \leq 2pn$.

4. Find a set of trace identities that generate $\mathrm{id}(M_n(G))$. (De Concini, Papi, and Procesi [DiPP13] showed that the superidentities z^{2n} and $z\,\mathrm{tr}(z^{2n-1}) = \sum_{i=1}^{n-1} z^{2i} \wedge \mathrm{tr}(z^{2(n-i)-1}) + nz^{2n-1}$ generate as superalgebras.)

   In positive characteristic we still are far from the full picture. For example, the following question surprisingly remains open:

5. Is the T-ideal (in an infinite number of indeterminates) of $M_n(F)$ finitely based, when F is an infinite field of characteristic p?

6. Does Specht's conjecture hold for T-ideals generated by multilinear polynomials? (This is true in characteristic p.)

7. Classify the prime T-ideals over an infinite field of characteristic > 0. (This is extremely difficult, with the best results provided by Kemer [Kem96]).

8. (Kemer) Is Remark 7.4.12 true in positive characteristic?

9. (L'vov) Is every PI-algebra a homomorphic image of a torsion-free PI-algebra? (This would enable one to transfer several theorems directly from characteristic 0 to characteristic p.)

10. (Procesi's problem) Let $C\{Y\}_n$ denote the algebra of generic matrices over C, cf. Definition 1.8.11. There is a natural homomorphism $\Phi_p\, \mathbb{Z}\{Y\}_n \to (\mathbb{Z}/p\mathbb{Z})\{Y\}_n$, obtained by taking mod p, and Schelter showed that $\ker \Phi_2$ properly includes $2\mathbb{Z}\{Y\}_2$. Kemer [Kem03] found examples for all n, with suitable $p < n$.

Fixing n, is $\ker \Phi_p = p\mathbb{Z}\{Y\}_n$ for large enough p? Averyanov and Kemer [KemA06] proved it for $n = 3$ and $p > 3$. (Belov has showed this is true in the affine case, i.e., when Y is finite.)

11. What are the prime ideals of the group algebra of the infinite symmetric group, in positive characteristic?

12. What are the T-prime varieties in the class of algebras having a formal trace function for which $\mathrm{tr}(1) = 0$?

Cocharacters

1. What are the cocharacters of $\mathrm{id}(M_n(\mathbb{Q}))$? for $\mathrm{id}(M_n(G))$? $\mathcal{M}_{k,\ell}$?

   This is perhaps the most studied question in the theory, and considerable advances have been made using multivariate Hilbert series, as described at the end of Chapter 11. Nevertheless, the solution still seems far away.

2. What is the best asymptotic formula for the codimensions of a T-ideal? (Some major advances have been made by Berele, Giambruno, Regev, and Zaicev, cf. the bibliography.)

GK-Dimension and Essential Height

1. Is there an example of a prime Goldie ring of GK dimension 2 that is neither primitive nor PI?

2. Is the GK dimension of any left Noetherian PI-algebra A an integer? More precisely, is $\mathrm{GKdim}(A) = \mathrm{GKdim}(A/N)$ where N is the nilpotent radical of A?

3. Under what conditions is there a faithful representation $\rho : A \to M_n(C)$ satisfying $\mathrm{GKdim}(A) = \mathrm{GKdim}(C)$?

4. Find the best possible formulas for GK dimension of affine PI-algebras.

5. Find the best possible formulas for the essential (Shirshov) height of affine PI-algebras. (Recall this is the same as for GK dim for representable PI-algebras.)

6. Is the essential height polynomially bounded?

7. Is the difference between the essential height and the GK dim bounded? By 1?

8. Is there any natural criterion weaker than representability that will guarantee that a PI-algebra has finite GK dimension?

9. What can one say about the (minimal) kernel of a PI-algebra into a representable algebra?

Nonassociative Algebras

Volume II is to cover the nonassociative theory in detail. The theory of Lie PI-algebras in particular is an intriguing area of research. Often the results are more negative than for the associative theory. For example, Question 5 of T-ideals is known to have a negative answer for f.d. Lie algebras.

1. Does every affine alternative algebra satisfy all identities of a f.d. alternative algebra? What about Jordan algebras? Find some workable (perhaps asymptotic) analog of Lewin's Theorem.

2. Does the Specht problem have an affirmative answer for affine Lie algebras? (Iltyakov [Ilt91] obtained an affirmative answer in characteristic 0 for Lie algebras satisfying a Capelli identity.)

Bibliography

[AB10]   Aljadeff, E. and Kanel-Belov, A., *Representability and Specht problem for G-graded algebras*, Adv. Math. **225** (2010), no. 5, 2391–2428.

[AB12]   Aljadeff, E. and Kanel-Belov, A., *Hilbert series of PI relatively free G-graded algebras are rational functions*, Bull. Lond. Math. Soc. **44** (2012), no. 3, 520–532.

[AH14]   Aljadeff, E. and Haile, D., *Simple G-graded algebras and their polynomial identities*, Trans. Amer. Math. Soc. **366** (2014), no. 4, 1749–1771.

[AlDF85]   Almkvist, G., Dicks, W., and Formanek, E., *Hilbert series of fixed free algebras and noncommutative classical invariant theory*, J. Algebra **93** (1985), no. 1, 189–214.

[Am57]   Amitsur, S.A., *A generalization of Hilbert's Nullstellensatz*, Proc. Amer. Math. Soc. **8** (1957), 649–656.

[Am66]   Amitsur, S.A., *Rational identities and applications to algebra and geometry*, J. Algebra **3** (1966), 304–359.

[Am70]   Amitsur, S.A., *A noncommutative Hilbert basis theorem and subrings of matrices*, Trans. Amer. Math. Soc. **149** (1970), 133–142.

[Am71]   Amitsur, S.A., *A note on PI-rings*, Israel J. Math **10** (1971), 210–211.

[Am72]   Amitsur, S.A., *On central division algebras*, Israel J. Math **12** (1972), 408–420.

[Am74]   Amitsur, S.A., *Polynomial identities*, Israel J. Math **19** (1974), 183–199.

[AmL50]   Amitsur, S.A. and Levitzki J., *Minimal identities for algebras*, Proc. Amer. Math. Soc. **1** (1950), 449–463.

[AmPr66]   Amitsur, S.A. and Procesi C., *Jacobson-rings and Hilbert algebras with polynomial identities*, Ann. Mat. Pura Appl. **71** (1966), 61–72.

[AmRe82]   Amitsur, S.A. and Regev A., *PI-algebras and their cocharacters*,
            J. Algebra **78** (1982), no. 1, 248–254.

[AmSm93]   Amitsur, S.A. and Small, L., *Affine algebras with polyno-
            mial identities*, Recent developments in the theory of algebras
            with polynomial identities (Palermo, 1992), Rend. Circ. Mat.
            Palermo **31** (1993), 943.

[An77]   Anan'in, A.Z., *Locally residually finite and locally representable
            varieties of algebras*, Algebra i Logika **16** (1977), no. 1, 3–23.

[An92]   Anan'in, A.Z., *The representability of finitely generated algebras
            with chain condition*, Arch. Math. **59** (1992), 1–5.

[Ani82a]   Anick, D., *The smallest singularity of a Hilbert series*, Math.
            Scand. **51** (1982), 35–44.

[Ani82b]   Anick, D., *Non-commutative graded algebras and their Hilbert
            series*, J. Algebra **78** (1982), no. 1, 120–140.

[Ani88]   Anick, D., *Recent progress in Hilbert and Poincare series*, Lect.
            Notes Math, vol. 1318 (1988), 1–25.

[ArSc81]   Artin, M. and Schelter, W., *Integral ring homomorphisms*, Adv.
            Math. **39** (1981), 289–329.

[Ba74]   Bakhturin, Yu.A., *Identities in the universal envelopes of Lie
            algebras*, Australian Math. Soc. **18** (1974), 10–21.

[Ba87]   Bakhturin, Yu.A., Identical relations in Lie algebras, Trans-
            lated from the Russian by Bakhturin, VNU Science Press, b.v.,
            Utrecht, (1987).

[Ba89]   Bakhturin, Yu.A. and Ol'shanskiĭ, A.Yu., *Identities. Current
            problems in mathematics*, Fundamental Directions **18**, Itogi
            Nauki i Tekhniki, Akad. Nauk SSSR, Vsesoyuz. Inst. Nauchn.
            i Tekhn. Inform., Moscow, (1988), 117–240, in Russian.

[Ba91]   Bakhturin, Yu.A., *Identities*, Encyclopedia of Math. Sci. **18**
            Algebra II, Springer-Verlag, Berlin-New York (1991), 107–234.

[BaSZ01]   Bahturin, Yu.A., Sehgal, S.K., and Zaicev, M.V., *Group grad-
            ings on associative algebras*, J. Algebra **241** (2001), no. 2, 677–
            698.

[BaZa02]   Bahturin, Yu. A., and Zaicev, M.V., *Group gradings on matrix
            algebras*, (Dedicated to Robert V. Moody, Canad. Math. Bull.
            45 (2002), no. 4, 499–508.

[BaZa03]   Bahturin, Yu. A., and Zaicev, M.V., *Graded algebras and graded identities*, Polynomial identities and combinatorial methods (Pantelleria, 2001), Lecture Notes in Pure and Appl. Math. **235**, Dekker, New York, (2003), 101–139.

[BaZS08]   Bakhturin, Yu.A., Zaitsev, M.V., and Segal, S.K., *Finite-dimensional simple graded algebras*, (Russian) Mat. Sb. 199 (2008), no. 7, 21–40; translation in Sb. Math. **199** (2008), 965–983.

[BGGKPP]   Bandman, T., Grunewald, F., Gruel G.M., Kunyavski B., Pfister G., and Plotkin E., *Two-variable identities for finite solvable groups*, C.R. Acad. Sci. Paris, Ser. I **337** (2003), 581–586.

[Bei86]   Beĭdar, K. I., *On A. I. Mal'tsev's theorems on matrix representations of algebras*, (Russian) Uspekhi Mat. Nauk **41** (1986), no. 5, (251), 161–162.

[BeMM96]   Beĭdar, K.I., Martindale, W.S., and Mikhalev, A.V., *Rings with generalized identities*, Monogr. Textbooks Pure Appl. Math. **196**, Dekker, New York (1996).

[Bell03]   Bell, J. P., *Examples in finite Gelfand-Kirillov dimension*, J. Algebra **263**, no. 1, (2003), 159–175.

[Bell04]   Bell, J. P., *Hilbert series of prime affine PI-algebras*, Israel J. Math. **139** (2004), 1–10.

[Bel88a]   Belov, A., *On Shirshov bases in relatively free algebras of complexity n*, Mat. Sb. **135** (1988), no. 3, 373–384.

[Bel88b]   Belov, A., *The height theorem for Jordan and Lie PI-algebras*, in: Tez. Dokl. Sib. Shkoly po Mnogoobr. Algebraicheskih Sistem, Barnaul (1988), pp. 12–13.

[Bel89]   Belov, A., *Estimations of the height and Gelfand-Kirillov dimension of associative PI-algebras*, In: Tez. Dokl. po Teorii Koletz, Algebr i Modulei. Mezhdunar. Konf. po Algebre Pamyati A.I.Mal'tzeva, Novosibirsk (1989), p. 21.

[Bel92]   Belov, A., *Some estimations for nilpotency of nil-algebras over a field of an arbitrary characteristic and height theorem*, Comm. Algebra **20** (1992), no. 10, 2919–2922.

[Bel97]   Belov, A., *Rationality of Hilbert series with respect to free algebras*, Russian Math. Surveys **52** (1997), no. 10, 394–395.

[Bel99]   Belov, A., *On non-Specht varieties*, Fundam. Prikl. Mat. **5** (1999), no. 6, 47–66.

[Bel00]   Belov, A., *Counterexamples to the Specht problem*, Sb. Math. **191** (3-4) (2000), 329–340.

[Bel02]   Belov, A., *Algebras with polynomial identities: Representations and combinatorial methods*, Doctor of Science Dissertation, Moscow (2002).

[Bel07]   Belov, A., *Burnside-type problems, height and independence theorems*, (Russian) Fund. Prikl. Matem. **13** (2007), no. 5, 19–79; translation in J. Math. Sci. (N. Y.) **156** (2009), no. 2, 219–260.

[BelBL97]   Belov, A., Borisenko, V.V., and Latyshev, V.N., *Monomial algebras. Algebra 4*, J. Math. Sci. (New York) **87** (1997), no. 3, 3463–3575.

[BelIv03]   Belov, A. Ya. and Ivanov, I. A., *Construction of semigroups with some exotic properties*, Comm. Algebra **31** (2003), no. 2, 673–696.

[BelK12]   Belov, A. Ya. and Kharitonov, M.I., *Subexponential estimates in Shirshov theorem on height*, Sb. Math. **203** (2012), no. 3-4, 534–553.

[BelRo05]   Belov, A. Ya. and Rowen, L.H., Computational sspects of polynomial rdentities, A.K. Peters (2005) (First edition of this volume).

[Ber82]   Berele, A., *Homogeneous polynomial identities*, Israel J. Math. **42** (1983), no. 3, 258–272.

[Ber85a]   Berele, A., *On a theorem of Kemer*, Israel J. Math. **51** (1985), no. 1–2, 121–124.

[Ber85b]   Berele, A., *Magnum PI*, Israel J. Math. **51** (1985), no. 1–2, 13–19.

[Ber88]   Berele, A., *Trace identities and $\mathbb{Z}/2\mathbb{Z}$ graded invariants*, Trans. Amer. Math. Soc. **309** (1988), 581–589.

[Ber93]   Berele, A., *Generic verbally prime PI-algebras and their GK-dimensions*, Comm. Algebra **21** (1993), no. 5, 1487–1504.

[Ber94]   Berele, A., *Rates of growth of p.i. algebras*, Proc. Amer. Math. Soc. **120** (1994), no. 4, 1047–1048.

[Ber05]   Berele, A., *Colength sequences for matrices*, J. Algebra **283** (2005), 700–710.

[BerBe99]   Berele, A., and Bergen., K., *PI algebras with Hopf algebra actions*, J. Algebra **214** (1999), 636–651.

[BerRe87]   Berele, A., and Regev, A., *Hook Young diagrams with applications to combinatorics and to representations of Lie superalgebras*, Adv. Math. **87**, no. 2, 118–175.

[BergP92]   Bergen, J., and Passman, D.S., *Delta methods in enveloping algebras of Lie superalgebras*, Trans. Amer. Math. Soc. **334** (1992), no. 1, 295–280.

[Bergm70]   Bergman, G., *Some examples in PI ring theory*, Israel J. Math. **18** (1970), 1–5.

[BergmD75]   Bergman, G., and Dicks, W., *On universal derivations*, J. Algebra **36** (1975), 193–211.

[Bir35]   Birkhoff, G., *On the structure of abstract algebras*, Proc. Cambridge Phil. Soc. **31** (1935), 433–431.

[Bog01]   Bogdanov I., *Nagata-Higman's theorem for hemirings*, Fundam. Prikl. Mat. **7** (2001), no. 3, 651–658 (in Russian).

[BLH88]   Bokut', L.A., L'vov, I.V., and Harchenko, V.K., *Noncommutative rings*, In: Sovrem. Probl. Mat. Fundam. Napravl. Vol. 18, Itogi Nauki i Tekhn., All-Union Institute for Scientific and Technical Information (VINITI), Akad. Nauk SSSR, Moscow (1988), 5–116.

[BoKr76]   Borho, W., and Kraft, H., *Uber die Gelfand-Kirillov dimension*, Math. Ann. **220** (1976), no. 1, 1–24.

[Br81]   Braun, A., *A note on Noetherian PI-rings*, Proc. Amer. Math. Soc. **83** (1981), 670–672.

[Br82]   Braun, A., *The radical in a finitely generated PI-algebra*, Bull. Amer. Math. Soc. **7** (1982), no. 2, 385–386.

[Br84]   Braun, A., *The nilpotency of the radical in a finitely generated PI-ring*, J. Algebra **89** (1984), 375–396.

[BrVo92]   Braun, A., and Vonessen, N., *Integrality for PI-rings*, J. Algebra **151** (1992), 39–79.

[BrPS12]   Bresar, M., Procesi, C., and Spenko, S., *Quasi-identities on matrices and the Cayley-Hamilton polynomial*, arXiv:1212.4597 (2012).

[Che94a]   Chekanu, G.P., *On independence in algebras*, Proc. Xth National Conference in Algebra (Timişoara, 1992), Timişoara (1994), 25–35.

[Che94b]   Chekanu, G.P., *Independency and quasiregularity in algebras*,
              Dokl. Akad. Nauk **337** (1994), no. 3, 316-319; translation: Rus-
              sian Acad. Sci. Dokl. Math. **50** (1995), no. 1, 84–89.

[ChKo93]   Chekanu, G.P. and Kozhuhar', E.P., *Independence and nilpo-
              tency in algebras*, Izv. Akad. Nauk Respub. Moldova Mat. **2**
              (1993), 51–62, 92–93, 95.

[ChUf85]   Chekanu, G.P., and Ufnarovski'i, V.A., *Nilpotent matrices*,
              Mat. Issled. no. 85, Algebry, Kotsa i Topologi (1985), 130–141,
              155.

[Ch01]   Chibrikov, Ye. S. *On Shirshov height of a finitely generated as-
              sociative algebra satisfying an identity of degree four* (Russian),
              Izvestiya Altaiskogo gosudarstvennogo universiteta **1** (2001),
              52–56.

[Co46]   Cohen, I.S., *On the structure and ideal theory of complete local
              rings*, Trans. Amer. Math. Soc. **59** (1946), 54–106.

[CuRei62]   Curtis, C.W. and Reiner, I., Representation theory of finite
              groups and of associative algebras, Interscience Publishers, New
              York, London, Wiley (1962).

[De22]   Dehn, M., *Uber die Grundlagen der projektiven Geometrie und
              allgemeine Zahlsysteme*, Math. Ann. **85** (1922), 184–193.

[DiPP13]   De Concini, C., Papi, P., and Procesi, C., *The adjoint repre-
              sentation inside the exterior algebra of a simple Lie algebra*,
              arXiv:1311.4338 (2013)

[Don92]   Donkin, S., *Invariants of several matrices*, Invent. Math. **110**
              (1992), no. 2, 389–401.

[Don93]   Donkin, S., *Invariant functions on matrices*, Math. Proc. Cam-
              bridge Philos. Soc. **113** (1993), no. 1, 23–43.

[Don94]   Donkin, S., *Polynomial invariants of representations of quivers*,
              Comment. Math. Helv. **69** (1994), no. 1, 137–141.

[Dor12]   Dor, G., *The extended Grassmann algebra and its applications
              in PI-theory*, Doctoral dissertation, Bar-Ilan University, 2012.

[DorKV15]   Dor, G., Kanel-Belov, A., and Vishne, U., *The Grassmann
              algebra in arbitrary characteristic and generalized sign*,
              arXiv:1501.02464

[Dr74]   Drensky, V., *About identities in Lie algebras*, Algebra and Logic
              13 (1974), no. 3,

[Dr81a]   Drensky, V., *Representations of the symmetric group and varieties of linear algebras*, Mat. Sb. **115** (1981), no. 1, 98-115.

[Dr81b]   Drensky, V., *Codimensions of T-ideals and Hilbert series of relatively free algebras*, C.R. Acad. Bulgare Sci. **34** (1981), no. 9, 1201–1204.

[Dr81c]   Drensky, V., *A minimal basis for identities of a second order matrix over a field of characteristic 0*, Algebra and Logic **20** (1981), 282-290.

[Dr84]   Drensky, V., *Codimensions of T-ideals and Hilbert series of relatively free algebras*, J. Algebra **91** (1984), no. 1, 1–17.

[Dr95]   Drensky, V., *New central polynomials for the matrix algebra*, Israel J. Math. **92** (1995), no. 1-3, 235–248.

[Dr00]   Drensky, V., Free algebras and PI-algebras: Graduate course in algebra, Springer-Verlag, Singapore (2000).

[DrFor04]   Drensky, V. and Formanek, E., Polynomials identity rings, CRM Advanced Courses in Mathematics, Birkhäuser, Basel (2004).

[DrKa84]   Drensky, V. and Kasparian, A., *A new central polynomial for* 3×3 *matrices*, Comm. Algebra **13** (1984), no. 3, 745–752.

[DrRa93]   Drensky, V. and Rashkova, T.G., *Weak polynomial identities for the matrix algebras*, Comm. Algebra **21** (1993), no. 10, 3779–3795.

[Fay05]   Fayers, M., *Irreducible Specht modules for Hecke algebras of type A*, Adv. Math. **193** (2005), 438–452.

[FaLM06]   Fayers, M., Lyle, S., and Martin, S., *p-restriction of partitions and homomorphisms between Specht modules*, J. Algebra **306** (2006), 175–190.

[Fl95]   Flavell, P., *Finite groups in which every two elements generate a soluble group*, Invent. Math. **121** (1995), 279–285.

[For72]   Formanek, E., *Central polynomials for matrix rings*, J. Algebra **23** (1972), 129–132.

[For82]   Formanek, E., *The polynomials identities of matrices*, Algebraists' Homage: Papers in Ring Theory and Related Topics (New Haven, Conn., 1981), Contemp. Math. **13** (1982), 41–79.

[For84]   Formanek, E., *Invariants and the ring of generic matrices*, J. Algebra **89** (1984), no. 1, 178–223.

[For87]   Formanek, E., *The invariants of $n \times n$ matrices*, Invariant The-
             ory, Lecture Notes in Math. **1278**, Springer, Berlin-New York
             (1987), 18–43.

[For91]   Formanek, E., *The polynomial identities and invariants of $n \times n$
             matrices*, CBMS Regional Conference Series in Mathematics
             78, Amer. Math. Soc. (1991).

[For02]   Formanek, E., *The ring of generic matrices*, J. Algebra **258**
             (2002), no. 1, 310–320.

[ForL76]   Formanek, E. and Lawrence J., *The group algebra of the infinite
             symmetric group*, Israel J. Math. **23** (1976), no. 3–4, 325–331.

[GeKi66a]   Gelfand, I.M., and Kirillov, A.A., *Sur les corps lies aux algebras
             envelopantes des algebras de Lie*, Publs. Math. Inst. Hautes
             Etud. Sci. **31** (1966), 509–523.

[GeKi66b]   Gelfand, I.M., and Kirillov, A.A., *On division algebras related
             to enveloping algebras of Lie algebras*, Dokl. Akad. Nauk SSSR
             167 (1966), no. 3, 503–505.

[GRZ00]   Giambruno, A., Regev, A., and Zaicev, M., *Simple and
             semisimple Lie algebras and codimension growth*, Trans. Amer.
             Math. Soc. **352** (2000), no. 4, 1935–1946.

[GiZa98]   Giambruno, A. and Zaicev, M., *On codimension growth of
             finitely generated associative algebras*, (English summary) Adv.
             Math. Sb. Math. 203 (2012), no. 3-4, 534–553 **140** (1998), no. 2,
             145–155.

[GiZa03a]   Giambruno, A. and Zaicev, M., *Codimension growth and mini-
             mal superalgebras*, Trans. Amer. Math. Soc. **355** (2003), no. 12,
             5091–5117.

[GiZa03b]   Giambruno, A. and Zaicev, M., *Asymptotics for the standard
             and the Capelli identities*, Israel J. Math. Soc. **135** (2003), 125–
             145.

[GiZa03c]   Giambruno, A. and Zaicev, M., *Minimal varieties of algebras
             of exponential growth*, Adv. Math. **174** (2003), no. 2, 310–323.

[GiZa05]   Giambruno, A. and Zaicev, M., *Polynomial identities and
             asymptotic methods*, A.M.S. Mathematical Surveys and Mono-
             graphs **122** (2005).

[Gol64]   Golod, E.S., *On nil-algebras and residually finite p-groups*, Izv.
             Akad. Nauk SSSR **28** (1964), no. 2, 273–276.

[GoWa89] Goodearl, K.R., and Warfield, R.B., *An introduction to non-commutative Noetherian rings*, London Math. Soc. Student Texts, vol. 16 (1989).

[Gov72] Govorov, V.E., *On graded algebras*, Mat. Zametki **12** (1972), no. 2, 197–204.

[Gov73] Govorov, V.E., *On dimension of graded algebras*, Mat. Zametki **14** (1973) no. 2, 209–216.

[Gre80] Green, J.A., Polynomial representations of GL_n, Lecture Notes in Math., vol. 830, Springer-Verlag, Berlin-New York (1980).

[Gri87] Grishin, A.V., *The growth index of varieties of algebras and its applications*, Algebra i Logika **26** (1987), no. 5, 536–557.

[Gri99] Grishin, A.V., *Examples of T-spaces and T-ideals in characteristic 2 without the finite basis property*, Fundam. Prikl. Mat. **5** (1) (1999), no. 6, 101–118 (in Russian).

[GuKr02] Gupta, C.K., and Krasilnikov, A.N., *A simple example of a non-finitely based system of polynomial identities*, Comm. Algebra **36** (2002), 4851–4866.

[Ha43] Hall, M., *Projective planes*, Trans. Amer. Math. Soc. **54** (1943), 229–277.

[Hel74] Helling, H., *Eine Kennzeichnung von Charakteren auf Gruppen und assoziativen Algebren*, Comm. Alg. **1** (1974), 491–501.

[Her68] Herstein, I.N., *Noncommutative rings*, Carus Math. Monographs **15**, Math. Assoc. Amer. (1968).

[Her71] Herstein, I.N., *Notes from a ring theory conference*, CBMS Regional Conference Series in Math., no. 9, Amer. Math. Soc. (1971).

[HerSm64] Herstein, I. and Small, L., *Nil rings satisfying certain chain conditions*, Canadian J. Math. (1964), 771–776.

[Hig56] Higman, G., *On a conjecture of Nagata*, Proc. Cam. Phil. Soc. **52** (1956), 1–4.

[Ilt91] Iltyakov, A.V., *Finiteness of basis identities of a finitely generated alternative PI-algebra*, Sibir. Mat. Zh. **31** (1991), no. 6, 87–99; English translation: Sib. Math. J. **31** (1991), 948–961.

[Ilt92] Iltyakov, A.V., *On finite basis identities of identities of Lie algebra representations*, Nova J. Algebra Geom. **1** (1992), no. 3, 207—259.

[Ilt03]   Iltyakov, A.V., *Polynomial Identities of Finite Dimensional Lie Algebras*, monograph (2003).

[IsPa64]   Isaacs, I.M., and Passman, D.S., *Groups with representations of bounded degree*, Canadian J. Math **16** (1964), 299-309.

[Jac75]   Jacobson, N., *PI-Algebras, an Introduction*, Springer Lecture Notes in Math. **441**, Springer, Berlin-New York (1975).

[Jac80]   Jacobson, N., *Basic Algebra II*, Freeman (1980).

[JamK80]   James, G. D. and Kerber, A., *The Representation theory of the symmetric group*, Encyclopedia of Mathematics and its Applications, Vol. 16, Addison–Wesley, Reading, MA (1981).

[JamMa99]   James, G. and Mathas, A., *The irreducible Specht modules in characteristic 2*, Bull. London Math. Soc. **31** (1999), 457–62.

[Kap48]   Kaplansky, I., *Rings with a polynomial identity*, Bull. Amer. Math. Soc. **54** (1948), 575–580.

[Kap49]   Kaplansky, I., *Groups with representations of bounded degree*, Canadian J. Math. **1** (1949), 105–112.

[Kap50]   Kaplansky, I., *Topological representation of algebras. II*, Trans. Amer. Math. Soc. **68** (1950), 62–75.

[Kap70a]   Kaplansky, I., *"Problems in the theory of rings" revisited*, Amer. Math. Monthly **77** (1970), 445–454.

[Kap70b]   Kaplansky, I., Commutative rings, Allyn and Bacon, Inc., Boston (1970).

[Kem78]   Kemer, A.R., *Remark on the standard identity*, Math. Notes **23** (5) (1978), 414–416.

[Kem80]   Kemer, A.R., *Capelli identities and the nilpotency of the radical of a finitely generated PI-algebra*, Soviet Math. Dokl. **22** (3) (1980), 750–753.

[Kem81]   Kemer, A.R., *Nonmatrix varieties with polynomial growth and finitely generated PI-algebras*, Ph.D. Dissertation, Novosibirsk (1981).

[Kem87]   Kemer, A.R., *Finite basability of identities of associative algebras (Russian)*, Algebra i Logika **26** (1987), 597–641; English translation: Algebra and Logic **26** (1987), 362–397.

[Kem88]   Kemer, A.R., *The representability of reduced-free algebras*, Algebra i Logika **27** (1988), no. 3, 274–294.

[Kem91]   Kemer, A.R., *Ideals of identities of associativealgebras*, Amer. Math. Soc. Translations of Monographs **87** (1991).

[Kem91b]   Kemer, A.R., *Identities of finitely generated algebras over an infinite field*, Math-USSR Izv.**37** (1991), 69–96.

[Kem93]   Kemer, A.R., *The standard identity in characteristic p: a conjecture of I.B. Volichenko*, Israel J. Math. **81** (1993), 343–355.

[Kem95]   Kemer, A.R., *Multilinear identities of the algebras over a field of characteristic p*, Internat. J. Algebra Comput. **5** (1995), no. 2, 189–197.

[Kem96]   Kemer, A.R., *Remarks on the prime varieties*, Israel J. Math. **96** (1996), part B, 341–356.

[Kem03]   Kemer, A.R, *On some problems in PI-theory in characteristic p connects with dividing by p*, Proc. of 3 Intern. Alg. Conf., Kluwer Acad. Publish. (2003), 53–67.

[KemA06]   Kemer, A.R. and Averyanov, I.V., *Conjecture of Procesi for 2-generated algebra of generic 3×3 matrices*, J. Algebra **299** (2006), 151–170.

[KemA08]   Kemer, A.R. and Averyanov, I.V., *Some problems in PI-theory*, Advances in Algebra and Combinatorics, ed. K.P. Shum et al. **96** (1996), part B, 189–204.

[Kh2011a]   Kharitonov, M. I., *Estimates for structure of piecewise periodicity in Shirshov height theorem* (Russian), Vestnik Moskovskogo universiteta, Ser. 1, Matematika. Mekhanika. arxiv.org/abs/1108.6295.

[Kh2011b]   Kharitonov, M., *Estimations of the particular periodicity in case of the small periods in Shirshov Height theorem*, arXiv: 1108.6295.

[Kh2015]   Kharitonov, M. I., *Estimates in Shirshov height theorem*, Chebyshevskii Sbornik **15**, no. 4 (2015), 55–120 (in Russian).

[Kl85]   Klein, A.A., *Indices of nilpotency in a PI-ring*, Archiv der Mathematik, **44** (1985), 323–329.

[Kl00]   Klein, A.A., *Bounds for indices of nilpotency and nility*, Archiv der Mathematik, **74** (2000), 6–10.

[Kn73]   Knuth, D.E., The art of computer programming. Volume 3. Sorting and searching, Addison-Wesley Pub. (1991).

[Kol81]   Kolotov, A.T., *Aperiodic sequences and growth functions in algebras*, Algebra i Logika **20** (1981), no. 2, 138–154.

[Kos58]   Kostant, B., *A theorem of Frobenius, a theorem of Amitsur-Levitzki and cohomology theory*, J. Math. Mech. **7** No. 2 (1958), 237–264.

[KrLe00]   Krause, G.R., and Lenagan, T.H., *Growth of algebras and Gelfand-Kirillov dimension*, Amer. Math. Soc. Graduate Studies in Mathematics **22** (2000).

[Kuz75]   Kuzmin, E.N., *About Nagata-Higman Theorem*, Proceedings dedicated to the 60th birthday of Academician Iliev, Sofia (1975), 101–107 (in Russian).

[Lat63]   Latyshev, V.N., *Two remarks on PI-algebras*, Sibirsk. Mat. Z. **4** (1963), 1120–1121.

[Lat73]   Latyshev, V.N., *On some varieties of associative algebras*, Izv. Akad. Nauk SSSR **37** (1973), no. 5, 1010–1037.

[Lat77]   Latyshev, V.N., *Nonmatrix varieties of associative rings*, Doctor of Science Dissertation, Moscow (1977).

[Lat88]   Latyshev, V.N., Combinatorial ring theory. Standard bases, Moscow University Press, Moscow (1988), (in Russian).

[LeBPr90]   Le Bruyn, L. and Procesi, C., *Semisimple representations of quivers*, Trans. Amer. Math. Soc. **317** (1990), no. 2, 585–598.

[Lev46]   Levitzki, J., *On a problem of Kurosch*, Bull. Amer. Math. Soc. **52** (1946), 1033–1035.

[Lev50]   Levitzki, J., *A theorem on polynomial identities*, Proc. Amer. Math. Soc. **1** (1950), 449–463.

[Lew74]   Lewin, J., *A matrix representation for associative algebras. I and II*, Trans. Amer. Math. Soc. **188** (1974), no. 2, 293–317.

[Lic89]   Lichtman, A.I., *PI-subrings and algebraic elements in enveloping algebras and their field of fractions*, J. Algebra. **121** (1989), no. 1, 139–154.

[Lop11]   Lopatin, A.A., *On the nilpotency degree of the algebra with identity $x^n = 0$*, arXiv:1106.0950v1.

[LoSm82]   Lorenz, M. and Small, L., *On the Gelfand-Kirillov dimension of Noetherian PI-algebras*, Contemporary Math. **13** (1982), 199–205.

[MacD95]   MacDonald, I.G., Symmetric functions and Hall polynomials, 2nd ed., with contributions by A. Zelevinsky, Oxford Mathematical Monographs, The Clarendon Press, Oxford University Press (1995).

[Mal36]   Mal'tsev, A.I., *Unterschungen ays tem Gebiete des mathemati-cischen Logik*, Mat. Sb. 1 **43** (1936), 323–336.

[Mar79]   Markov, V.T., *On generators of T-ideals in finitely generated free algebras*, Algebra i Logika **18** (1979), no. 5, 587–598.

[Mar88]   Markov, V.T., *Gelfand-Kirillov dimension: nilpotency, repre-sentability, nonmatrix varieties*, In: Tez.Dokl. Sib. Shkola po Mnogoobr. Algebraicheskih Sistem, Barnaul (1988), 43–45.

[Neu67]   Neumann, H., Varieties of groups, Springer-Verlag, Berlin-New York (1967).

[Oa64]   Oates, S., and Powell, M.B., *Identical relations in finite groups*, J. Algebra **1** (1964), no. 1, 11–39.

[PaSmWa]   Pappacena, Ch.J., Small, L.W., and Wald, J., *Affine semiprime algebras of GK dimension one are (still) p.i.*, Glasg. Math. J. **45** (2003), no. 2, 243–247.

[Pas71]   Passman, D.S., *Group rings satisfying a polynomial identity II*, Pacific J. Math. **39** (1971), 425-438.

[Pas72]   Passman, D.S., *Group rings satisfying a polynomial identity*, J. Algebra **20** (1972), 103-117.

[Pas77]   Passman, D.S., The algebraic structure of group rings, Wiley (1977), rev. ed. Krieger(1985), republished Dover (2011).

[Pas89]   Passman, D.S., *Crossed products and enveloping algebras satis-fying a polynomial identity*, Israel Math. Conf. Proc. **1** (1989), 61-73.

[Pchl84]   Pchelintzev, S.V., *The height theorem for alternate algebras*, Mat. Sb. **124** (1984), no. 4, 557–567.

[Plo04]   Plotkin, B., *Notes on Engel groups and Engel elements in groups. Some generalizations*, Izv. Ural. Univ. Ser. Mat. Mekh. **36** (7) (2005), 153–166.

[Pro73]   Procesi, C., Rings with polynomial identities, M. Dekker, New York (1973).

[Pro76]   Procesi, C., *The invariant theory of $n \times n$ matrices*, Adv. Math. **19** (1976), 306–381.

[Pro14]   Procesi, C., *On the theorem of Amitsur-Levitzki*, **207** (2015), no. 1, 151–154.

[Raz72]   Razmyslov, Yu.P., *On a problem of Kaplansky*, Math USSR. Izv. **7** (1972), 479–496.

[Raz74a]   Razmyslov, Yu.P., *The Jacobson radical in PI-algebras*, Algebra and Logic **13** (1974), no. 3, 192–204.

[Raz74b]   Razmyslov, Yu.P., *Trace identities of full matrix algebras over a field of characteristic zero*, Math. USSR Izv. **8** (1974), 724–760.

[Raz74c]   Razmyslov, Yu.P., *Existence of a finite basis for certain varieties of algebras*, Algera i Logika **13** (1974), 685–693; English translation: Algebra and Logic **13** (1974), 394–399.

[Raz78]   Razmyslov, Yu.P., *On the Hall-Higman problem*, Izv. Akad. Nauk SSSR Ser. Mat. **42** (1978), no. 4, 833–847.

[Raz87]   Razmyslov, Yu.P., *Trace identities and central polynomials in matrix superalgebras $M_{n;k}$* (Russian), Mat. Sb. (N.S.) **128** (1985), no. 2, 194–215, 287. English translation: Math. USSR-Sb. **56** (1987) 187–206

[Raz89]   Razmyslov, Yu.P., Identities of algebras and theirrepresentations, Nauka, Moscow (1989).

[RaZub94]   Razmyslov, Yu.P., and Zubrilin, K.A., *Capelli identities and representations of finite type*, Comm. Algebra **22** (1994), no. 14, 5733–5744.

[Reg78]   Regev, A., *The representations of S_n and explicit identities for P.I. algebras*, J. Algebra **51** (1978), no. 1, 25–40.

[Reg80]   Regev, A., *The Kronecker product of S_n-characters and an $A \otimes B$ theorem for Capelli identities*, J. Algebra **66** (1980), 505–510.

[Reg81]   Regev A., *Asymptotic values for degrees associated with strips of Young diagrams*, Adv. Math. **41** (1981), 115–136.

[Reg84]   Regev, A., *Codimensions and trace codimensions of matrices are asymptotically equal*, Israel J. Math. **47** (1984), 246–250.

[ReS93]   Resco, R. and Small, L., *Affine Noetherian algebras and extensions of the base field*, Bulletin London Math. Soc. **25** (1993), 549–552.

[ReSS82]   Richard Resco, Lance W. Small and J. T. Stafford, *Krull and global dimensions of semiprime Noetherian PI-rings*, Trans. Amer. Math. Soc. **274** (1982), 285–295.

[Ro80]   Rowen, L.H., *Polynomial identities in ring theory*, Acad. Press Pure and Applied Math. **84**, New York (1980).

[Ro88a]   Rowen, L.H., *Ring theory I*, Acad. Press Pure and Applied Math. **127**, New York (1988).

[Ro88b]   Rowen, L.H., *Ring theory II*, Acad. Press Pure and Applied Math. **128**, New York (1988).

[Ro05]   Rowen, L.H., *Graduate algebra: Commutative view*, Graduate Studies in Mathematics **73**, Amer. Math. Soc. (2006).

[Ro08]   Rowen, L.H., *Graduate algebra: Noncommutative view*, Graduate Studies in Mathematics **91**, Amer. Math. Soc. (2008).

[RoSm15] Rowen, L.H. and Small, L., *Representability of algebras finite over their centers*, J. Algebra (2015).

[Sag01]   Sagan, B.E., *The symmetric group: Representations, combinatorial algorithms, and symmetric functions*, 2nd ed., Graduate Texts in Mathematics **203**, New York, Springer, Berlin-New York (2001).

[Sch76]   Schelter, W., *Integral extensions of rings satisfying a polynomial identity*, J. Algebra **40** (1976), 245–257; errata op. cit. **44** (1977), 576.

[Sch78]   Schelter, W., *Noncommutative affine PI-algebras are catenary*, J. Algebra **51** (1978), 12–18.

[Shch00] Shchigolev, V.V., *Examples of infinitely basable T-spaces*, Mat. Sb. **191** (2000), no. 3, 143–160; translation: Sb. Math. **191** (2000), no. 3-4, 459–476.

[Shch01] Shchigolev, V.V., *Finite basis property of T-spaces over fields of characteristic zero*, Izv. Ross. Akad. Nauk Ser. Mat. **65** (2001), no. 5, 191–224; translation: Izv. Math. **65** (2001), no. 5, 1041–1071.

[Shes83] Shestakov, I.P., *Finitely generated special Jordan and alternate PI-algebras*, Mat. Sb. **122** (1983), no. 1, 31–40.

[Shir57a] Shirshov, A.I., *On some nonassociative nil-rings and algebraic algebras*, Mat. Sb. **41** (1957), no. 3, 381–394.

[Shir57b] Shirshov, A.I., *On rings with identity relations*, Mat. Sb. **43**, (1957), no. 2, 277–283.

[Sm71]   Small, L.W., *An example in PI rings*, J. Algebra **17** (1971), 434–436.

[SmStWa84] Small, L.W., Stafford J.T., and Warfield, R., *Affine algebras of Gelfand Kirillov dimension one are PI*, Math. Proc. Cambridge Phil. Soc. **97** (1984), 407–414.

[SmWa84] Small, L.W., and Warfield, R., *Prime affine algebras of Gelfand-Kirillov dimension one*, J. Algebra **91** (1984), 386–389.

[SmoVi03] Smoktunowicz A. and Vishne, U., *An affine prime non-semiprimitive monomial algebra with quadratic growth*, Adv. Appl. Math. **37** (2006), 511-513.

[Sp50] Specht, W., *Gesetze in Ringen I*, Math. Z. **52** (1950), 557–589.

[St99] Stanley, R.P. *Enumerative combinatorics. Vol. 2*, Cambridge Studies in Advanced Math. **62**, Cambridge Univ. Press (1999).

[Ufn78] Ufnarovski'i, V.A., *On growth of algebras*, Vestn. Moskovskogo Universiteta Ser. 1, no. 4 (1978) 59–65.

[Ufn80] Ufnarovski'i, V.A., *On Poincare series of graded algebras*, Mat. Zametki **27** (1980), no. 1, 21–32.

[Ufn85] Ufnarovski'i, V.A., *The independency theorem and its consequences*, Mat. Sb., **128** (1985), no. 1, 124–13.

[Ufn89] Ufnarovski'i, V.A., *On regular words in Shirshov sense*, In: Tez. Dokl. po Teorii Koletz, Algebr i Modulei. Mezhd. Konf. po Algebre Pamyati A.I. Mal'tzeva, Novosibirsk (1989), 140.

[Ufn90] Ufnarovski'i, V.A., *On using graphs for computing bases, growth functions and Hilbert series of associative algebras*, Mat. Sb. **180** (1990), no. 11, 1548–1550.

[VaZel89] Vais, A.Ja., and Zelmanov, E.I., *Kemer's theorem for finitely generated Jordan algebras*, Izv. Vyssh. Uchebn. Zved. Mat. (1989), no. 6, 42–51; translation: Soviet Math. (Iz. VUZ) **33** (1989), no. 6, 38–47.

[Vas99] Vasilovsky S. Yu., $\mathbb{Z}_n$-*graded polynomial identities of the full matrix algebra of order n*, Proc. Amer. Math. Soc. **127** (1999), 3517–3524.

[Vau70] Vaughan-Lee M.R., *Varieties of Lie algebras*, Quart. J. of Math. **21** (1970), 297–308.

[Vi99] Vishne, U., *Primitive algebras with arbitrary GK-dimension*, J. Algebra **211** (1999), 150–158.

[Vi02] Vishne, U., *Polynomial identities of 2×2 matrices over the Grassmannian*, Comm. Algebra **30(1)** (2002), 443–454.

[Vi11] Vishne, U., *Polynomial identities of $M_{2,1}(G)$*, Comm. Algebra, **39(6)** (2011), 2044–2050.

[Wag37] Wagner, W., *Uber die Grundlagen der projektiven Geometrie und allgemeine Zahlsysteme*, Math. Z. **113** (1937), 528–567.

[We76]   Wehrfritz, B.A.F., *Automorphism groups of Noetherian modules over commutative rings*, Arch. Math. **27** (1976), 276–281.

[Wil91]   Wilson, J.S., *Two-generator conditions for residually finite groups*, Bull. London Math. Soc. **23** (1991), 239–248.

[Za90]   Zalesskiĭ, A.E., *Group rings of inductive limits of alternating groups*, Algebra i Analiz **2** (1990), no. 6, 132–149; translation: Leningrad Math. J. **2** (1991), no. 6, 1287–1303.

[Za96]   Zalesskiĭ, A.E., *Modular group rings of the finitary symmetric group*, Israel J. Math. **96** (1996), 609–621.

[ZaMi73]   Zalesskiĭ, A.E., and Mihalev, A.V., *Group rings*, Current Problems in Mathematics, Vol. 2, Akad. Nauk SSSR Vseojuz. Inst. Naučn. i Tehn. Informacii, Moscow (1973), 5–118.

[ZelKos88]   Zelmanov, E.I., *On nilpotency of nilalgebras*, Lect. Notes Math. **1352** (1988), 227–240.

[ZelKos90]   Zelmanov, E.I., and Kostrikin, A.I., *A theorem about sandwich algebras*, Trudy Mat. Inst. Akad. Nauk SSSR **183** (1990), 106–111.

[ZSlShSh78] Zhelvakov, Slin'ko, A.M., Shestakov, I.P. and Shirshov, A.I., Rings, Close to associative, Nauka, Moscow (1978), in Russian.

[Zu96]   Zubkov, A.N., *On a generalization of the Razmyslov-Procesi theorem*, Algebra i Logika **35**, (1996), no. 4, 433–457, 498; transl. Algebra and Logic **35** (1996), no. 4, 241–254.

[Zub95a]   Zubrilin, K.A., *On the class of nilpotency of obstruction for the representability of algebras satisfying Capelli identities*, Fundam. Prikl. Mat. **1** (1995), 409–430.

[Zub95b]   Zubrilin, K.A., *Algebras satisfying Capelli identities*, Sb. Math. **186** (1995), no. 3, 53–64.

[Zub97]   Zubrilin, K.A., *On the largest nilpotent ideal in algebras satisfying Capelli identities*, Sb. Math. **188** (1997), 1203–1211.

Author Index

Subject Index

ACC, 11, 68, 198, 264, 267, 373, 381
  on ideals, 331
  on left annihilators, 10
  on principal annihilators, 51, 73
  on principal left annihilators, 354
ACC(Ann), 44, 68
action of S_n
  left, 124
  right, 124
K-admissible, 240
p-admissible, 300, 304–306, 309, 310, 318, 319
affine algebra, 7, 8, 40, 44–49, 71–73, 77, 87, 88, 92, 101, 108, 150, 174, 184, 197–236, 240, 336, 341, 353, 354, 378
  over a Noetherian ring, 174
  semiprimitive, 353
affine extension integral, 172
affine PI-algebra, 343, 345, 357, 379, 387
  semiprime, 342
affine superalgebra, 243
affinization, 339, 340
algebra
  affine, *see* affine
  basic, 205, 219
  finite dimensional, 90, 203
  free, *see* free algebra
  full, 208, 209
  of generic $n \times n$ matrices, 59
  PI-, *see* PI-algebra

  relatively free, *see* relatively free algebra
  representable, *see* representable algebra
  without 1, 5, 65, 68, 210, 238, 289, 292
algebraic, 5, 8, 72
  radical, 72
alphabet, 78
alternating, 20, 22, 23, 37, 40–42, 91, 93, 172, 193, 239, 373
  $(t_1, \ldots, t_\mu)$-alternating, 210
  (t_0, t_1)-alternating, 256, 258
  doubly, 157, 164, 165, 169
  polynomial, 212, 239
alternator, 23, 211
  $(t_1, \ldots, t_\mu)$-alternator, 211
  μ-alternator, 219
  n-alternator, 23
Amitsur-Levitzki Theorem, 21, 32, 36, 63, 71, 139, 286, 289, 373
annihilator
  ideal, 10
  left, 10
  principal (left), 51, 353
Artin-Tate Lemma, 10, 50
ascending chain condition, 9

basic, 224, 237
Bergman
  example, 47
  gap, 337, 338, 352, 384
bimodule, 7
Boolean algebra, 15
Branching Theorem, 132
Braun's Theorem, 172

411